Lecture Notes in Physics

Volume 870

Founding Editors

W. Beiglböck
J. Ehlers
K. Hepp
H. Weidenmüller

Editorial Board

B.-G. Englert, Singapore, Singapore
U. Frisch, Nice, France
P. Hänggi, Augsburg, Germany
W. Hillebrandt, Garching, Germany
M. Hjort-Jensen, Oslo, Norway
R. A. L. Jones, Sheffield, UK
H. von Löhneysen, Karlsruhe, Germany
M. S. Longair, Cambridge, UK
M. L. Mangano, Geneva, Switzerland
J.-F. Pinton, Lyon, France
J.-M. Raimond, Paris, France
A. Rubio, Donostia, San Sebastian, Spain
M. Salmhofer, Heidelberg, Germany
D. Sornette, Zurich, Switzerland
S. Theisen, Potsdam, Germany
D. Vollhardt, Augsburg, Germany
W. Weise, Garching, Germany and Trento, Italy

For further volumes:
www.springer.com/series/5304

The Lecture Notes in Physics

The series Lecture Notes in Physics (LNP), founded in 1969, reports new developments in physics research and teaching—quickly and informally, but with a high quality and the explicit aim to summarize and communicate current knowledge in an accessible way. Books published in this series are conceived as bridging material between advanced graduate textbooks and the forefront of research and to serve three purposes:

- to be a compact and modern up-to-date source of reference on a well-defined topic
- to serve as an accessible introduction to the field to postgraduate students and nonspecialist researchers from related areas
- to be a source of advanced teaching material for specialized seminars, courses and schools

Both monographs and multi-author volumes will be considered for publication. Edited volumes should, however, consist of a very limited number of contributions only. Proceedings will not be considered for LNP.

Volumes published in LNP are disseminated both in print and in electronic formats, the electronic archive being available at springerlink.com. The series content is indexed, abstracted and referenced by many abstracting and information services, bibliographic networks, subscription agencies, library networks, and consortia.

Proposals should be sent to a member of the Editorial Board, or directly to the managing editor at Springer:

Christian Caron
Springer Heidelberg
Physics Editorial Department I
Tiergartenstrasse 17
69121 Heidelberg/Germany
christian.caron@springer.com

Daniele Faccio · Francesco Belgiorno ·
Sergio Cacciatori · Vittorio Gorini ·
Stefano Liberati · Ugo Moschella
Editors

Analogue Gravity Phenomenology

Analogue Spacetimes and Horizons,
from Theory to Experiment

Editors
Daniele Faccio
School of Engineering and Physical
  Sciences, Institute of Photonics and
  Quantum Sciences, Scottish Universities
  Physics Alliance
Heriot-Watt University
Edinburgh, UK

Francesco Belgiorno
Dipartimento di Matematica
Politecnico di Milano
Milano, Italy

Sergio Cacciatori
Dipartimento di Scienza e Alta Tecnologia
Università dell'Insubria
Como, Italy

Vittorio Gorini
Dipartimento di Scienza e Alta Tecnologia
Università dell'Insubria
Como, Italy

Stefano Liberati
International School for Advanced Studies
  (SISSA)
Trieste, Italy

Ugo Moschella
Dipartimento di Scienza e Alta Tecnologia
Università dell'Insubria
Como, Italy

ISSN 0075-8450   ISSN 1616-6361 (electronic)
Lecture Notes in Physics
ISBN 978-3-319-00265-1   ISBN 978-3-319-00266-8 (eBook)
DOI 10.1007/978-3-319-00266-8
Springer Cham Heidelberg New York Dordrecht London

Library of Congress Control Number: 2013946369

© Springer International Publishing Switzerland 2013
This work is subject to copyright. All rights are reserved by the Publisher, whether the whole or part of the material is concerned, specifically the rights of translation, reprinting, reuse of illustrations, recitation, broadcasting, reproduction on microfilms or in any other physical way, and transmission or information storage and retrieval, electronic adaptation, computer software, or by similar or dissimilar methodology now known or hereafter developed. Exempted from this legal reservation are brief excerpts in connection with reviews or scholarly analysis or material supplied specifically for the purpose of being entered and executed on a computer system, for exclusive use by the purchaser of the work. Duplication of this publication or parts thereof is permitted only under the provisions of the Copyright Law of the Publisher's location, in its current version, and permission for use must always be obtained from Springer. Permissions for use may be obtained through RightsLink at the Copyright Clearance Center. Violations are liable to prosecution under the respective Copyright Law.
The use of general descriptive names, registered names, trademarks, service marks, etc. in this publication does not imply, even in the absence of a specific statement, that such names are exempt from the relevant protective laws and regulations and therefore free for general use.
While the advice and information in this book are believed to be true and accurate at the date of publication, neither the authors nor the editors nor the publisher can accept any legal responsibility for any errors or omissions that may be made. The publisher makes no warranty, express or implied, with respect to the material contained herein.

Printed on acid-free paper

Springer is part of Springer Science+Business Media (www.springer.com)

Preface

Reasoning by analogies is a natural inclination of the human brain that operates by associating new and unknown situations to a series of known and previously encountered situations. On the basis of these analogies, judgements and decisions are made: associations are the building blocks for predictive thought. It is therefore natural that analogue models are also a constant presence in the world of physics and an invaluable instrument in the progress of our knowledge of the world that surrounds us. It would be impossible to give a comprehensive list of these analogue models but a few recent and relevant examples are optical waveguide analogues of the relativistic Dirac equation (linking optics with quantum mechanics), photonic crystals (linking optical wave propagation in periodic lattices with electron propagation in metals) or, at a more profound level, the Anti-de Sitter/Conformal Field Theory correspondence (linking quantum systems in D dimensions to gravitational systems in $D+1$ dimensions). The purpose of this book is to give a general overview and introduction to the world of analogue gravity: the simulation or recreation of certain phenomena that are usually attributed to the effects of gravity but that can be shown to naturally emerge in a variety of systems ranging from flowing liquids to nonlinear optics.

Questions often arises regarding the implications of analogue models, particularly in the context of analogue gravity. So this appears to possibly be a good starting point for a Preface. The analogue models treated here can all be reconnected to some form of flowing medium. This flowing medium, under appropriate conditions is expected to reproduce or mimic the flow of space generated by a gravitational field. However, it is important to bear in mind that analogue models are always "analogies" and never, or hardly ever, "identities", meaning that we should not confuse the two systems under comparison. In the specific case of analogue gravity, the analogies do not usually attempt to reproduce the *dynamics* of a gravitational system, for example a black hole. The dynamics rely on Einstein's equations and require the presence of a gravitational source term. On the other hand, the analogies can reproduce to a large extent the *kinematics* of a black hole. The kinematics refer for example to photon or particle trajectories and these are determined by the system's spacetime metric. Whether the curved spacetime metric is the result of a gravitational field or of a flowing medium becomes irrelevant when the analysis is restricted to the description of wave propagation and evolution in this flowing medium: the kinematics are identical and the analogy is robust. The absence of a link between the dynamics of the analogue and gravitational systems may seem to be an inherent and even disappointing weakness of the ideas presented here. Questions often arise regarding the usefulness of analogue models for gravity if we cannot produce predictions regarding the evolution of, for example, binary black hole systems, the quantum or the purely geometrical nature of gravity or whatever may be the hot topic at the moment you are reading this book. These objections are certainly valid: unless things take an unexpected twist in the future, it seems rather unlikely that analogue gravity will provide us with answers to these questions. This is mainly because analogue gravity simply does not address these problems. Re-iterating once more, analogue gravity builds upon our knowledge in general relativity and condensed matter physics in order to build a deeper understanding of certain physical effects that rely solely on the kinematics of the two systems. This declaration contains within it a series of fundamental and outstanding problems in modern physics that fully justify the interest in the field. Moreover, the search for a deeper understanding of the laws that govern the universe is only one aspect of analogue gravity. We hope that in reading this book, you will appreciate how the strive to develop and understand both old and new analogue models is leading to innovation, technical advancement and new discoveries in a remarkable range of physical systems ranging from acoustics and gravity waves to optics, all linked by the common denominator that lies within the geometrical description of spacetime, as first introduced in the context of gravity.

So what are these "kinematical" effects addressed by analogue gravity? The most important and obvious member of this list is without doubt Hawking radiation. Hawking radiation is the spontaneous emission of blackbody radiation due to the distortion of the quantum vacuum in the vicinity of a black hole event horizon. This effect was first described in detail by Stephen Hawking in 1974 although the first hints that black holes should have non-zero equilibrium temperatures came from Bekenstein, a year earlier based on the analogy between the laws of thermodynamics and those of black hole mechanics (yet another fruitful analogy!). Hawking radiation has since attracted the imagination and efforts of countless scientists, all looking for a deeper understanding of why this radiation is emitted, how it is emitted and the implications of this emission with big questions concerning information loss and even the final fate of our universe. A remarkable point that needs to be remembered is that Hawking radiation in realistic gravitational systems such as stellar and galactic black holes, is remarkably weak—so weak that we have very little hope of directly observing it in a gravitational context. This is somewhat of a set-back for what is without doubt one of the most fascinating and prolific ideas of modern physics. Analogue gravity will probably not be able to claim that this unfortunate glitch will be overcome, simply because analogue gravity experiments do not deal with gravitational black holes, yet it certainly does give us the opportunity to study Hawking radiation from a fresh and rather unexpected perspective.

Bill Unruh first proposed an analogue for gravity in 1981 in the context of sound waves propagating in a flowing medium. The underlying idea is very simple: if we imagine a sound wave propagating against a counter-flowing medium it is easy to appreciate that if the medium flow velocity is smaller than the wave velocity, then the wave will be allowed to propagate upstream. If the wave then encounters a velocity gradient such that the flow accelerates up to supersonic speeds, then the wave will inevitably slow down until it is completely blocked by the counter-propagating medium. The sound wave cannot propagate upstream against a supersonic flow. The wave-blocking point lies at the transition from sub to super-sonic flow and, as such, represents the analogue of a gravitational wave-blocking horizon. This is just a pictorial description of the situation. The mathematics reveals a far more unexpected and revolutionary aspect: the sound wave propagation and trajectory is fully described by a spacetime metric that is distorted by the flowing medium close to the velocity gradient. This distortion can be formally identical to the metric close to the event horizon of a gravitational black hole and, taking one step further, Hawking radiation, now in the form of sound waves is predicted to be emitted from the "sonic" horizon. Most interestingly, the sonic horizon emits Hawking radiation that depends on the gradient of the medium velocity across the horizon and therefore can in principle be engineered and optimised in the laboratory. For several years these predictions were not fully appreciated and it was only later on that Bill re-proposed his idea and today, under the thrust of continuous technological improvements there is a thriving and expanding community dedicated to the search of new settings in which a flowing medium of some sort can be generated and controlled in a such a way as to recreate curved spacetime metrics with various applications.

The resulting theoretical models are becoming ever more sophisticated, spurred by and in turn spurring new laboratory tests and experimental success stories. Hawking-like radiation mechanisms have now been analysed in wide range of systems well beyond the original acoustical analogue, e.g. superfluids, flowing Bose-Einstein Condensates, ion-rings, electromagnetic waveguides, soliton-like pulses in optical media just to name a few. The last few years in particular have seen a sudden surge of experimental tests that are paving the way for greater things to come: horizons have been generated and observed with gravity waves in flowing water, phonon oscillations in flowing Bose-Einstein-Condensates and light scattering from laser pulse-induced flowing optical media. Negative frequency waves have been observed in water-wave and optical analogues and direct evidence of Hawking-like behaviour in a classically stimulated context has been observed using water-waves. The first evidence of spontaneous emission from an optically generated perturbation has also been observed. These are all remarkably important in light of the evident advancement of the whole field from a rather sidetrack idea to a fully fledged research area that is leading to innovation and important discoveries at both the theoretical and experimental level.

Clearly there still remain many challenges and hurdles. Notwithstanding the remarkable experimental progress in the last years, a clear and undisputed experimental indication of spontaneous Hawking emission is still lacking. This may come from one of the analogue models mentioned above or possibly from one of the many and new models that are currently emerging. An additional point that is requiring significant effort and is one of the main complications with respect to the original

Hawking description, is the presence of dispersion. On the one hand, dispersion provides a simple solution to the so-called transplanckian problem as it curbs any frequency divergence at the horizon. But on the other hand, it may significantly complicate the analogy or even modify the nature of the emission. This difficulty is certainly relevant in the optical analogues but plays a role also in other settings. Moreover, although Hawking radiation is certainly the most prominent and desired effect one would want to observe in these analogues, other effects are also equally noteworthy and are gradually gaining attention, such as the emission from analogues of a cosmological expansion, from superluminal flows, rapidly changing media or from rotating media in analogy with so-called superradiance from rotating black holes.

The scope of this book is to present an overview of the ideas underlying analogue gravity together with a description of some of the most promising and interesting systems in which analogues may be built. The aim is to provide scientists, independently of their background in any of the topics approached by analogue gravity, with a general understanding of the field. A full and deeper understanding will probably require some extra work beyond the reading of this book. However, it is our hope that these pages will encourage you to ponder the beauty of how wave propagation in flowing media is intricately connected to the geometry of space and time and to thus stimulate new ideas and new questions.

This book is divided into a number of chapters that start from a general overview of analogue gravity, providing also some background knowledge to black hole physics before moving on to consider in detail some specific analogue gravity models.

Chapter 1 gives an overview of black hole geometry and Hawking emission from gravitational and analogue black holes.

Chapter 2 delivers an overview of the field of analogue gravity as a whole with a brief description of the foundation pillars applied to a few specific examples.

Chapter 3 gives a broader description of some possibilities for observing diverse fundamental quantum effects in the laboratory, therefore going beyond Hawking radiation.

Motivated by recent experimental success, Chap. 4 gives a theoretical and general description of surface waves in fluids.

Chapters 4 to 8 give an in-depth description of surface waves on flowing media in terms of the analogue gravity perspective, rounding up with a description of recent experimental results in this area.

Chapter 9 treats the Bose-Einstein-Condensate analogue model and Chaps. 8 and 9 treat in detail the optical analogue first from a purely theoretical perspective and then introducing some recent experimental results with these analogues.

Chapters 10 and 11 deal with the so-called optical analogue model including an overview of some recent experimental results.

Chapter 12 also deals with a particular application of nonlinear optics in which light flows as if it were a fluid ("luminous liquid") and represents a promising experimental avenue.

Chapter 13 gives an overview of Lorentz-invariance breaking and possible observational tests in the context of analogue gravity.

Chapter 14 extends the concepts of analogue gravity to studies involving the topology of the vacuum—the topological constraints on the quantum vacuum structure determine some universal properties of our universe and these can be mimicked and studied in analogue systems.

Chapter 15 describes a further extension of analogue gravity, here in the realm of Einstein diffusion modified by a curved, or analogue curved spacetime background.

Finally, Chap. 16 is a return to the origins and provides the reader with an overview of the current observational evidence of event horizons in gravitational black holes.

In closing, I wish to acknowledge Fondazione Cariplo for funding and Centro Volta, Como, Italy for hosting the 2011 summer school on Analogue Gravity—the chapters in this book are mostly based on the lectures delivered during this summer school. We also gratefully acknowledge all of the authors who so kindly contributed with their knowledge to this publication and the students and collaborators who kindly assisted in proof-reading, Niclas Westerberg, Mihail Petev, Daniel Moss.

Edinburgh   Daniele Faccio
December 2012

Contents

1 Black Holes and Hawking Radiation in Spacetime and Its Analogues   1
Ted Jacobson
- 1.1 Spacetime Geometry and Black Holes 1
  - 1.1.1 Spacetime Geometry 1
  - 1.1.2 Spherical Black Hole 2
  - 1.1.3 Effective Black Hole and White Hole Spacetimes 7
  - 1.1.4 Symmetries, Killing Vectors, and Conserved Quantities . 8
  - 1.1.5 Killing Horizons and Surface Gravity 11
- 1.2 Thermality of the Vacuum 14
- 1.3 Hawking Effect 17
  - 1.3.1 Mode Solutions 19
  - 1.3.2 Positive Norm Modes and the Local Vacuum 20
  - 1.3.3 Stimulated Emission of Hawking Radiation 23
- 1.4 The Trans-Planckian Question 23
- 1.5 Short Wavelength Dispersion 24
  - 1.5.1 Stimulated Hawking Radiation and Dispersion 26
  - 1.5.2 White Hole Radiation 26
- References 28

2 Survey of Analogue Spacetimes 31
Matt Visser
- 2.1 Introduction 31
- 2.2 Optics: The Gordon Metric and Its Generalizations 32
- 2.3 Non-relativistic Acoustics: The Unruh Metric 35
- 2.4 Horizons and Ergo-Surfaces in Non-relativistic Acoustics 37
- 2.5 Relativistic Acoustics 41
- 2.6 Bose–Einstein Condensates 43
- 2.7 Surface Waves and Blocking Horizons 44
- 2.8 Optical Fibres/Optical Glass 46
- 2.9 Other Models 47
- 2.10 Discussion 47

	References	47
3	**Cosmological Particle Creation in the Lab?**	51
	Ralf Schützhold and William G. Unruh	
	3.1 Introduction	51
	3.2 Scattering Analogy	52
	3.3 WKB Analysis	53
	3.4 Adiabatic Expansion and Its Breakdown	56
	3.5 Example: Inflation	57
	3.6 Laboratory Analogues	60
	References	61
4	**Irrotational, Two-Dimensional Surface Waves in Fluids**	63
	William G. Unruh	
	4.1 Introduction	63
	4.1.1 $v_0 = 0$ Limit	68
	4.1.2 Formal General Solution	70
	4.2 Fluctuations	70
	4.3 Shallow Water Waves	73
	4.4 Deep Water Waves	74
	4.5 General Linearized Waves	75
	4.6 Blocking Flow	77
	4.7 Conversion to δy	78
	4.8 Waves in Stationary Water over Uneven Bottom	79
	References	80
5	**The Basics of Water Waves Theory for Analogue Gravity**	81
	Germain Rousseaux	
	5.1 Introduction	81
	5.2 A Glimpse of Dimensional Analysis	81
	5.2.1 Shallow Waters	82
	5.2.2 Deep Waters	83
	5.2.3 Arbitrary Water Depth	83
	5.2.4 The Capillary Length	83
	5.3 Long Water Waves on a Current as a Gravity Analogue	84
	5.4 Fluid Particles' Trajectories	87
	5.5 A Plethora of Dispersive Effects	90
	5.6 Hydrodynamic Horizons	92
	5.6.1 Non-dispersive Horizons	92
	5.6.2 Dispersive Horizons	94
	5.6.3 Natural and Artificial Horizons	97
	5.6.4 Zero Modes	99
	5.7 The "Norm"	101
	5.8 Conclusion	106
	References	106

6	**The Čerenkov Effect Revisited: From Swimming Ducks to Zero Modes in Gravitational Analogues** 109		
	Iacopo Carusotto and Germain Rousseaux		
	6.1	Introduction 109	
	6.2	Generic Model 111	
		6.2.1	The Wave Equation and the Source Term 111
		6.2.2	Qualitative Geometrical Study of the Wake Pattern 113
		6.2.3	Generalization to Higher-Order Wave Equations 115
	6.3	Čerenkov Emission by Uniformly Moving Charges 116	
		6.3.1	Non-dispersive Dielectric 116
	6.4	Moving Impurities in a Superfluid 119	
		6.4.1	The Bogoliubov Dispersion of Excitations 120
		6.4.2	Superfluidity vs. Bogoliubov-Čerenkov Wake 121
	6.5	Parabolic Dispersion: Conics in the Wake 124	
	6.6	Surface Waves on a Liquid 127	
		6.6.1	Dispersion of Surface Waves 127
		6.6.2	Deep Fluid 132
		6.6.3	Shallow Fluid 134
	6.7	Čerenkov Processes and the Stability of Analogue Black/White Holes 136	
		6.7.1	Superfluid-Based Analogue Models 137
		6.7.2	Analogue Models Based on Surface Waves 139
	6.8	Conclusions 141	
	References 141		
7	**Some Aspects of Dispersive Horizons: Lessons from Surface Waves** 145		
	Jennifer Chaline, Gil Jannes, Philpe Maïssa, and Germain Rousseaux		
	7.1	Introduction 145	
	7.2	Preliminaries 147	
	7.3	Experimental Setup 151	
	7.4	Experimental Results 152	
	7.5	Airy Interference and Gravity-Wave Blocking 155	
	7.6	The Trans-Planckian Problem 159	
	Appendix Airy Stopping Length 162		
	References 164		
8	**Classical Aspects of Hawking Radiation Verified in Analogue Gravity Experiment** 167		
	Silke Weinfurtner, Edmund W. Tedford, Matthew C.J. Penrice, William G. Unruh, and Gregory A. Lawrence		
	8.1	Motivation 167	
	8.2	Black & White Hole Evaporation Process 169	
	8.3	Experimental Setup 170	
	8.4	Quasi-Particle Excitations 172	
	8.5	Experimental Procedure 175	
	8.6	Data Analysis and Results 176	

	8.7	Conclusions and Outlook	179
	References	179

9 Understanding Hawking Radiation from Simple Models of Atomic Bose-Einstein Condensates ... 181
Roberto Balbinot, Iacopo Carusotto, Alessandro Fabbri,
Carlos Mayoral, and Alessio Recati
- 9.1 Introduction ... 181
- 9.2 The Theory of Dilute Bose-Einstein Condensates in a Nutshell ... 185
  - 9.2.1 The Gross-Pitaevskii Equation and the Bogoliubov Theory ... 185
  - 9.2.2 Analogue Gravity in Atomic BECs ... 187
- 9.3 Stepwise Homogeneous Condensates ... 189
- 9.4 Subsonic-Subsonic Configuration ... 193
  - 9.4.1 The Bogoliubov Modes and the Matching Matrix ... 193
  - 9.4.2 The "In" and "Out" Basis ... 195
  - 9.4.3 Bogoliubov Transformation ... 199
  - 9.4.4 Density-Density Correlations ... 200
- 9.5 Subsonic-Supersonic Configuration ... 201
  - 9.5.1 The Modes and the Matching Matrix ... 201
  - 9.5.2 The "In" and "Out" Basis ... 204
  - 9.5.3 Bogoliubov Transformation ... 207
  - 9.5.4 Density-Density Correlations ... 209
  - 9.5.5 Remarks ... 211
- 9.6 Supersonic-Supersonic Configuration ... 212
- 9.7 Conclusions ... 215
- Appendix ... 216
- References ... 217

10 Transformation Optics ... 221
Ulf Leonhardt
- 10.1 Introduction ... 221
- 10.2 Maxwell's Electromagnetism ... 223
  - 10.2.1 Maxwell's Equations ... 223
  - 10.2.2 The Medium of a Geometry ... 224
  - 10.2.3 The Geometry of a Medium ... 226
- 10.3 Spatial Transformations ... 226
  - 10.3.1 Invisibility Cloaking ... 227
  - 10.3.2 Transformation Media ... 228
  - 10.3.3 Perfect Imaging with Negative Refraction ... 230
  - 10.3.4 Quantum Levitation ... 232
- 10.4 Curved Space ... 233
  - 10.4.1 Einstein's Universe and Maxwell's Fish Eye ... 233
  - 10.4.2 Perfect Imaging with Positive Refraction ... 235
  - 10.4.3 Casimir Force ... 237
- 10.5 Spacetime Media ... 239
  - 10.5.1 Spacetime Geometries ... 239

	10.5.2 Magneto-Electric Media 240
	10.5.3 Moving Media 241
	10.5.4 Spacetime Transformations 242
References 244

11 Laser Pulse Analogues for Gravity 247
Eleonora Rubino, Francesco Belgiorno, Sergio Luigi Cacciatori, and Daniele Faccio
- 11.1 Introduction 247
  - 11.1.1 White Holes 248
- 11.2 Analogue Gravity with Optics in Moving Media 249
- 11.3 Dielectric Metrics and Hawking Radiation 251
  - 11.3.1 The Role of Dispersion 255
- 11.4 Numerical Simulations of One-Dimensional Dielectric White Holes 259
- 11.5 Stimulated Hawking Emission and Amplification 261
- 11.6 Creating an Effective Moving Medium with a Laser Pulse . . . 265
- 11.7 Experiments: *Spontaneous Emission from a Moving Perturbation* 267
- 11.8 Conclusions and Perspectives 271
- References 271

12 An All-Optical Event Horizon in an Optical Analogue of a Laval Nozzle 275
Moshe Elazar, Shimshon Bar-Ad, Victor Fleurov, and Rolf Schilling
- 12.1 Introduction 275
- 12.2 Transonic Flow of a Luminous Fluid 277
- 12.3 An Experimental Horizon 280
- 12.4 Fluctuations 285
  - 12.4.1 Classical Straddled Fluctuations 285
  - 12.4.2 Regularization of Fluctuations Near the Mach Horizon . . 288
  - 12.4.3 Quantization and the Hawking Temperature 290
- 12.5 Discussion 294
- 12.6 Summary 295
- References 295

13 Lorentz Breaking Effective Field Theory and Observational Tests . . 297
Stefano Liberati
- 13.1 Introduction 297
- 13.2 A Brief History of an Heresy 298
  - 13.2.1 The Dark Ages 298
  - 13.2.2 Windows on Quantum Gravity 299
- 13.3 Bose–Einstein Condensates as an Example of Emergent Local Lorentz Invariance 301
  - 13.3.1 The Acoustic Geometry in BEC 302
  - 13.3.2 Lorentz Violation in BEC 305
- 13.4 Modified Dispersion Relations and Their Naturalness 307
  - 13.4.1 The Naturalness Problem 308

- 13.5 Dynamical Frameworks ... 310
  - 13.5.1 SME with Renormalizable Operators ... 310
  - 13.5.2 Dimension Five Operators SME ... 312
  - 13.5.3 Dimension Six Operators SME ... 313
  - 13.5.4 Other Frameworks ... 315
- 13.6 Experimental Probes of Low Energy LV: Earth Based Experiments ... 319
  - 13.6.1 Penning Traps ... 319
  - 13.6.2 Clock Comparison Experiments ... 319
  - 13.6.3 Cavity Experiments ... 319
  - 13.6.4 Spin Polarized Torsion Balance ... 320
  - 13.6.5 Neutral Mesons ... 320
- 13.7 Observational Probes of High Energy LV: Astrophysical QED Reactions ... 320
  - 13.7.1 Photon Time of Flight ... 320
  - 13.7.2 Vacuum Birefringence ... 322
  - 13.7.3 Threshold Reactions ... 324
  - 13.7.4 Synchrotron Radiation ... 327
- 13.8 Current Constraints on the QED Sector ... 328
  - 13.8.1 mSME Constraints ... 328
  - 13.8.2 Constraints on QED with $O(E/M)$ LV ... 329
  - 13.8.3 Constraints on QED with $O(E/M)^2$ LV ... 332
- 13.9 Other SM Sectors Constraints ... 333
  - 13.9.1 Constraints on the Hadronic Sector ... 334
  - 13.9.2 Constraints on the Neutrino Sector ... 335
- 13.10 Summary and Perspectives ... 336
- References ... 338

14 The Topology of the Quantum Vacuum ... 343
Grigorii E. Volovik
- 14.1 Introduction ... 344
  - 14.1.1 Symmetry vs Topology ... 344
  - 14.1.2 Green's Function vs Order Parameter ... 344
  - 14.1.3 The Fermi Surface as a Topological Object ... 345
- 14.2 Vacuum in a Semi-metal State ... 347
  - 14.2.1 Fermi Points in $3+1$ Vacua ... 347
  - 14.2.2 Emergent Relativistic Fermionic Matter ... 348
  - 14.2.3 Emergent Gauge Fields ... 349
  - 14.2.4 Emergent Gravity ... 349
  - 14.2.5 Topological Invariant Protected by Symmetry in the Standard Model ... 350
  - 14.2.6 Higgs Mechanism vs Splitting of Fermi Points ... 351
  - 14.2.7 Splitting of Fermi Points and Problem of Generations ... 351
- 14.3 Exotic Fermions ... 353
  - 14.3.1 Dirac Fermions with Quadratic Spectrum ... 354

Contents   xvii

		14.3.2 Dirac Fermions with Cubic and Quadratic Spectrum ... 356
	14.4	Flat Bands 359
		14.4.1 Topological Origin of Surface Flat Band 359
		14.4.2 Dimensional Crossover in Topological Matter: Formation of the Flat Band in Multi-layered Systems 361
	14.5	Anisotropic Scaling and Hořava Gravity 363
		14.5.1 Effective Theory Near the Degenerate Dirac Point 363
		14.5.2 Effective Electromagnetic Action 365
	14.6	Fully Gapped Topological Media 366
		14.6.1 2 + 1 Fully Gapped Vacua 367
	14.7	Relativistic Quantum Vacuum and Superfluid ^{3}He-B 369
		14.7.1 Superfluid ^{3}He-B 370
		14.7.2 From Superfluid Relativistic Medium to ^{3}He-B 371
		14.7.3 Topology of Relativistic Medium and ^{3}He-B 373
	14.8	Fermions in the Core of Strings in Topological Materials 374
		14.8.1 Vortices in ^{3}He-B and Relativistic Strings 374
		14.8.2 Flat Band in a Vortex Core: Analogue of Dirac String Terminating on Monopole 376
	14.9	Discussion 377
	References 378

15 Einstein^2: Brownian Motion Meets General Relativity 385
Matteo Smerlak
    15.1 Introduction 385
    15.2 Preliminaries 386
       15.2.1 The $D + 1$ Formalism 387
       15.2.2 Markov Processes 388
    15.3 Master and Fokker-Planck Equations in Curved Spacetimes ... 389
       15.3.1 Markovian Setup 389
       15.3.2 Master Equation 390
       15.3.3 Diffusive Limit: The Generalized Fokker-Planck Equation 391
    15.4 The Case of Brownian Motion 392
       15.4.1 The Generalized Diffusion Equation 392
       15.4.2 Gravitational Corrections to the Mean Square Displacement 393
    15.5 Application: Tailored Diffusion in Gravitational Analogues ... 394
       15.5.1 Dissipation in Gravitational Analogues 394
       15.5.2 From Diffusion to Antidiffusion 395
       15.5.3 On the Second Law of Thermodynamics 395
    15.6 Conclusion 397
    References 398

16 Astrophysical Black Holes: Evidence of a Horizon? 399
Monica Colpi
    16.1 The Black Hole Hypothesis 399
    16.2 Gravitational Collapse 402

	16.3	Gravitational Equilibria and Stability	402
	16.4	Neutron Stars and Stellar-Mass Black Holes in the Realm of Observations	407
	16.5	Black Holes of Stellar Origin: A Maximum Mass?	412
	16.6	Black Holes: The Other Flavor	415
		16.6.1 Weighing Active Supermassive Black Holes	416
		16.6.2 Dormant Black Holes in the Local Universe and Their Demography	418
		16.6.3 Supermassive Black Holes: How Do They Form?	420
		16.6.4 The Supermassive Black Hole at the Galactic Center	421
		16.6.5 Testing the Black Hole Hypothesis?	422
	16.7	Black Holes: Are They Spinning?	424
	16.8	Black Holes: Evidence of an Event Horizon?	427
	16.9	Event Horizons: A New Perspective	429
		16.9.1 Testing the Kerr-ness of the Spacetime	431
	16.10	Conclusions	433
	References		434
Index			437

Contributors

Roberto Balbinot Dipartimento di Fisica, Università di Bologna and INFN sezione di Bologna, Bologna, Italy

Shimshon Bar-Ad Raymond and Beverly Sackler School of Physics and Astronomy, Tel-Aviv University, Tel-Aviv, Israel

Francesco Belgiorno Dipartimento di Matematica, Politecnico di Milano, Milano, Italy

Sergio Luigi Cacciatori Dipartimento di Scienza e Alta Tecnologia, Università dell'Insubria, Como, Italy

Iacopo Carusotto INO-CNR BEC Center and Dipartimento di Fisica, Università di Trento, Povo, Trento, Italy; INO-CNR BEC Center and Dipartimento di Fisica, Università di Trento, Povo, Italy

Jennifer Chaline Laboratoire J.-A. Dieudonné, UMR CNRS-UNS 6621, Université de Nice-Sophia Antipolis, Nice Cedex 02, France

Monica Colpi Dipartimento di Fisica G. Occhialini, Università degli Studi di Milano Bicocca, Milano, Italy

Moshe Elazar Raymond and Beverly Sackler School of Physics and Astronomy, Tel-Aviv University, Tel-Aviv, Israel

Alessandro Fabbri Departamento de Física Teórica and IFIC, Universidad de Valencia-CSIC, Burjassot, Spain

Daniele Faccio School of Engineering and Physical Sciences, SUPA, Heriot-Watt University, Edinburgh, UK

Victor Fleurov Raymond and Beverly Sackler School of Physics and Astronomy, Tel-Aviv University, Tel-Aviv, Israel

Ted Jacobson Center for Fundamental Physics, Department of Physics, University of Maryland, College Park, MD, USA

Gil Jannes Laboratoire J.-A. Dieudonné, UMR CNRS-UNS 6621, Université de Nice-Sophia Antipolis, Nice Cedex 02, France

Gregory A. Lawrence Department of Civil Engineering, University of British Columbia, Vancouver, Canada

Ulf Leonhardt School of Physics and Astronomy, University of St. Andrews, St. Andrews, UK

Stefano Liberati SISSA, Trieste, Italy

Philppe Maïssa Laboratoire J.-A. Dieudonné, UMR CNRS-UNS 6621, Université de Nice-Sophia Antipolis, Nice Cedex 02, France

Carlos Mayoral Departamento de Física Teórica and IFIC, Universidad de Valencia-CSIC, Burjassot, Spain

Matthew C.J. Penrice Department of Physics and Astronomy, University of Victoria, Victoria, Canada

Alessio Recati INO-CNR BEC Center and Dipartimento di Fisica, Università di Trento, Povo, Trento, Italy

Germain Rousseaux Laboratoire J.-A. Dieudonné, UMR CNRS-UNS 6621, Université de Nice-Sophia Antipolis, Nice Cedex 02, France

Eleonora Rubino Dipartimento di Scienza e Alta Tecnologia, Università dell'Insubria, Como, Italy

Rolf Schilling Johannes Gutenberg University, Mainz, Germany

Ralf Schützhold Fakultät für Physik, Universität Duisburg-Essen, Duisburg, Germany

Matteo Smerlak Max-Planck-Institut für Gravitationsphysik (Albert-Einstein-Institut), Am Mühlenberg 1, Golm, Germany

Edmund W. Tedford Department of Civil Engineering, University of British Columbia, Vancouver, Canada

William G. Unruh Canadian Institute for Advanced Research Cosmology and Gravity Program, Department of Physics and Astronomy, University of British Columbia, Vancouver, BC, Canada

Matt Visser School of Mathematics, Statistics, and Operations Research, Victoria University of Wellington, Wellington, New Zealand

Grigorii E. Volovik Low Temperature Laboratory, Aalto University, Aalto, Finland; L.D. Landau Institute for Theoretical Physics, Moscow, Russia

Silke Weinfurtner SISSA—International School for Advanced Studies, Trieste, Italy

Chapter 1
Black Holes and Hawking Radiation in Spacetime and Its Analogues

Ted Jacobson

Abstract These notes introduce the fundamentals of black hole geometry, the thermality of the vacuum, and the Hawking effect, in spacetime and its analogues. Stimulated emission of Hawking radiation, the trans-Planckian question, short wavelength dispersion, and white hole radiation in the setting of analogue models are also discussed. No prior knowledge of differential geometry, general relativity, or quantum field theory in curved spacetime is assumed. The discussion attempts to capture the essence of these topics without oversimplification.

1.1 Spacetime Geometry and Black Holes

In this section I explain how black holes are described in general relativity, starting with the example of a spherical black hole, and followed by the $1+1$ dimensional generalization that figures in many analogue models. Next I discuss how symmetries and conservation laws are formulated in this setting, and how negative energy states arise. Finally, I introduce the concepts of Killing horizon and surface gravity, and illustrate them with the *Rindler* or *acceleration horizon*, which forms the template for all horizons.

1.1.1 Spacetime Geometry

The *line element* or *metric* ds^2 assigns a number to any infinitesimal displacement in spacetime. In a flat spacetime in a Minkowski coordinate system it takes the form

$$ds^2 = c^2 dt^2 - (dx^2 + dy^2 + dz^2), \tag{1.1}$$

where t is the time coordinate, x, y, z are the spatial Cartesian coordinates, and c is the speed of light. Hereafter I will mostly employ units with $c = 1$ except when discussing analogue models (for which c may depend on position and time when using

---

T. Jacobson (✉)
Center for Fundamental Physics, Department of Physics, University of Maryland, College Park, MD 20742-4111, USA
e-mail: jacobson@umd.edu

D. Faccio et al. (eds.), *Analogue Gravity Phenomenology*,
Lecture Notes in Physics 870, DOI 10.1007/978-3-319-00266-8_1,
© Springer International Publishing Switzerland 2013

Fig. 1.1 The light cone at an event p. The event A is future timelike related to p, while B, C, D, and E respectively are future lightlike, spacelike, past lightlike, and past timelike related to p

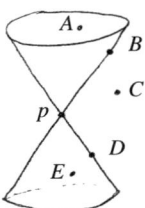

the Newtonian t coordinate). When $ds^2 = 0$ the displacement is called *lightlike*, or *null*. The set of such displacements at each event p forms a double cone with vertex at p and spherical cross sections, called the *light cone* or *null cone* (see Fig. 1.1). Events outside the light cone are *spacelike* related to p, while events inside the cone are either *future timelike* or *past timelike* related to p. For timelike displacements, ds^2 determines the square of the corresponding *proper time interval*.

The metric also defines the spacetime inner product $g(v, w)$ between two 4-vectors v and w, that is,

$$g(v, w) = ds^2(v, w)$$
$$= c^2 dt(v) dt(w) - \left[dx(v) dx(w) + dy(v) dy(w) + dz(v) dz(w) \right]. \quad (1.2)$$

Here $dt(v) = v^a \partial_a t = v^t$ is the rate of change of the t coordinate along v, etc.

In a general curved spacetime the metric takes the form

$$ds^2 = g_{\alpha\beta}(x) dx^\alpha dx^\beta, \quad (1.3)$$

where $\{x^\alpha\}$ are coordinates that label the points in a patch of a spacetime (perhaps the whole spacetime), and there is an implicit summation over the values of the indices α and β. The *metric components* $g_{\alpha\beta}$ are functions of the coordinates, denoted x in (1.3). In order to define a metric with Minkowski signature, the matrix $g_{\alpha\beta}$ must have one positive and three negative eigenvalues at each point. Then *local inertial coordinates* can be chosen in the neighborhood any point p such that (i) the metric has the Minkowski form (1.1) at p and (ii) the first partial derivatives of the metric vanish at p. In two spacetime dimensions there are 9 independent second partials of the metric at a point. These can be modified by a change of coordinates $x^\mu \to x^{\mu'}$, but the relevant freedom resides in the third order Taylor expansion coefficients $(\partial^3 x'^\mu / \partial x^\alpha \partial x^\beta \partial x^\gamma)_p$, of which only 8 are independent because of the symmetry of mixed partials. The discrepancy $9 - 8 = 1$ measures the number of independent second partials of the metric that cannot be set to zero at p, which is the same as the number of independent components of the *Riemann curvature tensor* at p. So a single curvature scalar characterizes the curvature in a two dimensional spacetime. In four dimensions the count is $100 - 80 = 20$.

1.1.2 Spherical Black Hole

The Einstein equation has a unique (up to coordinate changes) spherical solution in vacuum for each mass, called the *Schwarzschild spacetime*.

Fig. 1.2 Gravitational redshift. Two lightrays propagating from r_a to r_b, separated by a coordinate time $\delta \bar{t}$. The corresponding proper time at r_a is less than that at r_b

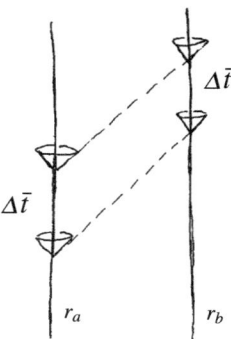

1.1.2.1 Schwarzschild Coordinates

The line element in so-called *Schwarzschild coordinates* is given by

$$ds^2 = \left(1 - \frac{r_s}{r}\right)d\bar{t}^2 - \left(1 - \frac{r_s}{r}\right)^{-1} dr^2 - r^2\left(d\theta^2 + \sin^2\theta d\phi^2\right). \quad (1.4)$$

Here $r_s = 2GM/c^2$ is the *Schwarzschild radius*, with M is the mass, and c is set to 1. Far from the black hole, M determines the force of attraction in the Newtonian limit, and Mc^2 is the total energy of the spacetime.

The spherical symmetry is manifest in the form of the line element. The coordinates θ and ϕ are standard spherical coordinates, while r measures $1/2\pi$ times the circumference of a great circle, or the square root of $1/4\pi$ times the area of a sphere. The value $r = r_s$ corresponds to the *event horizon*, as will be explained, and the value $r = 0$ is the "center", where the gravitational tidal force (curvature of the spacetime) is infinite. Note that r should *not* to be thought of as the radial distance to $r = 0$. That distance isn't well defined until a spacetime path is chosen. (A path at constant $\bar{t}$ does not reach any $r < r_s$.)

The coordinate $\bar{t}$ is the *Schwarzschild time*. It measures proper time at $r = \infty$, whereas at any other fixed r, θ, ϕ the proper time interval is $\Delta \tau = \sqrt{1 - r_s/r} dt$. The coefficients in the line element are independent of $\bar{t}$, hence the spacetime has a symmetry under $\bar{t}$ translation. This is ordinary time translation symmetry at $r = \infty$, but it becomes a lightlike translation at $r = r_s$, and a space translation symmetry for $r < r_s$, since the coefficient of $d\bar{t}^2$ is negative there. The defining property of the Schwarzschild time coordinate, other than that it measures proper time in the rest frame of the black hole at infinity, is that surfaces of constant $\bar{t}$ are orthogonal, in the spacetime sense, to the direction of the time-translation symmetry, i.e. to the lines of constant (r, θ, ϕ): there are no off-diagonal terms in the line element. But this nice property is also why $\bar{t}$ is ill behaved at the horizon.

Redshift and Horizon Suppose a light wave is generated with coordinate period $\Delta \bar{t}$ at some radius r_a, and propagates to another radius r_b (see Fig. 1.2). Because of the time translation symmetry of the spacetime, the coordinate period of the wave

Fig. 1.3 A null surface is tangent to the local light cone

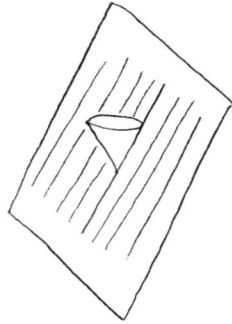

at r_b will also be $\Delta \bar{t}$. The ratio of the proper time periods will thus be $\Delta \tau_a / \Delta \tau_b = \sqrt{1 - r_s/r_a}/\sqrt{1 - r_s/r_b}$, and the ratio of the frequencies will the reciprocal. This is the *gravitational redshift*. Note that as $r_a \to r_s$, the redshift is infinite. The infinite redshift surface $r = r_s$ of the spherical black hole is the (stationary) event horizon. The same is true of the $1+1$ dimensional black holes we focus on later in these notes.

It is worth emphasizing that for a *non-spherical* stationary black hole, for instance a rotating black hole, the infinite redshift surface, where the time-translation symmetry becomes lightlike, is generally *not* the event horizon, because it is a timelike surface. A timelike surface can be crossed in either direction. In order to be a horizon, a surface must be tangent to the local light cone at each point, so that it cannot be crossed from inside to outside without going faster than light. At each point of such a *null surface* there is one null tangent direction, and all other tangent directions are spacelike and orthogonal to the null direction (see Fig. 1.3). Therefore the null tangent direction is orthogonal to all directions in the surface, i.e. the null tangent is also the normal. If the horizon is a constant r surface, then the gradient $\nabla_\alpha r$ is also orthogonal to all directions in the surface, so it must be parallel to the null normal. This means that it is a null (co)vector, hence $g^{\alpha\beta} \nabla_\alpha r \nabla_\beta r = g^{rr} = 0$ at the horizon.

1.1.2.2 Painlevé-Gullstrand Coordinates

A new time coordinate t that is well behaved at the horizon can be defined by $t = \bar{t} + h(r)$ for a suitable function $h(r)$ whose bad behavior at r_s cancels that of $\bar{t}$. This property of course leaves a huge freedom in $h(r)$, but a particularly nice choice is defined by

$$dt = d\bar{t} + \frac{\sqrt{r}}{r-1} dr, \quad \text{i.e.} \quad t = \bar{t} - 2\sqrt{r} + \ln\left(\frac{\sqrt{r}+1}{\sqrt{r}-1}\right) \tag{1.5}$$

where now I have adopted units with $r_s = 1$. It is easy to see that the t-r part of the Schwarzschild line element takes the form

$$ds^2 = dt^2 - \left(dr + \sqrt{\frac{1}{r}}dt\right)^2 - r^2(d\theta^2 + \sin^2\theta d\phi^2) \tag{1.6}$$

$$= \left(1 - \frac{1}{r}\right)dt^2 - \frac{2}{\sqrt{r}}dtdr - dr^2 - r^2(d\theta^2 + \sin^2\theta d\phi^2). \tag{1.7}$$

The new coordinate t is called the *Painlevé-Gullstrand* (PG) time. At $r = 1$ the metric coefficients are all regular, and indeed the coordinates are all well behaved there. According to (1.7), we have $ds^2 = 0$ along a line of constant $(r = 1, \theta, \phi)$, so such a line is lightlike. Such lines generate the event horizon of the black hole. The PG time coordinate has some remarkable properties:

- the constant t surfaces are flat, Euclidean spaces;
- the radial worldlines orthogonal to the constant t surfaces are timelike geodesics (free-fall trajectories) along which dt is the proper time.

For some practice in spacetime geometry, let me take you through verifying these properties. Setting $dt = 0$ in the line element we see immediately that $\{r, \theta, \phi\}$ are standard spherical coordinates in Euclidean space. To find the direction orthogonal to a constant t surface we could note that the gradient $\nabla_\alpha t$ has vanishing contraction with any vector tangent to this surface, which implies that the contravariant vector $g^{\alpha\beta}\nabla_\beta t$, formed by contraction with the inverse metric $g^{\alpha\beta}$, is orthogonal to the surface. Alternatively, we need not compute the inverse metric, since the form of the line element (1.6) allows us to read off the orthogonal direction "by inspection" as follows. Consider the inner product of two 4-vectors v and w in this metric,

$$g(v,w) = dt(v)dt(w) - \left(dr + \sqrt{\frac{1}{r}}dt\right)(v)\left(dr + \sqrt{\frac{1}{r}}dt\right)(w)$$
$$- r^2 d\theta(v)d\theta(w) - r^2 \sin^2\theta d\phi(v)d\phi(w), \tag{1.8}$$

using the notation of Eq. (1.2). If the vector v is tangent to the constant t surface, then $dt(v) = 0$, so the first term vanishes. The remaining terms will vanish if $(dr + \sqrt{\frac{1}{r}}dt)(w) = d\theta(w) = d\phi(w) = 0$. Thus radial curves with $dr + \sqrt{1/r}dt = d\theta = d\phi = 0$ are orthogonal to the surface, and along them $ds^2 = dt^2$, i.e. dt measures proper time along those curves. Moreover, any other timelike curve connecting the same two spacetime points will have shorter proper time, because the negative terms in ds^2 will contribute. The proper time is thus stationary with respect to first order variations of the curve, which is the defining property of a geodesic.[1]

---

[1] Even if the other terms in the line element (1.6) had not been negative, they would not contribute to the first order variation in the proper time away from a path with $(dr + \sqrt{1/r}dt) = d\theta = d\phi^2 = 0$, since the line element is quadratic in these terms. Thus the curve would still have been a geodesic (although the metric signature would not be Lorentzian).

Fig. 1.4 Painlevé-Gullstrand coordinate grid for Schwarzschild black hole. *Vertical lines* have constant r, *horizontal lines* have constant t. Shown are one ingoing radial light ray and three outgoing ones. The one outside the horizon escapes to larger radii, the one on the horizon remains at r_s, and the one inside the horizon falls to smaller radii and into the singularity at $r = 0$

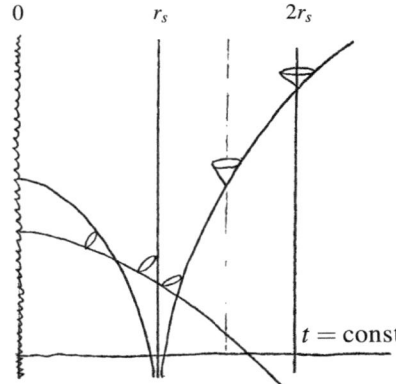

1.1.2.3 Spacetime Diagram of the Black Hole

The nature of the unusual geometry of the black hole spacetime can be grasped rather easily with the aid of a spacetime diagram (see Fig. 1.4). For the Schwarzschild black hole, we may exploit the spherical symmetry and plot just a fixed value of the spherical angles (θ, ϕ), and we may plot the lines of constant r vertically and the lines of constant PG time t horizontally. Then the time translation symmetry corresponds to a vertical translation symmetry of the diagram.

The diagram comes alive when the light cones are plotted. At a given event, the light cone is determined by $ds^2 = 0$, which for radial displacements corresponds to the two slopes

$$dt/dr = \frac{1}{\pm 1 - \sqrt{1/r}} \quad \text{(radial lightrays)}. \tag{1.9}$$

Far from the horizon these are the outgoing and incoming lightrays $dt/dr \to \pm 1$. The ingoing slope is negative and gets smaller in absolute value as r decreases, approaching 0 as $r \to 0$. The outgoing slope grows as r decreases, until reaching infinity at the horizon at $r = 1$. Inside the horizon it is negative, so an "outgoing" lightray actually propagates to smaller values of r. The outgoing slope also approaches 0 as $r \to 0$.

1.1.2.4 Redshift of Outgoing Waves Near the Horizon

An outgoing wave is stretched as it climbs away from the horizon. The lines of constant phase for an outgoing wave satisfying the relativistic wave equation are just the outgoing lightrays (1.9). The rate of change of a wavelength λ is given by the difference of dr/dt of the lightrays on the two ends of a wavelength, hence $d\lambda/dt = (d/dr)(dr/dt)\lambda$. The relative stretching rate is thus given by

$$\kappa \equiv \frac{d\lambda/dt}{\lambda} = \frac{d}{dr}\frac{dr}{dt} = \frac{c}{2r_s}, \tag{1.10}$$

where in the second step the expression is evaluated at the horizon, and the dimensionful constants are restored to better illustrate the meaning. This rate is called the "surface gravity" κ of the horizon. Later I will explain different ways in which the surface gravity can defined and calculated.

We can go further and use the lightray equation (1.9) to obtain an approximate expression for the wave phase near the horizon. Consider an outgoing wave of the form $e^{i\phi}$, with $\phi = -\omega t + \int^r k(r')dr'$. (This simple harmonic t dependence is exact because the metric is independent of t.) Along an outgoing lightray the phase is constant: $0 = d\phi = -\omega dt + k(r)dr$, so

$$k(r) = \frac{\omega}{1 - r^{-1/2}} \sim \frac{2\omega}{r-1} = \frac{\omega/\kappa}{r - r_s}, \qquad (1.11)$$

where in the second step a near horizon approximation is used, and in the last step the dimensionful constants are again restored. The wave thus has the near-horizon form

$$e^{-i\omega t} e^{i(\omega/\kappa)\ln(r-r_s)}. \qquad (1.12)$$

Note that the surface gravity appears in a ratio with the wave frequency, and there is a logarithmic divergence in the outgoing wave phase at the horizon.

1.1.3 Effective Black Hole and White Hole Spacetimes

Many black hole analogues can be described with one spatial dimension, and I will focus on those here. They are simple generalizations of the radial direction for a spherical black hole.

Waves or quasiparticles in a stationary $1 + 1$ dimensional setting can often be described by a relativistic field in an effective spacetime defined by a metric of the form

$$ds^2 = c(x)^2 dt^2 - [dx - v(x)dt]^2 = [c(x)^2 - v(x)^2]dt^2 - 2v(x)dtdx - dx^2. \qquad (1.13)$$

In fact, any stationary two dimensional metric can be put in this form, with $c(x) = 1$, by a suitable choice of coordinates (see e.g. Appendix A in Ref. [1] for a proof of this statement). If $c(x) = 1$ this corresponds to the PG metric, with $x \leftrightarrow r$ and $v(x) \leftrightarrow -1/\sqrt{r}$. if $|v(x)| > |c(x)|$ somewhere.

The metric (1.13) would arise for example in a Newtonian setting of a fluid, with velocity $v(x)$ in a "laboratory frame", with $c(x) = c$ a constant speed of sound. In that example, the coordinate x would measure distance in the lab at fixed Newtonian time t, and the metric would describe the effective spacetime for waves in the fluid that propagate at speed c relative to the local rest frame of the fluid. If the wave speed in the frame of the medium depends on some ambient local conditions then $c(x)$ will depend on position.

Fig. 1.5 Black hole horizon on the right and white hole horizon on the left. The *vertical arrows* depict the Killing vector, which is spacelike in the ergoregion between the horizons and timelike outside

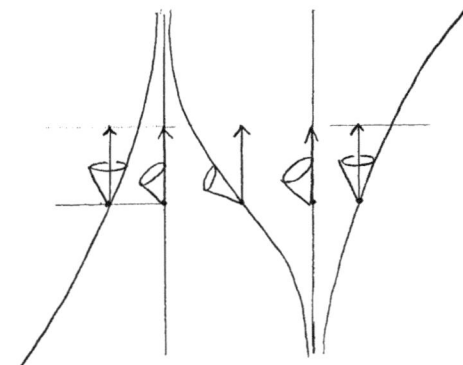

Moving Texture In some models the medium may be at rest in the lab, but the local conditions that determine the wave speed may depend on both time and space in a "texture" that moves. (If the motion is uniform then in the frame of the texture this is equivalent to the previous case.) An example of a line element of this sort is $[c(y - wt)]^2 dt^2 - dy^2$. Here again y measures proper distance in the lab at Newtonian time t, and the texture moves in the y direction with constant speed w. The line element may not look stationary, but it has a symmetry under $t \to t + \Delta t$ combined with $y \to y + w \Delta t$.

Black Hole–White Hole Pair An example that often arises has $v(x) < -c(x) < 0$ in a finite interval (x_-, x_+). Then x_+ is a black hole horizon, analogous to the one previously discussed for the PG spacetime, and x_- is a *white hole* horizon: no waves can escape from the region $x < x_+$ into the region $x > x_+$, and no waves can enter the region $x > x-$ from the region $x < x_-$. The region between the horizons is of finite size and nonsingular. Figure 1.5 is a spacetime diagram of this scenario. Black hole horizon on the right and white hole horizon on the left. The vertical arrows depict the Killing vector, which is spacelike in the ergoregion between the horizons and timelike outside.

1.1.4 Symmetries, Killing Vectors, and Conserved Quantities

Each symmetry of the background spacetime and fields leads to a corresponding conservation law. The most transparent situation is when the metric and any other background fields are simply independent of some coordinate. This holds for example with the Schwarzschild metric (1.4), which is independent of both t and ϕ. Of course the spherical symmetry goes beyond just ϕ translations, but the other rotational symmetries are not manifest in this particular form of the line element. They could be made manifest by a change of coordinates however, but not all at once. To be able to talk about symmetries in a way that is independent of whether or not they are manifest it is useful to introduce the notion of a *Killing vector field*. The flow of the spacetime along the integral curves of a Killing vector is a symmetry of the spacetime.

Suppose translation by some particular coordinate $x^{\hat{\alpha}}$ ($\hat{\alpha}$ indicates one particular value of the index α) is a manifest symmetry. The metric components satisfy $g_{\mu\nu,\hat{\alpha}} = 0$, where the comma notation denotes partial derivative with respect to $x^{\hat{\alpha}}$. The corresponding Killing vector, written in these coordinates, is $\chi^\mu = \delta^\mu_{\hat{\alpha}}$, i.e. the vector with all components zero except the $\hat{\alpha}$ component which is 1. Then the symmetry is expressed by the equation $g_{\mu\nu,\alpha}\chi^\alpha = 0$. This holds only in special coordinate systems adapted to the Killing vector. It is not a tensor equation, since the partial derivative of the metric is not a tensor.

It may be helpful to understand that this condition is equivalent to the covariant, tensor equation for a Killing vector,

$$\chi_{\alpha;\beta} + \chi_{\beta;\alpha} = 0, \tag{1.14}$$

where the semicolon denotes the covariant derivative. This is called *Killing's equation*. One way to see the equivalence is to use the fact that in a local inertial coordinate system at a point p, the covariant derivative reduces to the partial derivative, and the partials of the metric are zero. Thus Killing's equation at the point p becomes $\chi^\sigma{}_{,\beta}\eta_{\sigma\alpha} + \chi^\sigma{}_{,\alpha}\eta_{\sigma\beta} = 0$, where $\eta_{\sigma\tau}$ is the Minkowski metric. This implies that the infinitesimal flow $x^\sigma \to x^\sigma + \varepsilon\chi^\sigma(x)$ generated by χ^α is, to lowest order, a translation plus a Lorentz transformation, i.e. a symmetry of the metric.[2]

A simple example is the Euclidean plane with line element $ds^2 = dx^2 + dy^2 = dr^2 + r^2 d\phi^2$ in Cartesian and polar coordinates respectively. The rotation Killing vector about the origin in polar coordinates is just ∂_ϕ, with components δ^α_ϕ, as the metric components are independent of ϕ. The same Killing vector in Cartesian coordinates is $x\partial_y - y\partial_x$. This satisfies Killing's equation since $\chi_{x,x} = 0 = \chi_{y,y}$, and $\chi_{x,y} + \chi_{y,x} = -1 + 1 = 0$.

1.1.4.1 Ergoregions

It is of paramount importance in black hole physics that a Killing field may be timelike in some regions and spacelike in other regions of a spacetime. For example in the Schwarzschild spacetime, say in PG coordinates (1.6), or the $1+1$ dimensional generalization (1.13) the Killing vector ∂_t is timelike outside the horizon, but it is lightlike on the horizon and spacelike inside. For the black hole-white hole pair discussed above, it is the region between the black and white hole horizons (see

---

[2]For a more computational proof, note that since Killing's equation is a tensor equation it holds in all coordinate systems if it holds in one. In a coordinate system for which $\chi^\mu = \delta^\mu_{\hat{\alpha}}$ we have $\chi_{\alpha;\beta} = g_{\alpha\mu}\chi^\mu{}_{;\beta} = g_{\alpha\mu}\Gamma^\mu{}_{\beta\sigma}\chi^\sigma = \frac{1}{2}(g_{\alpha\beta,\sigma} + g_{\alpha\sigma,\beta} - g_{\beta\sigma,\alpha})\chi^\sigma$. If χ^σ is a Killing vector the first term vanishes in this adapted coordinate system, and the remaining expression is antisymmetric in α and β, so adding $\chi_{\beta;\alpha}$ yields zero. Conversely, if Killing's equation holds, the entire expression is antisymmetric in α and β, so the first term must vanish.

Fig. 1.5). This is evident because the coefficient of dt^2 in the line element becomes negative.

A region where an otherwise timelike Killing vector becomes spacelike is called an *ergoregion*. (The reason for the name will become clear below.) The boundary of this region is called the *ergosurface*, and it is a surface of infinite redshift, since the norm of the time translation Killing vector vanishes there. An ergoregion need not lie behind a horizon. For instance it occurs outside the horizon (as well as inside) of a spinning black hole. In analogue models, ergoregions can arise for example around a vortex [2] or in a moving soliton in superfluid ^{3}He-A [3]. For the Schwarzschild black hole, and the 1 + 1 dimensional generalization (1.13), however, the ergoregion always corresponds to the region inside the horizon.

1.1.4.2 Conserved Quantities

Particle trajectories (both timelike and lightlike) can be determined by the variational principle $\delta \int L d\lambda = 0$ with Lagrangian $L = \frac{1}{2} g_{\mu\nu}(x) \dot{x}^\mu \dot{x}^\nu$. Here λ is a path parameter and the dot denotes $d/d\lambda$. The Euler-Lagrange equation is the geodesic equation for motion in the metric $g_{\mu\nu}$ with affine parameter λ. If the metric is independent of $x^{\hat{\alpha}}$ then the corresponding conjugate momentum $p_{\hat{\alpha}} = \partial L / \partial \dot{x}^{\hat{\alpha}} = g_{\mu\hat{\alpha}} \dot{x}^\mu$ is a constant of motion. Note that this momentum can also be expressed as the inner product of the 4-velocity $u^\nu = \dot{x}^\nu$ with the Killing field, $u \cdot \chi = g_{\mu\nu} \dot{x}^\mu \chi^\nu = g_{\mu\hat{\alpha}} \dot{x}^\mu$.

Killing Energy and Ergoregions The conserved momentum conjugate to a particular timelike Killing field is called *Killing energy*. For a particle with rest mass m, the physical 4-momentum would be $p = mu$, so the Killing energy as defined above is actually the Killing energy per unit rest mass. For a massless particle, the physical 4-momentum is proportional to the lightlike 4-velocity, scaled so that the time component in a given frame is the energy in that frame. In both cases, the true Killing energy is the inner product of the 4-momentum and the Killing vector,

$$E_{\text{Killing}} = p \cdot \chi. \tag{1.15}$$

The 4-momentum of a massive particle is timelike, while that of a massless particle is lightlike. In both cases, for a physical state (i.e. an allowable excitation of the vacuum), stability of the local vacuum implies that the energy of the particle is positive as measured locally in any rest frame. This is equivalent to the statement that p is a *future pointing* 4-vector.

The importance of ergoregions stems from the fact that negative Killing energy physical states exist there. This happens because a future pointing 4-momentum can of course have a negative inner product with a spacelike vector (see Fig. 1.6). In an ergoregion, the Killing energy is what would normally be called a linear momentum component, and there is of course no lower limit on the linear momentum of a physical state.

Penrose [4, 5] realized that the existence of an ergoregion outside a spinning black hole implies that energy can be extracted from the black hole by a classical

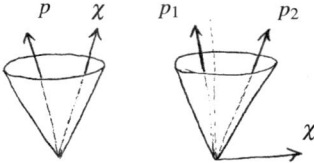

Fig. 1.6 Killing energy. On the *left* the Killing vector χ is timelike, hence all future causal (timelike or lightlike) 4-momenta have positive χ-energy. On the *right* χ is spacelike, hence future causal 4-momenta like p_2 can have negative χ-energy, while others like p_1 have positive χ-energy

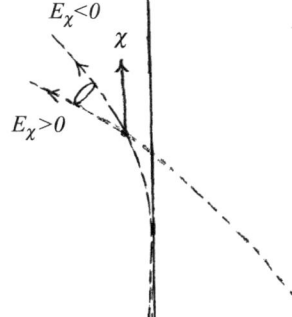

Fig. 1.7 Inside the horizon, the Killing vector χ is spacelike, outgoing radial particles have negative χ-energy, and infalling ones have positive χ-energy (Since the latter come from outside the ergoregion, and Killing energy is conserved, they must have positive Killing energy)

process, at the cost of lowering the angular momentum. This is the *Penrose process*, whose existence led to the discovery of black hole thermodynamics. For a non-spinning black hole the ergoregion lies inside the horizon, so no classical process can exploit it to extract energy, but the *Hawking effect* is a quantum process by which energy is extracted.

What do the negative Killing energy states "look like"? A particle with negative Killing energy cannot escape from the ergoregion, nor can it have fallen freely into the ergoregion, because Killing energy is conserved along a geodesic and it must have positive Killing energy if outside the ergoregion. For example, in the $1 + 1$ black hole, or in the radial direction of the Schwarzschild solution, a massless particle with negative Killing energy inside the horizon must be "outgoing" as seen by a local observer (see Fig. 1.7).

1.1.5 Killing Horizons and Surface Gravity

An event horizon can be defined purely in terms of the causal structure of a spacetime, and is meaningful even when the spacetime is not stationary, i.e. has no time translation symmetry. A *Killing horizon* on the other hand is a lightlike hypersurface (surface of one less dimension than the whole spacetime) generated by the flow of a Killing vector. This is sometimes called the *horizon generating Killing vector*.

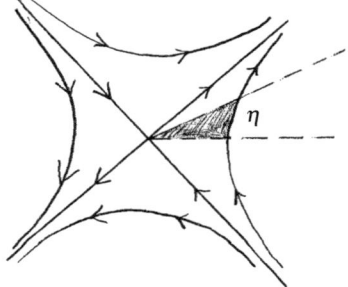

Fig. 1.8 Boost killing flow in Minkowski space (1.17). Curves of constant ℓ are hyperbolic flow lines. Lines of constant η are radial from the origin, and η measures the hyperbolic opening angle of the *shaded wedge*

The Schwarzschild event horizon is a Killing horizon with respect to the Killing vector ∂_t, as is the horizon of the $1+1$ black hole. A distinction arises in the case of a stationary black hole with spin. Then the Killing vector ∂_t that is a time translation at spatial infinity becomes lightlike at the boundary of the ergoregion, which lies outside the event horizon. However that boundary is timelike, so the ergosurface is not a Killing horizon. The event horizon of a spinning black hole is nevertheless a Killing horizon, but for a Killing vector $\partial_t + \Omega_H \partial_\phi$ that is a linear combination of the time translation and rotation Killing vectors, Ω_H being the angular velocity of the horizon. In the effective spacetime of a moving texture in superfluid ^{3}He-A, the horizon generating Killing vector has the similar form $\partial_t + w\partial_x$, where ∂_t and ∂_x are time and space translation Killing vectors, and the constant w can be thought of as the transverse velocity of the horizon [3].

Rindler (Acceleration) Horizon A simple yet canonical example of a Killing horizon is the Rindler horizon in Minkowski spacetime. The relevant Killing symmetry here is Lorentz boosts is a certain direction. Geometrically, these are just hyperbolic rotations. For example, using the Minkowski coordinates of (1.1) a *boost Killing vector* is

$$\chi_B = x\partial_t + t\partial_x. \tag{1.16}$$

This has covariant components $(\chi_B)_\alpha = \eta_{\alpha\beta}\chi_B^\beta = (x, -t)$ and so obviously satisfies Killing's equation (1.14). It can also be made manifest by changing from Minkowski to polar coordinates:

$$dt^2 - dx^2 = \ell^2 d\eta^2 - d\ell^2. \tag{1.17}$$

Then the boost symmetry is just rotation of the hyperbolic angle η, i.e.

$$\chi_B = \partial_\eta. \tag{1.18}$$

The flow lines of the Killing field are hyperbolas (see Fig. 1.8). Note that the polar coordinate system covers only one "Rindler wedge", e.g. $x > |t|$ of the Minkowski spacetime. The full Killing horizon is the set $|x| = |t|$.

1.1.5.1 Surface Gravity

Associated to a Killing horizon is a quantity κ called the *surface gravity*. There are many ways to define, calculate, and think of the surface gravity. It was already introduced in Sect. 1.1.2.4, as the relative rate of stretching of outgoing wavelengths near the horizon. I will mention here several other definitions, which are given directly in terms of the geometry of the horizon.

Geometrically, the simplest definition of surface gravity may be via

$$\left[\chi_{[\alpha,\beta]}\chi^{[\alpha,\beta]}\right]_H = -2\kappa^2, \tag{1.19}$$

horizon the square bracket on indices denotes antisymmetrization, and the subscript H indicates that the quantity is evaluated on the horizon. That is, κ is the magnitude of the infinitesimal Lorentz transformation generator. However the meaning of this is probably not very intuitive.

The conceptually simplest definition might be the rate at which the norm of the Killing vector vanishes as the horizon is approached from outside. That is,

$$\kappa = \left||\chi|_{,\alpha}\right|_H, \tag{1.20}$$

the horizon limit of the norm of the gradient of the norm of χ. Notice that if the Killing vector is rescaled by a constant multiple $\chi \to \alpha\chi$, then it remains a Killing vector, and the surface gravity for this new Killing vector is $\alpha\kappa$. This illustrates the important point that the intrinsic structure of a Killing horizon alone does not suffice to define the surface gravity. Rather, a particular normalization of the Killing vector is required. The symmetry implies that κ is constant along a particular null generator of the horizon, but in general it need not be the same on all generators. For a discussion of conditions under which the surface gravity can be proved to be constant see [6].

The surface gravity (1.20) has the interesting property that it is *conformally invariant*. That is, it is unchanged by a conformal rescaling of the metric $g_{ab} \to \Omega^2 g_{ab}$, provided the conformal factor Ω is regular at the horizon [7]. This follows simply because $|\chi|$ is rescaled by Ω, while the norm of its gradient is rescaled by Ω^{-1}, and the contribution from $d\Omega$ vanishes since it is multiplied by $|\chi|_H$ which vanishes.

For the metric (1.13) and the Killing vector $\chi = \partial_t$ we have $|\chi| = \sqrt{c^2 - v^2}$, which depends on x and not t. Thus $\kappa = (-g^{xx}\partial_x|\chi|\partial_x|\chi|)_H^{1/2}$, and the minus sign arises because the gradient is spacelike outside the horizon. At a horizon where $v = c$ this evaluates to $|\partial_x(v-c)|_H$, while at a horizon where $v = -c$ it would instead be $|\partial_x(v+c)|_H$.

In case $c = constant$, the surface gravity is thus just the gradient of the flow speed at the horizon. A covariant and more general version of this can be formulated. Any observer falling freely across a horizon can define the velocity of the static frame relative to himself, and can evaluate the spatial gradient of this velocity in his frame. If he has unit Killing energy ($u \cdot \chi = 1$) then it can be shown that this gradient, evaluated at the horizon, agrees with the surface gravity [8]. Another interesting observation is that this velocity gradient has a sort of "cosmological" interpretation as the local fractional rate of expansion ("Hubble constant") of the distances separating a family of freely falling observers stretched along the direction of the Killing frame velocity [8]. At the horizon, for unit energy observers, this expansion rate is the same as the surface gravity.

Computationally, a somewhat simpler definition of surface gravity is via

$$[\partial_\alpha(\chi^2) = -2\kappa \chi_\alpha]_H. \tag{1.21}$$

This is at least well-defined: since χ^2 vanishes everywhere on the Killing horizon, its gradient has zero contraction with all vectors tangent to the horizon. The same is true for $\chi_a = g_{\alpha\beta}\chi^\beta$, so these two co-vectors must be parallel. If using a coordinate component of this equation to evaluate κ, it is important that the coordinate system be regular at the horizon. For the metric (1.13), we may just evaluate the x component of this equation: $\partial_x(c^2 - v^2) = -2\kappa\chi_x = -2\kappa g_{xt} = -2\kappa v$, which on the horizon $v = c$ yields $\kappa = [\partial_x(v-c)]_H$ as before. (Note that this definition does not come with an absolute value. At a horizon $v = -c$ it yields $\kappa = [\partial_x(v+c)]_H$.)

Surface Gravity of the Rindler Horizon The surface gravity of the Rindler horizon can be computed for example using the polar coordinates to evaluate (1.20). Then the norm of the Killing vector is just ℓ, so $\partial_\alpha|\chi_B| = \delta_\alpha^\ell$, which has norm 1. Thus the boost Killing vector has unit surface gravity. Alternatively, we may use the x component of (1.21): $\partial_x \chi_B^2 = x^2 - t^2 = 2x$, and $-2\kappa(\chi_B)_x = 2\kappa t$, so $\kappa = (x/t)_H = \pm 1$. On the future horizon $x = t$ and this is positive, while on the past horizon it is negative. Usually one is only interested in the absolute value.

Finally, it is sometimes of interest to use the proper time along a particular hyperbola rather than the hyperbolic angle as the coordinate. On the hyperbola located at $\ell = \ell_0$ the proper time is $d\tau = \ell_0 d\eta$. The Minkowski line element can be written in terms of the time coordinate $\tau = \ell_0\eta$ as $ds^2 = (\ell/\ell_0)^2 d\tau^2 - d\ell^2$. The scaling of the Killing field $\partial_\tau = (1/\ell_0)\partial_\eta$ that generates proper time flow on this particular hyperbola has surface gravity $\kappa = 1/\ell_0$. This is also equal to the *acceleration* of the hyperbolic worldline. The relation between the surface gravity and acceleration can be shown quite generally using coordinate free methods, but here let's just show it by direct computation using Cartesian coordinates. The 4-velocity of the hyperbola is the unit vector $u = \ell_0^{-1}(x, t, 0, 0)$, and the acceleration of this worldline is $(u \cdot \nabla)u = \ell_0^{-2}(x\partial_t + t\partial_x)(x, t, 0, 0) = \ell_0^{-2}(t, x, 0, 0)$. The norm of the spacelike vector $(t, x, 0, 0)$ is ℓ_0, so the norm of the acceleration is $1/\ell_0$.

1.2 Thermality of the Vacuum

The subject of the rest of these notes is the Hawking effect, i.e. the emission of thermal radiation from a black hole. The root of the Hawking effect is the thermality of the vacuum in flat spacetime. This thermality is known as the Unruh, or Fulling-Davies-Unruh, effect [9]. In its narrowest form, this is the fact that a probe with uniform proper acceleration a, moving through the vacuum of a quantum field in flat spacetime, is thermally excited at the Unruh temperature

$$T_U = \hbar a/2\pi c. \tag{1.22}$$

(I've restored c here to show where it enters, but will immediately revert to units with $c = 1$.) When described this way, however, too much attention is focused on the probe and its acceleration.

Underlying the response of the probe is a rather amazing general fact: when restricted to a Rindler wedge, the vacuum of a relativistic quantum field is a canonical thermal state with density matrix

$$\rho_R \propto \exp(-2\pi H_\eta/\hbar), \tag{1.23}$$

where H_η is the "boost Hamiltonian" or "Rindler Hamiltonian" generating shifts of the hyperbolic angle coordinate η defined in (1.17). In terms of Minkowski coordinates (t, x, y, z), H_η is given on a $t = 0$ surface of the Rindler wedge by

$$H_\eta = \int_{\Sigma_R} T_{ab}\chi_B^a d\Sigma^b = \int x T_{tt} dx dy dz, \tag{1.24}$$

where T_{ab} is the energy-momentum tensor. The "temperature" of the thermal state (1.23) is

$$T_R = \hbar/2\pi. \tag{1.25}$$

Like a rotation angle, the hyperbolic angle is dimensionless, so the boost generator and temperature have dimensions of angular momentum.

Note that the thermal nature of the vacuum in the wedge does not refer to any particular acceleration, and it characterizes the state even on a single time slice. Nevertheless it does directly predict the Unruh effect. A localized probe that moves along a particular hyperbolic trajectory at proper distance ℓ_0 from the vertex of the wedge has proper time interval $d\tau = \ell_0 d\eta$ (cf. (1.17)). When scaled to generate translations of this proper time the field Hamiltonian is thus $H_\tau = \ell_0^{-1} H_\eta$, and the corresponding temperature is $T_0 = \ell_0^{-1}\hbar/2\pi$. The proper acceleration of that hyperbola is ℓ_0^{-1}, so the probe will be excited at the Unruh temperature (1.22).

The thermality of the vacuum in one wedge is related to entanglement between the quantum states in the right and left wedges. It can be understood using a simple, but abstract and formal, argument that employs the path integral expression for the ground state. Because the result is so central to the subject, I think this argument deserves to be explained.

The vacuum $|0\rangle$ is the ground state of the field Hamiltonian H, and can therefore be projected out of any state $|\chi\rangle$ as $|0\rangle \propto \lim_{t\to\infty} e^{-tH}|\chi\rangle$, as long as $\langle 0|\chi\rangle \neq 0$. The operator e^{-tH} can be thought of as the time evolution operator for an imaginary time $-it$, and its matrix elements can be represented by a path integral over fields ϕ on Euclidean space. This yields a path integral representation for the vacuum wave functional,

$$\Psi_0[\phi] \propto \lim_{t\to\infty} \langle\phi|e^{-tH}|\chi\rangle \propto \int_{\phi(t=-\infty)=\chi}^{\phi(t=0)=\phi} \mathcal{D}\phi \, e^{-S/\hbar}, \tag{1.26}$$

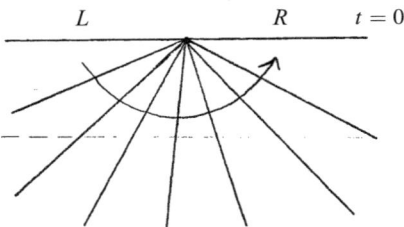

Fig. 1.9 Euclidean Minkowski space with boundary at $t = 0$. When the path integral (1.26) is sliced by constant t surfaces it presents the vacuum wave-functional. When sliced by constant angle surfaces, it presents matrix elements of the operator $\exp(-\pi H_\eta)$, where H_η is the Lorentz boost generator

where S is the Euclidean action corresponding to the Hamiltonian H. The standard demonstration of this path integral expression for matrix elements of e^{-tH} proceeds by slicing the Euclidean space into steps of constant Euclidean time, and exploits the time translation invariance of the Hamiltonian. If the original Hamiltonian is also Lorentz boost invariant, then the Euclidean action is also rotationally invariant. This extra symmetry leads to an alternate interpretation of the path integral.

Fixing a particular rotational symmetry, e.g. around the origin in the Euclidean tx plane, we may choose to slice the Euclidean space into steps of constant angle around the corresponding vertex (see Fig. 1.9). This vertex divides the time slice $t = 0$ into two halves, and the final field configuration ϕ restricts to some ϕ_L and ϕ_R on the left and right sides respectively. These configurations define Dirac "bras" $\langle \phi_L |$ and $\langle \phi_R |$ in the duals of the left and right side Hilbert spaces $\mathcal{H}_L$ and $\mathcal{H}_R$. The full Hilbert space is the tensor product $\mathcal{H}_L \otimes \mathcal{H}_R$.

With this angular slicing, (and not worrying about boundary conditions at the vertex), we can think of the path integral as producing the matrix element of the operator $\exp(-\pi H_\eta)$ between ϕ_L, regarded now as an *initial* state, and the final state ϕ_R,

$$\Psi_0[\phi_L, \phi_R] \propto \langle \phi_R | e^{-\pi H_\eta} J | \phi_L \rangle. \tag{1.27}$$

Here H_η is the boost Hamiltonian, which is the generator of angle shifts, and π is the rotation angle in the Euclidean plane. (The rotation angle is to the boost angle as the Euclidean time is to the Minkowski time.) The final state bra $\langle \phi_L |$ is replaced by a "corresponding" initial state ket $J | \phi_L \rangle$ that can be identified with a state in $\mathcal{H}_R$. Here $J = CTP^1$ is the operator of charge conjugation, time reversal, and reflection across the Rindler plane, which is a symmetry of all Lorentz invariant quantum field theories.[3]

---

[3]For a configuration eigenstate of a real field, the ket $J | \phi_L \rangle$ can just be identified with the same function ϕ_L, reflected by an operator P^1 across the Rindler plane. More generally, J includes CT to undo the conjugation of the $\langle bra | \to | ket \rangle$ duality.

The vacuum wave-functional (1.27) can also be represented as a vector in the Hilbert space $\mathscr{H}_L \otimes \mathscr{H}_R$, by multiplying the amplitudes (1.27) by the corresponding kets and integrating over the fields:

$$|0\rangle \propto \int \mathscr{D}\phi_L \mathscr{D}\phi_R |\phi_L\rangle |\phi_R\rangle \langle \phi_R | e^{-\pi H_\eta} J \, \phi_L\rangle \tag{1.28}$$

$$= \int \mathscr{D}\phi_L |\phi_L\rangle e^{-\pi H_\eta} J |\phi_L\rangle \tag{1.29}$$

$$= \sum_n e^{-\pi E_n} |n\rangle_L |\bar{n}\rangle_R. \tag{1.30}$$

In the last line the state is expressed in terms of eigenstates $|n\rangle$ of the boost Hamiltonian with boost energy E_n (with additional implicit quantum numbers). It is obtained via $J|\phi_L\rangle = \sum_n J|n\rangle\langle n|\phi_L\rangle = \sum_n \langle \phi_L|n\rangle J|n\rangle$, using the anti-linearity of J. Then the integral over ϕ_L yields the identity operator, and the result follows since H_η commutes with J. The state $|\bar{n}\rangle$ stands for the "antiparticle state" $J|n\rangle$.

This exhibits the precise sense in which the quantum field degrees of freedom in the left and right Rindler wedges are entangled in the vacuum state. This entanglement is the origin of the correlations between the Hawking quanta and their partners, and it produces the entanglement entropy for quantum fields outside a horizon. Tracing over the state in the left wedge we obtain the reduced density matrix for the state restricted to right wedge,

$$\rho_R = \mathrm{Tr}_L |0\rangle\langle 0| \propto \sum_n e^{-2\pi E_n} |n\rangle\langle n|. \tag{1.31}$$

This is the canonical thermal state (1.23) mentioned above.[4] The horizon entanglement entropy is the entropy of this thermal state. It diverges as the horizon area times the square of the momentum cutoff.

1.3 Hawking Effect

The essence of the Hawking effect [10] is that the correlated vacuum fluctuations described in the previous section exist near the horizon of a black hole, which is locally equivalent to a Rindler horizon. The crucial difference from flat space is that tidal effects of curved spacetime peel apart the correlated partners. The outside quanta sometimes escape to infinity and sometimes fall backwards into the black hole, while the inside ones fall deeper into the black hole. The escaping quanta have a thermal spectrum with respect to the analogue of the boost Hamiltonian, that is, with respect to the Hamiltonian for the horizon-generating symmetry. If the horizon generating Killing vector is normalized to have unit surface gravity, like the boost Killing vector, the temperature is again the Rindler temperature $T_R = \hbar/2\pi$

---

[4] Its matrix elements could also have been obtained directly using the wave functional (1.27), via $\int \mathscr{D}\phi_L \Psi_0[\phi_L, \phi_R] \Psi_0^*[\phi_L, \phi_R'] \propto \langle \phi_R | e^{-2\pi H_\eta} | \phi_R' \rangle$.

(1.25). However, for a quantum that escapes from the black hole region, the natural definition of energy is the generator of asymptotic time translations. For defining this energy we normalize the time translation Killing vector at infinity. Then the black hole horizon has a surface gravity κ, and the temperature is the Hawking temperature,

$$T_H = \hbar\kappa/2\pi. \tag{1.32}$$

Note that the Unruh temperature (1.22) can be expressed in exactly the same way as the Hawking temperature since, as explained in Sect. 1.1.5.1, when the boost Killing field is normalized to unity on a given hyperbola the surface gravity of the Rindler horizon is precisely the acceleration of that hyperbola.

For a rotating black hole, as explained in Sect. 1.1.5, the horizon generating Killing vector is $\partial_t + \Omega_H \partial_\phi$. The eigenvalues of the Hamiltonian corresponding to this Killing vector are[5] $E - \Omega_H L$, where E and L are the energy "at infinity" and angular momentum respectively. Thus the Boltzmann factor for the Hawking radiation is $e^{-(E-\Omega_H L)/T_H}$. The angular velocity Ω_H plays the role of a chemical potential for the angular momentum.

Missing from this explanation of the Hawking effect is the specification of the incoming state. In principle, there are two places where the state can "come in" from: spatial infinity, and the horizon. The state coming from the horizon is determined to be the local vacuum by a regularity condition, since anything other than the vacuum would be singular as a result of infinite blueshift when followed backwards in time toward the horizon. This is what accounts for the universality of the thermal emission. However the state coming in from infinity has freedom. If it is the vacuum, the state is called the "Unruh state", while if it is a thermal state, as appropriate for thermal equilibrium of a black hole with its surroundings, it is the "Hartle-Hawking" state. In the neighborhood of the intersection of past and future horizons, the Hartle-Hawking state is close to the local Minkowski vacuum.

For black holes in general relativity, the above description of the Hawking effect is, in a sense, the complete story. For analogue models, however, one wants a derivation that does not assume Lorentz invariance, and that shows the way to the modifications brought about by the lack thereof. Also, it is important to be able to allow for experimental conditions that determine different incoming states. Moreover, in the analogue case the horizon state need not be the vacuum, since in the presence of Lorentz violating dispersion a different state can exist without entailing anything singular on the horizon. Thus we now take a very different viewpoint, analyzing the vacuum "mode by mode". It is this approach that Hawking originally followed when he discovered black hole radiation. It should be emphasized at the outset however that, unlike the previous treatment, this approach will apply only to free field theory, with uncoupled modes satisfying a linear field equation.

---

[5]The sign of the L term is opposite to that of the E term because ∂_ϕ is spacelike while ∂_t is timelike.

1.3.1 Mode Solutions

My aim here is to convey the essence of the Hawking effect, using a language that is easily adapted to analogue models in which dispersive effects play a role. Hence I will discuss only a system with one spatial dimension, and will highlight the role of the dispersion relation, using WKB methods.

Consider a scalar field φ that satisfies the wave equation $\partial_\alpha(\sqrt{-g}g^{\alpha\beta}\partial_\beta\varphi) = 0$. For the metric (1.13) we have $\sqrt{-g} = c$ and $g^{tt} = 1/c^2$, $g^{tx} = v/c^2$, $g^{xx} = (v^2 - c^2)/c^2$. Since the metric is independent of t we can find solutions with definite Killing frequency, $\varphi = e^{-i\omega t}u(x)$. Because of the redshift effect an outgoing solution has very rapid spatial oscillations of $u(x)$ near the horizon. We can thus find an approximate solution near the horizon by neglecting all terms in which there is not at least one derivative of $u(x)$. This yields the equation

$$\partial_x[(v^2/c - c)\partial_x u] = (2i\omega v/c)\partial_x u. \tag{1.33}$$

Near a horizon $x = x_H$ where $v = -c$ we have the expansions $v/c = -1 + O(x - x_H)$ and $v^2/c - c = -2\kappa(x - x_H) + O[(x - x_H)^2]$. Thus at the lowest order in $x - x_H$ the near horizon approximation of (1.33) becomes

$$\partial_x[(x - x_H)\partial_x u] = (i\omega/\kappa)\partial_x u, \tag{1.34}$$

whose solutions have the form

$$u \sim (x - x_H)^{i\omega/\kappa} = e^{i(\omega/\kappa)\ln(x - x_H)}. \tag{1.35}$$

The logarithmic divergence in the phase justifies the dominance of spatial derivatives of φ near the horizon. Note that this mode has the same form as (1.12), which we inferred in Sect. 1.1.2.4 using the equation of outgoing lightrays to propagate the phase of the wave in the near horizon region.

Now let's see how to arrive at the same approximate solution using the dispersion relation with the fluid picture. First, a mode solution in a homogeneous fluid has the form $\varphi \sim e^{-i\omega t}e^{ikx}$, where x is the position in the fluid frame and the dispersion relation is $\omega^2 = F(k)^2$ for some function $F(k)$. For instance, for a nondispersive wave with speed c we have simply $F(k) = ck$. If the fluid is flowing with speed v relative to the "lab" then $x = x_f + vt$, where x_f is at rest with respect to the fluid. In terms of x_f the mode is $e^{-i(\omega - vk)t}e^{ikx_f}$, which allows us to read off the frequency as measured in the fluid frame, $\omega_f = \omega - vk$. The dispersion relation holds in the fluid frame, so we have $\omega - vk = \pm F(k)$.

If the flow velocity $v(x)$ is not uniform, $\omega_f = \omega - v(x)k$ is *locally* accurate provided the change of $v(x)$ over a wavelength is small compared to $v(x)$ itself. The local dispersion relation then becomes

$$\omega - v(x)k = \pm F(k), \tag{1.36}$$

which for a fixed Killing frequency yields a position-dependent wavevector, $k_\omega(x)$. It should be emphasized that the Killing frequency ω is a well-defined global constant for a solution, even if the Killing vector is not everywhere timelike.

An approximate, WKB mode solution, taking into account only the phase factor, is thus

$$u(x) \sim \exp\left(i \int^x k_\omega(x')dx'\right). \quad (1.37)$$

Finally, if the local wave velocity $c(x)$ also depends on position in the fluid (but is time independent in the lab frame), then the function $F(k, x)$ also depends on position. If $c(x)$ changes slowly over a wavelength, then the mode of the same form is again a good approximation. For the case of relativistic dispersion $F(k, x) = c(x)k$ we obtain $k_\omega = \omega/(c + v)$ for the outgoing mode. Expanding around the horizon this yields $k_\omega(x) = (\omega/\kappa)(x - x_H)^{-1}$, and so the mode takes the same form as (1.35) derived above.

1.3.2 Positive Norm Modes and the Local Vacuum

When the field is quantized, the Hilbert space is constructed as a Fock space built from single particle states corresponding to (complex) solutions to the field equation with positive conserved "norm". The norm can be identified using a conserved inner product, the existence of which follows from global phase invariance of the action. Here I will not attempt to explain the details of this construction, which can be found in many expositions,[6] but instead will try to provide a simple argument that captures the essence of the story. In this section the relativistic case will be explained, and in the last section I will make some brief comments about what happens when there is Lorentz violating dispersion for short wavelengths. The quantum field is taken to be a hermitian scalar, which arises from quantization of a real scalar field.

Positive norm modes that are localized can be recognized as those that have positive frequency in the fluid frame. In the relativistic case, this amounts to positive frequency in any freely falling frame. The time derivative in the fluid frame is $(\partial_t)_f = \partial_t + v\partial_x$. For a mode of the form (1.35) near the horizon, this is dominated by the second term, and $v \approx -c$, hence for such modes positive frequency with respect to t in the fluid frame is the same as positive frequency with respect to x. (There are two minus signs that cancel: $v = -c < 0$ at the horizon, but the conventional definition of "positive frequency" is $\sim e^{-i\omega t}$ with $\omega > 0$ for temporal frequency, and $\sim e^{+ikx}$ with $k > 0$ for spatial frequency.)

The mode (1.35) with logarithmic phase divergence at the horizon can be analytically continued across the horizon to make either a positive or a negative frequency solution. To see how this works, let's first simplify the notation a bit and set $x_H = 0$, so the horizon lies at $x = 0$. Now a positive x-frequency function has the form $\int_0^\infty dk f(k)e^{ikx}$, which is analytic in the upper-half complex x-plane since addition of a positive imaginary part to x leaves the integral convergent. Similarly, a negative

---

[6]For a pedagogical introduction see, e.g. [10], or references therein.

x-frequency function is analytic in the lower half x-plane. The argument of the logarithm is $x = e^{i\theta}|x|$, so $\ln x = i\theta + \ln|x|$. Continuing to $-x$ in the upper or lower half plane thus gives $(\ln x)_\pm = \pm i\pi + \ln|x|$ respectively, hence

$$e^{i(\omega/\kappa)\ln x} \to e^{\mp\pi\omega/\kappa} e^{i(\omega/\kappa)\ln|x|}. \tag{1.38}$$

We can thus write down positive and negative frequency continuations,

$$q_+ = u + e^{-\pi\omega/\kappa}\tilde{u}, \tag{1.39}$$

$$q_- = e^{-\pi\omega/\kappa}u + \tilde{u}, \tag{1.40}$$

where $u = \theta(x)e^{i(\omega/\kappa)\ln x}$ and $\tilde{u} = \theta(-x)e^{i(\omega/\kappa)\ln|x|}$, and N is a normalization factor. (The negative frequency continuation q_- has been multiplied by $e^{-\pi\omega/\kappa}$ to better reflect the symmetry and thus simplify the following discussion.)

We can now express u as a superposition of positive and negative norm parts,

$$u = u_+ + u_- \propto q_+ - e^{-\pi\omega/\kappa} q_-. \tag{1.41}$$

From the symmetry of the construction, the norms of q_+ and q_- are equal up to a sign, hence the ratio of the squared norms (denoted $\langle,\rangle$) of the negative and positive norm parts of u is

$$\frac{|\langle u_-, u_-\rangle|}{\langle u_+, u_+\rangle} = e^{-2\pi\omega/\kappa} = e^{-E/T_H}. \tag{1.42}$$

In the last equality I've defined the energy $E = \hbar\omega$, and $T_H = \hbar\kappa/2\pi$ is the Hawking temperature. This "thermal ratio" is the signature of the Hawking effect, as indicated via the mode u outside the horizon. Note that this ratio is a property of the classical solution to the wave equation, and is determined by the ratio of the frequency to the surface gravity. Planck's constant enters only when we express the result in terms of the energy quantum $\hbar\omega$. Note also that if the Killing vector is rescaled, then the Killing frequency ω and surface gravity κ are rescaled in the same way, so that the ratio ω/κ is unchanged.

The presence of the negative frequency part u_- in u (1.41) is unexpected from the WKB viewpoint. It corresponds to a negative wavevector, whereas when we solved the local dispersion relation we found $k_\omega(x) = (\omega/\kappa)(x - x_H)^{-1}$. Since the support of u lies outside the horizon at $x > x_H$, it might seem that this dispersion relation implies that $k_\omega(x)$ is positive, and thus that the frequency is purely positive. However this is a misconception, because a function with support on a half line cannot have purely positive frequency. The concept of a definite local wavevector must therefore have broken down. Indeed, if we examine the change of k over a wavelength we find $(dk/dx)/k \sim (\kappa/\omega)k$, which is *not* much smaller than k unless $\omega \gg \kappa$. This resolves the puzzle.[7]

---

[7]However, it raises another one: why did the WKB type mode $\sim \exp(i\int^x k_\omega(x')dx')$ agree so well with the mode function (1.35)? The answer is that (1.34) is a first order equation, not a second order one, once an overall ∂_x derivative is peeled off.

The Local Outgoing Vacuum   The local outgoing vacuum contains no outgoing excitations. More precisely, it is the ground state in the Fock space of outgoing positive norm modes. The outgoing modes we have been discussing are not themselves localized, but one can form localized wavepackets from superpositions of them with different frequencies. Hence we may characterize the local outgoing vacuum by the requirement that it be annihilated by the annihilation operators[8] $a(q^+)$ and $a(q_-^*)$ for all positive norm modes.

These operators can be expressed in terms of the annihilation and creation operators corresponding to u and $\tilde{u}$ using (i) linearity, (ii) Eqs. (1.39) and (1.40), and (iii) the relation $a(f) = -a^\dagger(f^*)$ which should be used if f has negative norm.[9] For example, $a(q_+) = a(u) + e^{-\pi\omega/\kappa} a(\tilde{u}) = a(u) - e^{-\pi\omega/\kappa} a(\tilde{u}^*)$. The vacuum conditions

$$a(q_+)|0\rangle = 0, \qquad (1.43)$$

$$a(q_-^*)|0\rangle = 0 \qquad (1.44)$$

thus amount to

$$a(u)|0\rangle = e^{-\pi\omega/\kappa} a^\dagger(\tilde{u}^*)|0\rangle, \qquad (1.45)$$

$$a(\tilde{u}^*)|0\rangle = e^{-\pi\omega/\kappa} a^\dagger(u)|0\rangle. \qquad (1.46)$$

If we normalize the mode u, then the commutation relation $[a(u), a^\dagger(u)] = 1$ holds and implies that, in effect, $a(u) = \partial/\partial a^\dagger(u)$, and similarly for $\tilde{u}$. Thus (1.45) can be solved to find the vacuum state for these particular modes of frequency ω,

$$|0\rangle \propto \exp\left(e^{-\pi\omega/\kappa} a^\dagger(u) a^\dagger(\tilde{u}^*)\right)|0_L 0_R\rangle, \qquad (1.47)$$

where $|0_L 0_R\rangle$ is the state with no u or $\tilde{u}^*$ excitations on either side of the horizon, $a(u)|0_L 0_R\rangle = 0 = a(\tilde{u}^*)|0_L 0_R\rangle$. In flat space $|0_L 0_R\rangle$ is called the (outgoing factor of the) "Rindler vacuum", while in a black hole spacetime it is the "Boulware vacuum".

Expanding the exponential in (1.47) we obtain another expression for the vacuum

$$|0\rangle \propto \sum e^{-n\pi\omega/\kappa} |n_L n_R\rangle, \qquad (1.48)$$

where n_L and n_R are the number of particles in the given mode.[10] Taking the product over all frequencies, we then arrive at an expression for the local vacuum of a free field theory near the horizon that has the same form as the general thermal result (1.30) obtained earlier using the path integral. The results look different only because here the energies of free field states with n quanta are given by $n\hbar\omega$, and because here the Killing vector is not normalized to unit surface gravity.

---

[8] What I am calling the annihilation operator here is related to the field operator ϕ by $a(f) = \langle f, \phi \rangle$, where f is a positive norm mode. If f is not normalized this is actually $\langle f, f \rangle^{1/2}$ times a true annihilation operator.

[9] The minus sign comes from the conjugation of a factor of i in the definition of the norm, which I will not explain in detail here.

[10] Here I've use the relation $(a^\dagger)^n |0\rangle = \sqrt{n!}|n\rangle$.

1.3.3 Stimulated Emission of Hawking Radiation

So far I spoke only of the Hawking effect arising from the local vacuum at the horizon. For a real black hole this is probably the only relevant condition, but for analogue models it is possible, and even unavoidable because of thermal fluctuations, noise, or coherent excitations, that the in-state is *not* the vacuum. Then what arises is stimulated emission of Hawking radiation [11], just as the decay of an excited atomic state can be stimulated by the presence of a photon.

To quantify this process, instead of imposing the vacuum condition (1.43) we can assume the quantum field is in an excited state,

$$a^\dagger(\hat{q}_+)a(\hat{q}_+)|\Psi\rangle = n_+|\Psi\rangle, \qquad (1.49)$$

$$a^\dagger(\hat{q}_-^*)a(\hat{q}_-^*)|\Psi\rangle = n_-|\Psi\rangle, \qquad (1.50)$$

where the $\hat{q}_\pm$ are normalized versions of (1.39,1.40). A simple way to diagnose the emission is via the expectation value of the occupation number of the normalized mode u. Using (1.41) and (1.42) we find

$$\langle\Psi|a^\dagger(u)a(u)|\Psi\rangle = \langle\Psi|a^\dagger(u_+)a(u_+) + a(u_-^*)a^\dagger(u_-^*)|\Psi\rangle \qquad (1.51)$$

$$= \langle u_+, u_+\rangle\left[n_+ + e^{-2\pi\omega/\kappa}(n_- + 1)\right] \qquad (1.52)$$

$$= n_+ + \frac{n_+ + n_- + 1}{e^{2\pi\omega/\kappa} - 1} \qquad (1.53)$$

where $\langle u_+, u_+\rangle = 1/(1 - e^{-2\pi\omega/\kappa})$. Thus both n_+ and n_- stimulate Hawking emission, while only n_+ shows up in the non-thermal spectrum. Had the state been a coherent state, the occupation numbers would be replaced by squared amplitudes. Something analogous to this occurs in the surface wave white hole radiation experiments [12], although those waves do not have a relativistic dispersion relation. In the case of a Bose condensate, the appropriate in-state would presumably be more like a thermal state [13].

1.4 The Trans-Planckian Question

The sonic black hole was originally conceived by Unruh [14] in part to address what has come to be called the trans-Planckian question: Can the derivation of Hawking radiation be considered reliable given that it refers to arbitrarily high frequency field modes? If one assumes local Lorentz invariance at arbitrarily large boosts, then any high frequency mode can be Doppler shifted to low frequency, so one might argue that there is nothing to be concerned about. Sometimes the point is raised that there is an arbitrarily large invariant center of mass energy in the collision between ingoing and outgoing modes in the vacuum outside a horizon. However, this is true even in flat spacetime. We never see the effects of such collisions because they concern the "internal structure" of the ground state. We could presumably see this quantum gravity structure of the vacuum only with probes that have Planckian invariant energy. Hence it is not clear to me that there is anything to worry about in the derivation, provided one is willing to assume local Lorentz symmetry at boost factors arbitrarily far beyond anything that will ever be tested.

Even without assuming exact Lorentz symmetry, one can infer the Hawking effect by assuming that the outgoing modes are in their local ground state near the horizon for free-fall frequencies high compared to, say, the light-crossing time of the black hole, but small compared to the Planck frequency [15]. Validity of this assumption is highly plausible since the black hole formation, and field propagation in the black hole background, is very slow compared to frequencies much higher than the light crossing time. One would thus expect that whatever is happening in the vacuum, it remains unexcited, and the outgoing modes would emerge in their ground state in the near horizon region. The sonic model and other analogues allow this hypothesis to be tested in well-understood material systems that break Lorentz symmetry.

Thus one is led to consider Hawking radiation in the presence of high frequency/short wavelength dispersion, both because of the possibility that spacetime is Lorentz violating (LV), and because of the fact that analogue models are LV. However, given the very strong observational constraints on Lorentz violation [16], as well as the difficulty of accounting for low energy Lorentz symmetry in a theory that is LV in the UV [17], the possibility of fundamental LV seems rather unlikely. Hence the main motivations for considering LV dispersion are to understand condensed matter analogues, and to have an example—probably unphysical from a fundamental viewpoint—in which the vacuum has strong UV modifications and the existence of Hawking radiation can be checked.

The central issue in my view is the origin of the outgoing modes [18]. In a condensed matter model with a UV cutoff these must arise from somewhere other than the near horizon region, either from "superluminal" modes behind the horizon, from "subluminal" modes that are dragged towards the horizon and then released, or from no modes at all. The last scenario refers to the possibility that modes "assemble" from microscopic degrees of freedom in the near horizon region. This seems most likely the closest to what happens near a spacetime black hole, and for that reason deserves to be better understood. Other than a linear model that has been studied in the cosmological context [19], and a linear model of quantum field theory on a $1+1$ dimensional growing lattice [20], I don't know of any work focusing on how to characterize or study such a process.

1.5 Short Wavelength Dispersion

In this concluding section, I discuss what becomes of the Hawking effect when the dispersion relation is Lorentz invariant ("relativistic") for long wavelengths but not for short wavelengths, as would be relevant for many analogue models. First I summarize results on the robustness of the "standard" black hole radiation spectrum, and then I describe the phenomena of stimulated emission and white hole radiation.

Dispersion relations of the form $\omega^2 = c^2(k^2 \pm k^4/\Lambda^2)$ have been exhaustively studied. The plus sign gives "superluminal" propagation at high wavevectors, while the minus sign gives "subluminal" propagation. Roughly speaking, a horizon (for long wavelengths) will emit thermal Hawking radiation in a given mode provided that there is a regime near the horizon in which the mode is relativistic and in the locally defined vacuum state. This much was argued carefully in Ref. [15], and much subsequent work has gone into determining the precise conditions under which this will happen, and the size of the deviations from the thermal spectrum, for specific types of dispersion relations. The dispersion determines how the outgoing modes arise, that is whether they come from inside or outside the horizon, and what quantum state they would be found in if the initial state were near the ground state of the field, as in Hawking's original calculation.

The most recent and most complete analysis of the effects of dispersion on the spectrum can be found in Ref. [21], in which many references to earlier work can also be found. The basic technique used there is that of matched asymptotic expansions, pioneered in Refs. [22, 23] as applied to Hawking radiation for dispersive fields. The dispersive modes have associated eikonal trajectories with a turning point outside or inside the horizon for the sub- and super-luminal cases respectively. Away from the turning point approximate solutions can be found using WKB methods. If the background fluid velocity (or its analogue) has a linear form $v(x) = -1 + \kappa x$ to a good approximation out beyond the turning point, then one can match a near horizon solution to WKB solutions, and use this to find the Hawking radiation state and correlation functions. The near horizon solution is most easily found in k space, because while the mode equation is of higher order in x derivatives, $v(x) = -1 + i\kappa \partial_k$ is linear in k derivatives, so the mode equation is second order in ∂_k. Further simplifications come about because a linear $v(x)$ in fact corresponds to de Sitter spacetime, which has an extra symmetry that produces factorized modes. One factor is independent of the dispersion and has a universal ω dependence, while the other factor is independent of ω and captures the dispersion dependence.

The result, for dispersion relations of the form $\omega^2 = c^2(k \pm k^{2n+1}/\Lambda^{2n})^2$ (chosen for convenience to be a perfect square), is that the relative deviations from the thermal spectrum are no greater than of order $(\kappa/\Lambda)(\kappa x_{\text{lin}})^{-(1+1/2n)}$ times a polynomial in ω/κ.[11] Here the horizon is at $x = 0$, and x_{lin} is the largest x for which $v(x)$ has the linear form to a good approximation. Thus while it is important that the Lorentz violation wavevector scale Λ be much greater than the surface gravity κ, this may not be good enough to ensure agreement with the relativistic Hawking spectrum if the linear regime of the velocity extends over a distance much shorter than the inverse surface gravity.

At the other extreme, when the surface gravity is much larger than the largest frequency for which the turning point falls in the linear region, the spectrum of created excitations has been found to be proportional to $1/\omega$, at least for dispersion relations of the form $\omega^2 = c^2(k^2 \pm k^4/\Lambda^2)$. This is the low frequency limit

---

[11] For frequencies of order the surface gravity, this quantity can also be expressed as $(x_{\text{tp}}/x_{\text{lin}})^{1+1/2n}$, where x_{tp} is the (ω-dependent) WKB turning point.

of a thermal spectrum, but the temperature is set not by the surface gravity but by $\sim \Lambda(\kappa x_{\text{lin}})^{3/2}$. This result applies even in the limit of an abrupt "step" at which the velocity changes discontinuously from sub- to supersonic [24, 25].

1.5.1 Stimulated Hawking Radiation and Dispersion

For a relativistic free field, the ancestors of Hawking quanta can be traced backwards in time along the horizon to the formation of the horizon, and then out to infinity. They are thus exponentially trans-Planckian. In the presence of dispersion, blueshifting is limited by the scale of dispersion, so that ancestors can be traced back to incoming modes with wave vectors of order Λ. If the dispersion is subluminal, those modes come from outside the black hole horizon, while if it is superluminal, they come from behind the horizon. Either way, they are potentially accessible to the control of an experiment. Instead of being in their ground state, they might be intentionally populated in an experiment, or they might be inadvertently thermally populated. Either way, they can lead to stimulated emission of Hawking radiation, as discussed in Sect. 1.3.3.

This opportunity to probe the dependence of the emitted radiation on the incoming state is useful to experiments, and it can amplify the Hawking effect, making it easier to detect. Note however that when the Hawking radiation is stimulated rather than spontaneous, it is less quantum mechanical, and if the incoming mode is significantly populated it is essentially purely classical.

1.5.2 White Hole Radiation

A white hole is the time reverse of a black hole. Just as nothing can escape from a black hole horizon without going faster than light, nothing can *enter* a white hole horizon without going faster than light. Einstein's field equation is time reversal invariant, so it admits white hole solutions. In fact the Schwarzschild solution is time reversal symmetric: when taken in its entirety it includes a white hole. A black hole that forms from collapse is of course not time reversal invariant, but the time reverse of this spacetime is also a solution to Einstein's equation. It is not a solution we expect to see in Nature, however, both because we don't expect the corresponding initial condition to occur, and because, even if it did, the white hole would be gravitationally unstable to forming a black hole due to accretion of matter [26, 27]. Moreover, even if there were no matter to accrete, the horizon would be classically and quantum mechanically unstable due to an infinite blueshift effect, as will be explained below.

White hole analogues, on the other hand, can be engineered in a laboratory, and are amenable to experimental investigation. For example, one could be realized by a fluid flow with velocity decreasing from supersonic to subsonic in the direction of the flow. Sound waves propagating against the flow would slow down and blueshift as they approach the sonic point, but the blueshifting would be limited by short wavelength dispersion, so the white hole horizon might be stable. If the horizon is stable, then the time reverse of the Hawking effect will take place on a white hole background, and the emitted radiation will be thermal, at the Hawking temperature of the white hole horizon [24] (see also Appendix D of Ref. [13]). Underlying this relation is the fact that the modes on the white hole background are the time reverse of the modes on the time-reversed black hole background. Note that this means that the roles of the in and out modes are swapped. In particular, the incoming vacuum relevant to the Hawking radiation consists of low wavenumber modes propagating against the flow.

When such a mode with positive norm approaches the white hole horizon, it is blocked and begins blueshifting. At this stage, it has become a superposition of positive and negative co-moving frequency (and therefore positive and negative norm) parts. If it were relativistic at all scales, it would continue blueshifting without limit. It would also be unentangled with the other side of the horizon, so would evidently be in an excited state, not the co-moving ground state. Hence there would be a quantum instability of the vacuum in which the state becomes increasingly singular on the horizon. A classical perturbation would behave in a similarly unstable fashion.

In the presence of dispersion, however, the blueshifting is arrested when the it reaches the dispersion scale. At that stage, if the mode becomes superluminal, it accelerates and both parts propagate across the horizon. If instead it becomes subluminal, then it slows down and both parts get dragged back out with the flow. In either case, the positive and negative norm parts are in an entangled, excited state that is thermal when tracing over one of the pair. Thus, a dispersive wave field exhibits Hawking radiation from a white hole horizon, but with two marked differences when compared to black hole radiation: the Hawking quanta have high wavevectors even when the Hawking temperature is low, and the entangled partners propagate on the same side of the horizon (inside for superluminal, outside for subluminal dispersion). While on the same side, the partners can separate, since in general they have different group velocities.

There is an important potential complication with this story of white hole radiation. Although the singularity that would arise in the relativistic case is cured by dispersion, an avatar of it emerges in the form of a zero Killing frequency standing wave. This has been shown to arise from the zero frequency limit of the Hawking radiation. In that limit, the emission rate diverges as $1/\omega$, leading to a state with macroscopic occupation number that grows in time [21, 28]. This process can also be seeded by classical perturbations, and it grows until nonlinear effects saturate the growth. The resulting standing wave, which is a well-known phenomenon in other contexts, is referred to in the white hole setting as an "undulation". It is composed of short wavelengths that are well into the dispersive regime. Depending on the nature of the flow and the saturation mechanism, it could disrupt the flow and prevent a smooth horizon from forming.

To conclude, I will now describe what was seen in the Vancouver experiment [12]. That experiment involved a flow of water in a flume tank with a velocity profile that produced a white hole horizon for long wavelength, shallow water, surface waves (which are dispersionless over a uniform bottom). When blueshifted those waves convert to deep water waves, with a lower group velocity, which behave like the "subluminal" case described above. In the experiment coherent, long waves with nine different frequencies were launched from downstream, propagating back upstream towards the white hole horizon, and the resulting conversion to short waves was observed. The squared norm ratio of the negative and positive norm components of the corresponding frequency eigenmode was consistent with the thermal ratio (1.42).[12] This can be understood as coherently stimulated emission of Hawking radiation (see Appendix C of Ref. [13] for a general discussion of this process). It is strictly classical, but it is governed by the same mode conversion amplitudes that would produce spontaneous emission if the system could be prepared in the ground state.

Acknowledgements I am grateful to Renaud Parentani for many helpful discussions on the material presented here, as well as suggestions for improving the presentation. Thanks also to Anton de la Fuente for helpful discussions on the path integral derivation of vacuum thermality. This work was supported in part by the National Science Foundation under Grant Nos. NSF PHY09-03572 and NSF PHY11-25915.

References

1. Corley, S., Jacobson, T.: Lattice black holes. Phys. Rev. D **57**, 6269 (1998). hep-th/9709166
2. Barcelo, C., Liberati, S., Visser, M.: Analogue gravity. Living Rev. Relativ. **8**, 12 (2005). Living Rev. Relativ. **14**, 3 (2011). gr-qc/0505065
3. Jacobson, T.A., Volovik, G.E.: Event horizons and ergoregions in He-3. Phys. Rev. D **58**, 064021 (1998)
4. Penrose, R.: Gravitational collapse: the role of general relativity. Riv. Nuovo Cimento **1**, 252 (1969). Gen. Relativ. Gravit. **34**, 1141 (2002)
5. Penrose, R., Floyd, R.M.: Extraction of rotational energy from a black hole. Nature **229**, 177 (1971)
6. Wald, R.M.: The thermodynamics of black holes. Living Rev. Relativ. **4**, 6 (2001). gr-qc/9912119
7. Jacobson, T., Kang, G.: Conformal invariance of black hole temperature. Class. Quantum Gravity **10**, L201 (1993). gr-qc/9307002
8. Jacobson, T., Parentani, R.: Horizon surface gravity as 2d geodesic expansion. Class. Quantum Gravity **25**, 195009 (2008). arXiv:0806.1677 [gr-qc]
9. Crispino, L.C.B., Higuchi, A., Matsas, G.E.A.: The Unruh effect and its applications. Rev. Mod. Phys. **80**, 787 (2008). arXiv:0710.5373 [gr-qc]
10. Jacobson, T.: Introduction to quantum fields in curved space-time and the Hawking effect. gr-qc/0308048 (2003)

---

[12] The relevant norm is determined by the action for the modes. This has been worked out assuming irrotational flow [29], which is a good approximation although the flow does develop some vorticity. The predicted Hawking temperature was estimated, but it is difficult to evaluate accurately because of the presence of the undulation, the fact that it depends on the flow velocity field that was not precisely measured, and the presence of vorticity which has not yet been included in an effective metric description.

11. Wald, R.M.: Stimulated emission effects in particle creation near black holes. Phys. Rev. D **13**, 3176 (1976)
12. Weinfurtner, S., Tedford, E.W., Penrice, M.C.J., Unruh, W.G., Lawrence, G.A.: Measurement of stimulated Hawking emission in an analogue system. Phys. Rev. Lett. **106**, 021302 (2011). arXiv:1008.1911 [gr-qc]
13. Macher, J., Parentani, R.: Black hole radiation in Bose-Einstein condensates. Phys. Rev. A **80**, 043601 (2009). arXiv:0905.3634 [cond-mat.quant-gas]
14. Unruh, W.G.: Experimental black hole evaporation. Phys. Rev. Lett. **46**, 1351 (1981)
15. Jacobson, T.: Black hole radiation in the presence of a short distance cutoff. Phys. Rev. D **48**, 728 (1993). hep-th/9303103
16. Mattingly, D.: Modern tests of Lorentz invariance. Living Rev. Relativ. **8**, 5 (2005). gr-qc/0502097
17. Collins, J., Perez, A., Sudarsky, D.: Lorentz invariance violation and its role in quantum gravity phenomenology. In: Oriti, D. (ed.) Approaches to Quantum Gravity, pp. 528–547 (2006). hep-th/0603002
18. Jacobson, T.: On the origin of the outgoing black hole modes. Phys. Rev. D **53**, 7082 (1996). hep-th/9601064
19. Parentani, R.: Constructing QFT's wherein Lorentz invariance is broken by dissipative effects in the UV. arXiv:0709.3943 [hep-th] (2007)
20. Foster, B.Z., Jacobson, T.: Quantum field theory on a growing lattice. J. High Energy Phys. **0408**, 024 (2004). hep-th/0407019
21. Coutant, A., Parentani, R., Finazzi, S.: Black hole radiation with short distance dispersion, an analytical S-matrix approach. Phys. Rev. D **85**, 024021 (2012). arXiv:1108.1821 [hep-th]
22. Brout, R., Massar, S., Parentani, R., Spindel, P.: Hawking radiation without trans-Planckian frequencies. Phys. Rev. D **52**, 4559 (1995). hep-th/9506121
23. Corley, S.: Computing the spectrum of black hole radiation in the presence of high frequency dispersion: an analytical approach. Phys. Rev. D **57**, 6280 (1998). hep-th/9710075
24. Macher, J., Parentani, R.: Black/white hole radiation from dispersive theories. Phys. Rev. D **79**, 124008 (2009). arXiv:0903.2224 [hep-th]
25. Finazzi, S., Parentani, R.: Hawking radiation in dispersive theories, the two regimes. Phys. Rev. D **85**, 124027 (2012). arXiv:1202.6015 [gr-qc]
26. Eardley, D.M.: Death of white holes in the early universe. Phys. Rev. Lett. **33**, 442 (1974)
27. Barrabès, C., Brady, P.R., Poisson, E.: Death of white holes. Phys. Rev. D **47**, 2383 (1993)
28. Mayoral, C., et al.: Acoustic white holes in flowing atomic Bose-Einstein condensates. New J. Phys. **13**, 025007 (2011). arXiv:1009.6196 [cond-mat.quant-gas]
29. Unruh, W.G.: Irrotational, two-dimensional Surface waves in fluids. In this book. arXiv:1205.6751 [gr-qc]

v

Chapter 2
Survey of Analogue Spacetimes

Matt Visser

Abstract Analogue spacetimes (and more boldly, analogue models both of and for gravity), have attracted significant and increasing attention over the last decade and a half. Perhaps the most straightforward physical example, which serves as a template for most of the others, is Bill Unruh's model for a dumb hole,(mute black hole, acoustic black hole), wherein sound is dragged along by a moving fluid—and can even be trapped behind an acoustic horizon. This and related analogue models for curved spacetimes are useful in many ways: analogue spacetimes provide general relativists with extremely concrete physical models to help focus their thinking, and conversely the techniques of curved spacetime can sometimes help improve our understanding of condensed matter and/or optical systems by providing an unexpected and countervailing viewpoint. In this chapter, I shall provide a few simple examples of analogue spacetimes as general background for the rest of the contributions.

2.1 Introduction

While the pre-history of analogue spacetimes is quite long and convoluted, with optics-based contributions dating as far back as the Gordon metric of 1923 [1], significant attention from within the general relativity community dates back to Bill Unruh's PRL concerning acoustic black holes (dumb holes) published in 1981 [2]. Even then, it is fair to say that the investigation of analogue spacetimes did not become mainstream until the late 1990's. (See the recently updated Living Review article on "Analogue gravity" for a summary of the historical context [3].)

In all of the analogue spacetimes, the key idea is to take some sort of "excitation", travelling on some sort of "background", and analyze its propagation in terms of the tools and methods of differential geometry. The first crucial technical distinction one has to make is between "rays" and "waves".

- The rays of ray optics (geometrical optics), ray acoustics (geometrical acoustics), or indeed any more general ray-like phenomenon, are only concerned with the

---

M. Visser (✉)
School of Mathematics, Statistics, and Operations Research, Victoria University of Wellington, PO Box 600, Wellington 6140, New Zealand
e-mail: matt.visser@msor.vuw.ac.nz

"light cones", "sound cones", or more generally the purely geometrical "propagation cones" defined by the ray propagation speed relative to the appropriate background. Physically, in this approximation one should think photons/phonons/quasi-particles following some well-localized trajectory, rather than the more diffuse notion of a wave. Mathematically, we will soon see that it is appropriate to construct some metric g_{ab}, and some tangent vector k^a to the particle trajectory, such that:

$$g_{ab} k^a k^b = 0. \qquad (2.1)$$

Here indices such as $a, b, c, \ldots$, take on values in $\{0, 1, 2, 3\}$, corresponding to both time and space, whereas indices such as $i, j, k, \ldots$, will take on values in $\{1, 2, 3\}$, corresponding to space only. Of course we could multiply the metric by any scalar quantity without affecting this equation; this is known as a conformal transformation of the metric. (So distances change but angles are unaffected.) In the language of differential geometry, ray phenomena are sensitive only to a conformal class of Lorentzian geometries.

- In contrast, for waves one needs to write down some PDE—some sort of wave equation. For example, for a scalar excitation Ψ one needs to construct a wave equation in terms of a d'Alembertian [2–7]:

$$\frac{1}{\sqrt{-g}} \partial_a \left(\sqrt{-g} g^{ab} \partial_b \Psi \right) = 0. \qquad (2.2)$$

This d'Alembertian, (and in fact very many of the different possible types of wave equation), depends on all the components of the metric g_{ab}, not just the conformal class. (And conformal wave equations, of which the Maxwell electromagnetic wave equations are the most common, have their own somewhat different issues.)

In short, depending on exactly what one is trying to accomplish, one may sometimes be able to get away with ignoring an overall multiplicative conformal factor—but for other applications knowledge of the conformal factor is utterly essential.

What I shall now do is to present some elementary examples—and a few not so elementary implications—that will hopefully serve as a pedagogical introduction to the more specific physics problems addressed in the other contributions to this volume.

2.2 Optics: The Gordon Metric and Its Generalizations

The original Gordon metric [1] of 1923 was limited to ray optics in a medium with a position-independent refractive index, and with some position-independent velocity. Let

$$\eta_{ab} = \left[\begin{array}{c|c} -1 & 0 \\ \hline 0 & \delta_{ij} \end{array} \right], \qquad (2.3)$$

denote (as usual) the special relativistic Minkowski metric, and correspondingly set the zeroth coordinate to $x^0 = t = ct_{\text{physical}}$. Denote the refractive index by n and the 4-velocity of the medium by $V^a = \gamma(1; \beta\mathbf{n})$. Then we have $V_a = \gamma(-1; \beta\mathbf{n})$. Now define

$$g_{ab} = (\eta_{ab} + V_a V_b) - \frac{V_a V_b}{n^2} = \eta_{ab} + \left(1 - \frac{1}{n^2}\right) V_a V_b. \tag{2.4}$$

In the rest frame of the medium $V^a \to (1; \mathbf{0})$ and

$$g_{ab} \to \begin{bmatrix} -1/n^2 & 0 \\ 0 & \delta_{ij} \end{bmatrix}. \tag{2.5}$$

Therefore in this rest frame the null cones of the medium are exactly what we want:

$$0 = ds^2 = -\frac{dt^2}{n^2} + \|d\mathbf{x}\|^2 \quad \Longrightarrow \quad \left\|\frac{d\mathbf{x}}{dt}\right\| = \frac{1}{n}. \tag{2.6}$$

But more generally, for non-zero velocity, $\beta \neq 0$, the metric g_{ab} provides a perfectly good special relativistic model for the light cones in a homogeneous moving medium. Let us agree to raise and lower the indices on the 4-velocity V using the Minkowski metric η, then the contravariant Gordon metric is

$$g^{ab} = \left(\eta^{ab} + V^a V^b\right) - n^2 V^a V^b. \tag{2.7}$$

The first and most obvious generalization is to note that one can easily make the refractive index and 4-velocity both space and time dependent. (Physically, this will certainly work as long as the wavelength and period of the light wavicle is short compared to the spatial and temporal scales over which changes of the background refractive index and 4-velocity are taking place.)

A second generalization is to note that from the point of view of ray optics one might as well take

$$g_{ab} = \Omega^2 \left[(\eta_{ab} + V_a V_b) - \frac{V_a V_b}{n^2}\right]; \quad g^{ab} = \Omega^{-2}\left[\left(\eta^{ab} + V^a V^b\right) - n^2 V^a V^b\right]. \tag{2.8}$$

The conformal factor Ω will simply drop out when determining the light cones. With the quantities $\Omega(x)$, $n(x)$, and $V(x)$ all being space and/or time dependent, this is the most general (but still physically natural) form of the (special relativity based) Gordon metric one can write.

One can certainly calculate the Einstein tensor for this optical metric, but there is *a priori* no really compelling reason to do so—there is *a priori* no good reason to attempt to enforce the Einstein equations for this optical metric, the physics is just completely different. (That being said, if one merely views this as an ansatz for interesting metrics to look at, then many of the standard spacetimes of general relativity can certainly be put into this form. For example, the Schwarzschild and Reissner–Nordström spacetimes, and the FLRW cosmologies, can certainly be put in this form [8].)

Example Let us take $\Omega = n$ and write

$$g_{ab} = n^2(\eta_{ab} + V_a V_b) - V_a V_b. \tag{2.9}$$

This corresponds to

$$ds^2 = -dt^2 + n^2 \|d\mathbf{x}^2\|. \tag{2.10}$$

Now pick the specific refractive-index profile

$$n = \frac{n_0}{1 + r^2/a^2}, \tag{2.11}$$

so

$$ds^2 = -dt^2 + \frac{n_0^2[dr^2 + r^2\{d\theta^2 + \sin^2\theta d\phi^2\}]}{(1 + r^2/a^2)^2}. \tag{2.12}$$

A theoretical cosmologist should recognize this as the Einstein static universe in isotropic coordinates [9–11]. A theoretician working in optics should recognize this as the Maxwell fish-eye lens [12–14]. This is simply the first of many cross-connections between optics and general relativity. This becomes (or should become) a two-way street for information exchange.

The Maxwell fish-eye above is an example of a Lüneburg lens [15], and has now become the canonical example which helped initiate much of the recently developed field of "transformation optics" [16–19]. In particular, if one looks at this from the perspective of a theoretical cosmologist then the prefect focussing properties are utterly trivial—after all, the spatial slices of the Einstein static universe are just the hyper-sphere S^3 in suitable coordinates—the geodesics are obviously just great circles, which by symmetry must meet at the antipodes of the emission event, and so perfect focussing in the ray optics approximation is trivial.

Limitations Perhaps the greatest limitation of the Gordon metric is its inability (in its original 1923 formulation) to deal with wave properties of light. There is a rather non-trivial generalization to the full Maxwell equations [3], but for technical reasons the generalization requires the very specific constraint

$$[\text{magnetic permittivity}] \propto [\text{electric permeability}]. \tag{2.13}$$

For ordinary physical media this constraint is somewhat unphysical [3], but there is hope that suitably designed metamaterials [20] may be designed to at least approximately satisfy this constraint.

Foreground-Background Version So far, the Gordon metric has been based on a optical medium in Minkowski space described by the flat metric η_{ab}. But now suppose we have a non-trivial background metric f_{ab} arising from standard general relativity, and place a flowing optical medium on top of that. It now becomes interesting to consider the generalized Gordon metric

$$g_{ab} = \Omega^2\left[(f_{ab} + V_a V_b) - \frac{V_a V_b}{n^2}\right]; \qquad g^{ab} = \Omega^{-2}\left[(f^{ab} + V^a V^b) - n^2 V^a V^b\right]. \tag{2.14}$$

We now have the possibility of a non-trivial general relativistic background $f_{ab}(x)$, a position-dependent refractive index $n(x)$, a position-dependent 4-velocity $V^a(x)$, and a position-dependent conformal factor $\Omega(x)$. Note that the 4-velocity $V^a(x)$ now has to be a timelike unit vector with respect to the background metric $f_{ab}(x)$, and the indices on V are raised and lowered using f. In particular, note $g_{ab} V^a V^b = -\Omega^2/n^2$ and $g^{ab} V_a V_b = -n^2/\Omega^2$, so it makes sense to define

$$\tilde{V}^a = \frac{n}{\Omega} V^a; \quad \text{and} \quad \tilde{V}_a = \frac{\Omega}{n} V_a. \tag{2.15}$$

Then $\tilde{V}$ is a timelike unit vector with respect to g, and its indices should be raised and lowered using g. Then we can adapt the generalized Gordon metric to also write the background f in terms of the foreground g as:

$$f_{ab} = \Omega^{-2}\left[(g_{ab} + \tilde{V}_a \tilde{V}_b) - n^2 \tilde{V}_a \tilde{V}_b\right]; \qquad f^{ab} = \Omega^2\left[(g^{ab} + \tilde{V}^a \tilde{V}^b) - \frac{\tilde{V}^a \tilde{V}^b}{n^2}\right]. \tag{2.16}$$

For a relativist the generalized Gordon metric provides an interesting ansatz for a potentially intriguing class of spacetimes to consider. From a theoretical optics perspective, one might view this procedure as an extremely general way of "composing" and/or "inverting" the transformations of transformation optics—for example, one might first design some metamaterial [20] to generate the background $f_{ab}(x)$, and then impose some flowing optical medium on top of that. Various interesting possibilities come to mind.

2.3 Non-relativistic Acoustics: The Unruh Metric

Bill Unruh's 1981 PRL article [2], and much of the follow up work [3–5], was explicitly and intrinsically based on non-relativistic acoustics. Let us explore the basic features of this particular model.

Geometric Acoustics From the acoustic ray perspective the derivation is trivial: Let c_s be the speed of sound, and let **v** be the velocity of the fluid. Then sound rays (phonon trajectories) satisfy [3–5]

$$\|\mathrm{d}\mathbf{x} - \mathbf{v}\mathrm{d}t\| = c_s \mathrm{d}t. \tag{2.17}$$

Let us, already anticipating the possibility of an arbitrary conformal factor, define

$$\mathrm{d}s^2 = \Omega^2\left\{-c_s^2\mathrm{d}t^2 + (\mathrm{d}\mathbf{x} - \mathbf{v}\mathrm{d}t)^2\right\} = \Omega^2\left\{-(c_s^2 - v^2) - 2\mathbf{v} \cdot \mathrm{d}\mathbf{x}\mathrm{d}t + \|\mathrm{d}\mathbf{x}\|^2\right\}. \tag{2.18}$$

(The zeroth coordinate is now most naturally chosen to simply be $x^0 = t = t_{\text{physical}}$, without any explicit factor of c. The speed of sound c_s has the dimensions of a physical velocity.) Then the sound-ray condition is completely equivalent geometrically to the null-cone condition $ds^2 = 0$. In terms of a 4×4 matrix this is equivalent to defining the metric tensor [3–5]

$$g_{ab} = \Omega^2 \left[\begin{array}{c|c} -(c_s^2 - v^2) & -v_j \\ \hline -v_i & \delta_{ij} \end{array} \right]. \tag{2.19}$$

The corresponding inverse metric is

$$g^{ab} = \Omega^{-2} \left[\begin{array}{c|c} -1/c_s^2 & -v^j/c_s^2 \\ \hline -v^i/c_s^2 & \delta^{ij} - v^i v^j/c_s^2 \end{array} \right]. \tag{2.20}$$

It should be emphasized that in this situation the velocity **v** and speed of sound c_s will be inter-related in some (often quite complicated) manner—the background fluid flow must satisfy the Euler equation and the continuity equation [2–5].

It should again further be emphasized (*forcefully*) that while one can certainly calculate the Einstein tensor for this acoustic metric, there is *a priori* no really compelling reason to do so—there is *a priori* no good reason to attempt to enforce the Einstein equations for this acoustic metric, the physics is just completely different. That being said, if one again views this as an ansatz for interesting metrics to look at, then many of the standard spacetimes of general relativity (but certainly not all interesting spacetimes) can be put into this form. (For instance the Schwarzschild and Reissner–Nordström spacetimes can be put into this form by going to Painlevé–Gullstrand coordinates, but the Kerr and Kerr–Newman spacetimes *cannot* be put in this form [3, 21].)

To further develop the discussion, let us now introduce quantities

$$Q^{ab} = \left[\begin{array}{c|c} 0 & 0 \\ \hline 0 & \delta^{ij} \end{array} \right]; \qquad V^a = (1; v^i) = (1; \mathbf{v}). \tag{2.21}$$

Here the 4-velocity V^a is normalized non-relativistically—with the time component being unity. Then for the inverse metric

$$g^{ab} = \Omega^{-2} \left[Q^{ab} - \frac{V^a V^b}{c_s^2} \right]. \tag{2.22}$$

But what about the covariant metric g_{ab}? Let us now define

$$\overset{\flat}{Q}_{ab} = \left[\begin{array}{c|c} 0 & 0 \\ \hline 0 & \delta_{ij} \end{array} \right]. \tag{2.23}$$

Then $\overset{\flat}{Q}_{ab}$ is the Moore–Penrose pseudo-inverse of Q^{ab}, and the object

$$P^a{}_b = Q^{ac} Q_{cb} = \left[\begin{array}{c|c} 0 & 0 \\ \hline 0 & \delta^i{}_j \end{array} \right] \tag{2.24}$$

is a projection operator onto spatial slices. Let us now furthermore define the quantities $V_a^{\flat} = Q_{ab}^{\flat} V^b = (0; v^i) = (0; \mathbf{v})$, while $T_a = (1; \mathbf{0})$. The best we can do for the covariant metric g_{ab} is to now write the somewhat clumsy expression:

$$g_{ab} = \Omega^2 [Q_{ab}^{\flat} - (c_s^2 - v^2) T_a T_b - T_a V_b^{\flat} - V_a^{\flat} T_b]. \tag{2.25}$$

In view of the fact that, with these definitions, one has $T_a V^a = 1$ and $V_a^{\flat} V^a = v^2$, while $Q^{ab} T_b = 0$ and $Q^{ab} V_b^{\flat} = V^a - T^a$, it is easy to verify that (as required)

$$g^{ab} g_{bc} = \delta^a{}_c. \tag{2.26}$$

As we shall soon see, relativistic acoustics is in some sense actually somewhat simpler than the non-relativistic case.

Wave Acoustics If one goes beyond ray acoustics, then the parameter Ω is no longer arbitrary. One does have to make some additional (and rather stringent) technical assumptions—barotropic, irrotational, and inviscid (zero viscosity) flow [3–5]. Under those assumptions, by linearizing the Euler equation and continuity equation, after a little work one ultimately obtains a wave equation (a curved-spacetime d'Alembertian equation) for perturbations of the velocity potential specified in terms of the density of the fluid and the speed of sound—specifically one has $\Omega = \sqrt{\rho/c_s}$ in 3 space dimensions, $\Omega = \rho/c_s$ in 2 space dimensions, and technical problems arise in 1 space dimension. Generally, in d space dimensions, $\Omega = (\rho/c_s)^{1/(d-1)}$. (See for example Ref. [3].)

A specific feature of physical (wave) acoustics, not probed in the geometrical acoustics limit, is the behaviour of quasi-normal modes [22, 23]. Furthermore, if the flow is not irrotational, so one is dealing with both background vorticity and vorticity-bearing perturbations, then a considerably more complicated *system* of wave equations can be written down [24], but this system of PDEs has nowhere near as clean a geometrical interpretation as the irrotational case.

2.4 Horizons and Ergo-Surfaces in Non-relativistic Acoustics

One of the very nice features of non-relativistic acoustics is that it is very simple and straightforward to define horizons and ergo-surfaces [3–5]. To define these concepts, it is sufficient to work in the geometric acoustics limit; wave acoustics adds additional constraints not directly needed to define horizons and ergo-surfaces. Consider for simplicity a stationary (time independent) configuration.

- Ergo-surfaces are defined by the condition $\|\mathbf{v}\| = c_s$.
- Horizons are surfaces, located for definiteness at $f(\mathbf{x}) = 0$, that are defined by the 3-dimensional spatial condition $\nabla f \cdot \mathbf{v} = c_s \|\nabla f\|$.

So the ergo-surface bounds the region where one cannot stand still without generating a sonic boom, and corresponds to Mach one, ($M \equiv v/c_s = 1$). In contrast, on a horizon the *normal component* of the fluid velocity equals the speed of sound, thereby either trapping or anti-trapping the acoustic excitations.

Stationary Versus Static In general relativity the words "stationary" and "static" have precise technical meanings that may not be obvious to non-experts. So a few words of explanation are called for:

- Stationary: For all practical purposes this means "time independent". More precisely, mathematically there is a Killing vector (a symmetry of the system) which is timelike at spatial infinity. Physically there is a class of natural time coordinates (not quite unique) in which the metric is time-independent. In this coordinate system the Killing vector is naturally associated with invariance under time translations $t \to t + C$.
- Static: For all practical purposes this means "time independent and non-rotating". More precisely, mathematically there is a Killing vector (a symmetry of the system) which is both timelike at spatial infinity *and* "hypersurface orthogonal", meaning there exist functions $\xi(x)$ and $\tau(x)$ such that $K^a = \xi(x)g^{ab}\partial_b\tau(x)$. Physically there is then a unique natural time coordinate, (in fact τ, which is unfortunately not necessarily "laboratory time"), in which the metric is both time-independent *and* block-diagonal. That is, with vanishing time-space components $g_{ti} = 0$, in these coordinates the metric block diagonalizes into $(time) \oplus (space)$. The existence of a coordinate system with vanishing time-space metric components is sufficient in general relativity to imply zero angular momentum for the spacetime, and absence of "frame dragging", hence the sobriquet "non-rotating".

A word of warning: Just because one *can* always choose a coordinate system to block diagonalize a static spacetime does not mean this is always a good idea. Coordinates in which static spacetimes are block diagonal will break down at any horizon that might be present in the spacetime. (For instance, Schwarzschild geometry in the usual coordinates.) Permitting coordinates for static spacetimes which retain the manifest time independence, but do not explicitly force block diagonalization of the metric, has significant technical and physical advantages. For one thing, this is the most natural situation when one works with "laboratory time" and a time independent fluid flow. For another thing, once one allows off-diagonal elements for the metric one can easily construct "horizon penetrating" coordinates, which are well defined both at and across the horizon. (For instance, Schwarzschild geometry in Painlevé–Gullstrand or Eddington–Finklestein coordinates.) In particular, the acoustic metric as given above (in terms of laboratory time, speed of sound, and fluid velocities) is automatically in horizon-penetrating form, all the components of both the metric g_{ab} and its inverse g^{ab} remain finite as one crosses the horizon. Let us now see how these ideas are used in practice.

Static Configurations Suppose the background flow satisfies the integrability constraint

$$\frac{\mathbf{v}}{c_s^2 - v^2} = \nabla \Phi, \qquad (2.27)$$

and then consider the new time coordinate $\tau = t + \Phi$. (Here t is explicitly laboratory time, while τ is constructed for mathematical convenience rather than for direct physical purposes.) Note that this integrability condition implies (but is stronger than) the vanishing of local helicity

$$h \equiv \mathbf{v} \cdot (\nabla \times \mathbf{v}) = 0. \tag{2.28}$$

In terms of this new time coordinate

$$ds^2 = \Omega^2 \left\{ -(c_s^2 - v^2) d\tau^2 + \left[\delta_{ij} + \frac{v_i v_j}{c_s^2 - v^2} \right] dx^i dx^j \right\}. \tag{2.29}$$

The geometry is now in this form block-diagonal so it is *manifestly* static, not just stationary. (And so the ergo-surfaces and horizons will automatically coincide.) The time translation Killing vector is

$$K^a = (1; \mathbf{0}), \quad \text{so} \quad K_a = -\Omega^2 (c_s^2 - v^2; \mathbf{0}). \tag{2.30}$$

To explicitly verify that this is hypersurface orthogonal in the sense defined above, note

$$K_a = -\Omega^2(c_s^2 - v^2)\partial_a \tau, \quad \text{so} \quad K^a = -\Omega^2(c_s^2 - v^2) g^{ab} \partial_b \tau. \tag{2.31}$$

The norm of this Killing vector is given by

$$K^a K_a = -\Omega^2(c_s^2 - v^2) = -\Omega^2 c_s^2 \left(1 - \frac{v^2}{c_s^2}\right). \tag{2.32}$$

Using very standard techniques, the surface gravity is then calculable in terms of the gradient of this norm [3]. It is a standard result that for a Killing horizon the overall conformal factor drops out of the calculation [25]. Generalizing Unruh's original calculation [2], which corresponds to c_s being constant, one finds [3–5]

$$g_H = \frac{1}{2} \| \mathbf{n} \cdot \nabla (c_s^2 - v^2) \|_H = c_H \| \mathbf{n} \cdot \nabla (c_s - v) \|_H = c_H \left| \frac{\partial (c_s - v)}{\partial n} \right|_H, \tag{2.33}$$

which can also be compactly written in terms of the Mach number $M \equiv v/c_s$ as

$$g_H = c_H^2 \| \mathbf{n} \cdot \nabla (v/c_s) \|_H = c_H^2 \| \mathbf{n} \cdot \nabla M \|_H = c_H^2 \left| \frac{\partial M}{\partial n} \right|_H. \tag{2.34}$$

We emphasise that this already works in the geometric acoustics framework, and that there is no need to make the more restrictive assumptions corresponding to wave acoustics that were made in references [3] and [5]. If the integrability condition is not satisfied one must be a little more devious.

Stationary but Non-static Configurations If the acoustic horizon is stationary but not static there may or may not be additional symmetries, (in addition to the assumed time independence), so in particular the horizon may or may not be a Killing horizon. (A horizon is said to be a Killing horizon if and only if there exists *some* Killing vector such that the location of the horizon coincides with the vanishing of the norm of that Killing vector. So Killing horizons automatically satisfy nice symmetry properties.) For a Killing horizon the calculation of surface gravity is still relatively straightforward, for non-Killing horizons the situation is far more complex.

Note that in full generality, on the horizon we have $(\nabla f \cdot \mathbf{v})^2 = c_s^2 \|\nabla f\|^2$, which we can rewrite in 3-dimensional form as $g^{ij} \partial_i f \partial_j f = 0$. Since the configuration, and location of the horizon, is time independent this statement can be bootstrapped to $3+1$ dimensions to see that *on the horizon*

$$g^{ab} \nabla_a f \nabla_b f = 0. \tag{2.35}$$

That is, the 4-vector ∇f is null on the horizon. In fact, on the horizon, where in terms of the 3-normal $\mathbf{n}$ we can decompose $\mathbf{v}_H = c_s \mathbf{n} + \mathbf{v}_\|$, we can furthermore write

$$(\nabla f)_H^a = \left(g^{ab} \nabla_b f\right)_H = \frac{\|\nabla f\|}{\Omega_H^2 c_H} (1; \mathbf{v}_\|)_H. \tag{2.36}$$

That is, not only is the 4-vector ∇f null on the horizon, it is also a 4-tangent to the horizon—so (as in general relativity) the horizon is ruled by a set of null curves. Furthermore, extending the 3-normal $\mathbf{n}$ to a region surrounding the horizon (for instance by taking $\mathbf{n} = \nabla f / \|\nabla f\|$) we can quite generally write $\mathbf{v} = v_\perp \mathbf{n} + \mathbf{v}_\|$. Then away from the horizon

$$g^{ab} \nabla_a f \nabla_b f = \frac{(c_s^2 - v_\perp^2) \|\nabla f\|^2}{\Omega^2 c_s^2}. \tag{2.37}$$

That is, the 4-vector ∇f is spacelike outside the horizon, null on the horizon, and timelike inside the horizon.

Stationary but Non-static Killing Horizons If the stationary horizon is Killing, then even if we do not explicitly know what the relevant Killing vector $\tilde{K}^a$ is, we know that its norm has to vanish on the horizon, and so the norm of this horizon-generating Killing vector is of the form

$$\tilde{K}^a \tilde{K}_a = Q(c_s^2 - v_\perp^2) = -Q c_s^2 \left(1 - \frac{v_\perp^2}{c_s^2}\right), \tag{2.38}$$

for some unknown (but for current purposes irrelevant) function Q. Following closely the argument for the static case, *mutatis mutandis*, we have [3–5]

$$g_H = \frac{1}{2} \left\|\mathbf{n} \cdot \nabla (c_s^2 - v_\perp^2)\right\|_H = c_H \left\|\mathbf{n} \cdot \nabla (c_s - v_\perp)\right\|_H = c_H \left|\frac{\partial (c_s - v_\perp)}{\partial n}\right|_H, \tag{2.39}$$

which can also be compactly written in terms of the horizon-crossing Mach number, $M_\perp \equiv v_\perp/c_s$, as

$$g_H = c_H^2 \|\mathbf{n} \cdot \nabla(v_\perp/c_s)\|_H = c_H^2 \|\mathbf{n} \cdot \nabla M_\perp\|_H = c_H^2 \left|\frac{\partial M_\perp}{\partial n}\right|_H. \quad (2.40)$$

We again emphasise that this already works in the geometric acoustics framework, and that there is no need to make the more restrictive assumptions corresponding to wave acoustics that were made in Refs. [3, 5]. If the horizon is non-Killing then one must be even more devious.

Stationary but Non-static Non-Killing Horizons Such situations are, from a technical perspective, much more difficult to deal with. Such behaviour cannot occur in standard general relativity, where the Einstein equations stringently constrain the allowable spacetimes, but there seems no good reason to exclude it for acoustic horizons. Unfortunately, when it comes to explicit computations of the surface gravity there are still some unresolved technical issues. There is still a lot of opportunity for significant new physics hiding in these non-Killing horizons.

2.5 Relativistic Acoustics

Full relativistic acoustics (either special relativistic or general relativistic) adds a few other quirks which I briefly describe below. (See early astrophysical work by Moncrief [26], a more recent cosmological framework developed in [27], and a pedagogical exposition in reference [7] for details.) Note that the interest in, and need for, relativistic acoustics is driven by astrophysical and cosmological considerations, not by direct laboratory applications. There are at least three situations in which relativistic acoustics is important:

- Speed of sound comparable to that of light.
  In any ideal gas once $kT \gg m_0 c^2$ then $p \approx \frac{1}{3}\rho$ and so $c_s \approx \frac{1}{\sqrt{3}}c$.
  This is physically relevant, for instance, in various stages of big bang cosmology.
- Speed of fluid flow comparable to that of light.
  This is physically relevant, for instance, in some black hole accretion disks and/or the jets emerging from active galactic nuclei (AGNs).
- Tight binding: $p \lesssim \rho$ or $|\mu| \ll m_0 c^2$.
  Once the pressure is an appreciable fraction of the energy density, or the absolute value of the chemical potential is much smaller than the rest mass, then the usual derivation of the conformal factor appearing in the wave version of the acoustic metric must be significantly modified.
  This is physically relevant, for instance, in cores of neutron stars.

It is somewhat unclear at present as to whether relativistic acoustics can be made directly relevant for laboratory physics. Some first steps in this regard may be found in reference [28], where the possibility of experimentally constructing relativistic BECs is considered.

Geometric Acoustics If one works with special relativistic acoustics, rather than non-relativistic acoustics, then at the level of ray acoustics one will simply obtain an acoustic variant of the Gordon optical metric

$$g_{ab} = \Omega^2 \left[(\eta_{ab} + V_a V_b) - \frac{c_s^2}{c^2} V_a V_b \right]. \tag{2.41}$$

The only difference is that the refractive index has now been replaced by the ratio of the speed of sound to the speed of light: $n^{-1}(x) \to c_s(x)/c$. (The 4-velocity of the medium is still $V^a(x)$, and the conformal factor $\Omega(x)$ is still undetermined.) In general relativistic acoustics this would become

$$g_{ab} = \Omega^2 \left[(f_{ab} + V_a V_b) - \frac{c_s^2}{c^2} V_a V_b \right], \tag{2.42}$$

where $f_{ab}(x)$ is now the general relativistic physical background metric obtained by solving the Einstein equations, and g_{ab} is the acoustic metric for the acoustic perturbations in the fluid flow. Note that the 4-velocity $V^a(x)$ now has to be a timelike unit vector with respect to the background metric $f_{ab}(x)$. For ray acoustics this is all one can say.

Wave Acoustics One can again go to wave acoustics, deriving a wave equation by linearizing the general-relativistic version of the Euler equations. The same sort of technical assumptions must be made, (irrotational, barotropic, and inviscid), and one now obtains a slightly more complicated formula for the conformal factor [7]

$$\Omega = \left(\frac{n^2}{c_s(\rho + p)} \right)^{1/(d-1)}. \tag{2.43}$$

Here n is the number density of particles, and ρ is the energy density (rather than the mass density ρ), while c_s is the speed of sound. The quantity p is the pressure, and d is the number of space dimensions.

Non-relativistic Limit In the non-relativistic limit among other things we certainly have $p \ll \rho$. Also in terms of the average particle mass $\overline{m}$ one has

$$\rho = \rho c^2 \approx n\overline{m}c^2, \tag{2.44}$$

and so

$$\frac{n^2}{c_s(\rho + p)} \approx \frac{n^2}{c_s \rho} \approx \frac{n}{c_s(\overline{m}c^2)} = \frac{\rho}{c_s(\overline{m}^2 c^2)} \propto \frac{\rho}{c_s}, \tag{2.45}$$

thereby (as required for internal consistency) reproducing the correct limit for the conformal factor.

The correct limit for the tensor structure is more subtle. (A suitable discussion can be found in reference [7].) Formally taking the limit $c \to \infty$, but holding c_s and v fixed, a brief calculation yields:

$$g_{00} = \Omega^2 \left[-1 + \gamma^2 - \frac{c_s^2}{c^2} \gamma^2 \right] \to -\Omega^2 \frac{c_s^2 - v^2}{c^2} + \cdots, \qquad (2.46)$$

$$g_{0i} = -\Omega^2 \left[1 - \frac{c_s^2}{c^2} \right] \gamma^2 \beta_i \to -\Omega^2 \frac{v}{c} + \cdots, \qquad (2.47)$$

$$g_{0i} = \Omega^2 \left\{ \delta_{ij} - \left[1 - \frac{c_s^2}{c^2} \right] \gamma^2 \beta_i \beta_j \right\} \to \Omega^2 \delta_{ij} + \cdots . \qquad (2.48)$$

Then, switching from $(ct, \mathbf{x})$ coordinates to $(t, \mathbf{x})$ coordinates, the relativistic g_{ab} of this section correctly limits to the non-relativistic g_{ab} of the previous section.

2.6 Bose–Einstein Condensates

Bose–Einstein condensates (BECs) provide a particularly interesting analogue model because they are relatively easy to construct and manipulate in the laboratory, and specifically because the speed of sound is as low as a few centimetres per second. Most work along these lines has focussed on non-relativistic BECs. Suitable background references are [29–35]. See also the companion chapter by Balbinot et al. in the current volume [36]. In view of the coverage of this topic already provided in that chapter, I shall not have more to say about it here.

In contrast, I will briefly discuss the relativistic BEC model of Fagnocchi et al. that is presented in reference [28]. While relativistic BECs do not seem currently to be a realistic experimental possibility, the theoretical treatment introduces some new issues and effects. The relativistic BECs naturally lead to two quasiparticle excitations, one massless and one massive, with rather complicated excitation spectra. (In this sense the relativistic BECs are reminiscent of the "massive phonon" models obtained from multiple mutually interacting non-relativistic BECs [37–39].) In the relativistic BEC one obtains a 4th-order differential wave equation for the excitations, which is ultimately why one has two branches of quasiparticle excitations. In the limit where the relativistic generalization of the so-called quantum potential can be neglected, the wave equation simplifies to the d'Alembertian equation—for a relativistic acoustic metric of the generalized Gordon form discussed in the previous section. In the limit where both relativistic effects and the quantum potential can be neglected, one recovers the (wave acoustic version of) Unruh's non-relativistic acoustic metric.

Madelung Representation There is an important and non-obvious technical point to be made regarding the linearization of the Madelung representation in a BEC context, (or in fact in any situation where one is dealing with a non-linear

Schrödinger-like equation). For any complex field ψ the Madelung representation is

$$\psi = \sqrt{\rho}e^{i\phi}. \tag{2.49}$$

When linearizing, (which is the basis of separating the system into background plus excitation, or condensate plus quasiparticle), there are at least three things one might envisage doing:

1. Take $\psi = \psi_0 + \varepsilon\psi_1 + \mathcal{O}(\varepsilon^2)$.
2. Take $\rho = \rho_0 + \varepsilon\rho_1 + \mathcal{O}(\varepsilon^2)$, and $\phi = \phi_0 + \varepsilon\phi_1 + \mathcal{O}(\varepsilon^2)$.
3. Take $\psi = \psi_0\{1 + \varepsilon\chi + \mathcal{O}(\varepsilon^2)\}$.

Note that routes 2 and 3 are related by:

$$\frac{\rho_1}{\rho_0} = \frac{\chi + \chi^\dagger}{2}; \qquad \phi_1 = \frac{\chi - \chi^\dagger}{2i}. \tag{2.50}$$

Mathematically, all three routes must carry the same intrinsic physical information, but the clarity with which the information can be extracted varies widely depending on the manner in which the perturbative analysis is presented. When actually carrying out the linearization, it turns out that route 1 is never particularly useful, and that routes 2 and 3 are essentially equivalent for a non-relativistic BEC, ultimately leading to formally identical wave equations. In contrast, for relativistic BECs it is route 3 that leads to the cleanest representation [28], while route 2 leads to a bit of a mess [3]. (A mess involving integro-differential equations.) This is not supposed to be obvious, and will not be obvious unless one tries to carefully work through the relevant technical literature. With hindsight, route 3 appears to be the superior way of organizing the perturbative calculation.

2.7 Surface Waves and Blocking Horizons

Surface waves (water-air, or more generally waves on any fluid-fluid interface) are described by an incredibly complex and subtle theoretical framework—one of the major technical complications comes from the fact that surface waves are highly dispersive, with a propagation speed that is very strongly frequency dependent. Thus, insofar as one can put surface wave propagation into a Lorentzian metric framework, one will have to adopt a "rainbow metric" formalism with a frequency dependent metric. The trade-off is that this system is relatively easily amenable to laboratory investigation through "wave tank" technology [40–42].

Surface Waves in the Geometric Limit As long as the wavelength and period of the surface wave are small compared to the distances and timescales on which the depth of water is changing one can usefully work in the geometric (ray) limit.

Under those conditions one can write a 2 + 1 dimensional metric to describe ray propagation:

$$ds^2 = \Omega^2\{-c_{sw}^2 dt^2 + (d\mathbf{x} - \mathbf{v}dt)^2\} = \Omega^2\{-(c_{sw}^2 - v^2) - 2\mathbf{v}\cdot d\mathbf{x}dt + \|d\mathbf{x}\|^2\}. \quad (2.51)$$

Here c_{sw} is the speed of the surface waves in the comoving frame (that is, comoving with the surface of the fluid), and $\mathbf{v}$ is the (horizontal) velocity of the surface. Unfortunately the speed c_{sw} is a relatively complicated function of (comoving) frequency, depth of the water, density of the fluid, the acceleration due to gravity, the surface tension, *etcetera*. (See for instance references [43–46].)

Now in terms of a 3×3 matrix, this is equivalent to defining the metric tensor

$$g_{ab} = \Omega^2 \left[\begin{array}{c|c} -(c_{sw}^2 - v^2) & -v_j \\ \hline -v_i & \delta_{ij} \end{array} \right]. \quad (2.52)$$

The indices $a, b, c, \ldots$, take on values in $\{0, 1, 2\}$, corresponding to both time and (horizontal) space, whereas indices such as $i, j, k, \ldots$, take on values in $\{1, 2\}$, corresponding to (horizontal) space only. The corresponding inverse metric is

$$g^{ab} = \Omega^{-2} \left[\begin{array}{c|c} -1/c_{sw}^2 & -v^j/c_{sw}^2 \\ \hline -v^i/c_{sw}^2 & \delta^{ij} - v^i v^j/c_{sw}^2 \end{array} \right]. \quad (2.53)$$

In the fluid dynamics community, one most often restricts attention to 1 spatial dimension, then surface waves are said to be "blocked" whenever one has $\|\mathbf{v}\| > c_{sw}$, and one will encounter considerable attention paid to this concept of "wave blocking" in that community. This is what a general relativist would instead call "trapping", and consequently the mixed terminology phrase "blocking horizon" has now come into use within the analogue spacetime community. Note that instead of speaking of Mach number (appropriate for acoustic propagation through the bulk of a medium), in a surface wave context it is the Froude number that governs the formation of ergo-regions and horizons.

Surface Waves in the Physical Limit Moving beyond the geometric/ray approximation for surface waves is mathematically rather tricky. Within the fluid dynamics community relevant work is based on the Boussinesq approximation [47, 48], and its modern variants [49]. Within the analogue spacetime community, see particularly the basic theoretical work in Ref. [50], and in the related chapter [51] in this volume. (See also [52, 53] for a more applied perspective.) Physically, in addition to the presence of dispersion, a second complicating issue is this: The fluid at the surface is moving both vertically (the wave) and horizontally (the background flow), while at the base of the fluid (which may be at variable depth), the no-slip boundary condition enforces zero velocity.

Based on the three-dimensional Euler and continuity equations one then has to construct an interpolating model for the fluid flow that connects the surface motion to the zero-velocity motion at the (variable depth) base. Once this is achieved, one throws away the interpolating model and concentrates only on the physical observable: the motion of the surface. The analysis is mathematically and physically subtle, and (in the physical or wave limit) the theoretical framework for surface waves is nowhere near as clean and straightforward as for barotropic inviscid irrotational acoustic perturbations travelling through the bulk.

Experiments The key benefit of surface waves is that the propagation speed c_{sw} is easily controllable by adjusting the depth of fluid, that background flows are easily set up by simple mechanical pumps, and that "wave tank" and related technologies are well understood and well developed. (See for example, the early 1983 experiments by Badulin et al. [40].) This particular analogue spacetime has recently led to several very interesting experimental efforts [41, 42, 54]. For instance, Weinfurtner et al. have performed an experiment looking at the classical (stimulated) analogue of Hawking radiation from a blocking horizon, and have detected an approximately Boltzmann spectrum of Hawking-like modes [41], while Rousseaux et al. have experimentally investigated the related "negative-norm modes" [42]. The relation between the "hydraulic jump" and blocking horizons has been experimentally investigated by Jannes et al. [54]. Some related theoretical developments are reported in [45, 55]. Work on this topic is ongoing.

2.8 Optical Fibres/Optical Glass

In an optical context, related "optical blocking" phenomena occur when a "refractive index pulse" (RIP) is initiated in an optical fibre [56], or in optical glass [57–59]. The basic idea is that things are arranged so that while the RIP moves at some speed v_{RIP}, the velocity of light outside the RIP is greater than the velocity of the RIP $c_{\mathrm{outside}} > v_{\mathrm{RIP}}$, while inside the RIP we have the contrary situation $c_{\mathrm{inside}} < v_{\mathrm{RIP}}$. This, (certainly within the geometric optics framework), sets up a "black" horizon at the leading edge of the RIP, and a "white" horizon at the trailing edge. (For technical details see references [56–59].) Some subtleties of the theoretical analysis lie in the distinction between group and phase velocities—are we dealing with "phase velocity horizons" or "group velocity horizons"? Other technical subtleties have to do with the transition from geometric optics to wave optics—there are a number of complex and messy technical details involved in this step.

An intriguing experiment based on these ideas has been carried out by Belgiorno et al., with results reported in reference [60]. While it is clear that some form of quantum radiation has been detected, there is still some disagreement as to whether this is (analogue) Hawking radiation, or possibly some other form of quantum vacuum radiation [61–63]. Work on this topic is ongoing.

2.9 Other Models

A complete and exhaustive catalogue of other analogue models would be impractical. See the Living Review article on "Analogue gravity" for more details [3]. Selected models, (a necessarily incomplete list), include:

- Electromagnetic wave guides [64].
- Graphene [65, 66].
- Slow light [67–71].
- Liquid helium [72, 73].
- Fermi gasses [74, 75].
- Ion rings [76].

Beyond the issue of simply *developing* analogue models, there is the whole subject of *using* analogue models to probe, (either theoretically or more boldly experimentally), a whole raft of physics questions such as directly verifying the existence of Hawking radiation, the possibility of Lorentz symmetry violations [77], the nature of the quantum vacuum [78–80], *etcetera*. For more details, see Ref. [3], and other chapters in this volume.

2.10 Discussion

The general theme to be extracted from these considerations is this: The propagation of excitations (either particles or waves) over a background can often (not always) be given a geometric interpretation in therms of some "analogue spacetime". As such a geometric interpretation exists, there is a strong likelihood of significant cross-fertilization of ideas and techniques between general relativity and other branches of physics. Such possibilities have increasingly attracted attention over the last decade, for many reasons. The other chapters in these proceedings will explore these ideas in more specific detail.

Acknowledgements This research was supported by the Marsden Fund, and by a James Cook Research Fellowship, both administered by the Royal Society of New Zealand.

References

1. Gordon, W.: Zur Lichtfortpflanzung nach der Relativitätstheorie. Ann. Phys. (Leipz.) **72**, 421–456 (1923). doi:10.1002/andp.19233772202
2. Unruh, W.G.: Experimental black hole evaporation. Phys. Rev. Lett. **46**, 1351 (1981)
3. Barceló, C., Liberati, S., Visser, M.: Analogue gravity. Living Rev. Relativ. **8**, 12 (2005). Updated as Living Rev. Relativ. **14**, 3 (2011). gr-qc/0505065
4. Visser, M.: Acoustic propagation in fluids: an unexpected example of Lorentzian geometry. gr-qc/9311028
5. Visser, M.: Acoustic black holes: horizons, ergospheres, and Hawking radiation. Class. Quantum Gravity **15**, 1767 (1998). gr-qc/9712010

6. Visser, M., Barceló, C., Liberati, S.: Analog models of and for gravity. Gen. Relativ. Gravit. **34**, 1719 (2002). gr-qc/0111111
7. Visser, M., Molina-París, C.: Acoustic geometry for general relativistic barotropic irrotational fluid flow. New J. Phys. **12**, 095014 (2010). arXiv:1001.1310 [gr-qc]
8. Baccetti, V., Martin–Moruno, P., Visser, M.: Gordon and Kerr–Schild ansätze in massive and bimetric gravity. arXiv:1206.4720v1 (2013)
9. Tolman, R.C.: On thermodynamic equilibrium in a static Einstein universe. Proc. Natl. Acad. Sci. USA **17**, 153–160 (1931)
10. Robertson, H.P.: Relativistic cosmology. Rev. Mod. Phys. **5**, 62–90 (1933)
11. Tolman, R.C.: Static solutions of Einstein's field equations for spheres of fluid. Phys. Rev. **55**, 364–373 (1939)
12. Niven, W.D. (ed.): The Scientific Papers of James Clerk Maxwell, 1890, p. 76. Dover, New York (2003)
13. Anonymous: Problems (3). Camb. Dublin Math. J. **8**, 188 (1853)
14. Anonymous: Solutions of problems (prob. 3, vol. VIII. p. 188). Camb. Dublin Math. J. **9**, 9–11 (1854)
15. Lüneburg, R.K.: In: Mathematical Theory of Optics, pp. 189–213. Brown University, Providence (1944)
16. Leonhardt, U.: Optical conformal mapping. Science **312**, 1777–1780 (2006). doi:10.1126/science.1126493
17. Pendry, J.B., Schurig, D., Smith, D.R.: Controlling electromagnetic fields. Science **312**, 1780–1782 (2006). doi:10.1126/science.1125907
18. Leonhardt, U., Philbin, T.: General relativity in electrical engineering. New J. Phys. **8**, 247 (2006). cond-mat/0607418. doi:10.1088/1367-2630/8/10/247
19. Leonhardt, U., Philbin, T.: Transformation optics and the geometry of light. Prog. Opt. **53**, 69–152 (2009). arXiv:0805.4778
20. Pendry, J.: Optics: all smoke and metamaterials. Nature **460**, 579 (2009)
21. Visser, M., Weinfurtner, S.E.C.: Vortex geometry for the equatorial slice of the Kerr black hole. Class. Quantum Gravity **22**, 2493 (2005). gr-qc/0409014
22. Berti, E., Cardoso, V., Lemos, J.P.S.: Quasinormal modes and classical wave propagation in analogue black holes. Phys. Rev. **D70**, 124006 (2004). gr-qc/0408099
23. Lemos, J.P.S.: Rotating acoustic holes: quasinormal modes and tails, super-resonance, and sonic bombs and plants in the draining bathtub. In: Proceedings of the II Amazonian Symposium on Physics—Analogue Models of Gravity (in press)
24. Perez Bergliaffa, S.E., Hibberd, K., Stone, M., Visser, M.: Wave equation for sound in fluids with vorticity. Physica D **191**, 121 (2004). cond-mat/0106255
25. Jacobson, T., Kang, G.: Conformal invariance of black hole temperature. Class. Quantum Gravity **10**, L201 (1993). gr-qc/9307002
26. Moncrief, V.: Stability of stationary, spherical accretion onto a Schwarzschild black hole. Astrophys. J. **235**, 1038–1046 (1980)
27. Babichev, E., Mukhanov, V., Vikman, A.: k-Essence, superluminal propagation, causality and emergent geometry. J. High Energy Phys. **0802**, 101 (2008). arXiv:0708.0561 [hep-th]
28. Fagnocchi, S., Finazzi, S., Liberati, S., Kormos, M., Trombettoni, A.: Relativistic Bose-Einstein condensates: a new system for analogue models of gravity. New J. Phys. **12**, 095012 (2010). arXiv:1001.1044 [gr-qc]
29. Garay, L.J., Anglin, J.R., Cirac, J.I., Zoller, P.: Black holes in Bose-Einstein condensates. Phys. Rev. Lett. **85**, 4643 (2000). gr-qc/0002015
30. Barceló, C., Liberati, S., Visser, M.: Analog gravity from Bose-Einstein condensates. Class. Quantum Gravity **18**, 1137 (2001). gr-qc/0011026
31. Barceló, C., Liberati, S., Visser, M.: Towards the observation of Hawking radiation in Bose-Einstein condensates. Int. J. Mod. Phys. A **18**, 3735 (2003). gr-qc/0110036
32. Barceló, C., Liberati, S., Visser, M.: Probing semiclassical analog gravity in Bose-Einstein condensates with widely tunable interactions. Phys. Rev. A **68**, 053613 (2003). cond-mat/0307491

33. Lahav, O., Itah, A., Blumkin, A., Gordon, C., Steinhauer, J.: Realization of a sonic black hole analogue in a Bose-Einstein condensate. Phys. Rev. Lett. **105**, 240401 (2010). arXiv:0906.1337 [cond-mat.quant-gas]
34. Jannes, G.: Emergent gravity: the BEC paradigm. arXiv:0907.2839 [gr-qc]
35. Mayoral, C., Recati, A., Fabbri, A., Parentani, R., Balbinot, R., Carusotto, I.: Acoustic white holes in flowing atomic Bose-Einstein condensates. New J. Phys. **13**, 025007 (2011). arXiv:1009.6196 [cond-mat.quant-gas]
36. Balbinot, R., Carusotto, I., Fabbri, A., Mayoral, C., Recati, A.: Understanding Hawking radiation from simple BEC models (these proceedings)
37. Visser, M., Weinfurtner, S.: Massive Klein-Gordon equation from a BEC-based analogue spacetime. Phys. Rev. D **72**, 044020 (2005). gr-qc/0506029
38. Visser, M., Weinfurtner, S.: Massive phonon modes from a BEC-based analog model. cond-mat/0409639
39. Liberati, S., Visser, M., Weinfurtner, S.: Naturalness in emergent spacetime. Phys. Rev. Lett. **96**, 151301 (2006). gr-qc/0512139
40. Badulin, S.I., Pokazayev, K.V., Rozenberg, A.D.: A laboratory study of the transformation of regular gravity-capillary waves in inhomogeneous flows. Izv., Atmos. Ocean. Phys. **19**, 782–787 (1983)
41. Weinfurtner, S., Tedford, E.W., Penrice, M.C.J., Unruh, W.G., Lawrence, G.A.: Measurement of stimulated Hawking emission in an analogue system. Phys. Rev. Lett. **106**, 021302 (2011). arXiv:1008.1911 [gr-qc]
42. Rousseaux, G., Mathis, C., Maissa, P., Philbin, T.G., Leonhardt, U.: Observation of negative phase velocity waves in a water tank: a classical analogue to the Hawking effect? New J. Phys. **10**, 053015 (2008). arXiv:0711.4767 [gr-qc]
43. Visser, M., Weinfurtner, S.: Analogue spacetimes: toy models for quantum gravity. arXiv:0712.0427 [gr-qc]
44. Visser, M.: Emergent rainbow spacetimes: two pedagogical examples. In: Time & Matter, Lake Bled, Slovenia (2007). arXiv:0712.0810 [gr-qc]
45. Rousseaux, G., Maissa, P., Mathis, C., Coullet, P., Philbin, T.G., Leonhardt, U.: Horizon effects with surface waves on moving water. New J. Phys. **12**, 095018 (2010). arXiv:1004.5546 [gr-qc]
46. Rousseaux, G.: The basics of water waves theory for analogue gravity (these proceedings). arXiv:1203.3018v1 [physics.flu-dyn]
47. Boussinesq, J.: Théorie de l'intumescence liquide, appelée onde solitaire ou de translation, se propageant dans un canal rectangulaire. C. R. Acad. Sci. **72**, 755–759 (1871)
48. Boussinesq, J.: Théorie des ondes et des remous qui se propagent le long d'un canal rectangulaire horizontal, en communiquant au liquide contenu dans ce canal des vitesses sensiblement pareilles de la surface au fond. J. Math. Pures Appl., Sér. II **17**, 55–108 (1872)
49. Madsen, P.A., Schaffer, H.A.: Higher-order Boussinesq type equations for surface gravity waves: derivation and analysis. Philos. Trans. R. Soc., Math. Phys. Eng. Sci. **356**, 3123–3184 (1998). See http://www.jtsor.org/stable/55084
50. Schützhold, R., Unruh, W.G.: Gravity wave analogs of black holes. Phys. Rev. D **66**, 044019 (2002). gr-qc/0205099
51. Unruh, W.G.: Irrotational, two-dimensional surface waves in fluids (these proceedings). arXiv:1205.6751 [gr-qc]
52. Chaline, J., Jannes, G., Maissa, P., Rousseaux, G.: Some aspects of dispersive horizons: lessons from surface waves (these proceedings). arXiv:1203.2492 [physics.flu-dyn]
53. Carusotto, I., Rousseaux, G.: The Cerenkov effect revisited: from swimming ducks to zero modes in gravitational analogs (these proceedings). arXiv:1202.3494 [physics.class-ph]
54. Jannes, G., Piquet, R., Maissa, P., Mathis, C., Rousseaux, G.: Experimental demonstration of the supersonic-subsonic bifurcation in the circular jump: a hydrodynamic white hole. Phys. Rev. E **83**, 056312 (2011). arXiv:1010.1701 [physics.flu-dyn]
55. Volovik, G.E.: The hydraulic jump as a white hole. JETP Lett. **82**, 624 (2005). Pis'ma Zh. Eksp. Teor. Fiz. **82**, 706 (2005). physics/0508215

56. Philbin, T.G., Kuklewicz, C., Robertson, S., Hill, S., Konig, F., Leonhardt, U.: Fiber-optical analogue of the event horizon. Science **319**, 1367–1370 (2008). arXiv:0711.4796 [gr-qc]
57. Belgiorno, F., Cacciatori, S.L., Ortenzi, G., Sala, V.G., Faccio, D.: Quantum radiation from superluminal refractive-index perturbations. Phys. Rev. Lett. **104**, 140403 (2010). arXiv:0910.3508 [quant-ph]
58. Belgiorno, F., Cacciatori, S.L., Ortenzi, G., Rizzi, L., Gorini, V., Faccio, D.: Dielectric black holes induced by a refractive index perturbation and the Hawking effect. Phys. Rev. D **83**, 024015 (2011). arXiv:1003.4150 [quant-ph]
59. Cacciatori, S.L., Belgiorno, F., Gorini, V., Ortenzi, G., Rizzi, L., Sala, V.G., Faccio, D.: Space-time geometries and light trapping in travelling refractive index perturbations. New J. Phys. **12**, 095021 (2010). arXiv:1006.1097 [physics.optics]
60. Belgiorno, F., Cacciatori, S.L., Clerici, M., Gorini, V., Ortenzi, G., Rizzi, L., Rubino, E., Sala, V.G., et al.: Hawking radiation from ultrashort laser pulse filaments. arXiv:1009.4634 [gr-qc]
61. Schützhold, R., Unruh, W.G.: Comment on: Hawking radiation from ultrashort laser pulse filaments. Phys. Rev. Lett. **107**, 149401 (2011). arXiv:1012.2686 [quant-ph]
62. Belgiorno, F., Cacciatori, S.L., Clerici, M., Gorini, V., Ortenzi, G., Rizzi, L., Rubino, E., Sala, V.G., et al.: Reply to "Comment on: Hawking radiation from ultrashort laser pulse filaments". Phys. Rev. Lett. **107**, 149402 (2011). arXiv:1012.5062 [quant-ph]
63. Liberati, S., Prain, A., Visser, M.: Quantum vacuum radiation in optical glass. Phys. Rev. D (in press). arXiv:1111.0214 [gr-qc]
64. Schützhold, R., Unruh, W.G.: Hawking radiation in an electromagnetic waveguide? Phys. Rev. Lett. **95**, 031301 (2005)
65. Cortijo, A., Vozmediano, M.A.H.: Effects of topological defects and local curvature on the electronic properties of planar graphene. Nucl. Phys. B **763**, 293 (2007). Nucl. Phys. B **807**, 659 (2009). cond-mat/0612374
66. Vozmediano, M.A.H., Katsnelson, M.I., Guinea, F.: Gauge fields in graphene. Phys. Rep. **496**, 109 (2010)
67. Leonhardt, U., Piwnicki, P.: Relativistic effects of light in moving media with extremely low group velocity. Phys. Rev. Lett. **84**, 822 (2000).
68. Visser, M.: Comment on: relativistic effects of light in moving media with extremely low group velocity. Phys. Rev. Lett. **85**, 5252 (2000). gr-qc/0002011
69. Leonhardt, U., Piwnicki, P.: Reply to the comment on 'Relativistic effects of light in moving media with extremely low group velocity' by M. Visser. Phys. Rev. Lett. **85**, 5253 (2000). gr-qc/0003016
70. Leonhardt, U.: Space-time geometry of quantum dielectrics. physics/0001064
71. Leonhardt, U.: A primer to slow light. gr-qc/0108085
72. Volovik, G.E.: Superfluid analogies of cosmological phenomena. Phys. Rep. **351**, 195 (2001). gr-qc/0005091
73. Jacobson, T.A., Volovik, G.E.: Event horizons and ergoregions in He-3. Phys. Rev. D **58**, 064021 (1998)
74. Giovanzzi, S.: Hawking radiation in sonic black holes. Phys. Rev. Lett. **94**, 061302 (2005). physics/0411064
75. Giovanzzi, S.: Entanglement entropy and mutual information production rates in acoustic black holes. Phys. Rev. Lett. **106**, 011302 (2011). arXiv:1101.3272 [cond-mat.other]
76. Horstmann, B., Reznik, B., Fagnocchi, S., Cirac, J.I.: Hawking radiation from an acoustic black hole on an ion ring. Phys. Rev. Lett. **104**, 250403 (2010). arXiv:0904.4801 [quant-ph]
77. Liberati, S.: Lorentz breaking: effective field theory and observational tests (these proceedings). arXiv:1203.4105 [gr-qc]
78. Volovik, G.E.: Topology of quantum vacuum (these proceedings). arXiv:1111.4627 [hep-ph]
79. Jannes, G., Volovik, G.E.: The cosmological constant: a lesson from the effective gravity of topological Weyl media. arXiv:1108.5086 [gr-qc]
80. Finazzi, S., Liberati, S., Sindoni, L.: The analogue cosmological constant in Bose-Einstein condensates: a lesson for quantum gravity. In: Proceedings of the II Amazonian Symposium on Physics—Analogue Models of Gravity (in press). arXiv:1204.3039 [gr-qc]

Chapter 3
Cosmological Particle Creation in the Lab?

Ralf Schützhold and William G. Unruh

Abstract We give an overview of some fundamental quantum vacuum effects in curved space times that may be studied in earth based laboratories. In particular we review the concept of cosmological particle creation related to a contraction or expansion of the Universe.

3.1 Introduction

One of the most striking examples for the production of particles out of the quantum vacuum due to external conditions is cosmological particle creation, which is caused by the expansion or contraction of the Universe. Already in 1939, Schrödinger understood that the cosmic evolution could lead to a mixing of positive and negative frequencies and that this "would mean production or annihilation of matter, merely by the expansion" [9]. Intuitively speaking, the expansion of the universe tears apart the quantum vacuum fluctuations and thereby transforms them into pairs of real particles. More precisely, the quantum state of the field under consideration cannot follow the cosmic evolution[1] (breakdown of adiabaticity) and thus deviates from the ground state, i.e., turns into an excited state containing particles. Later this phenomenon was derived via more modern techniques of quantum field theory in curved spacetimes by Parker [8] (who apparently was not aware of Schrödinger's work) and subsequently has been studied in numerous publications, see, e.g., [3, 6, 13]. Even

---

[1] It is like the situation in which the parameters of a quantised harmonic oscillator (spring constant or mass) are changed at a rate faster than the period of oscillation. This causes the oscillator to become excited (creates a squeezed state). The expansion of the universe alters the spring constant of each of the modes of the quantum field.

R. Schützhold (✉)
Fakultät für Physik, Universität Duisburg-Essen, 47048 Duisburg, Germany
e-mail: ralf.schuetzhold@uni-due.de

W.G. Unruh
Canadian Institute for Advanced Research Cosmology and Gravity Program, Department of Physics and Astronomy, University of British Columbia, Vancouver, BC V6T 1Z1, Canada
e-mail: unruh@physics.ubc.ca

though cosmological particle creation typically occurs on extremely large length scales, it is one of the very few examples for such fundamental effects where we actually may have observational evidence: According to the inflationary model of cosmology, the seeds for the anisotropies in the cosmic microwave background (CMB) and basically all large scale structures stem from this effect, see Sect. 3.5. In this chapter, we shall provide a brief discussion of this phenomenon and sketch a possibility for an experimental realisation via an analogue in the laboratory.

3.2 Scattering Analogy

For simplicity, let us consider a massive scalar field Φ in the $1+1$ dimensional Friedmann-Robertson-Walker metric with scale factor $a(\tau)$

$$ds^2 = d\tau^2 - a^2(\tau)dx^2 = a^2(\eta)[d\eta^2 - dx^2], \tag{3.1}$$

where τ is the proper (co-moving) time and η the conformal time. The latter coordinate is more convenient for our purpose since the wave equation simplifies to ($\hbar = c = 1$)

$$\left(\frac{\partial^2}{\partial \eta^2} - \frac{\partial^2}{\partial x^2} - a^2(\eta)m^2\right)\Phi(\eta, x). \tag{3.2}$$

In the massless case $m = 0$, the scalar field is conformally invariant (in $1+1$ dimensions) and thus the solution $\Phi(\eta, x)$ is not affected by the cosmic evolution $a^2(\eta)$. As a result, there is no mixing between positive and negative frequencies (see below) in this case $m = 0$, i.e., the expansion does only create particles for $m > 0$. The same argument applies to the electromagnetic field in $3+1$ dimensions, which is also comformally invariant.

After a spatial Fourier transform, we find that each mode $\phi_k(\eta)$ behaves like a harmonic oscillator with a time-dependent potential

$$\left(\frac{d^2}{dt^2} + \Omega^2(t)\right)\phi(t) = 0, \tag{3.3}$$

with $k^2 + a^2(\eta)m^2 \to \Omega^2(t)$ and $\eta \to t$. There is yet another analogy which might be interesting to notice. If we compare the above equation to a Schrödinger scattering problem in one spatial dimension

$$\left(-\frac{1}{2m}\frac{d^2}{dx^2} + V(x)\right)\Psi(x) = E\Psi(x), \tag{3.4}$$

we find that is has precisely the same form after identifying $t \leftrightarrow x$, $\phi(t) \leftrightarrow \Psi(x)$, and $\Omega^2(t) \leftrightarrow 2m[E - V(x)]$. Now we may apply our knowledge of such one-dimensional scattering problems. Assuming a localised potential $V(|x| \to \infty) =$ const, the general asymptotic solutions for $\Psi(x \to \pm\infty)$ are linear combinations of plane waves e^{+ikx} and e^{-ikx}. The connection formulas between these solutions on the far left and the far right of the potential $V(x)$ then define the 2×2 scattering matrix $\underline{S}$. For example, a solution which behaves as $\Psi(x \to -\infty) = Te^{-ikx}$ on the far left is connected via Eq. (3.4) to the corresponding solution $\Psi(x \to +\infty) = e^{-ikx} + Re^{+ikx}$ on the far right, where R and T are the reflection and transmission coefficients. If $E > V(x)$ holds everywhere, we have propagation over the barrier, where the reflection coefficient R vanishes in the classical limit. On the other hand, if $E < V(x)$, we have a barrier penetration problem where the transmission coefficient T due to quantum tunnelling vanishes in the classical limit.

Note that Ω^2 is always greater than zero in our case—which corresponds to propagation over the barrier $E > V(x)$. If Ω^2 were less than zero over some region in time, one would have a barrier penetration (i.e., tunnelling) problem $E < V(x)$. In both cases, an initial solution of the form $e^{i\Omega_{in}t}$ evolves into a future solution $\alpha e^{i\Omega_{out}t} + \beta e^{-i\Omega_{out}t}$ due to scattering from the region where $\Omega^2 \neq$ const. This would correspond to particle creation with probability proportional to $|\beta|^2$.

Assuming that Ω is constant asymptotically, we have a static wave equation (3.3) in that regime and thus positive and negative pseudo-norm solutions correspond to positive and negative frequencies, see, e.g., [3, 6, 13]. In order to derive the cosmological particle creation, we can study an initial positive pseudo-norm solution $e^{-i\Omega_{in}t}$ of Eq. (3.3) and see how it finally evolves into a mixture of positive and negative pseudo-norm solutions $\alpha e^{-i\Omega_{out}t} + \beta e^{+i\Omega_{out}t}$. In the Schrödinger scattering problem, the initial solution $e^{-i\Omega t}$ could be identified with a left-moving wave on the left-hand side of the potential "barrier" while the final solution $\alpha e^{-i\Omega t} + \beta e^{+i\Omega t}$ would then correspond to a mixture of left-moving $\alpha e^{-i\Omega t}$ and right-moving $\beta e^{+i\Omega t}$ waves on the right-hand-side. As a consequence, the Bogoliubov coefficients α and β are related to the reflection R and transmission T coefficients via $\alpha = 1/T$ and $\beta = R/T$. In this way, the Bogoliubov relation $|\alpha|^2 - |\beta|^2 = 1$ is equivalent to the conservation law $|R|^2 + |T|^2 = 1$ for the Schrödinger scattering problem. The probability for particle creation can be inferred from the expectation value of the number of final particles in the initial vacuum state which reads $\langle 0_{in}|\hat{n}_{out}|0_{in}\rangle = |\beta|^2$.

3.3 WKB Analysis

In order to actually calculate or estimate the Bogoliubov coefficients, let us re-write Eq. (3.3) in a first-order form via introducing the phase-space vector $\mathbf{u}$ and the matrix $\mathbf{M}$

$$\frac{d}{dt}\begin{pmatrix}\phi\\ \dot\phi\end{pmatrix} = \dot{\mathbf{u}} = \begin{pmatrix} 0 & 1 \\ -\Omega^2(t) & 0 \end{pmatrix} \cdot \begin{pmatrix}\phi\\ \dot\phi\end{pmatrix} = \mathbf{M} \cdot \mathbf{u}. \qquad (3.5)$$

If we define an inner product via

$$(\mathbf{u}|\mathbf{u}') = i\left(u_2^* u_1' - u_1^* u_2'\right), \qquad (3.6)$$

we find that the inner product of two solutions $\mathbf{u}$ and $\mathbf{u}'$ of Eq. (3.5) is conserved

$$\frac{d}{dt}(\mathbf{u}|\mathbf{u}') = 0. \tag{3.7}$$

The split of a solution into positive and negative frequencies (i.e., positive and negative pseudo-norm) corresponds to a decomposition in the instantaneous eigen-basis of the matrix

$$\mathbf{M} \cdot \mathbf{u}_{\pm} = \pm i \Omega \mathbf{u}_{\pm}. \tag{3.8}$$

Choosing the usual normalisation $\mathbf{u}_{\pm} = (1, \pm i\Omega)^T/\sqrt{2\Omega}$ known from the definition of the creation and annihilation operators $(\hat{x} \pm i\Omega \hat{p})/\sqrt{2\Omega}$ of the harmonic oscillator, we find

$$(\mathbf{u}_+|\mathbf{u}_+) = 1, \qquad (\mathbf{u}_-|\mathbf{u}_-) = -1, \qquad (\mathbf{u}_+|\mathbf{u}_-) = 0. \tag{3.9}$$

At each time t, we may expand a given solution $\mathbf{u}(t)$ of Eq. (3.5) into the instantaneous eigen-vectors

$$\mathbf{u}(t) = \alpha(t) e^{i\varphi(t)} \mathbf{u}_+(t) + \beta(t) e^{-i\varphi(t)} \mathbf{u}_-(t), \tag{3.10}$$

where the pre-factors are now defined as time-dependent Bogoliubov coefficients $\alpha(t)$ and $\beta(t)$. It is useful to separate out the oscillatory part with the WKB phase

$$\varphi(t) = \int_{-\infty}^{t} dt' \Omega(t'). \tag{3.11}$$

Now we may insert the expansion (3.10) into the equation of motion (3.5), i.e., $\dot{\mathbf{u}} = \mathbf{M} \cdot \mathbf{u}$. Using Eq. (3.8), we find that the eigen-values $\pm i\Omega$ are cancelled by the $\dot{\varphi}$-derivatives. If we project the remaining equation with the inner product (3.6) onto the eigen-vectors $\mathbf{u}_{\pm}$ and use the properties $(\mathbf{u}_+|\dot{\mathbf{u}}_+) = (\mathbf{u}_-|\dot{\mathbf{u}}_-) = 0$, $(\mathbf{u}_-|\dot{\mathbf{u}}_+) = \dot{\Omega}/(2\Omega)$ and $(\mathbf{u}_+|\dot{\mathbf{u}}_-) = -\dot{\Omega}/(2\Omega)$, as well as (3.9), we find

$$\dot{\alpha} = \frac{\dot{\Omega}}{2\Omega} e^{-2i\varphi} \beta, \qquad \dot{\beta} = \frac{\dot{\Omega}}{2\Omega} e^{2i\varphi} \alpha. \tag{3.12}$$

These equations (3.12) are still exact but very hard to solve analytically—except in very special cases. They can be solved formally by a iterative integral equation

$$\alpha_{n+1} = \alpha_{\text{in}} + \int_{-\infty}^{t} dt' \frac{\dot{\Omega}(t')}{2\Omega(t')} e^{-2i\varphi(t')} \beta_n(t'),$$
$$\beta_{n+1} = \beta_{\text{in}} + \int_{-\infty}^{t} dt' \frac{\dot{\Omega}(t')}{2\Omega(t')} e^{-2i\varphi(t')} \alpha_n(t'). \tag{3.13}$$

It can be shown that this iteration converges to the exact solution for well-behaved $\Omega(t)$ [1]. Standard perturbation theory would then correspond to cutting off this iteration at a finite order, which can be justified if $\Omega(t)$ changes only very little. For the scalar field in Eq. (3.2) this perturbative treatment should be applicable in the ultra-relativistic limit, i.e., as long as the mass is much smaller than the wave-number.

In many cases, however, another approximation—the WKB method—is more useful. This method can be applied if the rate of change of $\Omega(t)$, e.g., the expansion of the universe, is much slower than the internal frequency $\Omega(t)$ itself. Formally, we may write

$$\Omega(t) = \Omega_0 f(\omega t), \qquad (3.14)$$

where Ω_0 denotes the overall magnitude of $\Omega(t)$, i.e., the internal frequency, and ω its rate of change, i.e., the external frequency, while f is some dimensionless function f of order one. In this scaling, the WKB limit corresponds to $\Omega_0 \gg \omega$. In terms of the reflection coefficient $R = \beta/\alpha$ mentioned earlier, we have $\dot{R} = (\alpha \dot{\beta} - \dot{\alpha}\beta)/\alpha^2$ and inserting (3.12) we can combine these two Eqs. (3.12) into one simple equality via [14]

$$\dot{R} = \frac{\dot{\Omega}}{2\Omega}\left(e^{2i\varphi} - R^2 e^{-2i\varphi}\right), \qquad (3.15)$$

which is known as Riccati equation. Again, this equation is still exact but unfortunately non-linear. Neglecting the quadratic term R^2 would bring us back to perturbation theory. In the WKB-limit, the phase factors $e^{\pm 2i\varphi}$ are rapidly oscillating and the magnitude of R can be estimated by going to the complex plane. Re-writing the Riccati equation (3.15) as

$$\frac{dR}{d\varphi} = \frac{1}{2}\left(e^{2i\varphi} - R^2 e^{-2i\varphi}\right)\frac{d\ln\Omega}{d\varphi}, \qquad (3.16)$$

we may use an analytic continuation $\varphi \to \varphi + i\chi$ to see that R becomes exponentially suppressed $R \sim e^{-2\chi}$. How strongly it is suppressed depends on the point where the analytic continuation breaks down. Since $e^{\pm 2i\varphi}$ is analytic everywhere, this will be determined by the term $\ln\Omega$. Typically, the first non-analytic points t_* encountered are the zeros of Ω, i.e., where $\Omega(t_*) = 0$. In the case of barrier reflection, these points t_* where $\Omega(t_*) = 0$, i.e., where $V = E$, lie on the real axis and correspond to the classical turning points in WKB. In our case, we have scattering above the barrier and thus these points become complex—but are still analogous to the classical turning points in WKB. If we analytically continue R in Eq. (3.16) into the upper complex plane $\varphi \to \varphi + i\chi$ until we hit the first turning point t_*, the exponential $e^{2i\varphi}$ in Eq. (3.16) contains an oscillating factor from the real part $\Re[\varphi]$ and an exponentially suppressed factor $e^{-2\chi_*}$ from the imaginary part $\chi_* = \Im[\varphi(t_*)]$. Consequently, we find[2]

$$R = \frac{\beta}{\alpha} \sim e^{-2\chi_*} = \exp\left\{-2\Im\left[\int_0^{t_*} dt' \Omega(t')\right]\right\}. \qquad (3.17)$$

If there is more than one turning point, the one with the smallest $\chi_* > 0$, i.e., closest to the real axis (in the complex φ-plane) dominates. If these multiple turning points have similar $\chi_* > 0$, there can be interference effects between the different contributions, see, e.g., [5].

---

[2]In fact, it can be shown that Eq. (3.17) becomes exact in the adiabatic limit $\omega/\Omega \downarrow 0$, i.e., the pre-factor in front of the exponent tends to one, see, e.g., [4, 7].

3.4 Adiabatic Expansion and Its Breakdown

Note that we could repeat steps (3.5) till (3.12) and expand the solution $\mathbf{u}(t)$ into the first-order adiabatic eigen-states instead of the instantaneous eigen-vectors $\mathbf{u}_\pm$. To this end, let us re-write (3.12) as

$$\frac{d}{dt}\begin{pmatrix}\alpha(t)e^{+i\varphi(t)}\\ \beta(t)e^{-i\varphi(t)}\end{pmatrix} = \dot{\mathbf{w}} = \begin{pmatrix} i\Omega & \dot{\Omega}/(2\Omega) \\ \dot{\Omega}/(2\Omega) & -i\Omega \end{pmatrix} \cdot \begin{pmatrix}\alpha(t)e^{+i\varphi(t)}\\ \beta(t)e^{-i\varphi(t)}\end{pmatrix} = \mathbf{N}\cdot\mathbf{w}. \tag{3.18}$$

The eigen-vectors of the matrix $\mathbf{N}$ are the first-order adiabatic eigen-states $\mathbf{w}_\pm$ and the eigen-frequencies $\mathbf{N}\cdot\mathbf{w}_\pm = \pm i\Omega_{\mathrm{ad}}\mathbf{w}_\pm$ are renormalised to

$$\Omega_{\mathrm{ad}} = \Omega\sqrt{1 - \frac{\dot{\Omega}^2}{4\Omega^4}}. \tag{3.19}$$

Assuming $\alpha_{\mathrm{in}} = 1$ and $\beta_{\mathrm{in}} = 0$, the system stays in the adiabatic eigen-state $\mathbf{w}_+$ to lowest order in ω/Ω and we get

$$\alpha(t) = 1 + \mathscr{O}\left(\frac{\omega^2}{\Omega^2}\right), \qquad \beta(t) = -\frac{i}{4}\frac{\dot{\Omega}}{\Omega^2} + \mathscr{O}\left(\frac{\omega^2}{\Omega^2}\right). \tag{3.20}$$

This adiabatic expansion into powers of ω/Ω can be continued and gives terms like $\dot{\Omega}^2/\Omega^4$ and $\ddot{\Omega}/\Omega^3$ to the next order in ω/Ω (see below). One should stress that this expansion is *not* the same as in (3.13) since it is local—i.e., only contains time-derivatives—while (3.13) is global—i.e., contains time-integrals. Since all terms of the adiabatic expansion (3.20) are local, they cannot describe particle creation—which depends on the whole history of $\Omega(t)$. In terms of the adiabatic expansion into powers of ω/Ω, particle creation is a non-perturbative effect, i.e., it is exponentially suppressed: If we estimate the exponent in Eq. (3.17), we find that the turning point t_* scales with $t \propto 1/\omega$ whereas the integrand $\Omega(t)$ obviously scales with Ω. Thus we find the following exponential scaling

$$R \sim \exp\left\{-\mathscr{O}\left(\frac{\Omega}{\omega}\right)\right\}, \tag{3.21}$$

which does not admit a Taylor expansion into powers of ω/Ω and thus is non-perturbative in terms of the adiabatic expansion. For any finite ratio of ω/Ω, this also means that the adiabatic expansion (into powers of ω/Ω) must break down at some order. To make this argument more precise, let us re-write Eq. (3.18) in yet another form

$$\frac{d\mathbf{w}}{dt} = \mathbf{N}\cdot\mathbf{w} = \Lambda\begin{pmatrix} i\cosh(2\xi) & \sinh(2\xi) \\ \sinh(2\xi) & -i\cosh(2\xi) \end{pmatrix}\cdot\mathbf{w}. \tag{3.22}$$

In this representation, the eigen-values of $\mathbf{N}$ are given by $\pm i\Lambda$ and the eigen-vectors read

$$\mathbf{w}_+ = \begin{pmatrix}\cosh\xi\\ -i\sinh\xi\end{pmatrix}, \qquad \mathbf{w}_- = \begin{pmatrix}\sinh\xi\\ -i\cosh\xi\end{pmatrix}. \tag{3.23}$$

Decomposing the solution $\mathbf{w}(t)$ into these eigen-vectors

$$\mathbf{w}(t) = a(t)\mathbf{w}_+(t) + b(t)\mathbf{w}_-(t), \qquad (3.24)$$

and using $\dot{\mathbf{w}}_+ = \dot{\xi}\mathbf{w}_-$ as well as $\dot{\mathbf{w}}_- = \dot{\xi}\mathbf{w}_+$, we find

$$\frac{d}{dt}\begin{pmatrix} a \\ b \end{pmatrix} = \begin{pmatrix} i\Lambda & -\dot{\xi} \\ -\dot{\xi} & -i\Lambda \end{pmatrix} \cdot \begin{pmatrix} a \\ b \end{pmatrix}. \qquad (3.25)$$

This is the same form as Eq. (3.22) if we change Λ and ξ accordingly. Thus, by repeating this procedure, we get the iteration law

$$\Lambda_{n+1} = \sqrt{\Lambda_n^2 - \dot{\xi}_n^2}, \qquad \xi_{n+1} = -\frac{1}{2}\mathrm{arctanh}\left(\frac{\dot{\xi}_n}{\Lambda_n}\right). \qquad (3.26)$$

By this iteration, we go higher and higher up in the adiabatic expansion since ξ_n always acquires an additional factor of ω/Ω. Thus, for $\omega \ll \Omega$, the values of ξ_n quickly decay with a power-law $\xi_n = \mathcal{O}([\omega/\Omega]^n)$ initially. As we go up in this expansion, however, the effective rate of change of ξ_n increases. For example, if $\Omega(t)$ has one global maximum (or minimum) and otherwise no structure, the time-derivative $\dot{\Omega}/(2\Omega^2) = \tanh(2\xi_1)$ has two extremal points and a zero in between. By taking higher and higher time derivatives, more and more extremal points and a zeros arise and thus the effective frequency ω_n^{eff} of $\xi_n(t)$ increases roughly linearly with the number n of iterations $\omega_n^{\mathrm{eff}} = \mathcal{O}(n\omega)$. Furthermore, the adiabatically renormalised eigen-values Λ_n decrease with each iteration. Thus, after approximately $n = \mathcal{O}(\Omega/\omega)$ iterations, the effective frequency ω_n^{eff} becomes comparable to the internal frequency Λ_n. At that point, the adiabatic expansion starts to break down. Estimating the order of magnitude of ξ_n at that order gives

$$\xi_n = \mathcal{O}\left(\left[\frac{\omega}{\Omega}\right]^n\right) = \mathcal{O}\left(\left[\frac{\omega}{\Omega}\right]^{\mathcal{O}(\Omega/\omega)}\right). \qquad (3.27)$$

Since the effective external ω_n^{eff} and internal Λ_n frequencies are comparable and ξ_n is very small, we may just use perturbation theory to estimate β and we get $\beta = \mathcal{O}(\xi_n)$, i.e., the same exponential suppression as in Eq. (3.21). If we would continue the iteration beyond that order, the ξ_n would start to increase again—which the usual situation in an asymptotic expansion, see Fig. 3.1. Carrying on the iteration too far beyond this point, the $\dot{\xi}_n^2$ exceed the Λ_n^2 and thus we have barrier penetration instead of propagation over the barrier (as occurs for all orders below this value of n). In this procedure, it is this barrier penetration which gives the mixing of positive and negative pseudo-norm, and the creation of particles. Were the system to remain as propagation over the barrier for all orders n in this adiabatic expansion, one would have no particle creation.

3.5 Example: Inflation

As an illustrative example, let us consider a minimally coupled massive scalar field in $3+1$ dimensions—which could be the inflaton field (according to our standard

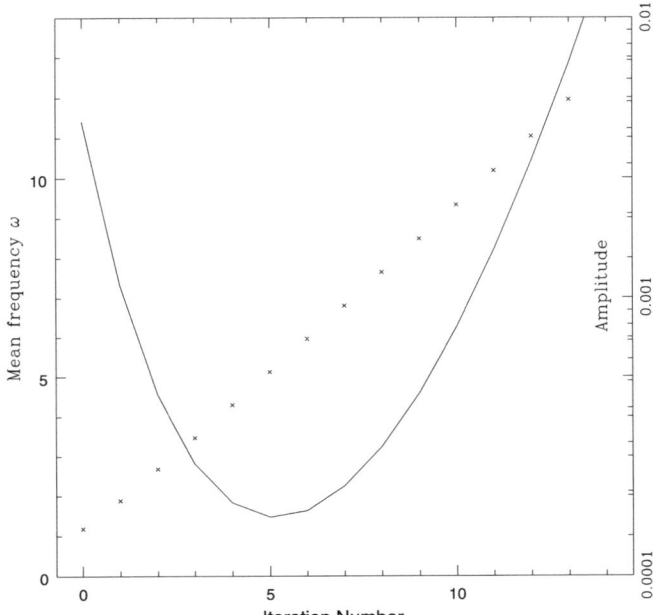

Fig. 3.1 Sketch of the effective external frequencies ω_n^{eff} (*crosses*) and amplitudes ξ_n (*solid line*) depending on the iteration number n obtained numerically for a concrete example. One can observe that ω_n^{eff} grows approximately linearly with n while ξ_n first decreases but later (for $n > 5$) increases again

model of cosmology). Again, we start with the Friedmann-Robertson-Walker metric (3.1) with a scale factor $a(\tau)$ and obtain the equation of motion

$$\left(\frac{1}{a^3(\tau)} \frac{\partial}{\partial \tau} a^3(\tau) \frac{\partial}{\partial \tau} - \frac{1}{a^2(\tau)} \nabla^2 + m^2 \right) \Phi = 0. \tag{3.28}$$

Rescaling the field $\phi(\tau, \mathbf{r}) = \eth(\tau) \Phi(\tau, \mathbf{r})$ with $\eth(\tau) = a^{3/2}(\tau)$ and applying a spatial Fourier transform, we obtain the same form as in Eq. (3.3)

$$\left(\frac{d^2}{d\tau^2} + \frac{\mathbf{k}^2}{a^2(\tau)} + m^2 - \frac{1}{\eth(\tau)} \frac{d^2 \eth(\tau)}{d\tau^2} \right) \phi_k = 0. \tag{3.29}$$

In the standard scenario of inflation, the spacetime can be described by the de Sitter metric $a(\tau) = \exp\{H\tau\}$ to a very good approximation, where H is the Hubble parameter. In this case, the effective potential $\ddot{\eth}/\eth$ just becomes a constant $(3H/2)^2$ and the frequency $\Omega(\tau)$ reads

$$\Omega^2(\tau) = \frac{\mathbf{k}^2}{a^2(\tau)} + m^2 - \frac{9H^2}{4}. \tag{3.30}$$

Inserting $a(\tau) = \exp\{H\tau\}$, we see that modes with different k-values follow the same evolution—just translated in time. (This fact is related to the scale invariance of the created k spectrum.) Initially, this frequency is dominated by the $\mathbf{k}^2$ term and we have $\dot{\Omega}/\Omega = -H$ which means that we are in the WKB regime $\dot{\Omega}/\Omega \ll \Omega$. However, due to the cosmological red-shift, this $\mathbf{k}^2$ term decreases with time until the other terms become relevant. Then the behaviour of the modes depends on the ratio m/H. For $m \gg H$, the modes remain adiabatic (i.e., stay in the WKB regime) and thus particle creation is exponentially suppressed. If m and H are not very different, but still $m > 3H/2$ holds, the modes are adiabatic again for large times—but for intermediated times, the WKB expansion breaks down, leading to a moderate particle creation. For $m < 3H/2$, on the other hand—which is (or was) supposed to be the case during inflation —the frequency $\Omega(\tau)$ goes to zero at some time and becomes imaginary afterwards. This means that we get a barrier penetration (tunnelling) problem where the modes $\phi_k(\tau)$ do not oscillate but evolve exponentially in time $\phi_k(\tau) \propto \exp\{\pm\tau\sqrt{9H^2/4 - m^2}\}$. Here one should remember that the original field does not grow exponentially due to the re-scaling with the additional factor $\mho(\tau) = a^{3/2}(\tau)$. This behaviour persists until the barrier vanishes, i.e., the expansion slows down (at the end of the inflationary period) and thus the effective potential $\ddot{\mho}/\mho$ drops below the mass term. After that, the modes start oscillating again. However, in view of the barrier penetration (tunnelling) over a relatively long time (distance), we get reflection coefficients R which are not small but extremely close to unity $R \approx 1$. This means that the Bogoliubov coefficients α and β are huge—i.e., that we have created a tremendous amount of particles out of the initial vacuum fluctuations. According to our understanding, precisely this effect is responsible for the creation of the seeds for all structures in our Universe. Perhaps the most direct signatures of this effect are still visible today in the anisotropies of the cosmic microwave background radiation.

An alternative picture of the mode evolution in terms of a damped harmonic oscillator can be obtained from the original field in Eq. (3.28)

$$\left(\frac{d^2}{d\tau^2} + 3H\frac{d}{d\tau} + e^{-2H\tau}\mathbf{k}^2 + m^2\right)\Phi_k = 0. \quad (3.31)$$

Initially, the term $e^{-2H\tau}\mathbf{k}^2$ dominates and the modes oscillate. Assuming $m \ll H$ (which is related to the slow-roll condition of inflation), the damping term dominates for late times and we get a strongly over-damped oscillator, whose dynamics is basically frozen (like a pendulum in a very sticky liquid). The transition happens when $H \sim ke^{-H\tau}$, i.e., when the physical wavelength $\lambda = 2\pi e^{H\tau}/k$ exceeds the de Sitter horizon $\propto 1/H$ due to the cosmological expansion $e^{H\tau}$. After that, crest and trough of a wave lose causal contact and cannot exchange energy any more—that's why the oscillations effectively stops.

As a final remark, we stress that this enormous particle creation effect is facilitated by the rapid (here: exponential) expansion and the resulting stretching of wavelengths over many many orders of magnitude (i.e., the extremely large red-shift). Therefore, a final mode with a moderate wavelength originated from waves with extremely short wavelengths initially. Formally, these initial wavelengths could be easily far shorter than the Planck length. However, on these scales one would expect deviations from the theory of quantum fields in classical spacetimes we used to derive these effects. On the other hand, this problem is not only negative—it might open up the possibility to actually see signatures of new (Planckian) physics in high-precision measurements of the cosmic microwave background radiation, for example.

3.6 Laboratory Analogues

Apart from the observation evidence in the anisotropies of the cosmic microwave background radiation mentioned above, one may study the phenomenon of cosmological particle creation experimentally by means of suitable laboratory analogues, see, e.g., [2, 12]. The are two major possibilities to mimic the expansion or contraction of the Universe—a medium at rest with time-dependent properties (such as the propagation speed of the quasi-particles) or an expanding medium, see, e.g., [2, 11]. Let us start with the former option and consider linearised and scalar quasi-particles (e.g., sound waves) with low energies and momenta propagating in a spatially homogeneous and isotropic medium. Under these conditions, their dynamics is governed by the low-energy effective action [2, 11]

$$\mathscr{L}_{\text{eff}} = \frac{1}{2}\left(a^2(t)\dot{\phi}^2 + b^2(t)\phi^2 + c^2(t)[\nabla\phi]^2\right) + \mathscr{O}(\phi^3) + \mathscr{O}(\partial^3). \quad (3.32)$$

Here we assume positive a^2 and non-negative b^2 and c^2 for stability. The factor $a^2(t)$ can be eliminated by suitable re-scaling of the time co-ordinate. Then, after a spatial Fourier transform, we obtain the same form as in Eq. (3.3). The quasi-particle excitations ϕ in such a medium behave in the same way as a scalar field in an expanding or contracting Universe with a possibly time-dependent potential (mass) term $\propto b^2(t)\phi^2$. In order to avoid this additional time-dependence of the potential (mass) term, the factors b and c must obey special conditions. For example, Goldstone modes with $b = 0$ correspond to a massless scalar field in $3 + 1$ dimensions—whereas the case of constant c is analogous to a massive scalar field in $1 + 1$ dimensions.

As one would intuitively expect, the expansion or contraction of the Universe can also be mimicked by an expanding or contracting medium. Due to local Galilee invariance, such a medium can also be effectively spatially homogeneous and isotropic as in Eq. (3.32) when described in terms of co-moving co-ordinates. For a quite detailed list of references, see [2].

There are basically three major experimental challenges for observing the analogue of cosmological particle creation in the laboratory. First, the initial temperature should be low enough such that the particles are produced due to quantum rather than thermal fluctuations. Second, one must be able to generate a time-dependence (e.g., expansion of the medium) during which the effective action in Eq. (3.32) remains valid (in some sense) but which is also sufficiently rapid to create particles. Third, one must be able to detect the created particles and to distinguish them from the radiation stemming from other sources. For trapped ions, for example (see, e.g.,

[10]), the first and third point (i.e., cooling and detection) is experimental state of the art, while a sufficiently rapid but still controlled expansion/contraction of the ion trap presents difficulties. For Bose-Einstein condensates (see, e.g., [2, 11] and references therein), on the other hand, the first and third points are the main obstacles.

Acknowledgements The authors benefited from fruitful discussions, especially with R. Parentani, during the *SIGRAV Graduate School in Contemporary Relativity and Gravitational Physics*, IX Edition "Analogue Gravity", at the Centro di Cultura Scientifica "A. Volta", Villa Olmo, in Como (Italy, 2011). R.S. acknowledges support from DFG and the kind hospitality during a visit at the University of British Columbia where part of this research was carried out. W.G.U. thanks the Natural Sciences and Engineering Research Council of Canada and the Canadian Institute for Advanced Research, for research support, and the University of Duisburg-Essen for their hospitality while part of this research was carried out.

References

1. Baird, L.C.: New integral formulation of the Schrödinger equation. J. Math. Phys. **11**, 2235 (1970)
2. Barceló, C., Liberati, S., Visser, M.: Analogue gravity. Living Rev. Relativ. **14**, 3 (2011)
3. Birrell, N.D., Davies, P.C.W.: Quantum Fields in Curved Space. Cambridge University Press, Cambridge (1982)
4. Davis, J.P., Pechukas, P.: Nonadiabatic transitions induced by a time-dependent Hamiltonian in the semiclassical/adiabatic limit: the two-state case. J. Chem. Phys. **64**, 3129 (1976)
5. Dumlu, C.K., Dunne, G.V.: Stokes phenomenon and Schwinger vacuum pair production in time-dependent laser pulses. Phys. Rev. Lett. **104**, 250402 (2010)
6. Fulling, S.A.: Aspects of Quantum Field Theory in Curved Space-Time. Cambridge University Press, Cambridge (1989)
7. Massar, S., Parentani, R.: Particle creation and non-adiabatic transitions in quantum cosmology. Nucl. Phys. B **513**, 375 (1998)
8. Parker, L.: Particle creation in expanding universes. Phys. Rev. Lett. **21**, 562 (1968)
9. Schrödinger, E.: The proper vibrations of the expanding universe. Physica **6**, 899 (1939)
10. Schützhold, R., Uhlmann, M., Petersen, L., Schmitz, H., Friedenauer, A., Schätz, T.: Analogue of cosmological particle creation in an ion trap. Phys. Rev. Lett. **99**, 201301 (2007)
11. Schützhold, R.: Emergent horizons in the laboratory. Class. Quantum Gravity **25**, 114011 (2008)
12. Unruh, W.G.: Experimental black hole evaporation? Phys. Rev. Lett. **46**, 1351 (1981)
13. Wald, R.M.: Quantum Field Theory in Curved Spacetime and Black Hole Thermodynamics. University of Chicago Press, Chicago (1994)
14. Wasow, W.: Adiabatic invariance of a simple oscillator. SIAM J. Math. Anal. **4**, 78 (1973)

Chapter 4
Irrotational, Two-Dimensional Surface Waves in Fluids

William G. Unruh

Abstract The equations for waves on the surface of an irrotational incompressible fluid are derived in the coordinates of the velocity potential/stream function. The low frequency shallow water approximation for these waves is derived for a varying bottom topography. Most importantly, the conserved norm for the surface waves is derived, important for quantisation of these waves and their use in analogue models for black holes.

4.1 Introduction

One of the most fascinating predictions of Einstein's theory of general relativity is the potential existence of black holes—i.e. spacetime regions from which nothing is able to escape. Perhaps no less interesting are their antonyms: white holes which nothing can penetrate. Both are described by solutions of the Einstein equations and are related to each other via time-inversion, see e.g. [1, 2].

It is equally fascinating that some of the predictions for fields in a black hole spacetime can be modelled by waves in a variety of other situations, with the interior of the black hole or white hole horizons that can be mimicked by fluid flow which exceeds the velocity of the waves in some regions. One of these is the use of surface waves on a incompressible fluid [3]. One can alter the flow properties of the fluid by placing obstacles into the bottom of a flume (a long tank along which the water flows) to speed up and slow down the fluid over these obstacles.

One of the difficulties in the theoretical treatment of such systems is the complicated boundary conditions on the bottom of the tank (where the fluid velocities must be tangential to the bottom) and the top (where the pressure of the fluid must be zero or at a constant atmospheric pressure). In fact, as we will see, the equations for the fluid itself are remarkably simple. The interesting physics arises entirely from those boundary conditions.

W.G. Unruh (✉)
Department of Physics and Astronomy, University of British Columbia, Vancouver, BC V6T 1Z1, Canada
e-mail: unruh@physics.ubc.ca

We will be interested in irrotational, incompressible flow. While both are certainly approximations for water flow (the former assumes no turbulence, and no viscosity which would create vorticity at the shear layer along the bottom, while the latter assumes that the velocity of sound in the fluid is far higher than any other velocities in the problem). While this problem has been investigated before [4, 5], this is in general in the three dimensional context (which is more difficult) and using approximations and expansions for the shape of the bottom.

I will assume that the fluid flow is a two dimensional flow—i.e. is uniform across the tank and that the tank maintains a constant width throughout. This is much simpler case than three dimensional flow, which allows the coordinate transformations I use.

The usual spatial coordinates are x, y with x being the horizontal direction in which the fluid flows, and y is the vertical direction parallel to the gravitational acceleration, g, directed in the negative y direction.

The Euler-Lagrange equations are

$$\partial_t \mathbf{v} + \mathbf{v} \cdot \nabla \mathbf{v} = -g \mathbf{e}_y - \nabla \frac{p}{\rho}, \tag{4.1}$$

$$\nabla \cdot \mathbf{v} = 0, \tag{4.2}$$

where the second equation is the incompressibility condition. In the usual way, if we assume that the flow is irrotational, then

$$\mathbf{v} = \nabla \tilde{\phi}, \tag{4.3}$$

and the above equation can be written as

$$\nabla \left(\partial_t \tilde{\phi} + \frac{1}{2} v^2 + gy + \frac{\tilde{p}}{\rho} \right) = 0, \tag{4.4}$$

$$\nabla^2 \tilde{\phi} = 0, \tag{4.5}$$

where $\tilde{p}$ is the pressure. Let me define the specific pressure, $p = \frac{\tilde{p}}{\rho}$.

In the following I will consider only flows in the $x - y$ directions. Everything is assumed to be independent of z. Consider the vector $\mathbf{w} = \mathbf{e}_z \times \mathbf{v}$. This vector also obeys

$$\nabla \cdot \mathbf{w} = -\mathbf{e}_z \cdot \nabla \times \mathbf{v} = 0, \tag{4.6}$$

$$\nabla \times \mathbf{w} = \mathbf{e}_z \nabla \cdot \mathbf{v} - (\mathbf{e}_z \cdot \nabla) \mathbf{v} = 0, \tag{4.7}$$

since nothing depends on z.

Thus we can define

$$\mathbf{w} = \nabla \tilde{\psi}, \tag{4.8}$$

where $\tilde{\psi}$ also obeys $\nabla^2 \tilde{\psi} = 0$ and where

$$\nabla \tilde{\psi} \cdot \nabla \tilde{\phi} = 0, \tag{4.9}$$

$$\nabla \tilde{\phi} \cdot \nabla \tilde{\phi} = \mathbf{v} \cdot \mathbf{v} = v^2, \tag{4.10}$$

$$\nabla \tilde{\psi} \cdot \nabla \tilde{\psi} = \mathbf{w} \cdot \mathbf{w} = v^2. \tag{4.11}$$

4 Irrotational, Two-Dimensional Surface Waves in Fluids

Let me now define a new coordinate system. I could use $\tilde\phi$ and $\tilde\psi$, but I will be interested in fluid flows where the velocity approaches a constant value $v_x = v_0$, $v_y = 0$ at large distances. I will thus instead use the functions ψ, ϕ defined by

$$\phi = \frac{\tilde\phi}{v_0}, \tag{4.12}$$

$$\psi = \frac{\tilde\psi}{v_0}, \tag{4.13}$$

as the new coordinates. This choice will also allow me to take the limit as the velocity v_0 goes to zero, where the potentials $\tilde\phi$, $\tilde\psi$ are undefined. Thus at large distances, $\phi = x$ and $\psi = y$. The spatial metric in the xy coordinates is

$$ds^2 = dx^2 + dy^2 = g_{ij} dz^i dz^j \tag{4.14}$$

(where the Einstein summation convention has been used where a repeated index implies summation over that index, and where $z^1 = x$, $z^2 = y$). Do not confuse z^i with the horizontal direction z which nothing depends on. The Laplacian for a general metric function of $g_{ij}(z^k)$ is

$$\nabla^2 = \frac{1}{\sqrt{|g|}} \partial_i \sqrt{|g|} g^{ij} \partial_j, \tag{4.15}$$

where g^{ij} are the components of the matrix which is the inverse to the matrix of coefficients g_{ij} and where g is the determinant of the matrix with coefficients g_{ij}. For a reference regarding metrics and the coordinate independent equations see almost any book on General Relativity [6].

In two dimensions, if $g_{ij} = f \tilde g_{ij}$ where f is some function of the coordinates z^i, then since $g^{ij} = \frac{1}{f} \tilde g^{ij}$ and $g = \det(g_{ij}) = f^2 \det(\tilde g_{ij}) = f^2 \tilde g$, we have $\nabla^2 = \frac{1}{f} \tilde\nabla^2$. Metrics such as g_{ij} and $\tilde g_{ij}$ are said to be conformally related.

Recalling that the change in the metric components from one coordinate system z^i to a new system $\hat z^j$ are given by

$$\hat g^{kl} = \frac{\partial \hat z^K}{\partial z^i} \frac{\partial \hat z^l}{\partial z^j} g^{ij}, \tag{4.16}$$

where the Einstein summation convention has been used, the upper components of the usual flat space metric in this new $\hat z^1 = \phi, \hat z^2 = \psi$ coordinate system are

$$\hat g^{\phi\phi} = \nabla\phi \cdot \nabla\phi = \frac{v^2}{v_0^2}, \tag{4.17}$$

$$\hat g^{\psi\psi} = \nabla\psi \cdot \nabla\psi = \frac{v^2}{v_0^2}, \tag{4.18}$$

$$\hat g^{\phi\psi} = \nabla\phi \cdot \nabla\psi = 0, \tag{4.19}$$

i.e., the new metric (the inverse of this upper form metric) in these new coordinates is a conformally flat metric

$$\hat{g}_{ij} = \frac{v_0^2}{v^2} \begin{pmatrix} 1 & 0 \\ 0 & 1 \end{pmatrix}. \tag{4.20}$$

Since in the xy coordinates the metric is flat, this metric is also flat in ψ, ϕ coordinates, (the curvature is not changed by a coordinate transformation) and the scalar curvature in this new coordinate system is zero. Using the equation for the scalar curvature of a metric (and in two dimensions, the scalar curvature is the only independent component of the curvature) one gets

$$(\partial_\phi^2 + \partial_\psi^2) \ln\left(\frac{v^2}{v_0^2}\right) = 0 \tag{4.21}$$

(This is valid as long as $\frac{v^2}{v_0^2}$ is not equal to zero anywhere.)

I define

$$\tilde{\nabla}^2 = \partial_\phi^2 + \partial_\psi^2. \tag{4.22}$$

Since the metric in ψ, ϕ coordinates is conformally flat, the Laplacian

$$\frac{1}{\sqrt{(\hat{g})}} \partial_i \sqrt{\hat{g}} \hat{g}^{ij} \partial_j \Phi \tag{4.23}$$

is just

$$\frac{v^2}{v_0^2} \tilde{\nabla}^2 \Phi \tag{4.24}$$

for any scalar function Φ.

Since in x, y coordinates, the Laplacian of both the scalar functions x and y are zero, they must also be zero in ϕ, ψ coordinates (since the Laplacian is an invariant scalar operator), and, as functions of ϕ, ψ, we have

$$\tilde{\nabla}^2 x(\phi, \psi, t) = \tilde{\nabla}^2 y(\phi, \psi, t) = 0, \tag{4.25}$$

as the equations of motion obeyed by x and y in these new coordinates.

ψ is the stream function, and the vector $\mathbf{v}$ is tangent to the surfaces of constant ψ: $\mathbf{v} \cdot \nabla \psi = \mathbf{v} \cdot \mathbf{w} = 0$. The bottom of the flow must be tangent to the flow vector (no flow can penetrate the bottom), and thus must be a surface of constant ψ, which I will take to be $\psi = 0$. Similarly, if the flow is stationary, the top of the water, no matter how convoluted, must also lie along a streamline, since a particle of the fluid which is at the top, must flow along the top (the velocity of the particles must be parallel to the top surface). This means that the top of a stationary flow (but not a time dependent flow) also is at a constant value of ψ which I will label ψ_T.

We also have

$$\partial_x \phi = \partial_y \psi = \frac{v_x}{v_0}, \tag{4.26}$$

$$\partial_y \phi = -\partial_x \psi = \frac{v_y}{v_0}, \tag{4.27}$$

4 Irrotational, Two-Dimensional Surface Waves in Fluids

$$\partial_\phi x = \partial_\psi y = \frac{v_x v_0}{v^2}, \tag{4.28}$$

$$-\partial_\psi x = \partial_\phi y = \frac{v_y v_0}{v^2} \tag{4.29}$$

and thus

$$\frac{v^2}{v_0^2} = \frac{1}{(\partial_\phi y)^2 + (\partial_\psi y)^2} \tag{4.30}$$

$$= \frac{1}{(\partial_\phi x)^2 + (\partial_\psi x)^2}. \tag{4.31}$$

Solving for x and y as a function of ψ, ϕ, which is just solving the Laplacian in terms of ψ, ϕ, gives us the velocity at all points.

The boundary condition along the bottom for these functions must be that the velocity along the bottom be parallel to the bottom. If the bottom has the functional form $y = F(x)$ then $y(\phi, 0) = F(x(\phi, 0))$. On the top of the flow, we have the boundary condition that $p = 0$. The Bernoulli equation for a stationary flow is

$$\frac{1}{2}v^2 + gy + p = \text{const}, \tag{4.32}$$

which, if the flow has constant velocity u over a constant depth bottom of height h far away from the obstacle, gives the equation for the top of the flow

$$\frac{1}{2}v(\phi, \psi_T)^2 + gy(\phi, \psi_T) = \frac{1}{2}v_0^2 + gh. \tag{4.33}$$

Writing this in terms of ϕ, ψ we have the upper boundary condition of

$$\frac{v_0^2}{2((\partial_\phi y(\phi, \psi_T))^2 + (\partial_\phi x(\phi, \psi_T))^2)} + gy(\phi, \psi_T) = \frac{1}{2}v_0^2 + gh. \tag{4.34}$$

This is a complicated, non-linear, boundary condition. Thus while the equations of motion of x, y are simple (Laplacian equals zero), the physics is all contained in the boundary conditions.

If we are given $y(x)$ as the equation for the bottom, the solution of the above non-linear boundary value problem is difficult. However if, instead of specifying the lower boundary, one specifies the shape of the upper boundary $y(\phi, \psi_T)$, one can use Bernoulli's equation in these new coordinates to determine the ψ derivative of y. Since

$$\partial_\psi y = \frac{v_y}{v^2}, \tag{4.35}$$

$$\partial_\phi y = \frac{v_x}{v^2}, \tag{4.36}$$

we have

$$v^2 = \frac{1}{(\partial_\psi y)^2 + (\partial_\phi y)^2}, \tag{4.37}$$

and Bernoulli's equation is $v^2 + gy = const$ along the top surface of the fluid where $p = 0$. Solving for $\partial_\psi y$ we get

$$\partial_\psi y(\phi, \psi_T) = -\sqrt{-(\partial_\phi y(\phi, \psi_T))^2 + \frac{1}{v_0^2 + g(y(\infty, \psi_T) - y(\phi, \psi_T))}}. \quad (4.38)$$

Any function $H(\psi\phi)$ which is a solution of $\partial_\psi^2 H + \partial_\phi^2 H = 0$ can be expanded in exponentials $e^{ik\phi}$. We see immediately that the dependence of these modes of ψ must be in terms of $e^{\pm k\psi}$ or equivalently in terms of $\cosh(k\psi)$ and $\sinh(k\psi)$ for the ψ dependence. Thus, since y obeys that equation, we have

$$y(\phi, \psi) = \int e^{ik\phi}\left(\alpha_k \cosh(k(\psi - \psi_T)) + \beta_k \sinh(k(\psi - \psi_T))\right) dk, \quad (4.39)$$

with

$$\alpha_k = \frac{1}{2\pi} \int y(\phi, \psi_T) e^{-ik\phi} d\phi, \quad (4.40)$$

$$\beta_k = \frac{1}{2\pi} \int \frac{1}{k} \partial_\psi y(\phi, \psi_T) e^{-ik\phi} d\phi. \quad (4.41)$$

Then at the lower boundary,

$$y(\phi, 0) = \int \left[\hat{y}(k)\cosh(k\psi_T) - \hat{\partial}y(k)\frac{\sinh(k\psi_T)}{k}\right] e^{ik\phi} dk, \quad (4.42)$$

$$x(\phi, 0) = \int \left[\hat{\partial}y(k) k\cosh(k\psi_T) + \hat{y}(k)\sinh(k\psi_T)\right] e^{ik\phi} dk, \quad (4.43)$$

where

$$\hat{y}(k) = \frac{1}{2\pi} \int y(\phi, \psi_T) e^{-ik\phi} d\phi = \alpha_k, \quad (4.44)$$

$$\hat{\partial}y(k) = \frac{1}{2\pi} \int \partial_\psi y(\phi, \psi_T) e^{-ik\phi} d\phi = k\beta_k. \quad (4.45)$$

This gives the bottom as a parametric set of functions of ϕ.

In Fig. 4.1 we have an example of sub to supercritical flow over an obstacle, calculated as above. Note that the obstacle is a reasonable function $y(x)$.

4.1.1 $v_0 = 0$ Limit

The boundary condition equations are easily solved in the limit as $v_0 \to 0$. The upper boundary condition becomes simply $y = h$ and $\partial_\phi y = 0$. This can be solved (in terms of the unknown lower boundary solutions $y(\phi, 0)$, $x(\phi, 0)$ by

$$y(\phi, \psi) = \int \alpha_k e^{ik\phi} \frac{\sinh(k(\psi_T - \psi))}{\sinh(k\psi_T)} dk, \quad (4.46)$$

$$x(\phi, \psi) = i \int \alpha_k e^{ik\phi} \frac{\cosh(k(\psi_T - \psi))}{\sinh(k\psi_T)} dk, \quad (4.47)$$

4 Irrotational, Two-Dimensional Surface Waves in Fluids

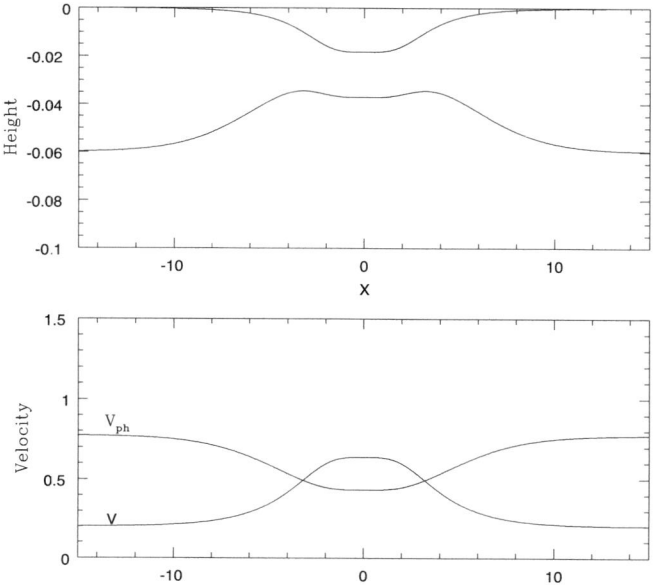

Fig. 4.1 The *upper graph* gives the top and bottom ($y(\psi_T)$ and $y(0)$) of a symmetric flume flow with $v_0 = 0.3$ m/s. The top of the flow was specified with $y(\phi, \psi_T) = 0.015(e^{(\psi-.5)^2/2} + e^{(\psi+0.5)^2/2})$. Note that the bottom of the flume is a reasonable function of x. The *lower graph* gives the velocity of the fluid flow, ($v(\phi)$) as a function of x and the phase velocity of long wavelength waves $\sqrt{g(y(\phi, \psi_T) - y(\phi, 0))}$ as a function of x. The ratio of these two velocities is the Froude number, which is greater than unity over the obstacle

where

$$\alpha_k = \frac{1}{2\pi} \int y(\phi, 0) e^{-ik\phi} d\phi. \qquad (4.48)$$

Of course, we are not given $y(\phi, 0)$ but rather $y(\phi, 0) = F(x(\phi, 0))$. However, one can get rapid convergence by iteration

$$x_0(\phi, 0) = \phi, \qquad (4.49)$$

$$y_{i+1}(\phi, 0) = F\big(x_i(\phi, 0)\big), \qquad (4.50)$$

which gives, via the above equations, the solution $y_{i+1}(\phi, \psi)$ and thus

$$x_{i+1}(\phi, 0) = \int \partial_\psi y_{i+1}(\phi, 0) d\phi. \qquad (4.51)$$

For small v_0, one can get a first order correction for the surface value of $y(\phi, \psi_T)$ by taking

$$y(\phi, \psi_T) = h - v_0^2 \frac{1}{(\partial_\psi y_{v_0=0}(\phi, \psi))^2|_{\psi=\psi_T}}. \qquad (4.52)$$

I.e., for slow flow over a bottom boundary, the stationary solution for that flow is easy to find.

4.1.2 Formal General Solution

The general solution to the equation $\tilde{\nabla}^2 F = 0$ can be written as

$$F = f(\phi + i\psi) + g(\phi - i\psi). \tag{4.53}$$

If F is real, then $g(\phi - i\psi) = (f(\phi + i\psi))^*$. We then have

$$x(\phi, \psi) = \hat{x}(\phi + i\psi) + \hat{x}^*(\phi + i\psi), \tag{4.54}$$

$$y(\phi, \psi) = i(\hat{x}(\phi + i\psi) - \hat{x}^*(\phi + i\psi)). \tag{4.55}$$

Given the boundary conditions along the bottom, we have

$$\hat{x}(\phi) = \frac{1}{2}(x_0(\phi, 0) - iy_0(\phi, 0)). \tag{4.56}$$

This of course still leaves the highly non-linear boundary conditions at the top to solve, to find x and y everywhere.

4.2 Fluctuations

Let us assume that we have a background solution to the stationary equation, $x_0(\phi, \psi)$, $y_0(\phi, \psi)$, or equivalently, $\phi_0(x, y)$, $\psi_0(x, y)$. We want to find the equations for small perturbations around this background flow. Let us also consider a solution to the full time dependent equations, $\phi(x, y, t)$, $\psi(x, y, t)$ together with their inverses, $x(\phi, \psi, t)$, $y(\phi, \psi, t)$, such that $y(\phi(x, y, t), \psi(x, y, t), t) = y$ and $x(\phi(x, y, t), \psi(x, y, t), t) = x$. Define the small deviations from the background by

$$\delta\phi = \phi(x, y, t) - \phi_0(x, y), \tag{4.57}$$

$$\delta\psi = \psi(x, y, t) - \psi_0(x, y), \tag{4.58}$$

$$\delta x = x(\psi, \phi, t) - x_0(\phi, \psi), \tag{4.59}$$

$$\delta y = y(\phi, \psi, t) - y_0(\phi, \psi). \tag{4.60}$$

Then, we have

$$y = y(\phi_0(x, y) + \delta\phi(x, y, t), \psi_0(x, y) + \delta\psi(x, y, t), t), \tag{4.61}$$

$$= y_0(\phi_0(x, y) + \delta\phi(x, y, t), \psi_0(x, y) + \delta\psi(x, y, t), t)$$
$$+ \delta y(\phi_0(x, y) + \delta\phi(x, y, t), \psi_0(x, y) + \delta\psi(x, y, t), t). \tag{4.62}$$

Keeping terms only to first order in "δ", we have

4 Irrotational, Two-Dimensional Surface Waves in Fluids

$$y = y_0\big(\phi_0(x,y), \psi_0(x,y)\big) + \partial_\phi y_0\big(\phi_0(x,y), \psi_0(x,y)\big)\delta\phi$$
$$+ \partial_\psi y_0\big(\phi_0(x,y), \psi_0(x,y)\big)\delta\psi + \delta y\big(\phi_0(x,y), \psi_0(x,y)\big), \quad (4.63)$$

or

$$\delta y\big(\phi_0(x,y), \psi_0(x,y)\big) = -\frac{v_0 v_y}{v^2}\delta\phi(x,y) - \frac{v_0 v_x}{v^2}\delta\psi(x,y) \quad (4.64)$$

(where all velocity components are those in the background flow).
Similarly

$$\delta x = -\frac{v_0 v_x}{v^2}\delta\phi(x,y) + \frac{v_0 v_y}{v^2}\delta\psi(x,y) \quad (4.65)$$

and

$$\delta\phi\big(x_0(\phi,\psi), y_0(\phi,\psi)\big) = \frac{1}{v_0}\big(v_x\delta x(\phi,\psi) + v_y\delta y(\phi,\psi)\big), \quad (4.66)$$

$$\delta\psi\big(x_0(\phi,\psi), y_0(\phi,\psi)\big) = \frac{1}{v_0}\big(-v_y\delta x(\phi,\psi) + v_x\delta y(\phi,\psi)\big). \quad (4.67)$$

The Bernoulli equation is

$$v_0\partial_t\phi\big(x(\phi,\psi,t), y(\phi,\psi,t), t\big) + \frac{v_0^2}{2}\frac{1}{(\partial_\phi x(\phi,\psi,t))^2 + (\partial_\phi y(\phi,\psi,t))^2}$$
$$+ gy(\phi,\psi,t) + p = const, \quad (4.68)$$

where the first ∂_t is defined as the derivative keeping x, y fixed, not ϕ, ψ fixed. Here p is the specific pressure.

Writing this equation perturbatively, we have

$$-v_x\partial_t\delta x - v_y\partial_t y - \frac{v_0^2}{((\partial_\phi x)^2 + (\partial_\phi y)^2)^2}\left(\frac{v_0 v_x}{v^2}\partial_\phi \delta x + \frac{v_0 v_y}{v^2}\partial_\phi y\right)$$
$$+ g\delta y + \delta p = 0, \quad (4.69)$$

where all of the velocities are the values of the background velocities at the location ϕ, ψ. I.e., $v_x(\phi,\psi) = v_{0x}(x_0(\phi,\psi), y_0(\phi,\psi))$.

We can now rewrite this equation in terms of $\delta\phi = \delta\phi(x_0(\phi,\psi), y_0(\phi,\psi))$ to get

$$v_0\tilde{\partial}_t\delta\phi + v^2\left(v_x\partial_\phi\left(\frac{v_x}{v^2}\delta\phi - \frac{v_y}{v^2}\delta\psi\right) + v_y\partial_\phi\left(\frac{v_y}{v^2}\delta\phi + \frac{v_x}{v^2}\delta\psi\right)\right)$$
$$- g\left(\frac{v_0 v_x}{v^2}\delta\psi + \frac{v_0 v_y}{v^2}\delta\phi\right) + \delta p = 0. \quad (4.70)$$

Recalling that $\partial_\phi \frac{v_x}{v^2} = \partial_\phi \partial_\psi y_0 = \partial_\psi \frac{v_0 v_y}{v^2}$ and $\partial_\phi \frac{v_0 v_y}{v^2} = -\partial_\psi \frac{v_0 v_x}{v^2}$, we finally get

$$v_0\tilde{\partial}_t\delta\phi + v^2\partial_\phi\delta\phi + \partial_\phi\left(\frac{1}{2}v^2 + gy_0\right)\delta\phi - \partial_\psi\left(gy_0 + \frac{1}{2}v^2\right)\delta\psi + \delta p = 0. \quad (4.71)$$

The boundary conditions at the bottom are that δx and δy must be parallel to the bottom, or $v_x\delta y - v_y\delta x = 0$ which is just

$$\delta\psi(\phi, 0) = 0. \tag{4.72}$$

At the top, the pressure at the surface must be 0. However the surface is no longer simply $\psi = \psi_T$ because of the time dependence of the equations. Let us assume that the surface is defined by

$$\psi = \Psi(\phi, t) + \psi_T. \tag{4.73}$$

Since a particle of the fluid which starts on the surface, remains on the surface, we can define the fluid coordinates η, ζ. Then the velocity of the fluid is

$$v^\phi = \frac{d}{dt}\phi(\zeta, \eta, t), \tag{4.74}$$

$$v^\psi = \frac{d}{dt}\psi(\zeta, \eta, t). \tag{4.75}$$

Along the surface, we therefore have

$$v^\psi = \partial_t \Psi + v^\phi \partial_\phi \Psi. \tag{4.76}$$

But,

$$v^\phi = \frac{d}{dt}\phi\big(x(\eta, \zeta, t), y(\eta, \zeta, t), t\big) \tag{4.77}$$

$$= v_x \partial_x \phi + v_y \partial_y \phi + \partial_t \phi \tag{4.78}$$

$$= \frac{v^2}{v_0} + \partial_t \phi(x, y, t), \tag{4.79}$$

$$v^\psi = \partial_t \psi(x, y, t). \tag{4.80}$$

Thus, assuming that Ψ is also small (the same order as the other "δ" terms), we have

$$v_0 \partial_t \Psi + \frac{v^2}{v_0}\partial_\phi \Psi = v_0 \partial_t \delta\psi. \tag{4.81}$$

On the surface, we have the Bernoulli equation, which to first order is

$$\frac{1}{2}v^2(\phi, \psi_T + \Psi) + gy_0(\phi, \psi_T + \Psi) - \frac{1}{2}v^2(\phi, \psi_T + \Psi)$$
$$+ gy_0(\phi, \psi_t + \Psi) + \tilde{\partial}_t \delta\phi + v^2 \partial_\phi \delta\phi$$
$$- \partial_\phi\left(\frac{1}{2}v^2 + gy\right)\delta\phi - \partial_\psi\left(\frac{1}{2}v^2 + gy\right)\delta\psi + p - p_0 = 0. \tag{4.82}$$

But along the surface $\psi = \psi_T$, the background $\frac{1}{2}v^2 + gy$ is constant, so the ϕ derivative is 0. We have

$$(\tilde{\partial}_t + v^2 \partial_\phi)\delta\phi + \partial_\psi\left(\frac{1}{2}v^2 + gy\right)(\Psi - \delta\psi) = 0. \tag{4.83}$$

Dividing by $G = \partial_\psi(\frac{1}{2}v^2 + gy)$ and taking the derivative $\tilde{\partial}_t + \frac{v^2}{v_0}\partial_\phi$ we get

$$\left(\tilde{\partial}_t + \frac{v^2}{v_0}\partial_\phi\right)\left[\frac{1}{G}\left(\tilde{\partial}_t + \frac{v^2}{v_0}\partial_\phi\right)\right]\delta\phi - \frac{v^2}{v_0}\partial_\phi \delta\psi = 0, \tag{4.84}$$

› 4 Irrotational, Two-Dimensional Surface Waves in Fluids

as the equation of motion for the surface wave. $\delta\phi$ and $\delta\psi$ are related by the boundary condition $\delta\phi = 0$ along the bottom.

Since both $\delta\phi$ and $\delta\psi$ obey $\nabla^2 \delta\psi = \nabla^2 \delta\phi = 0$, we have

$$\tilde{\nabla}^2 \delta\psi = \tilde{\nabla}^2 \delta\phi = 0. \tag{4.85}$$

Furthermore, since

$$\partial_x \delta\phi = \partial_y \delta\psi, \tag{4.86}$$
$$\partial_y \delta\phi = -\partial_x \delta\psi, \tag{4.87}$$

so

$$\partial_\phi \delta\phi = \partial_\phi x_0 \partial_x \delta\phi + \partial_\phi y_0 \partial_y \delta\phi\phi \tag{4.88}$$
$$= \partial_\psi y_0 \partial_y \delta\psi - \partial_\psi x_0 (-\partial_x \delta\psi) = \partial_\psi \delta\psi, \tag{4.89}$$
$$\partial_\psi \delta\phi = -\partial_\phi \delta\psi. \tag{4.90}$$

For irrotational time-independent flow, the acceleration of a parcel of fluid is $\mathbf{v} \cdot \nabla \mathbf{v} = \nabla(\frac{1}{2}v^2)$ and the orthogonal component of this, the centripetal acceleration is

$$\frac{1}{|\nabla\psi|^2} \nabla\psi \cdot \nabla\left(\frac{1}{2}v^2\right) = \frac{1}{v}\partial_\psi\left(\frac{1}{2}v^2\right). \tag{4.91}$$

Also $g\partial_\psi y = g\frac{v_x v_0}{v^2} \approx gv_0/v$ so Gv/v_0 is the effective gravitational field orthogonal to the flow lines (including the centripetal acceleration).

However it is important to note that it is the effective force of gravity only at the surface of the fluid, not at the obstacle to the flow along the bottom, that is important for the equations of motion.

4.3 Shallow Water Waves

Since ϕ, ψ are real functions, the solutions can be written as

$$\delta\phi(\phi,\psi) = Z(\phi + i\psi) + \left(Z(\phi + i\psi)\right)^*, \tag{4.92}$$
$$\delta\psi(\phi,\psi) = i\left(Z(\phi + i\psi) - \left(Z(\phi + i\psi)\right)^*\right), \tag{4.93}$$

for some function Z. These functions clearly satisfy the Laplacian equation for, and furthermore also satisfy the differential relations on the derivatives of x, y with respect to ϕ, ψ. This gives

$$0 = \delta\psi(\phi,0) = i\left(Z(\phi) - Z^*(\phi)\right), \tag{4.94}$$

i.e., Z is a real function of a real arguments, which gives

$$\delta\phi(\phi,\psi) = \left(Z(\phi + i\psi) + Z(\phi - i\psi)\right) \approx 2Z(\phi) + Z''(\phi)\psi^2, \tag{4.95}$$
$$\delta\psi = 2\psi Z'(\phi), \tag{4.96}$$

or, to first order in ψ_T

$$\delta\psi = \psi_T \partial_\phi \delta\phi. \tag{4.97}$$

The equation for the waves then becomes

$$(\tilde{\partial}_t + v^2 \partial_\phi)\frac{1}{G}(\tilde{\partial}_t + v^2 \partial_\phi)\delta\phi - v^2 \psi_T \partial_\phi^2 \delta\phi = 0. \tag{4.98}$$

We note that this is not a Hermitian operator acting on $\delta\phi$. Recall that a Hermitian operator is one such that

$$\int \delta\hat{\phi}\mathcal{H}\phi d\phi dt = \int (\mathcal{H}\delta\hat{\phi})\delta\phi d\phi dt, \tag{4.99}$$

if we assume that all of the boundary terms in the integration by parts are zero. We can rewrite the equation for $\delta\phi$ by dividing by v^2 as

$$(\tilde{\partial}_t + \partial_\phi v^2)\frac{1}{v^2 G}(\tilde{\partial}_t + v^2 \partial_\phi)\delta\phi - \psi_T \partial_\phi^2 \delta\phi = 0. \tag{4.100}$$

This is a symmetric equation, derivable from an action,

$$\int \left[\frac{1}{v^2 G}(\tilde{\partial}_t + v^2 \partial_\phi)\delta\phi^*(\tilde{\partial}_t + v^2 \partial_\phi)\delta\phi - \Psi_T \partial_\phi \delta\phi^* \partial_\phi \delta\phi \right] d\phi dt. \tag{4.101}$$

This action has the global symmetry $\delta\phi \to e^{i\mu}\delta\phi$ and thus has the usual Noether current associated with this symmetry. In particular it has the conserved norm

$$\langle \delta\phi, \delta\phi' \rangle = \frac{i}{2} \int \left\{ \delta\phi^* \frac{1}{Gv}(\partial_t + v^2 \partial_\phi)\frac{\delta\phi'}{v} - \delta\phi' \frac{1}{Gv}(\partial_t + v^2 \partial_\phi)\frac{\delta\phi^*}{v} \right\} d\phi. \tag{4.102}$$

4.4 Deep Water Waves

For deep water waves, we can assume that either $Z(\phi + i\psi_T) \gg Z(\psi - i\phi_T)$ or $Z(\phi + i\psi_T) \ll Z(\psi - i\phi_T)$. (I.e., we assume that as analytic functions, Z goes to zero either in the upper or lower half plane.)

Let us also assume it is the first case, and let us define $\hat{Z}(\phi) = Z(\phi + i\psi_T)$, and that $\tilde{\partial}_t \delta\phi = i\omega\delta\phi$. We then have

$$(i\omega + v^2 \partial_\phi)\frac{1}{G}(i\omega + v_\phi^2)\hat{Z} - (-i)v^2 \partial_\phi \hat{Z} = 0. \tag{4.103}$$

If we assume that $K = i(\partial_\phi \ln(\hat{Z}))$ is large and negative, such that $\hat{Z}$ varies faster than v^2 or G, we have approximately

$$\frac{(\omega + v^2(\phi)K)^2}{G} + Kv^2 = 0, \tag{4.104}$$

or

$$\omega = -v^2 K \pm \sqrt{v^2 G K}. \tag{4.105}$$

4.5 General Linearized Waves

The equation in general is

$$\left(\tilde{\partial}_t + v^2 \partial_\phi\right)\frac{1}{G}\left(\tilde{\partial}_t + v^2 \partial_\phi\right)\delta\phi - v^2 \partial_\phi \delta\psi = 0. \tag{4.106}$$

Fourier transforming with respect to ϕ and ψ, and using the fact that $\delta\psi = 0$ at $\psi = 0$, the functions $\delta\phi$, $\delta\psi$ then can be written as

$$\delta\phi(\phi, \psi, t) = \int A(k, t) e^{ik\phi} \cosh(k\psi) dk, \tag{4.107}$$

$$\delta\psi(\phi, \psi, t) = i \int A(k, t) e^{ik\phi} \sinh(k\psi) dk, \tag{4.108}$$

since again they obey the Laplacian equal to zero in these variables.

Defining $B(k, t) = A(k, t) \cosh(k\psi)$ this can be written as

$$\delta\phi(\phi, \psi_T, t) = \int B(k, t) e^{ik\phi} dk, \tag{4.109}$$

$$\delta\psi(\phi, \psi_T, t) = i \int B(k, t) e^{ik\phi_T} \tanh(k\psi) dk = i \tanh(-i\psi_T \partial_\phi) \int B(k, t) e^{ik\phi} dk$$
$$= i \tanh(-i\psi_T \partial_\phi) \delta\phi. \tag{4.110}$$

Thus the equation of the surface waves can be written as

$$0 = \left(\tilde{\partial}_t + \partial_\phi v^2\right)\frac{1}{v^2 G}\left(\tilde{\partial}_t + v^2 \partial_\phi\right)\delta\phi - i\partial_k \tanh(-i\psi_T \partial_\phi)\delta\phi. \tag{4.111}$$

I.e., we get the usual tanh dispersion relation for the transition from shallow to deep water waves.

This equation is symmetric and real, and thus if $\delta\phi$ is a solution, so is $\delta\phi^*$. Again this gives a conserved norm between two solutions to the equations of motion $\delta\phi$ and $\delta\phi'$ of

$$\langle\delta\phi, \delta\phi'\rangle = \int \frac{1}{v^2 G}[\phi^*(\tilde{\partial}_t + v^2 \partial_\phi)\delta\phi' - \delta\phi'(\tilde{\partial}_t + v^2 \partial_\phi)\delta\phi^*] d\phi. \tag{4.112}$$

We note that this equation depends only the conditions at the surface of the flow. It is defined entirely in terms of the factors v^2 and $G = \partial_\psi(gy + \frac{1}{2}v^2)$ defined at $\psi = \psi_T$, and is independent of the obstacles, or the flow throughout the rest of the stream except insofar as they affect the flow at the surface. This might well change if either vorticity or viscosity were introduced into the equations.

This norm is crucial to the analysis of the wave equation. It is conserved (in the absence of viscosity), and in the use of such waves as models for black holes, it is this norm which determines the Bogoliubov coefficients (or the amplification factor) for waves in the vicinity of a horizon (blocking flow in the hydrodynamics sense) and determines the quantum noise (Hawking radiation) emitted by such a horizon analogue. The quantum norm used in the quantization procedure is

$$\langle\delta\phi, \delta\phi\rangle_Q = \frac{i}{2}\langle\delta\phi, \delta\phi\rangle. \tag{4.113}$$

If we define a new coordinate $\hat{\phi} = \int \frac{1}{v^2} d\phi$, the norm becomes

$$\langle \delta\phi, \delta\phi' \rangle = \int \frac{1}{G} [\delta\phi^*(\tilde{\partial}_t - \partial_{\hat{\psi}})\delta\phi' - \delta\phi^*(\tilde{\partial}_t - \partial_{\hat{\psi}})\delta\phi] d\hat{\phi}. \quad (4.114)$$

If the surface of the flow is shallow ($\frac{dy_T}{dx} \ll 1$) then $\frac{d\phi}{dx} = v_x \approx v$ and $\hat{\phi} \approx \frac{dx}{v_x}$.

To relate this to the measured quantity, the vertical displacement at the surface of the waves, we must relate $\delta\phi$ to δy at the surface of the fluid. We have

$$\Psi(t, \phi) = \psi(t, x, y_T(t, x)) - \psi_T = \delta\psi(t, x(\phi, \psi_T), y(\phi, \psi_T)) + v_x \delta y_T, \quad (4.115)$$

or

$$\delta y_T = \frac{1}{v_x}(\Psi - \delta\psi) = \left(\frac{1}{Gv_x} \partial_t + v^2 \partial_\phi \right) \delta\phi. \quad (4.116)$$

Now, $Gv_x \approx g \frac{v_x^2}{v^2} \approx g$ (ignoring the centrifugal contribution to the effective gravity), so the norm becomes

$$\langle \delta y_t, \delta y_T \rangle = \int \frac{v^2}{g} [(\partial_t + \partial_{\hat{\psi}})^{-1} \delta y_T^* \delta y_T - (\partial_t + \partial_{\hat{\psi}})^{-1} \delta y_T \delta y_T^*] d\hat{\phi}$$

$$= \int \frac{1}{g} [((\partial_t + \partial_{\hat{\psi}})^{-1} \sqrt{v} \delta y_T^*) \sqrt{v} \delta y_T$$

$$- ((\partial_t + \partial_{\hat{\psi}})^{-1} \sqrt{v} \delta y_T) \sqrt{v} \delta y_T^*] d\hat{\phi}, \quad (4.117)$$

and $d\hat{\phi} \frac{d\phi}{v^2} \approx \frac{dx}{v}$.

If we assume that the incoming wave is at a set frequency ω and take the Fourier transform with respect to $t, \hat{x}$ of $\sqrt{v(\hat{\psi})} y_T(t, \hat{\phi})$ this becomes

$$\langle \delta y, \delta y \rangle = \int \frac{|(\sqrt{v} y_T)(\hat{k})|^2}{(\omega + \hat{k})} d\hat{k}. \quad (4.118)$$

We can also look at the norm current

$$\partial_t \int_{\phi_1}^{\phi_2} \frac{1}{v^2 G} [\phi^*(\tilde{\partial}_t + v^2 \partial_\phi) \delta\phi' - \delta\phi'(\tilde{\partial}_t + v^2 \partial_\phi) \delta\phi^*] d\phi$$

$$= \int_{\phi_1}^{\phi_2} \partial_x \left(\frac{1}{G}(\tilde{\partial}_t + v^2 \partial_\phi^*) \delta\phi - \partial_x \left(\frac{1}{G}(\tilde{\partial}_t + v^2 \partial_\phi) \delta\phi^* \right) \right)$$

$$+ [(-i \partial_\phi \tanh(-i \psi_T \partial_\phi) \delta\phi^*) \delta\phi - (i \partial_\phi \tanh(i \psi_T \partial_\phi) \delta\phi^*) \delta\phi] d\phi. \quad (4.119)$$

The integrand is a complete derivative. Although this is not obvious for the terms with the tanh in them, we can use

$$(\partial_\phi^{2n} \delta\phi^*)\delta\phi - \delta\phi^* \partial_\phi^{2n} \delta\phi = \partial_\phi \left(\sum_{r=0}^{2n-1} (-1)^r \partial_\phi^r \delta\phi^{*r} \partial_\phi^{2n-1-r} \delta\phi \right) \quad (4.120)$$

4 Irrotational, Two-Dimensional Surface Waves in Fluids

and the fact that $i\partial_\phi \tanh(i\psi_T \partial_\phi)$ can be expanded in a power series in ∂_ϕ^2 to show that they also a complete derivative.

Thus the integrand can be written in terms of a complete derivative of with respect to ∂_ψ and we can regard the term that is being taken the derivative of as a spatial norm current J^ϕ so that if J^t is the temporal part of the norm current, we have $\partial_t J^t + \partial_\phi J^\phi = 0$.

If we are in a regime where $\delta\phi = Ae^{-i\omega t - k\phi}$, (i.e., a regime where the velocity v and G are both constants), then we have

$$J^\phi = i|A|^2 \frac{(\omega + v^2 k)}{Gv^2} + \partial_k \left(k \tanh(\Psi_T k)\right) = i|A|^2 \omega \frac{(1 + v^2/v_p - 2v_g)}{Gv^2}, \quad (4.121)$$

where v_p and v_g are the phase and group velocity of the wave. In a situation in which one has a wave train with some definite frequency and wave number entering a region, then the sum of all the norm currents for each k at the boundary of the region must be zero.

4.6 Blocking Flow

Let us return to the static situation. Define $U = \partial_\phi \delta\phi$, we have the equation

$$\partial_\phi \frac{v^2}{G} U + i \tanh(i\psi_T \partial_\phi) U = 0. \quad (4.122)$$

As above, there is a solution if we assume that the derivatives are small, which gives

$$U = \frac{\text{const}}{\frac{v^2}{G} - \psi_T}. \quad (4.123)$$

For rapid variations, we have

$$U = \text{const} \frac{v^2}{G} e^{i \int \frac{G}{v^2} d\phi}, \quad (4.124)$$

with the transition from one to the other occuring roughly when the logarithmic derivatives of the two solutions are equal

$$\frac{(v^2/G)'}{\frac{v^2}{G} - \psi_T} \approx \frac{\sqrt{(v^2/G)'^2 + 1}}{v^2/G}. \quad (4.125)$$

Defining the Froude number by $F^2 = \frac{v^2}{G\psi_T}$ (the square of the velocity of the fluid over the velocity of the long wavelengths in the fluid in the WKB approximation), we have

$$\frac{(F^2)'}{F^2 - 1} \approx \sqrt{4(\ln(F)')^2 + \left(\frac{1}{F^2 \psi_T}\right)^2}. \quad (4.126)$$

Note that for a non-trivial rate of change of the bottom, the turning point occurs well before the horizon.

The $'$ denotes derivative with respect to ϕ not x. We can rewrite this approximately (assuming that $\frac{v_x}{v} \approx 1$ and that $\frac{2\ln(F)'<1}{F^2\psi_T}$ and $\psi_T \approx vd$ where d is the depth of the water at position x) as

$$\frac{dF^2}{dx} \approx \frac{(F^2-1)}{F^2 d}. \tag{4.127}$$

Note that this transition occurs before $G\psi_T = v^2$ or Froude number equals 1. The wave on the slope piles up and its frequency makes the transition to deep water wave before we hit the effective horizon.

The long wavelength equation,

$$\frac{1}{v^2}(\tilde{\partial}_t + v^2\partial_\phi)\frac{1}{G}(\tilde{\partial}_t + v^2\partial_\phi)\delta\phi - \psi_T\partial_\phi^2\delta\phi = 0, \tag{4.128}$$

is not that of a two dimension metric, which is always conformally flat, but can be written as a the wave equation for a three dimensional metric where all derivatives are equal to zero in the third ξ dimension for the variable $\delta\phi$. The metric is

$$ds^2 = \alpha\left(\left(1 - \frac{v^2}{G\psi_T}\right)dt^2 + 2\frac{1}{G\psi_T}dt d\phi - \frac{1}{v^2 G\psi_T}d\phi^2\right) - \frac{1}{v^2 G\psi_T}d\xi^2, \tag{4.129}$$

where α is an arbitrary function of ϕ, a two dimensional conformal factor which does not affect the two dimensional wave equation. This metric has surface gravity

$$\kappa = \frac{v^2}{2}\partial_\phi\left(\frac{v^2}{G\psi_T}\right) = \frac{1}{2}v^2\partial_\phi F^2. \tag{4.130}$$

(The surface gravity is the acceleration in the horizon as seen from far away. For a static time independent metric in a coordinate system which is regular across the horizon, it can be defined by $\kappa = \Gamma^t_{tt}$ at the horizon, where Γ^i_{jk} is the Christoffel symbol for the metric. Then $\Gamma^t_{tt} = -\frac{1}{2}g^{t\phi}(\partial_\phi g_{tt})$ at the horizon.)

4.7 Conversion to δy

Of course $\delta\phi$ is not what is actually measured in an experiment. That is the fluctuation $\Delta y(x)$ which is the difference in height between the stationary flow, and the height with the wave present. We can relate this to $\delta\psi$ and Ψ:

$$y_s(x,t) = y_0(x) + \Delta y(x,t) \tag{4.131}$$

where y_0 is the surface for the background,

$$\delta y = \frac{v_y}{v^2}\delta\phi + \frac{v_x}{v^2}\delta\psi. \tag{4.132}$$

Since $\delta\psi = \tanh(\Psi_H \partial_\phi)\delta\phi$, we have

$$v^2 \delta y = [v_y + v_x \tanh(\Psi_H \partial_\phi)]\delta\phi. \tag{4.133}$$

4 Irrotational, Two-Dimensional Surface Waves in Fluids

Inverting this for deep water waves,

$$\delta\phi = \frac{v^2}{v_y + v_x}\delta y, \tag{4.134}$$

while for shallow water waves

$$\delta\phi = \int e^{-\int \frac{v_y}{v_x}d\phi} \frac{v^2}{v_x}\delta y d\phi. \tag{4.135}$$

The integrand in the exponent is non-zero only in the region where the background flow is dimpled, and, since $\frac{v_y}{v_x}$ is in general very small, the exponential can be neglected in most situations.

In the intermediate region, where the wave changes from shallow to deep water wave, there is no easy solution to these equations, but they can be integrated numerically.

4.8 Waves in Stationary Water over Uneven Bottom

In the limit as v_0 goes to zero, so does v with the ratio being a finite function. y obeys the equation $(\partial_\phi^2 + \partial_\psi^2)y = 0$ with the boundary conditions along the bottom that $y = Y(x)$, with Y the given function of x of the bottom, and along the top, $y = H$, a constant. If we assume that we know $Y(\phi)$ (instead of $Y(x)$) along the bottom, this can be solved by

$$y(\phi, \psi) = H + \int \alpha(k) e^{ik\phi} \frac{\sinh(k(\psi - \psi_T))}{\sinh(k\psi_T)} dk, \tag{4.136}$$

where

$$\alpha(k) = \frac{1}{2\pi} \int Y(\phi) e^{ik\phi}, \tag{4.137}$$

and

$$x(\phi, \psi) = \phi + i \int \alpha(k) \frac{\cosh(k(\psi - \psi_T))}{\sinh(k(\psi_T))} dk. \tag{4.138}$$

One gets rapid convergence if one starts by taking $x = \phi$, substituting into $Y(x(\phi))$ to find $Y(\phi)$, finding the new $x(\phi)$ and substituting in again.

Then $\frac{v_y}{v_0}$ at the surface is zero, while

$$\frac{v_0}{v_x} = \frac{v_0}{v} = \partial_\psi y = \int k\alpha(k) \frac{1}{\sinh(k(\psi_T))} dk. \tag{4.139}$$

The equation for small perturbations becomes

$$\frac{v_0^2}{v^2 G} \partial_t^2 \delta\phi - i\partial_\phi \tanh(i\psi_T \partial_\phi) \delta\phi = 0, \tag{4.140}$$

where

$$\frac{v^2}{v_0^2} G = \frac{v^2}{v_0^2} g \partial_\psi y = g \frac{v_x}{v_0} = g \frac{\partial \phi}{\partial_x}. \quad (4.141)$$

If the depth is constant, the background $\psi = y$ and $\phi = x$ giving the usual equation, which allows us to write

$$\partial_t^2 \delta\phi + ig\partial_x \tanh(\psi_T \partial_\phi) \delta\phi. \quad (4.142)$$

For deep water waves, where the tanh is unity, this equation is exactly the same as the deep water equation for constant depth. The fact that the bottom varies makes no difference to the propagation of the waves, as one would expect.

For shallow water waves, where the tanh can be approximated as the linear function in its argument, the equation becomes

$$\partial_t^2 \delta\phi = \psi_T \partial_x \partial_\phi \delta\psi = g\psi_T \frac{v_0}{v} \partial_x^2 \delta\phi. \quad (4.143)$$

This allows us to determine the wave propagation over an arbitrarily defined bottom. Note that in the stationary limit, the background flow is certainly irrotational, implying that the assumptions made here should certainly be valid (of course neglecting the viscosity of the fluid).

Acknowledgements This work was supported by The Canadian Institute for Advanced Research (CIfAR) and by The Natural Science and Engineering Research Council of Canada (NSERC) and was completed while the author was a visitor at the Perimeter Institute.

References

1. Misner, C.M., Thorne, K.S., Wheeler, J.A.: Gravitation. Freemann, San Francisco (1973)
2. Hawking, S.W., Ellis, G.F.R.: The Large Scale Structure of Spacetime. Cambridge University Press, Cambridge (1973)
3. Schützhold, R., Unruh, W.G.: Phys. Rev. D **66**, 044019 (2002)
4. Keller, J.B.: Surface waves on water of non-uniform depth. J. Fluid Mech. **4**, 607 (1958)
5. Birkoff, J.C.W.: Computation of combined refraction-diffraction. In: Khaskhachikh, G.D., Plakida, M.É., Popov, I.Ya. (eds.) Proceedings 13th International Conference on Coastal Engineering, Vancouver, p. 471. Springer, Berlin (1972)
6. Wald, R.M.: General Relativity. University of Chicago Press, Chicago (1984)

Chapter 5
The Basics of Water Waves Theory for Analogue Gravity

Germain Rousseaux

Abstract This chapter gives an introduction to the connection between the physics of water waves and analogue gravity. Only a basic knowledge of fluid mechanics is assumed as a prerequisite.

5.1 Introduction

According to Pierre-Gilles de Gennes, *"the borders between great empires are often populated by the most interesting groups"*. Indeed, these people often speak several languages and are more open-minded due to cultural exchanges. A wonderful analogy exists between the propagation of hydrodynamic waves on a fluid flow and the propagation of light in the curved spacetime of a black hole. It allows us to test astrophysical predictions such as Hawking radiation and the effects of high frequency dispersion on it. It provides new insights in Fluid Mechanics thanks to the use of tools and concepts borrowed from Quantum Field Theory in curved spacetime and vice versa. General relativists speak with hydraulicians and this chapter is a testimony of their common language and relationships [1, 2].

Here, we provide the general background on water waves propagation for analogue gravity: we will try to explain how water waves propagate and how a flow current implies the existence of an effective spacetime; we will insist on the difference between propagation in deep and shallow waters on the dispersion relation; a generalized definition of a horizon will be given and which, in the particular case of shallow water, reduces to the usual habit of general relativists.

5.2 A Glimpse of Dimensional Analysis

The equations of fluid mechanics are known since several centuries but they still defy modern physics when we try to understand one of its outstanding mysteries

---

G. Rousseaux (✉)
Laboratoire J.-A. Dieudonné, UMR CNRS-UNS 6621, Université de Nice-Sophia Antipolis, Parc Valrose, 06108 Nice Cedex 02, France
e-mail: germain.rousseaux@unice.fr

D. Faccio et al. (eds.), *Analogue Gravity Phenomenology*,
Lecture Notes in Physics 870, DOI 10.1007/978-3-319-00266-8_5,
© Springer International Publishing Switzerland 2013

like turbulence. As they are non-linear and feature several effects such as pressure, gravity and external forces, practitioners have been forced to introduce a very useful way to grasp the relevant effects when dealing with a peculiar flow. This technique is the so-called dimensional analysis which is often the only rescue procedure to disentangle the relative magnitude of several processes at play. The reader will be referred to the book by Barenblatt on scaling and dimensional analysis for a thorough introduction [3]. Here, we will construct with simple arguments the relevant velocities of propagation of water waves depending on the water depth. Three regimes of propagation will be uncovered with corresponding dispersion relations. Then, the effect of a current will be added and this ingredient is going to be essential in order to have an analogue gravity system.

Waves are characterized by both their wavenumber k and their angular frequency ω. Since we are dealing with Newton's laws of motion applied to fluids, the second time derivative (namely inertia) will translate in a square term in the angular frequency and the dispersion relation has the general form $\omega^2 = F(k)$. Obviously, waves with both positive and negative angular frequency are thus described by the dispersion relation ($\omega = \pm\sqrt{F(k)}$). Usually, the negative root is dismissed since, when the propagation is free, it is a matter of convention to focus either on the right or left-propagating waves. Of course, in the presence of a current, the system will no longer be symmetric with respect to space reflection ($x \to -x$) and then both types of waves (positive and negative) turn to be important...

5.2.1 Shallow Waters

Let us assume that a train of continuous sinusoidal water waves with wavelength λ is propagating at the surface of a fluid at rest in a given depth h. We make the strong hypothesis that the wavelength is much longer than the depth, that is $kh \ll 1$ using the wavenumber $k = 2\pi/\lambda$. The fluid vertical extension h has the dimension of a length L. Since inertia is balanced by gravity, the gravity field is a relevant parameter and its intensity g has the dimension of an acceleration $L.T^{-2}$. We are looking for the typical scaling of the wave velocity. The crests of the water waves propagate with the so-called phase velocity and its value c has the following dimension $L.T^{-1}$. We are lead to the obvious scaling law up to a constant term:

$$c_{shallow} \approx \sqrt{gh}. \tag{5.1}$$

Using the definition of the phase velocity $c_\phi = \omega/k$, it is straightforward to infer the approximate dispersion relation for water waves propagating in shallow waters:

$$\omega^2 \approx ghk^2 \tag{5.2}$$

which is similar to the dispersion relation for light waves in empty flat spacetime. Analogue Gravity will emerge when effective curved spacetime is added as we will see...

Gravity waves in shallow waters are not dispersive since, whatever their wavelength, they do propagate with the same velocity.

5 The Basics of Water Waves Theory for Analogue Gravity

5.2.2 Deep Waters

Far from the sea shore or for very short gravity waves, the water depth is no more a relevant parameter and the only length scale left is the wavelength of the water waves. So, if we assume that $kh \gg 1$ and recalling that the wavenumber k has the dimension of an inverse length L^{-1}, we find the scaling for the phase velocity in deep waters:

$$c_{deep} \approx \sqrt{\frac{g}{k}} \qquad (5.3)$$

and the dispersion relation:

$$\omega^2 \approx gk. \qquad (5.4)$$

Newton derived this scaling in his Principia by applying the Galileo formula for the period of oscillation of a pendulum $T \simeq \sqrt{l/g}$ to the water waves. Indeed, if the length of the pendulum l is replaced by the wavelength, we do recover the same scaling.

Gravity waves in deep waters are dispersive since longer waves propagates faster than the short ones.

5.2.3 Arbitrary Water Depth

Without doing more calculations, we can anticipate that the general dispersion relation for waters on arbitrary depth will write according to the following form:

$$\omega^2 = gkH(k), \qquad (5.5)$$

bearing in mind that the following asymptotic limits must be fulfilled: $\lim_{kh \to \infty} H(k) = 1$ and $\lim_{kh \to 0} H(k) = kh$. It can been shown that one has rigorously $H(k) = \tanh(kh)$ after a lengthy calculation that we will avoid to the reader (see the books by Mei [4] or Dingemans [5] for a mathematical demonstration).

5.2.4 The Capillary Length

Water is made of molecules. In the bulk of the liquid, every molecule is surrounded by the same number of neighboring molecules whereas at the boundary with another substance (like air at the free surface of sea water), a water molecule is also surrounded by gas molecules and the resulting imbalance in the chemical interactions results in a pressure difference between the liquid and the gas. The Young-Laplace law states that the pressure jump is proportional both to the local curvature of the interface and to a phenomenological coefficient γ named surface tension which is a property of both media:

$$\Delta p_{Y-L} = \gamma C, \qquad (5.6)$$

where C is the curvature of the free surface and has the dimension of an inverse length (roughly the radius of curvature).

When the Young-Laplace pressure Δp_{Y-L} is balanced by the Pascal pressure drop $\Delta p_P = \rho g h$ due to the static gravity field, a scaling law is deduced easily for the so-called capillary length which is the typical size on which surface tension effects are acting:

$$\Delta p_{Y-L} \approx \frac{\gamma}{l_c} = \Delta p_P \approx \rho g l_c. \tag{5.7}$$

Hence, the capillary length writes:

$$l_c = \sqrt{\frac{\gamma}{\rho g}}. \tag{5.8}$$

Let us introduce the effective gravity field g^* induced by the capillarity and its dispersive scaling law in terms of the wavenumber k:

$$g^* \simeq \frac{\gamma}{\rho} \frac{1}{l_c^2} \approx \frac{\gamma}{\rho} k^2. \tag{5.9}$$

The dispersion relation for water waves taking into account the effect of surface tension becomes:

$$\omega^2 = (g + g^*) k \tanh(kh), \tag{5.10}$$

that is [4]:

$$\omega^2 = \left(gk + \frac{\gamma}{\rho} k^3\right) \tanh(kh). \tag{5.11}$$

5.3 Long Water Waves on a Current as a Gravity Analogue

In a letter to H. Cavendish in 1783, Reverend John Michell introduced the concept of what is named, in modern physics, a "black hole" (he was inspired by the corpuscular theory of light by I. Newton) [6]: "*If the semi-diameter of a sphere of the same density as the Sun in the proportion of five hundred to one, and by supposing light to be attracted by the same force in proportion to its [mass] with other bodies, all light emitted from such a body would be made to return towards it, by its own proper gravity*". According to him, this situation occurs when the "escape velocity" of a massive particle is equal to the velocity of light. Then, Pierre-Simon de Laplace introduced the term "*étoile sombre*" (dark star) in his "Exposition du Système du Monde" in 1796 to denote such an object. Laplace is also well known for having proposed in 1775 an analytical model to describe standing water waves in shallow water and for having derived the related dispersion relation [7].

Recently, Schützhold & Unruh derived the equation of propagation of water waves moving on a background flow in the shallow waters limit [8] and this equation describes also the behaviour of light near the event horizon of a black hole. Indeed, under the impulsion of the seminal work by Unruh [9], there has been more

5 The Basics of Water Waves Theory for Analogue Gravity

and more interest for analogue models in general relativity in order to understand the physics of wave propagation on an effective curved spacetime. Several systems exhibit a so-called "acoustic" metric similar to the metric describing a black hole when a wave is moving in a "flowing" medium [1, 2].

Here, we will reproduce the derivation of Schützhold & Unruh with some details for pedagogy [8]. Hence, we consider the propagation of small linear perturbations of a free surface between water and air in the presence of an underlying current. The current is uniform in depth z, time-independent and varies slowly in the longitudinal direction x. We are in the so-called WKB approximation such that the wavelength is smaller than the typical length on which the current varies ($\lambda \ll \frac{U(x)}{\frac{dU}{dx}}$). $\mathbf{U} = \mathbf{v}_B$ will denote the background flow current whereas $\mathbf{v}$ will stand for the velocity associated with the propagation of waves.

The liquid is inviscid, its density is constant ($\rho = \text{const}$) and the flow is incompressible.

We have the following equations of motion:

- The continuity equation which comes from the incompressibility condition: $\nabla \cdot \mathbf{v} = 0$;
- The Euler equation:

$$\frac{d\mathbf{v}}{dt} = \dot{\mathbf{v}} + (\mathbf{v} \cdot \nabla)\mathbf{v} = -\frac{\nabla p}{\rho} + \mathbf{g} + \frac{\mathbf{f}}{\rho}, \qquad (5.12)$$

with p the pressure, $\mathbf{g} = -g\mathbf{e_z}$ the gravitational acceleration and $\mathbf{f} = -\rho \nabla_\parallel V^\parallel$ a horizontal and irrotational force in the x direction ($\parallel$) driven by the potential $V^\parallel$ which is at the origin of the flow.

Since the flow is assumed to be vorticity-free $\nabla \times \mathbf{v} = 0$ (a crucial feature of water waves propagation), one has $(\mathbf{v} \cdot \nabla)\mathbf{v} = (\nabla \times \mathbf{v}) \times \mathbf{v} + \frac{1}{2}\nabla(v^2) = \frac{1}{2}\nabla(v^2)$, with $\mathbf{v} = \nabla \phi$ where ϕ stands for the velocity potential. The vectorial Euler equation reduces to the simpler scalar Bernoulli equation

$$\dot{\phi} + \frac{1}{2}(\nabla \phi)^2 = -\frac{p}{\rho} - gz - V^\parallel. \qquad (5.13)$$

The boundary conditions are such that:

- In $z = 0$, the vertical flow velocity must be null, i.e. $v^\perp(z=0) = 0$. $z = 0$ is by definition the bottom depth.
- The height variations of the fluid are determined by the very same velocity but computed on the free surface:

$$v^\perp(z=h) = \frac{dh}{dt} = \dot{h} + (\mathbf{v} \cdot \nabla)h, \qquad (5.14)$$

where $\frac{dh}{dt}$ is the velocity of a point on the air-water interface.
- The relative pressure with respect to the atmospheric pressure on the free surface cancels by definition: $p(z=h) = 0$.

Let us consider a velocity perturbation δv of the background flow $\mathbf{v}_B$ with a corresponding vertical displacement δh. We assume the background flow $\mathbf{v}_B$ to be stationary, irrotational and horizontal: $\nabla_\perp \mathbf{v}_B = 0$, $\mathbf{v}_B = \mathbf{v}_B^\parallel \to \nabla_\parallel \cdot \mathbf{v}_B = 0$.

The Bernoulli equation gives:

$$\frac{1}{2}v_B^2 = -\frac{p_B}{\rho} - gz - V^\parallel, \tag{5.15}$$

where p_B follows Pascal's law for static pressure distribution in the water column $p_B(z) = \rho g(h - z)$.

We assume that the velocity perturbation δv is also curl-less: hence, we can define a perturbed velocity potential $\delta \phi$. Using again the Bernoulli equation, we get:

$$\dot{\delta\phi} + \mathbf{v}_B^\parallel \cdot \nabla_\parallel \delta\phi = -\frac{\delta p}{\rho}. \tag{5.16}$$

By taking into account the condition $p_B(z = h) = 0$ and using the expression for p_B, we obtain the boundary condition for the pressure fluctuation at the free surface δp: $\delta p(z = h) = g\rho \delta h$.

The same procedure applies to the vertical velocity, using the following conditions (5.14) and $v^\perp(z = 0) = 0$: $\delta v^\perp(z = 0) = 0$ and $\delta v^\perp(z = h_B) = \dot{\delta h} + (\mathbf{v}_B^\parallel \cdot \nabla_\parallel)\delta h$.

We now develop the velocity potential $\delta\phi$ using a Taylor series:

$$\delta\phi(x, y, z) = \sum_{n=0}^{\infty} \frac{z^n}{n!} \delta\phi_{(n)}(x, y). \tag{5.17}$$

The boundary condition $v^\perp(z=0) = 0$, implies $\delta\phi_{(1)} = 0$. With the continuity equation, we find:

$$\nabla_\parallel^2 \delta\phi_{(0)} + \delta\phi_{(2)} + \cdots = 0. \tag{5.18}$$

We assume that the depth h is much longer than the wavelength λ of the free surface perturbation. Hence, the higher-order terms in the Taylor expansion are suppressed by powers of $h/\lambda \ll 1$ since we have $\nabla_\parallel^2 = \mathcal{O}(1/\lambda^2)$. Keeping only the two lowest terms in the Taylor series, we get:

$$\delta v^\perp(z) = \nabla_\perp \delta\phi$$
$$= \nabla_\perp \left(\delta\phi_{(0)} + \frac{z^2}{2} \delta\phi_{(2)} \right)$$
$$= z\delta\phi_{(2)}. \tag{5.19}$$

At the free surface $z = h$ and using (5.18), we then find:

$$\delta v^\perp(z = h) = -h\nabla_\parallel^2 \delta\phi_{(0)}. \tag{5.20}$$

In order to find a wave equation for $\delta\phi_{(0)}$, let us take the partial derivative with respect to time t of Eq. (5.16), then, with the help of the boundary conditions $\delta p(z = h)$, $\delta v^\perp(z = h)$ and of Eq. (5.20), we substitute $\dot{\delta h}$ with an equivalent expression. Finally, we end up with:

$$\partial_t^2(\delta\phi) + 2(\mathbf{v}_B^\parallel \cdot \nabla_\parallel)\partial_t(\delta\phi) + (\mathbf{v}_B^\parallel \otimes \mathbf{v}_B^\parallel - gh)\nabla^2(\delta\phi) = 0, \tag{5.21}$$

that is the so-called Beltrami-Laplace equation:

$$\Box \delta\phi_{(0)} = \frac{1}{\sqrt{-g}} \partial_\mu \left(\sqrt{-g} g^{\mu\nu} \partial_\nu \delta\phi_{(0)} \right) = 0, \qquad (5.22)$$

provided that we identify $g^{\mu\nu}$ as an inverse metric expressed in the following matrix form:

$$g^{\mu\nu} = \begin{pmatrix} 1 & \vdots & \mathbf{v}_B^{\parallel} \\ \dots\dots\dots & \cdot & \dots\dots\dots\dots \\ \mathbf{v}_B^{\parallel} & \vdots & \mathbf{v}_B^{\parallel 2} - gh I \end{pmatrix}. \qquad (5.23)$$

Using $g^{\mu\nu} g_{\mu\sigma} = \delta^\nu_\sigma$ we get the so-called acoustic metric $g_{\mu\nu}$ in its typical Painlevé-Gullstrand form:

$$g_{\mu\nu} = \frac{1}{c^2} \begin{pmatrix} gh - \mathbf{v}_B^{\parallel 2} & \vdots & \mathbf{v}_B^{\parallel} \\ \dots\dots\dots & \cdot & \dots\dots\dots\dots \\ \mathbf{v}_B^{\parallel} & \vdots & -1 \end{pmatrix}, \qquad (5.24)$$

with $c = \sqrt{gh}$ the velocity of water waves in shallow water which is the analogue of the velocity of light.

It is straightforward to show that the dispersion relation associated with Eq. (5.21) is:

$$(\omega - \mathbf{k}.\mathbf{U})^2 \approx c^2 k^2, \qquad (5.25)$$

which describes the propagation of long water waves on a given flow $\mathbf{U} = \mathbf{v}_B$ where we insist on its approximate nature ($kh \ll 1$). The flow induces a Doppler shift of the angular frequency: we refer the reader to the hydrodynamics literature where the effect of a current on water waves has been discussed extensively [10–15]. One speaks of blue-shifting (red-shifting) when the current encounters (follows) the waves and the wavenumber increases (decreases). When the flow vanishes, the Beltrami-Laplace operator reduces to the usual d'Alembertian operator.

The dispersion relation is solved graphically in Fig. 5.1. The so-called transplanckian problem arises when $U = -c$ that is when the wavenumber of the positive solution (in green) diverges at $+\infty$, disappears and then reappears as a new diverging negative solution (in blue) at $-\infty$ for increasing modulus of the flow velocity.

5.4 Fluid Particles' Trajectories

In this part, we recall (without demonstration) some classical results from water waves theory on the trajectories of the fluid particles beneath a water wave [4]. For small wave amplitudes ($ka \ll 1$), the non-linear terms of the Euler equation and of the boundary conditions can be neglected. In 1845, G.B. Airy derived within this approximation the fluid particles' trajectories compatible with the following

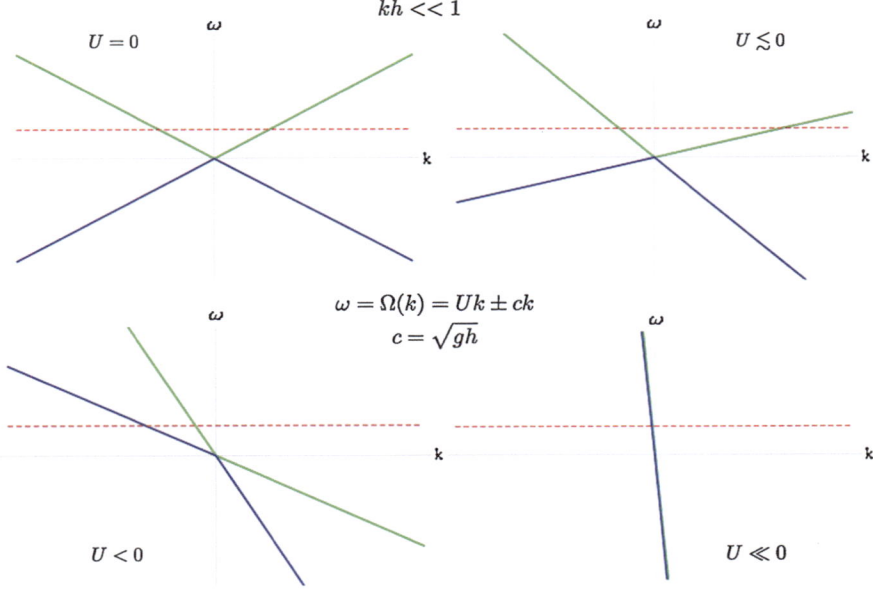

Fig. 5.1 Graphical solutions of the dispersion relation $(\omega - \mathbf{k}.\mathbf{U})^2 \approx c^2 k^2$: $\omega = \Omega(k)$ is plotted as a function of k for increasing modulus of the background flow $U < 0$. The conserved frequency ω is the *horizontal red dotted line*. The *green* (*blue*) color corresponds to the positive (negative) branches

dispersion relation valid for pure gravity waves without a background flow for a given depth:

$$\omega^2 = gk \tanh(kh). \tag{5.26}$$

We denote $z' = 0$ the mean water depth of the free surface without wave. Let us consider the following perturbation with respect to rest:

$$z' = \eta(x,t) = a \sin(\omega t - kx). \tag{5.27}$$

Airy computed the resulting velocity profile:

$$u(x,z',t) = a\omega \frac{\cosh(k(z'+h))}{\sinh(kh)} \sin(\omega t - kx), \tag{5.28}$$

and

$$w(x,z',t) = a\omega \frac{\sinh(k(z'+h))}{\sinh(kh)} \cos(\omega t - kx), \tag{5.29}$$

where u and w correspond to the projections of the perturbation velocity in the horizontal and vertical directions.

5 The Basics of Water Waves Theory for Analogue Gravity

Fig. 5.2 The flow generated beneath a surface waves: (**a**) deep water case; (**b**) shallow water case

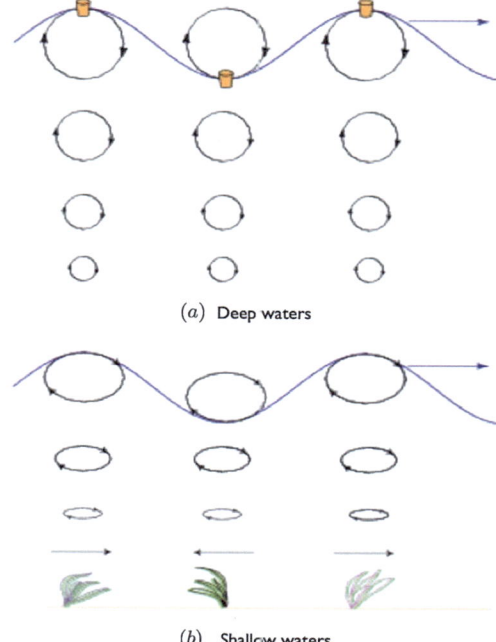

Under the hypothesis of small displacements, one deduces the horizontal motion of fluid particles:

$$X(x,z',t) - X_0 = -a\frac{\cosh(k(z'+h))}{\sinh(kh)}\cos(\omega t - kx), \qquad (5.30)$$

as well as the vertical motion:

$$Z(x,z',t) - Z_0 = a\frac{\sinh(k(z'+h))}{\sinh(kh)}\sin(\omega t - kx). \qquad (5.31)$$

In deep waters (far from the sea shore for example), the fluid particles' trajectories are circular with radius R. In shallow waters (close to the beach!), the trajectories flatten and the particles follow an ellipse of semi-axis A and B for respectively the horizontal and vertical motions (see Fig. 5.2). In practice, practitioners distinguish three zones:

- The deep water case ($h/\lambda > 1/2$):

$$A \sim B \sim ae^{kz'}. \qquad (5.32)$$

The trajectories are circles of radius $R \sim A \sim B$ which decreases exponentially with the depth z'.
- The intermediate case ($1/20 < h/\lambda < 1/2$):

$$A = a\frac{\cosh(k(z'+h))}{\sinh(kh)}, \qquad (5.33)$$

and

$$B = a\frac{\sinh(k(z'+h))}{\sinh(kh)}. \tag{5.34}$$

The trajectories are ellipses whose semi-axes diminish with depth. The decrease is slower than the exponential one of deep waters.
- The shallow water case ($h/\lambda < 1/20$):

$$A \sim \frac{a}{kh}, \tag{5.35}$$

and

$$B \sim \frac{ak(z'+h)}{kh}. \tag{5.36}$$

The trajectories are ellipses whose major semi-axis A is independent of the water depth z' and whose minor semi-axis B decreases linearly with z'. On the bottom ($z' = -h$), B cancels and the trajectories become a horizontal oscillation of amplitude A.

In the presence of a current **U**, the previous expressions for the velocity field keep the same form provided the dependence of the amplitude (and not of the phase) with the angular frequency ω is replaced by the relative angular frequency $\omega' = \omega -$ **k.U** [10–16]. Of course, u becomes $u' + U$ whereas w' is invariant. The particles' trajectories are thus similar to cycloids whose amplitude decreases with the water depth for waves following the current [16].

5.5 A Plethora of Dispersive Effects

One of the salient effects of analogue gravity is the possibility to solve the transplanckian problem thanks to the introduction of dispersion close to the horizon of an artificial black hole. As a matter of fact, a major drawback of the original calculation by Stephen Hawking of the black hole radiation is the necessity for the field to have a wavelength which goes to zero as one gets close to the event horizon. Water waves provide several regularization scales in a cascade such as the water depth, the capillary length or even a viscous scale in order to cope with a diverging wavenumber by counter-acting the continuous blue-shifting of the flow...

Let us consider the propagation of gravity waves (without surface tension $\gamma = 0$ for the moment) on a linear shear flow $U(z) = U_0 + \Omega z$ with constant plug flow U_0 and constant vorticity Ω. Here, one assumes that both the bottom depth and the flow velocity vary slowly such that $\frac{h}{\frac{dh}{dx}} \gg \lambda$ and $\frac{U}{\frac{dU}{dx}} \gg \lambda$. The dispersion relation between the frequency $\frac{\omega}{2\pi}$ and the wavenumber k writes either with its implicit expression due to Thompson [17]:

$$(\omega - U_0 k)^2 = [gk - \Omega(\omega - kU_0)]\tanh(kh), \tag{5.37}$$

5 The Basics of Water Waves Theory for Analogue Gravity

or with its explicit expression due to Biesel [18]:

$$\omega = U_0 k - \frac{\Omega}{2} \tanh(kh) \pm \sqrt{\left(\frac{\Omega}{2} \tanh(kh)\right)^2 + gk \tanh(kh)}. \quad (5.38)$$

With surface tension, Huang has derived recently the following dispersion relation with its implicit expression [19]:

$$(\omega - kU_0)^2 = \left[gk + \frac{\gamma}{\rho}k^3 - \Omega(\omega - kU_0)\right] \tanh(kh), \quad (5.39)$$

which can be written explicitly according to Choi [20] in the form:

$$\omega = U_0 k - \frac{\Omega}{2} \tanh(kh) \pm \sqrt{\left(\frac{\Omega}{2} \tanh(kh)\right)^2 + \left(gk + \frac{\gamma}{\rho}k^3\right) \tanh(kh)}. \quad (5.40)$$

It is interesting to notice that the "relativistic" dispersion relation $(\omega - U.k)^2 = c^2 k^2$ is recovered in the long wavelength limit $kh \ll 1$ whatever is the dispersive correction. It obvious when dealing with the surface tension since the capillary length is smaller that the long wavelength. It is less obvious for the dispersive effect of vorticity. Indeed, the long wavelength approximation of the Biesel's dispersion relation writes:

$$\omega \simeq U_0 k - \frac{h\Omega}{2}k \pm \sqrt{ghk^2 + \frac{\Omega^2 h^2}{4}k^2}. \quad (5.41)$$

Fortunately, it can be transformed into the usual dispersion relation associated with the acoustic metric $(\omega - U'.k)^2 = c'^2 k^2$ provided one introduces renormalized flow and waves velocities $U' = U_0 - \Omega h/2$ and $c' = \sqrt{gh + \Omega^2 h^2/4}$.

Assuming a uniform flow in the vertical direction ($\Omega = 0$), the dispersion relation becomes [10–15]:

$$(\omega - Uk)^2 \simeq \left(gk + \frac{\gamma}{\rho}k^3\right) \tanh(kh). \quad (5.42)$$

- In the shallow water limit $kh \ll 1$,

$$(\omega - Uk)^2 \simeq ghk^2 + \left(\frac{\gamma h}{\rho} - \frac{gh^3}{3}\right)k^4 + \mathcal{O}(k^6), \quad (5.43)$$

the dispersion relation is identical to a BEC-type phonons spectrum:

$$(\omega - Uk)^2 \simeq c^2 k^2 \pm c^2 \xi^2 k^4, \quad (5.44)$$

with the corresponding "healing length":

$$\xi = \sqrt{\left|l_c^2 - \frac{h^2}{3}\right|}. \quad (5.45)$$

An interesting observation is that the superluminal correction can have a negative sign in contrast to the BEC case if the capillary length is less than $h/\sqrt{3}$ or even null...

- In the deep water limit $kh \gg 1$, the dispersion relation looses its "relativistic/acoustic" branch:

$$(\omega - Uk)^2 \simeq gk + \frac{\gamma}{\rho}k^3. \qquad (5.46)$$

Viscosity has both a dissipative (imaginary term) and a dispersive (real term) contributions to the dispersion relation. By dimensional analysis, it is obvious that the typical viscous scale would be of the order of $\delta \approx \sqrt{\frac{\nu}{\omega}}$ otherwise known as the Stokes viscous length which is the scale of the viscous boundary layer [4].

5.6 Hydrodynamic Horizons

In this part, we propose a generalized definition of a horizon with respect to the usual custom in General Relativity. Condensed matter horizons and here, hydrodynamic horizons lead to a dispersive-like definition. What is a Horizon? The word horizon derives from the Greek "$οριζων$ $κυκλος$" (*horizon kyklos*), "separating circle", from the verb "$οριζω$" (*horizo*), "to divide, to separate", from the word "$ορος$" (oros), "boundary, landmark". In the Fable of Jean de la Fontaine recalled at the beginning of this chapter, will the Lamb be right to argue against the Wolf that the waves he creates as he is drinking at the river border will not climb against the current and reach the Wolf? Will a frontier separate the Lamb from the Wolf: will a horizon form? Will the position of the frontier depend on the period of the waves: will the horizon be dispersive or not?

5.6.1 Non-dispersive Horizons

The analogy between the propagation of light in a curved spacetime and the propagation of long gravity waves on a current features the so-called "acoustic/relativistic" dispersion relation $(\omega - \mathbf{k}.\mathbf{U})^2 \approx c^2 k^2$ as a common characteristic for both systems assuming $kh \ll 1$ and without surface tension. A simple dimensional analysis of it:

$$\omega^2 \approx U^2 k^2 \approx ghk^2, \qquad (5.47)$$

allows to infer scaling laws for the wavenumber:

$$k \approx \frac{\omega}{\sqrt{gh}}, \qquad (5.48)$$

and the blocking velocity:

$$U \approx \sqrt{gh}, \qquad (5.49)$$

which we confirm by solving the "relativistic" dispersion relation as a polynomial in k:

$$k_h = \frac{\omega}{U + \sqrt{gh}}, \qquad (5.50)$$

Fig. 5.3 Phase-space U versus T of the dispersion relation $(\omega - \mathbf{k}.\mathbf{U})^2 \approx c^2 k^2$

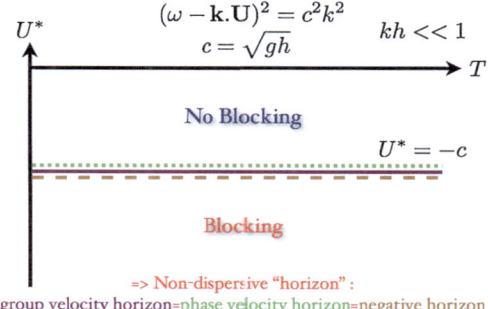

implying the transplanckian problem ($k_h \to \infty$) when the blocking velocity (U^* in modulus) of long gravity waves matches the current flow:

$$U_h = -\sqrt{gh}. \tag{5.51}$$

This non-dispersive definition of a horizon in hydrodynamics corresponds to the definition of General Relativity where the pure temporal matrix element of the Painlevé-Gullstrand metric is cancelled:

$$g_{00} = 0, \tag{5.52}$$

leading to:

$$U = -c = -\sqrt{gh}. \tag{5.53}$$

Here, it is crucial to understand that the relativistic horizon hides in fact three intricate horizons (Fig. 5.3: blocking velocity U^* versus the wave period T): a group velocity horizon ($c_g = \frac{\partial \omega}{\partial k} = U + c = 0$), a phase velocity horizon ($c_\phi = \frac{\omega}{k} = U + c = 0$) and a negative horizon (or negative energy mode horizon): negative relative frequencies $\omega - \mathbf{k}.\mathbf{U} < 0$ can appear. This last fact implies that Stimulated Hawking Radiation can be observed in Classical Physics using water waves and this is one of the major interests of the analogue gravity program for the Fluid Mechanics community [21–25]. As soon as there is a phase velocity horizon, this one is identical with a negative horizon. Because of dispersion, a phase velocity horizon can be absent whereas a negative group velocity horizon can be present (see below).

These "negative energy waves" are well known in Hydrodynamics. Werner Heisenberg discovered them in his PhD Thesis on the stability of the plane Couette flow. He showed that viscosity can have a destabilizing effect if negative energy waves (also named Tollmien-Schlichting waves) are present (at the so-called critical layer corresponding to a phase velocity horizon) in a unidirectional non-inflectional plane flow which is normally stable if inviscid according to the classical Rayleigh criterion [7, 14]!

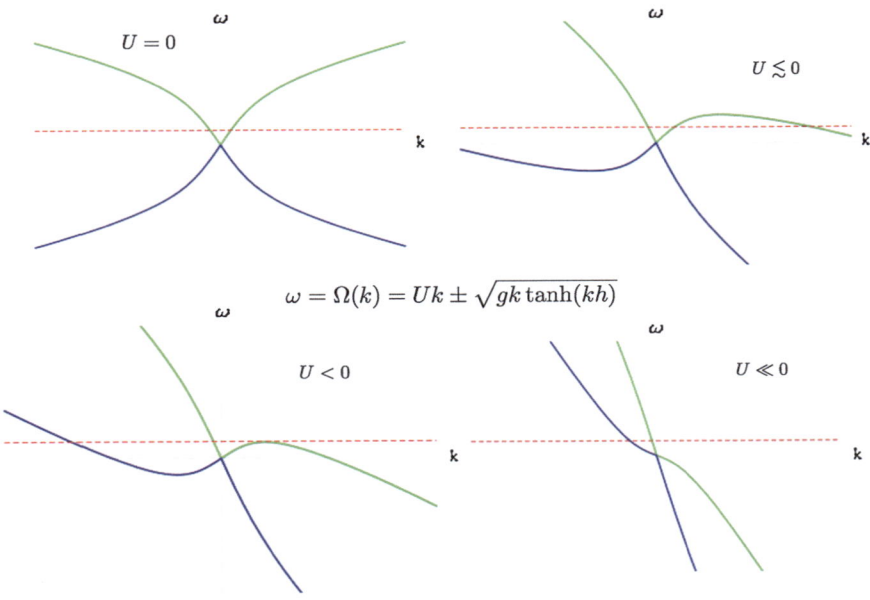

Fig. 5.4 Graphical solutions of the dispersion relation $(\omega - \mathbf{k}.\mathbf{U})^2 = gk\tanh(kh)$. The conserved frequency ω is the *horizontal red dotted line*. The *green* (*blue*) *color* corresponds to the positive (negative) branches

5.6.2 Dispersive Horizons

How is the definition of a horizon modified in the presence of dispersion? Wave blocking is a process where a flow separates a free surface into a flat and a deformed surface. The boundary defines a "horizon". A wave phenomenon implies the existence of a dispersion relation $\omega = \Omega(k)$. At the boundary, the energy flow of the system "waves + current" cancels:

$$c_{group}^{wave+current} = \frac{\partial \Omega}{\partial k} = 0. \tag{5.54}$$

This last criterion will define a hydrodynamic horizon as a group velocity horizon (or turning point using WKBJ terminology). Of course, one recovers the non-dispersive definition $U = -c$ for an "acoustic/relativistic" dispersion relation. We treat here the simple case of a white hole horizon which is the time reverse of a black hole horizon. As previously, the dispersion relation for water waves in arbitrary depth is solved graphically (Fig. 5.4). An extremum of the function $\omega = \Omega(k) = Uk \pm \sqrt{gk\tanh(kh)}$ corresponds to a horizon.

Some scaling laws can be derived in the high dispersive regime where $kh \gg 1$:

- Without surface tension. Dimensional analysis leads to:

$$\omega^2 \approx U^2 k^2 \approx gk, \tag{5.55}$$

5 The Basics of Water Waves Theory for Analogue Gravity

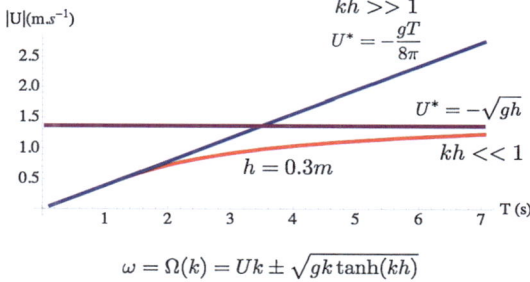

Fig. 5.5 Superposed phase-spaces U versus T for the shallow and deep water cases without surface tension

that is:

$$k \approx \frac{\omega^2}{g}, \tag{5.56}$$

and:

$$U \approx \frac{\omega}{k} \approx \frac{g}{\omega} \approx gT. \tag{5.57}$$

The rigorous mathematical treatment gives [22]:

$$k_g = \frac{4\omega^2}{g}, \tag{5.58}$$

and

$$U_g = -\frac{g}{4\omega} = -\frac{gT}{8\pi}. \tag{5.59}$$

The blocking velocity U^* depends now on the incoming period of the water waves (Fig. 5.5). Depending on the period, we have either $U^* = U_h$ for long waves or $U^* = U_g$ for short waves.

- With surface tension. Dimensional analysis leads to:

$$\omega^2 \approx U^2 k^2 \approx gk \approx \frac{\gamma}{\rho} k^3, \tag{5.60}$$

that is:

$$k \approx \left(\frac{\rho g}{\gamma}\right)^{1/2}, \tag{5.61}$$

and:

$$U \approx \left(\frac{\gamma g}{\rho}\right)^{1/4}. \tag{5.62}$$

The rigorous mathematical treatment gives [23]:

$$k_\gamma = \left(\frac{\rho g}{\gamma}\right)^{1/2}, \tag{5.63}$$

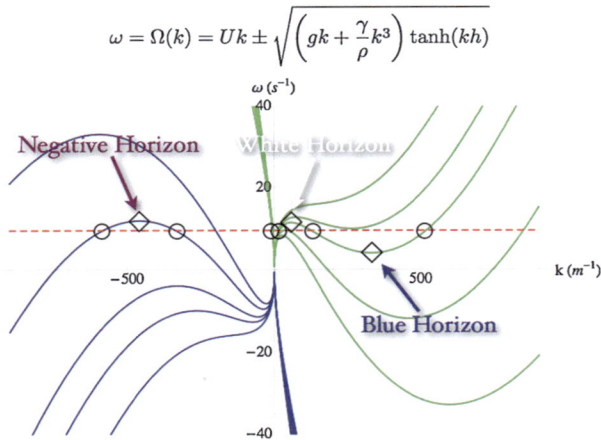

Fig. 5.6 Graphical solutions of the dispersion relation $(\omega - \mathbf{k}.\mathbf{U})^2 = (gk + \frac{\gamma}{\rho}k^3)\tanh(kh)$. The conserved frequency ω is the *horizontal red dotted line*. The *green* (*blue*) *color* corresponds to the positive (negative) branches

Fig. 5.7 Phase-space U versus T for the deep water case including surface tension

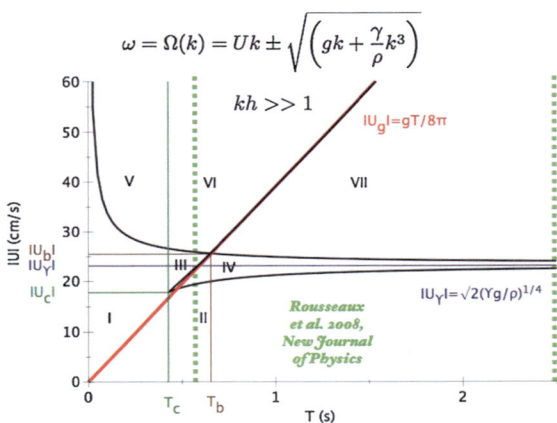

and:

$$U_\gamma = -\sqrt{2}\left(\frac{\gamma g}{\rho}\right)^{1/4}. \qquad (5.64)$$

A new horizon (in fact two) appears. Blue-shifted waves and negative energy waves can be reflected at a blue horizon and a negative horizon whose common asymptotic value is U_γ (see [23] for the details and the corresponding chapter in this book). Two maxima and a minimum appear in the graphical analysis of the dispersion relation (Fig. 5.6). A cusp where the white and blue horizons merge appears in the Phase-Space (Fig. 5.7).

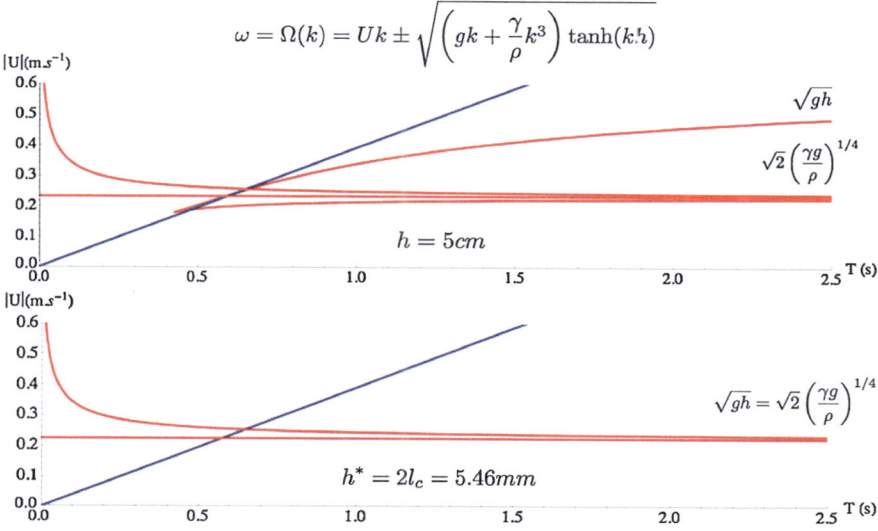

Fig. 5.8 Phase-spaces U versus T for a changing water depth including surface tension

When the water depth changes, the dispersion relation is either $(\omega - Uk)^2 \simeq c^2 k^2 \pm c^2 \xi^2 k^4$ for $kh \ll 1$ allowing dispersive corrections (only a negative horizon remains with the positive quartic correction) or $(\omega - \mathbf{k}.\mathbf{U})^2 = (gk + \frac{\gamma}{\rho} k^3) \tanh(kh)$ and three horizons are observed (Fig. 5.8). $h^* = 2l_c$ determines the transition depth between both behaviours.

- With vorticity. Similar arguments would lead to a new horizon replacing U_γ when including vorticity Ω with qualitatively the same behaviour in the limit $kh \gg 1$:

$$k_\Omega \approx \left(\frac{\rho \Omega^2}{\gamma}\right)^{1/3}, \quad (5.65)$$

and

$$U_\Omega \approx \left(\frac{\Omega \gamma}{\rho}\right)^{1/3}. \quad (5.66)$$

5.6.3 Natural and Artificial Horizons

In this part, we give some examples of water wave horizons. In hydrodynamics, white holes are more usual than black holes whose canonical example is the draining flow in the bathtub. A river mouth dying in the sea is a nice case of a natural white hole: the sea waves are blocked by the river flow. Figure 5.9 is an example found by the author when he used to walk on the Promenade des Anglais in Nice (France). This white hole inspired the following studies [21, 23, 24].

Fig. 5.9 A natural white hole of the French Riviera

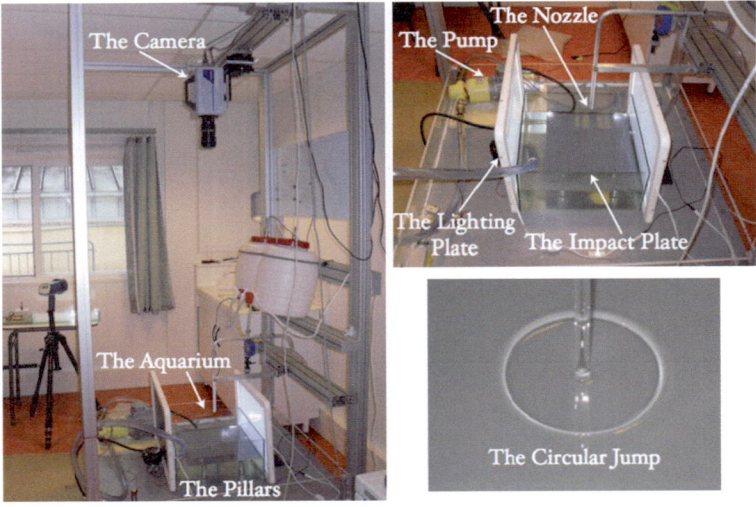

Fig. 5.10 A laboratory white hole in the kitchen sink

A more controlled white hole in the laboratory was suggested a few years ago by Volovik [26]: the circular jump in the kitchen sink (Fig. 5.10). We studied its related Mach cone and its dispersive properties in [25].

Fig. 5.11 A "biological" dispersive white hole

Recently, the author became aware of a biological-induced white hole with interesting dispersive properties, namely the whale fluke-print. As a whale swims or dives, it releases a vortex ring behind its fluke at each oscillation. The flow induced on the free surface is directed radially and forms a oval patch that gravity waves cannot enter whereas capillary waves are seen on its boundary (Fig. 5.11).

Dispersive and non-dispersive horizons are all encountered in nature and can be simulated in the laboratory. Dispersion has another intriguing consequence, namely the appearance of current-induced zero-frequency waves which appear spontaneously. These so-called "zero modes" have no counterpart in General Relativity so far...

5.6.4 Zero Modes

A flat interface can be considered as a wave with zero frequency $\omega = 0$ and zero wavenumber $k = 0$. When a spatially varying current is flowing under such an interface, the infinite wavelength of the interface can be reflected at a blocking line (creation of a group velocity horizon). This process produces a static ($\omega = 0$) jump through the interferences between the incident wave (flat surface with infinite wavelength $k = 0$) and the reflected one. This explains the formation of the circular [25, 26] and hydraulic [27] jumps. An undulation is observed which has a zero phase velocity but a non-zero group velocity and thus withdraws energy from the horizon towards infinity. When surface tension is present, the "gravity" jump is decorated by static capillary ripples inside the circular jump [28].

5.6.4.1 The Zero Mode (Static Undulation) for Gravity Waves

Two opposite wavenumbers are solutions of the dispersion relation for a zero frequency:

$$(0 - Uk)^2 = gk\tanh(kh). \tag{5.67}$$

Clearly, there is no threshold since there is always a solution whatever the velocity of the flow. Whatever the water depth, the slightest current flow induces a free surface deformation.

Let us take the extreme shallow waters limit ($kh \ll 1$):

$$U^2 k^2 \simeq ghk^2. \tag{5.68}$$

The threshold velocity for the zero mode appearance would correspond to:

$$U \simeq \sqrt{gh} = c_{phase} = c_{group}, \tag{5.69}$$

that is:

$$Fr = \frac{U}{\sqrt{gh}} = 1, \tag{5.70}$$

in terms of the dimensionless Froude number Fr. This latter constraint is well known in Hydraulics as the condition of appearance of the hydraulic jump when water flows over a bump:

$$Froude = Fr = \frac{U}{\sqrt{gh}} = \frac{U}{c_{phase}} = \frac{U}{c_{group}} = M = Mach, \tag{5.71}$$

which is similar to the supersonic-subsonic transition of air flows in aerodynamics described by the so-called Mach number M [25].

Then, one distinguishes in Hydraulics the following regimes:

- $Fr < 1$: (a) subcritical-to-subcritical flow over a bump. A group velocity horizon can appear but no phase velocity horizon [21]. No hydraulic jump is created but a static undulation is observed. The water depth decreases on average over the bump.
- $Fr > 1$: (b) supercritical-to-supercritical flow over a bump. The water depth increases on average over the bump.
- $Fr = 1$: (c) subcritical-to-supercritical flow over a bump. The group and phase velocity horizons are the same. A hydraulic jump appears as part of the static undulation [24, 27].

5.6.4.2 The Zero Mode (Static Undulation) for Capillo-Gravity Waves

If one takes into account the effect of surface tension:

$$(0 - Uk)^2 = \left(gk + \frac{\gamma}{\rho}k^3\right)\tanh(kh). \tag{5.72}$$

One is lead to the existence of a velocity threshold which corresponds to the minimum of the phase velocity with the wavenumber $U_\gamma = -\sqrt{2}(\frac{\gamma g}{\rho})^{1/4}$ [23].
For the case of the circular jump assuming $kh \ll 1$,

$$(0 - Uk)^2 = c^2 k^2 + \left(l_c^2 - \frac{h^2}{3}\right)k^4, \tag{5.73}$$

the following condition:

$$h < \sqrt{3}l_c \Rightarrow U > c = \sqrt{gh}, \tag{5.74}$$

implies the existence of static capillary undulations in the supersonic region of the circular jump [28].

We have seen how the dispersion relation explains the appearance of a horizon as well as the evolution of the wavelength of the converted modes. How does the amplitude of the modes evolve?

5.7 The "Norm"

> *If someone tells you that he knows what $E = \hbar\omega$ means,*
> *tell him that he is a liar.*
>
> Albert Einstein

In this final part, we will show how the so-called "norm" used by relativists in order to derive the Hawking spectrum is nothing else than the wave action, a pure classical concept.

In 1905, Albert Einstein pointed out that the four-momentum and the four-wave vector transform similarly under a Lorentz boost. This simple remark was fundamental in order to infer the existence of the light quantum whose energy is proportional to the frequency. The factor of proportionality was the Planck constant and Physicists soon realized that the latter constant of nature was measured in units of action. Quantum Mechanics then will shortly take its roots in Analytical Mechanics. In the famous 1911 Solvay conference [29], Lorentz wondered about the paradoxical behaviour of a harmonic oscillator like a pendulum whose frequency was made to change slowly with time by reducing its length. Indeed, the corresponding quantum behaviour of the oscillator would forbid a change in the quantum number describing the state of the oscillator since the frequency variation would not be high enough to allow transition to another state. Einstein pointed out that both the energy and the frequency of the pendulum would change with time but not their ratio as discovered by Rayleigh in 1902 [30]. Ehrenfest showed that the ratio of energy to frequency namely the action was an "adiabatic invariant". Adiabatic invariants of a given dynamical system are approximate constants of motion which are approximately preserved during a process where the parameters of the system change slowly on a time scale, which is supposed to be much larger than any typical dynamical time scale. They are the quantities to quantize when switching from Analytical Mechanics to Quantum Mechanics. A similar relation was discovered later by De Broglie between the momentum and the wavenumber. It should be borne in mind that the ratio of the action to the Planck constant is the number of photons which is another way to interpret the norm (as we will see) as the number of photons/phonons times the quantum of action in a quantum context.

Water waves are an example of a classical field. Thus, we can anticipate that the fluid system will have a corresponding wave action density defined as the ratio between a mean energy density (computed by averaging the instantaneous energy density on a spatial wave period) and the wave frequency. This action $J = E/\omega$ is assumed to be an adiabatic invariant (see [31] for a demonstration based on classical field theory). Then, if prime denotes a moving frame of reference with velocity $\mathbf{v}$, we must have $J' = J$ for a Galilean boost (recall that Einstein dealt with Lorentz transformations applied to light) that is [32, 33]:

$$\frac{E'}{\omega'} = \frac{E}{\omega}, \qquad (5.75)$$

which is valid if and only if we have the following transformations:

$$E' = E - \mathbf{v}.\mathbf{P}, \quad \mathbf{P}' = \mathbf{P}, \qquad (5.76)$$

and

$$\omega' = \omega - \mathbf{v}.\mathbf{k}, \quad \mathbf{k}' = \mathbf{k}. \qquad (5.77)$$

The latter formulae are just the usual Doppler effect whereas the former correspond to the change of energy/momentum for a classical wave and NOT a particle [34]. These would apply to a quasi-particle that is a collective excitation: phonon in acoustics or ripplons for water waves. In the following, momentum and energy would refer to quasi-momentum and quasi-energy if not specified.

It is now obvious that the energy in the moving frame will be given by $E' = E(1 - v/c_\phi)$ where $c_\phi = \omega/k$ is the phase velocity in the rest frame [32, 33]. In order to have negative energy waves ($E' < 0$), the so-called Landau criterion must be fulfilled $\omega - \mathbf{U}.\mathbf{k} < 0$ since the energy in the rest frame $E > 0$ is always positive. It is well known that superfluidity is lost when negative energy waves are created at the minimum of the roton spectrum [26]. It is similar to waves creation (Cerenkov-like effect) by an object in a flowing current U perforating the interface between water and air (capillary waves in the front and gravity waves in the rear) when $\min(c_\phi) < U$: the phase velocity $c_\phi = \omega/k$ features a minimum under which no waves are created (cf. Thomson and Helmholtz fishing line as described by Darrigol [7] and the corresponding chapter in this book). Let us recall that, in the direct space, the mean energy density (or pseudo-energy) for water waves (without any current) is proportional to the square of the amplitude $E = 1/2\rho g a^2$, where a is the amplitude of the wave [4].

Here, we must be careful when we want to evaluate the wave energy in the moving frame of the current because the velocity of the flow is changing with space. Hence, every part of the water waves wavelength will be "desynchronized" by the spatial-dependent Doppler effect due to the current. That is why Weinfurtner et al. [24] introduced a time shift $t_c = \int \frac{dx}{U(x)}$ which is reminiscent of Carroll kinematics for classical waves $t' = t - vx/c^2$ and $x' = x$ or $t' = t - x/V_0$ where $V_0 = c^2/v$ is the dual velocity associated with the wavefront [34]. The Carrollian time shift writes $dt' = dt - dx/V_0$ in differential form where V_0 is now a function of space x in the water waves problem. The t_c coordinate has dimension of time and its associated "wavenumber" f_c has units of a frequency. The usual convective derivative operator $\partial_t + U(x)\partial_x$ becomes $\partial_t + \partial_{t_c}$ and in Fourier transform space $f + f_c$. Then, when analysing data in the Carrollian coordinate system, the amplitude of the wave η is a function of both the normal time t and the Carrollian time t_c that is in the Fourier transform space $\tilde{\eta}(f, f_c)$. The wave action density in the Fourier space $\hat{J}_{wave}$ is by definition the integral on the different Carrollian times of the ratio between the Fourier transform of the wave energy density (E) and the Fourier transform of the relative angular frequency (ω'):

$$\hat{J}_{wave} = \int \frac{|\tilde{\eta}(f, f_c)|^2}{f + f_c} df_c. \tag{5.78}$$

The expression of J_{wave} is similar to the Zeldovich formula for the number of photons N when dealing with plane electromagnetic waves that are not monochromatic ($\omega = \pm c_L |\mathbf{k}|$):

$$N = \frac{1}{\hbar} \frac{1}{8\pi} \int d^3k \frac{|\hat{\mathbf{E}}(\mathbf{k}, t)|^2 + |\hat{\mathbf{B}}(\mathbf{k}, t)|^2}{\pm c_L |\mathbf{k}|}, \tag{5.79}$$

where c_L is the light velocity and $\hat{\mathbf{E}}(\mathbf{k}, t)$, $\hat{\mathbf{B}}(\mathbf{k}, t)$ are the Fourier transforms of the electric and magnetic fields [35]. The sign in the denominator comes from the dispersion relation of light which features both positive and negative branches. The Zeldovich formula and the "norm" used in [24] writes as the ratio between a wave energy (which scales with the square of an amplitude) and the wave frequency. The case of acoustics is discussed in [36] following the treatment by Landau and Lifschitz [37].

The equivalence between the norm and the wave action density can be formally proven as follows. First, wave packets on the free surface of water obey the Beltrami-Laplace equation in the Painlevé-Gullstrand metric as shown by Schutzhold and Unruh in 2002 (see [8] and the corresponding chapter in this book):

$$\partial_t(\partial_t\phi + U\partial_x\phi) + \partial_x(U\partial_t\phi + U^2\partial_x\phi) - c^2\partial_x^2\phi = 0, \tag{5.80}$$

where ϕ is the velocity potential fluctuation. The complete velocity potential featuring both the waves and the background flow U is such that its space derivative is by definition the flow velocity.

We can expect the conservation of two quantities due to the invariance of the corresponding action under (1) the transformation $\phi \to e^{i\alpha}\phi$, α constant, and (2) time translation (for time-independent U). The former invariance gives conservation of the Klein-Gordon norm (as demonstrated elsewhere in this book):

$$N = \frac{i}{2c^2} \int_{-\infty}^{\infty} dx \left[\phi^*(\partial_t \phi + U\partial_x \phi) - \phi(\partial_t \phi^* + U\partial_x \phi^*)\right], \tag{5.81}$$

whereas the latter gives conservation of (pseudo-)energy. For wave packets confined to a region where the flow velocity U is constant, the norm (5.81) can be written in k-space in terms of the Fourier transform $\tilde{\phi}(k)$ as:

$$N = \frac{1}{c^2} \int_{-\infty}^{\infty} dk (\omega - Uk) |\tilde{\phi}(k)|^2. \tag{5.82}$$

The typical interpretation of the Zeldovich formula is that it is a positive quantity: the number of photons. However, the Klein-Gordon norm used but the relativists is either positive or negative. Then, the Zeldovich formula encodes in general a different information than the Klein-Gordon norm. The latter counts the amount of charge, that is why for real fields it is zero. The complex solutions however do have charge. In fact, strictly speaking, the last equation is not correct as for a single k there can be modes with $\pm|k|$. Thus, apart from the integral there should be a sum in positive/negative branches:

$$N = \frac{1}{c^2} \int dk \left(c|k||a_k|^2 - c|k||b_k|^2\right), \tag{5.83}$$

with a_k, b_k the corresponding Fourier coefficients.

The Zeldovich formula (without the negative sign in the denominator) does not give zero for real fields. It really corresponds to $\frac{1}{c^2} \int dk(c|k||a_k|^2)$. It does not contain the second term that in the case of real fields ($a_k = b_k$), would combine to yield a total zero (this is the case also for photons). The point is that what corresponds to the norm is not one of the pieces individually but the addition of the two.

In general, the norm scales like the integral over the wavenumber of the amplitude square of the Fourier transformed velocity potential times the relative frequency in the moving frame. Hence, the norm scaling is $N \approx \int dk(\omega - Uk)\tilde{\phi}^2$ [24]. However, because the velocity potential is related to the free surface deformation η by the Bernoulli equation $\partial \phi/\partial t + g\eta = 0$ (here, without a flow to simplify), it follows that the velocity potential scales like $\tilde{\phi} \approx g\tilde{\eta}/(\omega - Uk)$ in the Fourier space [8]. We conclude that the norm behaves like $N \approx \int dk g^2 \tilde{\eta}^2/(\omega - Uk)$ as the wave action that is as the ratio between the square of the amplitude (the energy) and the relative frequency.

The norm is strictly conserved. Is this the case for the wave action? In the fluid mechanics literature, the wave action is the solution of a conservation equation which replaces obviously the conservation of energy of a closed system. Here the system is open since the waves interact with the flow and do exchange energy. Bretherton and Garrett have shown that the wave action conservation writes in the so-called WKBJ regime where the flow velocity varies on a length scale much larger that the wavelength [38, 39]:

$$\frac{\partial}{\partial t}\left(\frac{E'}{\omega'}\right) + \nabla \cdot \left(c_g \frac{E'}{\omega'}\right) = 0, \tag{5.84}$$

5 The Basics of Water Waves Theory for Analogue Gravity

where $\omega' = \omega - Uk$ and c_g is the total group velocity including the background flow. According to Bretherton and Garrett, *"because E' is an energy density, it is not constant down a ray, even if wave energy is conserved. However, in a time dependent and/or non-uniformly moving medium, ω' varies along a ray. If E'/ω' is the wave action density, total wave action is conserved, whereas total wave energy is not"*. For a stationary process, we deduce that $\eta^2 c_g/(\omega - kU) = \text{const}$ since the energy density in the moving E' is proportional to the square of the interface deformation η (as in the rest frame without current). As a consequence, the amplitude diverges to infinity if one gets close to a turning point where the group velocity vanishes and where the WKBJ approximation is no longer valid. Dispersion enters the game to avoid such a caustic.

The change of wave action $J = E/\omega$ of a slowly modulated oscillator is exponentially small in the non-adiabatic parameter (ω/(rate of change of the medium properties)): a mathematical theorem due to Meyer in 1973 [40]. For the linear pendulum of Rayleigh with a varying length, the rate of change is directly the inverse of the time lapse. Here, with water waves on a non-uniform flow, the property is the velocity U and its typical rate of change is its space gradient (the so-called surface gravity in General Relativity) whose dimension is the one of a frequency: Jacobson and Parentani defined the surface gravity as a local expansion rate seen by a freely falling observer when he crosses the horizon [41]. One is tempted to extrapolate the following behaviour for the change of wave action as the waves propagate against the flow:

$$\Delta J = J_0 \exp\left(-\text{constant}\frac{\omega}{\kappa}\right), \quad (5.85)$$

where $\kappa = dU/dx$ is the surface gravity for water waves.

The change of wave action would be very similar to the famous Hawking spectrum [2]:

$$\frac{\beta^2}{\alpha^2} = \exp\left(-2\pi\frac{\omega}{\kappa}\right). \quad (5.86)$$

The fact that the Bogoliubov coefficient behaves as an exponential has been discussed by Jacobson [42].

Let us introduce the following dimensionless numbers ω / (rate of change of the medium properties) with names of distinguished physicists:

$$\frac{\omega}{\kappa} = \text{Hawking number} = \mathcal{H}_w, \quad (5.87)$$

$$\frac{\omega}{\frac{dU}{dx}} = \text{Unruh number} = \mathcal{U}_n. \quad (5.88)$$

The validity of the WKBJ inequality $\frac{U}{\frac{dU}{dx}} \gg \lambda$ can be reassessed thanks to the Unruh number close to the horizon. As a matter of fact, $U^* \approx g/\omega$ and $\lambda^* \approx g/\omega^2$ then $\mathcal{U}_n \simeq O(1)$: close to the horizon, the WKBJ approximation breaks down. If $\mathcal{U}_n \gg 1$, then the process is adiabatic. In order not to have a vanishing spectrum, $\mathcal{U}_n \approx O(1)$, then vacuum radiation à la Hawking-Unruh is a non-adiabatic process

[43]. The case $\mathscr{U}_n \ll 1$ would imply a too small frequency, hence the amplitude of the energy spectrum (which scales with the cube of the frequency in 3D) would vanish.

5.8 Conclusion

This rapid tour of the field of analogue gravity through the prism of water waves theory has broadened our definition of a horizon and has deepened our understanding of the concept of norm as used by relativists. In a related chapter of this book, we study experimentally the influence of surface tension and the associated dispersive horizons. Moreover, we try to answer to the question "what is a particle close to a horizon?".

Acknowledgements I would like to thank Thomas Philbin, Gil Jannes, Carlos Barcelo and Iacopo Carusotto for very interesting remarks which improve the content of this chapter.

References

1. Schützhold, R.: Emergent horizons in the laboratory. Class. Quantum Gravity **25**, 114011 (2008)
2. Barcelo, C., Liberati, S., Visser, M.: Analogue gravity. Living Rev. Relativ. **8**, 12 (2011)
3. Barenblatt, G.I.: Scaling. Cambridge University Press, Cambridge (2003)
4. Mei, C.C., Stiassnie, M., Yue, D.K.P.: Theory and Applications of Ocean Surface Waves: Part I, Linear Aspects; Part II, Nonlinear Aspects. World Scientific, Singapore (2005)
5. Dingemans, M.W.: Water Wave Propagation over Uneven Bottoms. World Scientific, Singapore (1997)
6. Schaffer, S.: John Michell and black holes. J. Hist. Astron. **10**, 42–43 (1979)
7. Darrigol, O.: Worlds of Flow: a History of Hydrodynamics from the Bernoullis to Prandtl. Oxford University Press, Oxford (2005)
8. Schützhold, R., Unruh, W.G.: Gravity wave analogues of black holes. Phys. Rev. D **66**, 044019 (2002)
9. Unruh, W.G.: Experimental black-hole evaporation? Phys. Rev. Lett. **46**, 1351–1353 (1981)
10. Peregrine, D.H.: Interaction of water waves and currents. Adv. Appl. Mech. **16**, 9–117 (1976)
11. Hedges, T.S.: Combinations of waves and currents: an introduction. Proc., Inst. Civ. Eng. **82**, 567–585 (1987)
12. Jonsson, I.G.: Wave-current interactions. In: Le Mehaute, B., Hanes, D.M. (eds.) The Sea, pp. 65–120. Wiley, New York (1990)
13. Thomas, G.P., Klopman, G.: Wave-current interactions in the nearshore region. In: Hunt, J.N. (ed.) Gravity Waves in Water of Finite Depth. Advances in Fluid Mechanics, vol. 10. CMP, Southampton (1997)
14. Fabrikant, A.L., Stepanyants, Y.A.: Propagation of Waves in Shear Flows. World Scientific, Singapore (1998)
15. Lavrenov, I.: Wind-Waves in Oceans. Springer, Berlin (2003)
16. Umeyama, M.: Coupled PIV and PTV measurements of particle velocities and trajectories for surface waves following a steady current. J. Waterw. Port Coast. Ocean Eng. **137**(2), 85–94 (2011)

17. Thompson, P.D.: The propagation of small surface disturbance through rotational flow. Ann. N.Y. Acad. Sci. **5**, 463–474 (1949)
18. Biesel, F.: Etude théorique de la houle en eau courante. Houille Blanche **5**, 279–285 (1950)
19. Huang, H.: Linear surface capillary-gravity short-crested waves on a current. Chin. Sci. Bull. **53**, 3267 (2008)
20. Choi, W.: Nonlinear surface waves interacting with a linear shear current. Math. Comput. Simul. **80**, 29–36 (2009)
21. Rousseaux, G., Mathis, C., Maïssa, P., Philbin, T.G., Leonhardt, U.: Observation of negative phase velocity waves in a water tank: a classical analogue to the Hawking effect? New J. Phys. **10**, 053015 (2008)
22. Nardin, J.-C., Rousseaux, G., Coullet, P.: Wave-current interaction as a spatial dynamical system: analogies with rainbow and black hole physics. Phys. Rev. Lett. **102**, 124504-1/4 (2009)
23. Rousseaux, G., Maïssa, P., Mathis, C., Coullet, P., Philbin, T.G., Leonhardt, U.: Horizon effects with surface waves on moving water. New J. Phys. **12**, 095018 (2010)
24. Weinfurtner, S., Tedford, E.W., Penrice, M.C.J., Unruh, W.G., Lawrence, G.A.: Measurement of stimulated Hawking emission in an analogue system. Phys. Rev. Lett. **106**, 021302 (2011)
25. Jannes, G., Piquet, R., Maïssa, P., Mathis, C., Rousseaux, G.: Experimental demonstration of the supersonic-subsonic bifurcation in the circular jump: a hydrodynamic white hole. Phys. Rev. E **83**(5), 056312 (2011)
26. Volovik, G.E.: Horizons and ergoregions in superfluids. J. Low Temp. Phys. **145**, 337–356 (2006)
27. Unruh, W.G.: Dumb holes: analogues for black holes. Philos. Trans. R. Soc. Lond. A **366**, 2905–2913 (2008)
28. Rolley, E., Guthmann, C., Petersen, M.S.: Hydraulic jump and ripples in liquid helium-4. Physica B **394**, 46–55 (2007)
29. Langevin, P., de Broglie, M.: La Théorie du Rayonnement et des Quanta, Rapports et Discussions de la Réunion "Solvay" de 1911. Gauthier-Villars, Paris (1912)
30. Rayleigh, J.W.S.: On the pressure of vibrations. Philos. Mag. **3**, 338–346 (1902)
31. Sturrock, P.A.: Field-theory analogs of the Lagrange and Poincaré invariants. J. Math. Phys. **3**, 43 (1962)
32. Sturrock, P.A.: In what sense do slow waves carry negative energy? J. Appl. Phys. **31**, 2052 (1960)
33. Sturrock, P.A.: Energy-momentum tensor for plane waves. Phys. Rev. **121**, 18–19 (1961)
34. Houlrik, J.M., Rousseaux, G.: Non-relativistic kinematics: particles or waves? arXiv: 1005.1762
35. Avron, J.E., Berg, E., Goldsmith, D., Gordon, A.: Is the number of photons a classical invariant? Eur. J. Phys. **20**, 153–159 (1999)
36. Stone, M.: Acoustic energy and momentum in a moving medium. Phys. Rev. E **62**, 1341–1350 (2000)
37. Landau, L.D., Lifshitz, E.M.: Fluid Mechanics, vol. 6, 2nd edn. Butterworth-Heinemann, Stoneham (1987)
38. Bretherton, F.P., Garrett, C.J.R.: Wavetrains in inhomogeneous moving media. Proc. R. Soc. Lond. A **302**, 529–554 (1968)
39. Jonsson, I.G.: Energy flux and wave action in gravity waves propagating on a current. J. Hydraul. Res. **16**(3), 223–234 (1978)
40. Meyer, R.E.: Adiabatic variation, part II: action change for the simple oscillator. Z. Angew. Math. Phys. **24**, 517–524 (1973)
41. Jacobson, T., Parentani, R.: Horizon surface gravity as 2D geodesic expansion. Class. Quantum Gravity **25**, 195009 (2008)
42. Jacobson, T.: Introduction to quantum fields in curved spacetime and the Hawking effect. Lectures given at the CECS School on Quantum Gravity in Valdivia, Chile, January 2002. Available on arXiv:gr-qc/0308048v3
43. Massar, S., Parentani, R.: Particle creation and non-adiabatic transitions in quantum cosmology. Nucl. Phys. B **513**(1–2), 375–401 (1998)

Chapter 6
The Čerenkov Effect Revisited: From Swimming Ducks to Zero Modes in Gravitational Analogues

Iacopo Carusotto and Germain Rousseaux

Abstract We present an interdisciplinary review of the generalized Čerenkov emission of radiation from uniformly moving sources in the different contexts of classical electromagnetism, superfluid hydrodynamics, and classical hydrodynamics. The details of each specific physical systems enter our theory via the dispersion law of the excitations. A geometrical recipe to obtain the emission patterns in both real and wave-vector space from the geometrical shape of the dispersion law is discussed and applied to a number of cases of current experimental interest. Some consequences of these emission processes onto the stability of condensed-matter analogues of gravitational systems are finally illustrated.

6.1 Introduction

The emission of radiation by a uniformly moving source is a widely used paradigm in field theories to describe a number of very different effects, from the wake generated by a swimming duck on the surface of a quiet lake [1–12], to the Čerenkov emission by a charged particle relativistically moving through a dielectric medium [13], to the sound waves emitted by an object travelling across a fluid or a superfluid at super-sonic speed [14–18]. Of course, the radiated field has a different physical nature in each case, consisting e.g. of gravity or capillary waves at the water/air interface, or electromagnetic waves in a dielectric medium, or Bogoliubov excitations in the superfluid. In spite of this, the basic qualitative features of the emission process are very similar in all cases and a unitary discussion is possible.

In the present chapter, we shall present an interdisciplinary review of this *generalized Čerenkov effect* from the various points of view of classical electromagnetism,

---

I. Carusotto (✉)
INO-CNR BEC Center and Dipartimento di Fisica, Università di Trento, 38123 Povo, Italy
e-mail: carusott@science.unitn.it

G. Rousseaux
Laboratoire J.-A. Dieudonné, UMR CNRS-UNS 6621, Université de Nice-Sophia Antipolis, Parc Valrose, 06108 Nice Cedex 02, France
e-mail: Germain.Rousseaux@unice.fr

superfluid hydrodynamics, and classical hydrodynamics. The details of each specific physical systems enter our theory via the dispersion law $\Omega(\mathbf{k})$ of its excitations. In particular, the emission patterns in both real and wave-vector space can be extracted from the geometrical shape of the intersection of the $\Omega(\mathbf{k})$ dispersion law with the $\Omega = \mathbf{k} \cdot \mathbf{v}$ hyper-plane that encodes energy-momentum conservation. Once the dispersion law of a generic system is known, our geometrical algorithm provides an efficient tool to obtain the most significant qualitative features of wake pattern in a straightforward and physically transparent way.

In the last years, the interest of the scientific community on this classical problem of wave theory has been revived by several experiments which have started exploring the peculiar features that appear in new configurations made accessible by the last technological developments, e.g. the Čerenkov emission of electromagnetic radiation in resonant media [19, 20] and the Bogoliubov-Čerenkov emission of sound waves in bulk dilute superfluids [16–18]. Another reason for this renewed interest comes from the condensed-matter models of gravitational systems that are the central subject of the present book. In many of such analogue models, the presence of a horizon may be responsible for the emission of waves from the horizon by generalized Čerenkov processes. A full understanding of these classical effects is then required if one is to isolate quantum features such as the analogues of Hawking radiation, dynamical Casimir emission and anomalous Doppler effect.

The structure of the chapter is the following. In Sect. 6.2, we shall introduce the general field-theoretical formalism to calculate the real and momentum space emission patterns and the geometrical construction to obtain qualitative information on them. These methods will then be applied in the following sections to a few different systems of current interest. As a first example, in Sect. 6.3 we will review the main features of the Čerenkov emission of electromagnetic waves from relativistically moving charges in a dielectric medium. We shall restrict our attention to the simplest case of a non-dispersive dielectric with frequency-independent refractive index n, where the phase and group velocities are equal and constant. In this case, a Čerenkov emission takes place as soon as the charge speed exceeds the velocity of light in the medium $c = c_0/n$. Modern developments for the case of a strongly dispersive media [19–21], photonic crystals, and left-handed metamaterials [22] will be briefly mentioned. In Sect. 6.4 we shall review the emission of sound waves by a supersonically moving impurity in the bulk of a dilute superfluid. In addition to the Mach cone that appears in the wake behind the object, the presence of single-particle excitations in the excitation spectrum is responsible for the appearance of a series of curved wavefronts ahead of the impurity. On the other hand, a subsonically moving impurity will produce no propagating wave and the perturbation will remain localized in its vicinity: the resulting frictionless motion is one of the clearest examples of the class of phenomena that go under the name of superfluidity [24–26]. This physics is currently of high experimental relevance, as first real-space images of the density perturbation pattern induced by a moving impurity have been recently obtained using Bose-Einstein condensates of ultracold atoms [17] and of exciton-polaritons in semiconductor microcavities [16]. The case of a parabolic dispersion will be presented in Sect. 6.5: this specific functional form allows for an elementary analytical treatment of the wake pattern in both real and momentum space. On one hand, this discussion provides a useful guideline to understand the qualitative shape of the wake in superfluids and in surface waves. On the other hand, it is of central importance in view of the experimental realization of gravitational analogues based on magnon excitations in magnetic solids [27]. The physics of a material object such as a boat, a duck or a fishing line creating surface waves on the air/water interface of a lake will be considered in Sect. 6.6: not only does this example provide the most intuitive example of the generalized Čerenkov effect, but is perhaps also the richest one in terms of different behaviours that can be observed depending on the system parameters, e.g. the velocity of the object with respect to the fluid, the depth of the water, the surface tension of the fluid [1–12]. The concepts that have been laid down so far are finally applied in Sect. 6.7 to analogue models of gravity based on flowing superfluids or surface waves on flowing water. In both these cases, classical Čerenkov emission into the so-called *zero modes* at the horizon may disturb detection of the analogue Hawking radiation as well as affect the dynamical stability of the analogue black/white hole [28]. Conclusions are finally drawn in Sect. 6.8.

6.2 Generic Model

In this section, we introduce the generic model that will be used to study the different physical systems in the following sections. The model is based on a linear partial differential equation for a scalar $\mathbb{C}$-number field $\phi(\mathbf{r}, t)$: in most relevant cases, the multi-component physical field (i.e. the vector electromagnetic field or the Bogoliubov spinor) can in fact be reduced to a single scalar field upon straightforward algebraic manipulations under controlled approximations. We are also assuming that quantum fluctuations of the field $\phi(\mathbf{r}, t)$ can be fully neglected. The geometry under investigations consists of a spatially homogeneous system interacting with a spatially localized moving source describing the moving electric charge, or the interaction potential of the moving impurity with the fluid, or the extra pressure exerted on the fluid surface by the moving object. In this geometry, the microscopic information on the field dynamics is summarized in the dispersion law relating the frequency Ω of a plane wave to its wave-vector $\mathbf{k}$: different forms of dispersion laws corresponding to first- or higher-order partial differential equations are discussed in the subsections Sects. 6.2.1 and 6.2.3. The geometric construction of the wake pattern starting from the dispersion law $\Omega(\mathbf{k})$ is discussed in Sect. 6.2.2.

6.2.1 The Wave Equation and the Source Term

We start by considering a generic, d-dimensional classical complex field $\phi(\mathbf{r}, t)$ ($d = 2$ in the figures) that evolves according to the generic linear, first-order in time, partial differential equation:

$$i\partial_t \phi(\mathbf{r}, t) = \Omega(-i\nabla_\mathbf{r})\phi(\mathbf{r}, t) + S(\mathbf{r}, t) \qquad (6.1)$$

with a source term $S(\mathbf{r}, t)$.

The function $\Omega(\mathbf{k})$ defines the so-called dispersion law for free field propagation, that is the frequency of the plane wave solutions

$$\phi(\mathbf{r},t) = \phi_0 e^{i\mathbf{k}\mathbf{r}} e^{-i\Omega(\mathbf{k})t} \qquad (6.2)$$

as a function of the wave-vector $\mathbf{k}$ in the absence of sources, $S(\mathbf{r},t) = 0$.

Throughout this chapter we shall consider a spatially localized and uniformly moving source term of the form

$$S(\mathbf{r},t) = S_0(\mathbf{r} - \mathbf{v}t), \qquad (6.3)$$

with a spatial profile $S_0(\mathbf{r})$ concentrated in the vicinity of $\mathbf{r} = 0$ and moving at a speed $\mathbf{v}$.

Thanks to the translational invariance of the free field problem in both space and time, solution of the full wave equation (6.1) in the presence of the source term is easily obtained in Fourier space with respect to both space and time. Defining the Fourier transform in the usual way

$$\tilde{\phi}(\mathbf{k},\omega) = \int dt \int d^d\mathbf{r}\, \phi(\mathbf{r},t) e^{-i\mathbf{k}\cdot\mathbf{r}} e^{i\omega t}, \qquad (6.4)$$

the source term in Fourier space has the simple form

$$\tilde{S}(\mathbf{k},\omega) = 2\pi \tilde{S}_0(\mathbf{k})\delta(\omega - \mathbf{k}\cdot\mathbf{v}) \qquad (6.5)$$

in terms of the structure factor $\tilde{S}_0(\mathbf{k})$ defined as the Fourier transform of the source shape $S_0(\mathbf{r})$.

In Fourier space, the solution of (6.1) is then

$$\tilde{\phi}(\mathbf{k},\omega) = \frac{2\pi \tilde{S}_0(\mathbf{k})\delta(\omega - \mathbf{k}\cdot\mathbf{v})}{\omega - \Omega(\mathbf{k}) + i0^+}, \qquad (6.6)$$

where an infinitesimally small imaginary part is introduced in the denominator of (6.6) to specify the integration contour to be followed around the poles and, in this way, ensure causality of the solution. This trick dates back to Rayleigh [4, 5] and is equivalent to a infinitesimal shift of the dispersion law into the lower half-space, $\Omega(\mathbf{k}) \to \Omega(\mathbf{k}) - i0^+$. Physically, it corresponds to introducing a very weak damping of the plane wave solutions in time,

$$\phi(\mathbf{r},t) = \phi_0 e^{i\mathbf{k}\mathbf{r}} e^{-i\Omega(\mathbf{k})t} e^{-0^+ t} \qquad (6.7)$$

or to assume that the source term is slowly switched on in time [11].

The real-space pattern is obtained by an inverse Fourier transform of (6.6),

$$\phi(\mathbf{r},t) = -\int \frac{d^d\mathbf{k}}{(2\pi)^d} \frac{\tilde{S}_0(\mathbf{k}) e^{i\mathbf{k}(\mathbf{r}-\mathbf{v}t)}}{\Omega(\mathbf{k}) - \mathbf{k}\cdot\mathbf{v} - i0^+} = \phi(\mathbf{r} - \mathbf{v}t). \qquad (6.8)$$

Thanks to the $\delta(\omega - \mathbf{k}\cdot\mathbf{v})$ factor in (6.6), this expression only depends on the combination $\mathbf{r}' = \mathbf{r} - \mathbf{v}t$: as expected, the wake pattern is rigidly moving at the speed of the source. Within Galilean invariance, the $\mathbf{r}'$ coordinate corresponds to the spatial coordinate in the reference frame of the source in motion at velocity $\mathbf{v}$.

Evaluation of (6.8) can be performed with standard numerical tools. The result for some most interesting cases will be presented in the next sections. Now, we shall rather proceed with some analytical manipulations of (6.8) that allow to extract qualitative information on the emitted field pattern from the dispersion law $\Omega(\mathbf{k})$. The first step in this direction is to note that the integral in (6.8) is dominated by those $\mathbf{k}$ values for which the resonant denominator vanishes, that is

$$\Omega(\mathbf{k}) = \mathbf{k} \cdot \mathbf{v}. \tag{6.9}$$

This equation recovers the standard Čerenkov condition for emission of radiation [13] and can be geometrically interpreted as the intersection of the dispersion surface $\Omega(\mathbf{k})$ with the $\Omega = \mathbf{k} \cdot \mathbf{v}$ plane. In a quantum description of the Čerenkov emission by a massive charged particle, the condition (6.9) naturally appears when energy-momentum conservation is imposed to the photon emission process [13]. When reformulated in the reference frame of the moving source, the condition (6.9) reduces to $\Omega' = \gamma(\Omega - \mathbf{k} \cdot \mathbf{v}) = 0$, meaning that the perturbation pattern around the source at rest is stationary in time in the moving reference frame.

The locus Σ of $\mathbf{k} \neq 0$ modes that satisfy (6.9) is a central object in all the following discussion as it defines the modes in $\mathbf{k}$ space into which the Čerenkov emission will be peaked.[1] In particular, no emission of propagating waves takes place if the locus Σ is empty; the non-resonant contributions to (6.8) only provide a non-radiative perturbation that remains spatially localized in the close vicinity of the source and is not able to transport energy away. In spite of this, the momentum and energy that are stored in the localized moving pattern of the field ϕ are responsible for a sizeable renormalization of the particle mass [11, 30, 31].

6.2.2 Qualitative Geometrical Study of the Wake Pattern

Let us consider a generic point $\mathbf{k}_0 \in \Sigma$. Within a neighbourhood of $\mathbf{k}_0$, we introduce a new set of $\mathbf{k}$-space coordinates defined as follows: for each point $\mathbf{q}$, q_n is the distance of $\mathbf{k}$ from the Σ surface and the position of the closest point on the surface Σ is parametrized by the $(d-1)$-dimensional $\mathbf{q}_\parallel$ curvilinear coordinate system. A sketch of this coordinate system is indicated as a grid in Fig. 6.1(a).

In this new coordinate system, the Fourier integral giving the emitted field pattern in real space can be approximately rewritten as:

$$\phi(\mathbf{r}') = -\tilde{S}_0(\mathbf{k}_0) \int_\Sigma \frac{d^{d-1}\mathbf{q}_\parallel}{(2\pi)^{d-1}} e^{i\bar{\mathbf{k}}(\mathbf{q}_\parallel)\cdot\mathbf{r}'} \int \frac{dq_n}{2\pi} \frac{e^{iq_n\hat{\mathbf{n}}\cdot\mathbf{r}'}}{v'_g q_n - i0^+}, \tag{6.10}$$

where $\bar{\mathbf{k}}(\mathbf{q}_\parallel)$ is the position of the point on the surface Σ corresponding to the value $\mathbf{q}_\parallel$ of the $(d-1)$-dimensional coordinate and $\hat{\mathbf{n}}$ is the unit vector normal to the

---

[1] The $\mathbf{k} = 0$ mode corresponds to a spatially constant modulation that does not transport energy nor momentum. As discussed in [29], many other interesting features of wave propagation can be graphically studied starting from iso-frequency surfaces analogous to the locus Σ.

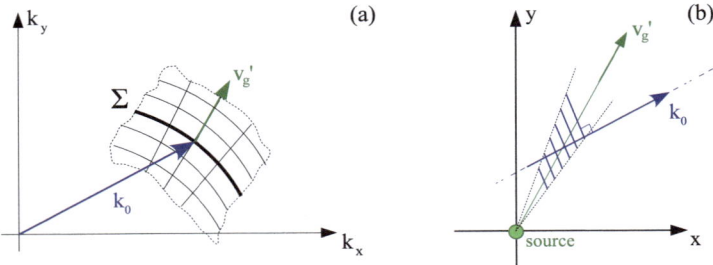

Fig. 6.1 (a) k-space sketch of a patch of the locus Σ around the wave-vector $\mathbf{k}_0$. The grid shows the $(\mathbf{q}_\parallel, q_n)$ coordinate system used in the geometrical construction of the wake pattern. (b) Sketch of the region of the wake pattern generated by the emission in the neighbourhood of $\mathbf{k}_0$: the *blue fringes* have wave-vector $\mathbf{k}_0$, the direction of propagation $\mathbf{v}'_g$ is determined by the normal to the locus Σ at the point $\mathbf{k}_0$

surface at $\mathbf{k}_0$ in the direction of growing $\Omega(\mathbf{k}) - \mathbf{k} \cdot \mathbf{v}$. As the surface Σ is defined by the zeros of $\Omega(\mathbf{k}) - \mathbf{k} \cdot \mathbf{v}$, the velocity

$$\mathbf{v}'_g = v'_g \hat{\mathbf{n}} = \nabla_\mathbf{k}[\Omega(\mathbf{k}) - \mathbf{k} \cdot \mathbf{v}] = \mathbf{v}_g - \mathbf{v} \qquad (6.11)$$

is directed along the normal $\hat{\mathbf{n}}$ and corresponds to the group velocity of the wave, as measured relative to the moving source at $\mathbf{v}$. For a non-relativistic source speed $v \ll c_0$, it can be interpreted as the group velocity observed from the source reference frame.

The integral over q_n can be performed by closing the contour on the complex plane. The only pole is located slightly above the real axis. Depending on the sign of $\hat{\mathbf{n}} \cdot \mathbf{r}'$, the contour has to be closed in the upper or lower half plane, which gives

$$\phi(\mathbf{r}') = -i S_0(\mathbf{k}_0) \int_\Sigma \frac{d^{d-1}\mathbf{q}_\parallel}{(2\pi)^{d-1}} \frac{e^{i\bar{\mathbf{k}}(\mathbf{q}_\parallel)\cdot\mathbf{r}'}}{v'_g} \Theta[\mathbf{v}'_g \cdot \mathbf{r}']. \qquad (6.12)$$

The expression (6.12) can be further simplified by performing the so-called stationary phase approximation, as first proposed by Thomson [4, 5]. For each value $\mathbf{r}'$ of the relative coordinate, the integral over $\mathbf{q}_\parallel$ is dominated by those points for which the phase is stationary, i.e. the variation of $\bar{\mathbf{k}}(\mathbf{q}_\parallel)$ on $\mathbf{q}_\parallel$ is orthogonal to $\mathbf{r}'$. In combination with the Heaviside-Θ function in (6.12), this is equivalent to requiring that the vector $\mathbf{r}'$ is parallel to the normal $\hat{\mathbf{n}}$ to the surface Σ at point $\mathbf{k}_0$ in the direction of growing $\Omega(\mathbf{k}) - \mathbf{k} \cdot \mathbf{v}$, i.e. parallel to the relative group velocity $\mathbf{v}'_g$.

For a generic relative position $\bar{\mathbf{r}}'$, there are only a few discrete points $\mathbf{k}_j$ on Σ such that this condition is met. As a consequence, for generic values of $\mathbf{r}'$ in a neighbourhood of $\bar{\mathbf{r}}'$, one can approximately write

$$\phi(\mathbf{r}') \approx -i \frac{S_0(\mathbf{k}_j)}{(2\pi)^{d-1}} \sum_j \frac{\Delta k_j^2}{v'_{g,j}} e^{i\mathbf{k}_j \cdot \mathbf{r}'}, \qquad (6.13)$$

where the sum is over the allowed $\mathbf{k}_j$ vectors: the numerical coefficient Δk_j^2 is inversely proportional to the curvature of Σ at $\mathbf{k}_j$ and $\mathbf{v}'_{g,j}$ is the group velocity of the $\mathbf{k}_j$ mode.

A physical understanding of this result can be easily obtained by looking at the diagram of Fig. 6.1(b). Within Galilean invariance, sitting in the moving reference frame of the source may facilitate building an intuitive picture of the emission process: every point on the surface Σ corresponds to a continuous plane wave of wave-vector $\mathbf{k}_0$ that is emitted from the source and propagates away from it at a group velocity $\mathbf{v}'_g$ (indicated by the green arrow in the figure). As a result, it is able to reach all points $\mathbf{r}'$ that lie in the vicinity of the straight line of direction $\mathbf{v}'_g$. While the group velocity $\mathbf{v}'_g$ is always along the radial direction, the wave-vector $\mathbf{k}_0$ (blue arrow in the figure) can have arbitrary direction: as a result, the wave-fronts (indicated by the blue fringes) are generally tilted and the emission pattern does not necessarily resemble a spherical wave. Of course, all this reasoning can be performed equally well in the laboratory frame if $\mathbf{v}'_g$ is interpreted as the relative group velocity of the wave with respect to the moving source.

6.2.3 Generalization to Higher-Order Wave Equations

The geometrical framework introduced in the previous subsections is not restricted to partial differential equations that are of first-order in time, but can be extended to more general wave equations of the form

$$P[i\partial_t, -i\nabla_\mathbf{r}]\phi(\mathbf{r},t) = S(\mathbf{r},t), \qquad (6.14)$$

where P is an arbitrary polynomial in two variables, a scalar variable and a d-component vectorial variable. The degree of the polynomial P corresponds to the order of the partial differential equation for $\phi(\mathbf{r},t)$: in the case of electromagnetic waves in a non-dispersive medium, it is of second order in both variables; in the case of Bogoliubov excitations in a superfluid, it is of second order in time and of fourth order in space; in the case of surface waves, it is of second order in time, but it involves arbitrarily high derivatives in the spatial coordinates. The different branches $\Omega(\mathbf{k})$ of the dispersion law are then defined by the roots of P via the equation

$$P[\Omega(\mathbf{k}), \mathbf{k}] = 0. \qquad (6.15)$$

In the presence of a source term of the form (6.3), the solution of (6.14) has the form

$$\tilde{\phi}(\mathbf{k}, \omega) = \frac{2\pi \tilde{S}_0(\mathbf{k})}{P(\mathbf{k}\cdot\mathbf{v} + i0^+, \mathbf{k})}, \qquad (6.16)$$

where the infinitesimally small imaginary part has been again added in order to enforce causality by shifting the real roots Ω of the dispersion law (6.15) into the lower half of the complex-plane.

The reasoning to extract from (6.16) the qualitative shape of the real-space pattern is then the same as before, the locus Σ in **k**-space being now defined by the zeros of the polynomial equation

$$P(\mathbf{k} \cdot \mathbf{v}, \mathbf{k}) = 0 \qquad (6.17)$$

with $\mathbf{k} \neq 0$. In the next sections, we shall discuss in full detail a few physical examples illustrating how the geometrical structure of Σ determines the shape of the emission pattern in both real and momentum spaces.

6.3 Čerenkov Emission by Uniformly Moving Charges

As a first application of the theory, in this section we shall review the basic features of the emission of electromagnetic radiation by a charged particle relativistically moving through a dielectric medium at speed higher than the phase velocity of light in the medium. This is the well-known Čerenkov effect (or, more precisely, Vavilov-Čerenkov effect) first observed by Marie Curie, then experimentally characterized by Vavilov and Čerenkov [32, 33] and finally theoretically understood by Frank and Tamm [34].

6.3.1 Non-dispersive Dielectric

In the simplest case of a non-dispersive dielectric with a frequency-independent refractive index n, the dispersion law satisfies the second-order equation

$$\Omega^2 = \frac{c_0^2}{n^2} k^2 : \qquad (6.18)$$

in the $(\Omega, \mathbf{k})$ space, the dispersion $\Omega(\mathbf{k})$ corresponds to a conical surface with vertex in $\Omega = k = 0$. A cut of this cone along the $k_y = 0$ line is shown in Fig. 6.2(a): the thick lines indicate the positive frequency part of the conical surface, the thin line indicate the negative frequency part. In the absence of dispersion, the phase and group velocities coincide and are equal to $c = c_0/n$.

The shape of the locus Σ of $\mathbf{k} \neq 0$ points satisfying $\Omega(\mathbf{k}) - \mathbf{k} \cdot \mathbf{v} = 0$ crucially depends on whether the charge is moving at a sub-luminal $v < c$ or super-luminal $v > c$ speed. In the former case, the locus Σ is empty and no radiation is emitted. The localized, non-radiative perturbation that is present around the charge due to the non-resonant excitation of the field modes contributes to the (velocity-dependent) Coulomb field around the charge.

The locus Σ in the case of a super-luminally moving charge in the positive x direction is illustrated in Fig. 6.2(b): it has the analytic form

$$k_y^2 = k_x^2 \left(\frac{v^2}{c^2} - 1 \right) \qquad (6.19)$$

Fig. 6.2 (**a**) Cut along $k_y = 0$ of the photon dispersion in a non-dispersive medium of frequency-independent refractive index n. The *dashed line* indicates the $\Omega = \mathbf{k} \cdot \mathbf{v}$ plane for a super-luminal charge speed $v > c$. (**b**) **k**-space locus Σ of resonant modes into which the Čerenkov emission occurs, the so-called Čerenkov cone; the *green arrows* indicate the normal to the Σ locus, that is the direction of the relative group velocity $\mathbf{v}'_g = \nabla_{\mathbf{k}}[\Omega(\mathbf{k}) - \mathbf{k} \cdot \mathbf{v}]$. (**c**) Real-space pattern of the electric field amplitude in the wake of the charge; the pattern is numerically obtained as the fast Fourier transform of the **k**-space perturbation (6.16). The Mach cone around the negative x axis is apparent with aperture ϕ

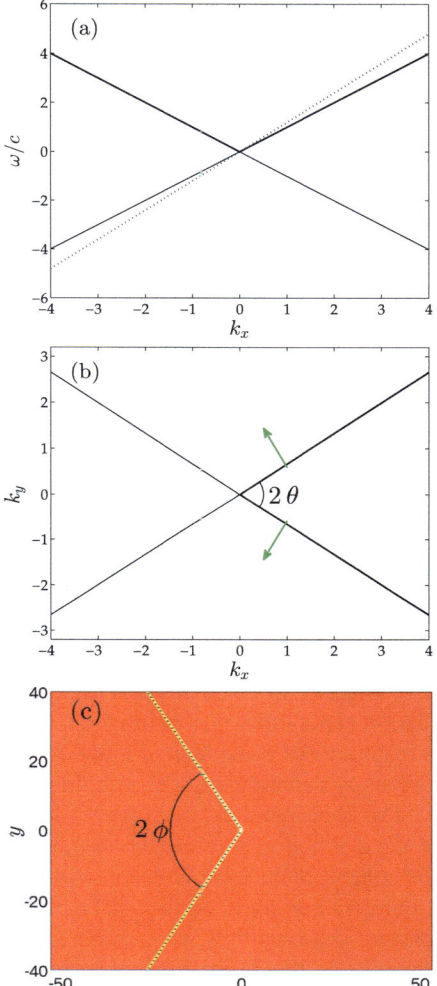

and consists of a pair of half straight lines originating from $\mathbf{k} = 0$ and symmetrically located with respect to the k_x axis at an angle θ such that $\cos\theta = c/v$. The higher the particle speed v/c, the wider the angle θ made by the direction of the Čerenkov emission with the direction of the charge motion.

The most peculiar feature of the locus Σ is that the normal vector to Σ indicating the direction of the relative group velocity $\mathbf{v}'_g = \nabla_{\mathbf{k}}[\Omega(\mathbf{k}) - \mathbf{k} \cdot \mathbf{v}] = \mathbf{v}_g - \mathbf{v}$ [indicated by the green arrows on Fig. 6.2(b)] is constant for all points **k** lying on each of the two straight lines forming Σ and points in the backward direction. This last feature is a straightforward consequence of the fact that the charge velocity is larger than the speed of light c in the medium. As a result, all modes on Σ propagate (as seen from the charge reference frame) in the same direction and the electromagnetic field radiated by the charge is spatially concentrated around the direction of $\mathbf{v}'_g$. This defines a single-sheet conical surface in real space, i.e. a pair of half straight lines in the two dimensional geometry considered here,

$$y^2 = \frac{c^2 x^2}{v^2 - c^2} \quad \text{with } x < 0. \tag{6.20}$$

Its aperture ϕ around the negative x axis[2] is determined by the condition $\sin\phi = c/v$: the faster the charge speed, the narrower the cone behind the charge. In the analogue y with the conical sonic wake generated by a super-sonically moving bullet in a bulk fluid, we will refer to this real space cone as the *Mach cone*. The very thin shape of the Mach cone results from the interference of the continuum of points on the Σ locus. For each $\mathbf{k}_0 \in \Sigma$, the fringes are orthogonal to the Mach cone and have different spacing: the interference is everywhere destructive but for the thin surface of the Mach cone. If the correct form of the structure factor $S_0(\mathbf{k})$ is included, the usual δ-shape for the Mach cone is recovered [13].

In view of the discussion of the next sections, it is crucial to clearly keep in mind the conceptual distinction between the Mach cone in real space on which the electric field intensity is spatially concentrated and the $\mathbf{k}$-space *Čerenkov cone* defining the directions into which the radiation does occur. The former was experimentally detected and characterized in [20, 35] by looking at the spatial profile of the electric field wake behind the charge.[3] The latter is observed in any standard Čerenkov radiation experiment measuring the far-field angular distribution of the radiation, that turns out to be concentrated in the forward direction on a conical surface making a Čerenkov radiation angle θ with the charge velocity.

The conceptual distinction between the Čerenkov and the Mach cones is related to the distinction between the so-called *phase* and *group cones*, first pointed out in the context of the Čerenkov emission in dispersive media in [36, 37]. Restricting for a moment our attention to a given emission frequency, the *wave cone* is defined as the real space conical wavefront passing through the source and orthogonal to the direction of the far-field emission in $\mathbf{k}$ space: its aperture ϕ_{ph} around the negative x axis is determined by the *phase* velocity as $\sin\phi_{ph} = v_{ph}/v$ and is related to the aperture of the Čerenkov cone by $\phi_{ph} = \pi/2 - \theta$. With some caveats, it can be interpreted as the wavefront on which the Čerenkov emission has a constant phase. On the other hand, the *group cone* is defined as the Mach cone for the given frequency and describes the spatial points on which the (spectrally filtered) electric field intensity is peaked. Its aperture ϕ depends on the *group* velocity of light v_{gr}

---

[2]The coefficients of the analytical form (6.20) can be understood from the Fourier transform of a delta function peaked on the conically-shaped locus Σ of Eq. (6.19).

[3]It is interesting to note that in both these experiments the moving charge responsible for the Čerenkov emission did not consist of a charged physical particle travelling through the medium, but rather consisted of a moving bullet of nonlinear optical polarization generated by a femtosecond optical pulse via the so-called inverse electro-optic effect.

as $\sin\phi = v_{\text{gr}}/v$. The distinction between the phase and group cones has been anticipated in [21] to be most striking in the case of ultra-slow light media where v_{gr} is reduced to the m/s range while v_{ph} remains of the order of the speed of light in vacuum $c_0 \simeq 3 \times 10^8$ m/s [38–41].

The study of the Čerenkov effect in strongly dispersive media where the refractive index $n(\omega)$ has a strong dependence on the frequency and/or the medium exhibits a non-trivial spatial patterning is still a very active domain of research from both the theoretical and the experimental points of view. For instance, the consequences of a strong resonance in $n(\omega)$ were theoretically investigated in [19]: the sub-linear dispersion of the photon in a resonant medium is responsible for the disappearance of the threshold velocity for the Čerenkov emission and for a non-trivial spatial patterning of the electric field wake behind the charge. These striking results were experimentally confirmed in [20] and bear a close resemblance to the surface wave pattern in the wake of a duck swimming on shallow water that will be discussed in Sect. 6.6.3. Another active and promising research line is addressing those new features of Čerenkov radiation that follow from the peculiar band dispersion of photons in spatially periodic media [22] and in negative refractive index metamaterials, the so-called left-handed media [23].

6.4 Moving Impurities in a Superfluid

A central concept in the theory of superfluids [24–26] is the so-called Landau criterion for superfluidity, that determines the maximum speed at which a weak impurity can freely travel across a superfluid without experiencing any friction force and without generating any propagating perturbation in the fluid. In terms of the dispersion $\Omega(\mathbf{k})$ of the excitations in the superfluid, the Landau critical velocity has the form

$$v_{\text{cr}} = \min_{\mathbf{k}}\left[\frac{\Omega(\mathbf{k})}{k}\right]. \tag{6.21}$$

This cornerstone of our theoretical understanding of quantum liquids has a simple interpretation in terms of the theory of the generalized Čerenkov effect reviewed in Sect. 6.2: the friction force experienced by the moving impurity is due to the emission of elementary excitations in the fluid by a mechanism that is a quantum fluid analogue of Čerenkov emission. The $v < v_{\text{cr}}$ condition for superfluidity corresponds to imposing that the locus Σ of excited modes is empty, while for faster impurities a characteristic wake pattern is generated around the impurity.

An experimental image of this wake using a dilute Bose-Einstein condensate of ultracold atoms hitting[4] the repulsive potential of a blue-detuned laser is reproduced in the left panel of Fig. 6.3; an analogous image for a condensate of exciton-polaritons in a semiconductor microcavity is reproduced in the middle panel. In both

---

[4]Needless to say that the configuration of a moving superfluid hitting an impurity at rest is fully equivalent modulo a Galilean transformation to the case of a moving impurity crossing a superfluid at rest.

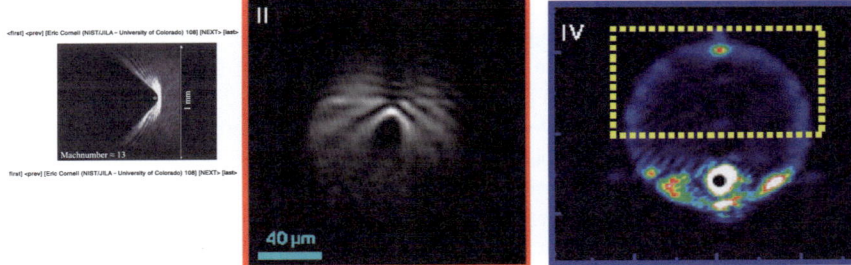

Fig. 6.3 *Left panel*: experimental image of the real-space wake pattern that appears in a Bose-Einstein condensate of ultracold atoms hitting the repulsive potential of a blue-detuned laser beam. The condensate motion is from right to left. Figure taken from [17], as published in [18]. *Middle* and *right panels*: experimental images of the real-space wake pattern (*middle*) and the momentum distribution (*right*) for a Bose-Einstein condensate of exciton-polaritons hitting a fabrication defect in the planar microcavity. The polariton flow is from top to bottom. The value of the density in the right panel is very small and interactions negligible. Figures taken from [16]

cases, the density wake extends both behind and ahead of the impurity. The geometrical shape of the Σ locus is instead clearly visible in the momentum distribution pattern shown in the right panel.

The situation is of course more complex when stronger impurities are considered, e.g. a finite-sized impenetrable object: in this case, the critical speed for frictionless flow was predicted in [42] to be limited by the nucleation of pairs of quantized vortices at the surface of the object, and therefore to be significantly lower than the speed of sound. This mechanism was recently confirmed in experiments for with atomic [43] and polariton [44–46] condensates. Furthermore, it is worth reminding that all our reasonings are based on a mean-field description of the condensate that neglects quantum fluctuations: more sophisticated Bethe ansatz calculations for a strongly interacting one-dimensional Bose gas [31] have anticipated the appearance of a finite drag force also at sub-sonic speed. Including higher order terms of the Bogoliubov theory led the authors of [47] to a similar claim for a three-dimensional condensate.

6.4.1 The Bogoliubov Dispersion of Excitations

The theoretical description of superfluids is simplest in the case of a dilute Bose gas below the transition temperature T_{BEC} for Bose-Einstein condensation [24]. For $T \ll T_{BEC}$ and sufficiently weak interactions, most of the atoms are accumulated in the same one-particle orbital, the so-called Bose-Einstein condensate. The elementary excitations in a dilute Bose-Einstein condensate are characterized by the Bogoliubov dispersion [24]

$$\hbar^2 \Omega^2 = \frac{\hbar^2 k^2}{2m}\left(\frac{\hbar^2 k^2}{2m} + 2\mu\right), \tag{6.22}$$

where m is the mass of the constituent particles and the chemical potential μ is given (at zero temperature) by

$$\mu = \frac{4\pi \hbar^2 a_0 n}{m} \quad (6.23)$$

in terms of the particle-particle low-energy collisional scattering length a_0 and the particle density n. In the standard three-dimensional case, the weak interaction (or diluteness) condition requires that $n a_0^3 \ll 1$.

The characteristic shape of the Bogoliubov dispersion (6.22) is illustrated in the (a, d) panels of Fig. 6.4. For small momenta $k\xi \ll 1$ (the so-called healing length ξ being defined as $\hbar^2/m\xi^2 = \mu$), the dispersion has a sonic behaviour

$$\Omega^2 \simeq c_s^2 k^2 \quad (6.24)$$

with a sound speed $c_s = \sqrt{\mu/m}$, while at high wave-vectors $k\xi \gg 1$ it grows at a super-sonic rate and eventually recovers the parabolic behaviour of single particles,

$$\Omega \simeq \pm \left[\frac{\hbar k^2}{2m} + \mu \right]. \quad (6.25)$$

An explicit calculation from (6.22) shows that the Landau critical velocity (6.21) in the dilute Bose gas is determined by the speed of sound $v_{cr} = c_s$. It is worth reminding that this is no longer true in more complex superfluids with strong interparticle interactions as liquid He-II, where v_{cr} is determined by the roton branch of the elementary excitations [24–26, 48, 49]. Remarkably, super-linear dispersions in the form (6.22) also appear in the theory of surface waves on shallow fluids when the fluid depth is lower than the capillary length, see Eqs. (6.42) in Sect. 6.6.

The effect of the moving impurity onto the superfluid can be described by a time-dependent external potential of the form $V(\mathbf{r}, t) = V_0(\mathbf{r} - \mathbf{v}t)$ coupled to the particles forming the superfluid. Inclusion of this external potential in the Bogoliubov theory requires including a classical source term in the Bogoliubov equations of motion for the two-component spinor describing the quantum field of the non-condensed particles: a complete theoretical discussion along these lines can be found in the recent works [14, 15, 18]. Here we shall use an approximate, yet qualitatively accurate model based on the simplified scalar theory of Sect. 6.2: the real and imaginary parts of the field $\phi(\mathbf{r}, t)$ correspond to the density and phase modulation of the condensate.

6.4.2 Superfluidity vs. Bogoliubov-Čerenkov Wake

As we have already mentioned, the locus Σ is empty for sub-sonic impurity speeds $v < c_s$: the impurity is able to cross the superfluid without resonantly exciting any propagating mode of the fluid. As a result, within mean-field theory it is not expected to experience any friction force. Modulo a Galilean transformation, this effect is equivalent to a frictionless flow along a containing pipe in spite of the roughness of the walls, which is one of the clearest signatures of superfluid behaviour [24–26].

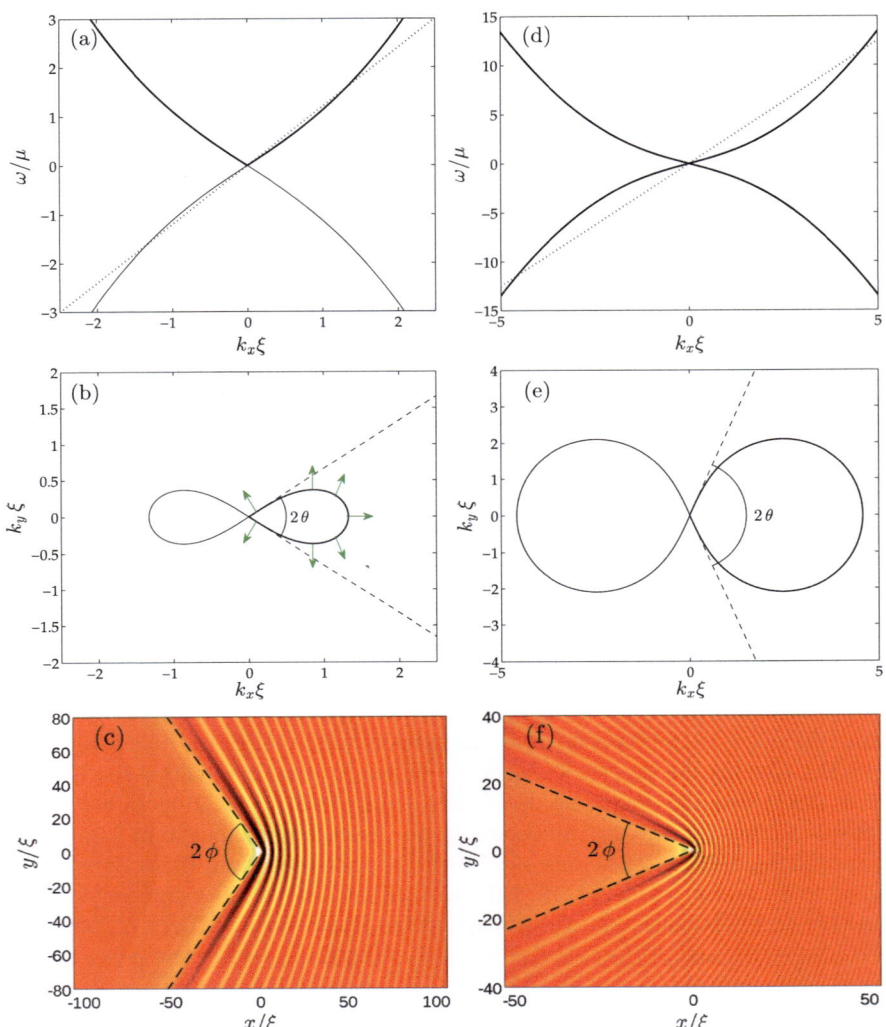

Fig. 6.4 *Top row*: Bogoliubov dispersion of excitations in a dilute Bose-Einstein condensate. The *dashed line* indicates the $\Omega = \mathbf{k} \cdot \mathbf{v}$ plane for two different impurity speeds $v/c_s = 1.2$ [panel (**a**)] and $v/c_s = 2.5$ [panel (**d**)]. *Middle row*: shape of the corresponding **k**-space locus Σ of resonantly excited modes. The *dashed lines* indicate the Čerenkov cone in the low wave-vector region $k\xi \ll 1$; the *green arrows* indicate the normal to the Σ locus, that is the direction of the relative group velocity $\mathbf{v}'_g$. *Bottom row*: real space pattern of the density modulation. All patterns are numerically obtained performing the integral via a fast Fourier transform of the **k**-space perturbation (6.16). The *black dashed lines* indicate the Mach cone

Still, the non-resonant excitation of the Bogoliubov modes by the moving impurity is responsible for a sizable density modulation in the vicinity of the impurity, that quickly decays to zero in space with an exponential law. An important consequence of this localized density perturbation is a sizable renormalization of the mass of the object [31]: the linear momentum that is associated to the moving impurity gets in fact a contribution from the portion of fluid that is displaced by it.

For super-sonic motion, the locus Σ consists of a conical region at small $k\xi \ll 1$ analogue ous to what was found in Sect. 6.3.1 for a purely linear dispersion: as in that case, the aperture angle θ of the **k**-space Čerenkov cone [dashed lines in Fig. 6.4(b, e)] defining the far-field angle at which phonons are emitted by the impurity is defined by the condition $\cos\theta = c_s/v$.

Correspondingly, the aperture ϕ of the Mach cone that is visible in the real-space density modulation pattern behind the impurity is defined by $\sin\phi = c_s/v$ (dashed black lines in [Fig. 6.4(c, f)]. This cone is the superfluid analogue of the Mach cone that is created in a generic fluid by a super-sonically moving object, e.g. an aircraft or a bullet. An experimental image of a Mach cone in a superfluid of exciton-polaritons is shown in the central panel of Fig. 6.3. As usual, the faster the impurity, the narrower the Mach cone.

Differently from sound waves in an ordinary fluid, the Bogoliubov dispersion of the excitations in a superfluid is characterized by a parabolic shape at large wave-vectors $k\xi \gg 1$ according to (6.25). This region of the Bogoliubov spectrum is responsible for the smooth arc in the high wave-vector region of Σ that connects the two straight lines emerging from the origin $\mathbf{k} = 0$. In experiments, the shape of Σ can be inferred following the peak of the momentum distribution of the particles in the superfluid: an example of experimental image using exciton-polaritons in the low-density regime is reproduced in the right panel of Fig. 6.3 analogous images for atomic gases can be found e.g. in [50].

In the low wave-vector region, the relative group velocity $\mathbf{v}'_g$ is oriented along the edges of the Mach cone. Along the high wave-vector part of Σ, the relative group velocity $\mathbf{v}'_g$ rotates in a continuous and monotonous way spanning all intermediate directions external to the Mach cone. As no point on Σ corresponds to a relative group velocity oriented in the backward direction inside the Mach cone, the density profile remains unperturbed in this region. On the other hand, the density perturbation shows peculiar features in front of the Mach cone, with a series of curved wavefronts extending all the way ahead of the impurity. These wavefronts are clearly visible in the experimental images that are shown in Fig. 6.3 for atomic (left panel) and polaritonic (middle panel) superfluids. In the **k**-space diagrams of Fig. 6.4(b, e), these waves correspond to the regions in the vicinity of the extreme points of Σ where $\mathbf{v}'_g$ is directed in the direction of the impurity speed along the positive x direction.

Physically, these curved wavefronts in the density modulation pattern can be understood as originating from the interference of the macroscopic coherent wave associated to the Bose-Einstein condensate with the atoms that are coherently scattered by the moving impurity. An analytic discussion of their shape is discussed in detail in [53]; their one-dimensional restriction was first mentioned in [51, 52]. In the next section we shall present analytical formulas for an approximated theory where the single particle region of the Bogoliubov dispersion is modelled with the parabolic dispersion of single-particle excitations.

6.5 Parabolic Dispersion: Conics in the Wake

Another example of dispersion that is fully amenable to analytic treatment is the parabolic one,

$$\Omega(\mathbf{k}) = \frac{\hbar k^2}{2m} + \mu. \tag{6.26}$$

In spite of its simplicity, this form of dispersion can be used to model a number of different physical configurations, from the large wave-vector $k\xi \gg 1$ region of the Bogoliubov dispersion (6.22) of superfluids, to the resonant Rayleigh scattering in planar microcavities [54–56], to magnons in solid-state materials [27, 57–59]. In particular, the results of this section will shine further light on the curved wavefronts observed in Fig. 6.4(c, f) ahead of the impurity.

For a generic dispersion of the parabolic form (6.26), simple analytical manipulations show that the locus Σ has a circular shape as shown in Fig. 6.5(b, e, h). Assuming again that the particle speed $\mathbf{v}$ is directed along the positive x axis, the center of the circle is located at $k_x = k_o = mv/\hbar$, $k_y = 0$ and has a radius $\bar{k}$ such that

$$\frac{\hbar \bar{k}^2}{2m} = \frac{mv^2}{2\hbar} - \mu. \tag{6.27}$$

Depending on the relative value of the velocity v and of the μ parameter, different regimes can be identified.

For positive μ (but such that the RHS of (6.27) is still positive), the radius $\bar{k}$ is smaller than k_o and the origin $\mathbf{k} = 0$ lies outside the circle. This is the typical case of large $\mathbf{k}$ excitations in superfluids, whose dispersion is approximated by Eq. (6.25). The usual resonant Rayleigh scattering ring [54–56] passing through the origin $\mathbf{k} = 0$ is recovered in the $\mu = 0$ case describing the case of an ideal gas of non-interacting particles: an experimental example of such a ring is visible in the momentum distribution shown in Fig. 6.3(c) for a low-density gas of (almost) non-interacting polaritons flowing against a localized impurity potential. For negative μ, the radius is instead larger $\bar{k} > k_0$ and the origin $\mathbf{k} = 0$ falls inside the circle.

The relative group velocity $\mathbf{v}'_g$ is directed in the outward radial direction. As a consequence of the smooth shape of Σ, $\mathbf{v}'_g$ spans all possible directions and the real-space perturbation shown in Fig. 6.5(c, f, i) extends to the whole plane. However, the wavefronts can have very different shapes depending on the relative value of v and μ.

A closed form for the real-space wake pattern is straightforwardly obtained by noting that the integral in the right-hand side of (6.8) is in this case closely related to the retarded Green's function for a free non-relativistic particle [60],

$$G_{\text{ret}}(\mathbf{r}, \omega) = \int \frac{d^d \mathbf{k}}{(2\pi)^d} \frac{e^{i\mathbf{k} \cdot \mathbf{r}}}{\hbar \mathbf{k}^2/2m - \omega - i0^+}. \tag{6.28}$$

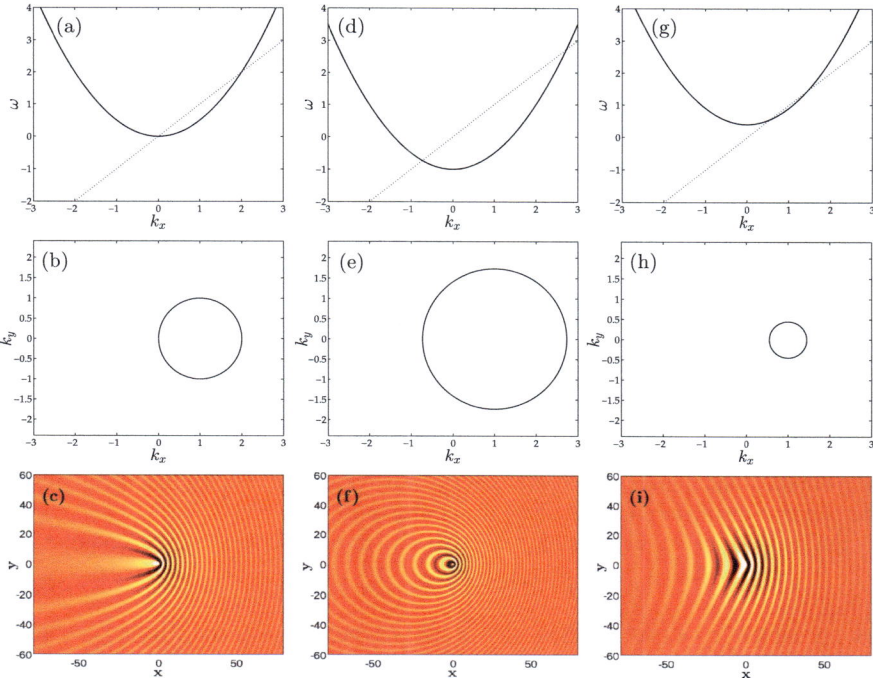

Fig. 6.5 *Top row*: parabolic dispersion of excitations $\omega = k^2/2 + \mu$ in the $\mu = 0$ (**a**), $\mu = -1 < 0$ (**b**), $\mu = 0.4 > 0$ (**c**) cases (for notational simplicity, we have set $m = \hbar = 1$). The *dashed line* indicates the $\Omega = \mathbf{k} \cdot \mathbf{v}$ plane for a generic particle speed $v = 1$ along the positive x direction. *Middle row*: circular shape of the **k**-space locus Σ of resonantly excited modes. *Bottom row*: real space patterns of the density modulation. All patterns are numerically obtained performing the integral in (6.8) via a fast Fourier transform algorithm

In a generic dimension d, the asymptotic form of G_{ret} at large **r** has the outgoing spherical wave form

$$G_{\text{ret}}(\mathbf{r}, \omega) = \frac{2\pi i C_d}{r^{(d-1)/2}} e^{i\bar{k}r} \tag{6.29}$$

with a wave-vector $\bar{k}$ such that

$$\frac{\hbar \bar{k}^2}{2m} = \omega. \tag{6.30}$$

Of course, for $\omega > 0$ (or $\omega < 0$), the solution such that $\bar{k} > 0$ (or $\text{Im}[\bar{k}] > 0$) must be considered. C_d is a dimension- and energy-dependent normalization constant.

Using this result, the expression (6.8) for the wake generated by a point-like source term can be simplified into

$$\phi(\mathbf{r}') = -\int \frac{d^2\mathbf{k}}{(2\pi)^2} \frac{e^{i\mathbf{k}\cdot\mathbf{r}'}}{\mu + \frac{\hbar k^2}{2m} - \mathbf{k}\cdot\mathbf{v} - i0^+}$$

$$= -\int \frac{d^2\mathbf{k}}{(2\pi)^2} \frac{e^{i\mathbf{k}\cdot\mathbf{r}'}}{\frac{\hbar k^2}{2m} - (\frac{mv^2}{2\hbar} - \mu) - i0^+} e^{i\frac{m}{\hbar}\mathbf{v}\cdot\mathbf{r}'} = -\frac{2\pi i C_d}{\sqrt{r'}} e^{i\bar{k}r'} e^{ik_o x'}. \tag{6.31}$$

The real part of this wave provides the wake pattern plotted in Fig. 6.5(c, f, i): The different panels correspond to the $\mu = 0$ (c), $\mu < 0$ (f) and $\mu > 0$ (i) cases, which correspond to $\bar{k} = k_o$ (c), $\bar{k} > k_o$ (f), and $\bar{k} < k_o$ (i), respectively.

The shape of the wavefronts is obtained as the constant phase loci of (6.31). For instance, the loci of points for which the phase of the field ϕ equals an integer multiple of 2π are described by

$$\bar{k}\sqrt{x^2 + y^2} + k_o x = 2\pi M \tag{6.32}$$

with M a generic integer. After moving the $k_o x$ term to the LHS and then taking the square of both members, this equation is straightforwardly rewritten as a quadratic equation in the spatial coordinates. The shape of the wavefronts in the two-dimensional plane is therefore described by conic curves: the specific nature of the conic in the different cases depends on the ratio $\bar{k}/k_o$.

For $k_o = \bar{k}$, the wavefronts have a parabolic shape described by the equation

$$4\pi^2 M^2 - 4\pi M k_o x = k_o y^2. \tag{6.33}$$

As the square root has by definition a non-negative value, the further condition $2\pi M - k_o x \geq 0$ has to be imposed to ensure that the RHS of (6.32) is non-negative. Combined with (6.33), this condition is equivalent to imposing that the integer $M \geq 0$. An example of these parabolic wavefronts is shown in Fig. 6.5(c).

For $\bar{k} > k_o$, the wavefronts have an elliptic shape described by the equation

$$\bar{k}^2 y^2 + (\bar{k}^2 - k_o^2)\left[x + \frac{2\pi M k_o}{\bar{k}^2 - k_o^2}\right]^2 = \frac{4\pi^2 M^2 \bar{k}^2}{\bar{k}^2 - k_o^2}. \tag{6.34}$$

An example of these elliptic wavefronts is shown in Fig. 6.5(f). The condition on the non-negativity of the RHS of (6.32) imposes that $M \geq 0$. The ellipticity of Σ is a function of the ratio $\bar{k}/k_o$: the closer this ratio is to 1, the more elongated the ellipse is. In the limit $\bar{k}/k_o \to 1$, the ellipse tends to a parabola, recovering the case $\bar{k} = k_o$ discussed above. The larger the ratio $\bar{k}/k_o$, the closer the wavefront shape to a series of concentric circles.

Finally, for $\bar{k} < k_o$, the wavefronts have hyperbolic shapes described by the equation

$$(k_o^2 - \bar{k}^2)\left[x - \frac{2\pi M k_o}{k_o^2 - \bar{k}^2}\right]^2 - \bar{k}^2 y^2 = \frac{4\pi^2 M^2 \bar{k}^2}{k_o^2 - \bar{k}^2}. \tag{6.35}$$

In this case, the condition on the RHS of (6.32) does not impose any condition on M that can have arbitrary positive or negative integer values. However, combining this condition with the equation defining the hyperbola, one finds that for each M only the left branch of the hyperbola at lower x has to be retained. Positive vs. negative values of M are responsible for the different periodicities that are visible in Fig. 6.5(i) in the $x > 0$ and $x < 0$ regions, respectively.

From the **k** space diagrams in Fig. 6.5(b, e, h), it is immediate to see that the points on Σ corresponding to the backward and forward propagating waves are the two intersections of the circle with the x axis: the wider spacing of the wavefronts in the backward direction is due to the smaller magnitude of the wave-vector at the intersection point at lower x. In the parabolic case, this point coincides with the origin, which explains the absence of density oscillations on the negative x axis.

The characteristic curvature of the forward propagating wavefronts provides a qualitative explanation for the shape of the density modulation experimentally observed ahead of the impurity and illustrated in the left and central panel of Fig. 6.3. Of course, the absence of backward propagating waves in the superfluid behind the impurity is due to the $\mathbf{k} = 0$ singularity of the Σ locus for the case of the Bogoliubov dispersion. It is worth reminding that, in contrast to previous works, the shape of the forward propagating wavefronts is exactly parabolic only in the $\mu = 0$ limit of non-interacting particles. For the generic $\mu > 0$ case of Bogoliubov theory, the Hartree potential in (6.25) makes their shape to be closer to (part of) an hyperbola.

6.6 Surface Waves on a Liquid

The discussion of the previous sections on the Čerenkov effect in classical electromagnetism and on the response of superfluids to moving impurities puts us in the position of getting an easy qualitative understanding of the surface waves that are generated by a duck steadily swimming on the surface of a quiet pond or, equivalently, a fishing line in a uniformly flowing river.[5] This system is by far the most accessible from the experimental point of view, but perhaps also the richest one for the variety of different behaviours that can be observed depending on the system parameters. A few examples of experimental pictures are shown in Fig. 6.6. For the sake of conciseness, we shall restrict ourselves to the case of the water-air interface and restrict to the linear regime of wave propagation described by the model equation (6.1). More complete treatments based on the full hydrodynamic equations including nonlinear effects can be found in the dedicated literature, see e.g. [1–12].

6.6.1 Dispersion of Surface Waves

The dispersion of surface waves on top of a fluid layer of height h and at rest has the form

$$\Omega(k)^2 = \left(gk + \frac{\gamma}{\rho}k^3\right)\tanh kh, \qquad (6.36)$$

---

[5]It is interesting to note that, as in the case of electromagnetic waves, accelerated objects emit surface waves independently from their speed [64].

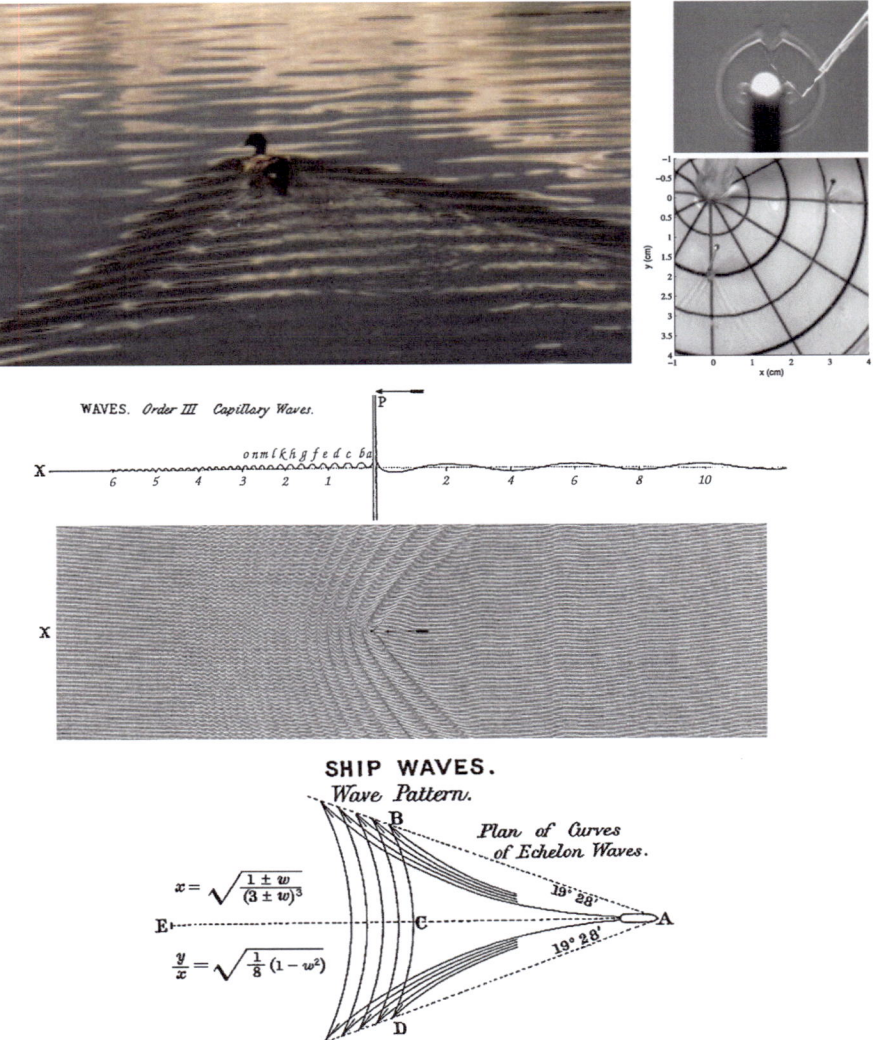

Fig. 6.6 *Upper panels*: picture of the Kelvin's ship-wave pattern behind a duck swimming at uniform speed on a quiet lake (*left*). Photograph courtesy of Fabrice Neyret, ARTIS-CNRS, France. Experimental picture of the Mach cone downstream of a wire immersed in radially flowing silicon oil (*upper right*). Capillary waves are not visible as they are quickly damped by the larger viscosity of silicon oil. Picture from [61]. Experimental picture of the Mach cone downstream of a pin immersed in very shallow flowing water: the surface wave dispersion is supersonic and the height modulation stays outside the Mach cone. Picture courtesy of Silke Weinfurtner (*lower right*). *Middle panel*: Original hand drawing by John Scott Russell [62] of the waves generated by a vertical rod (diameter = 1/16 inch) moving along the water surface with a uniform velocity. The rod moves in the leftward direction: the capillary waves are visible in front of the rod and the gravity waves behind it. A cut of the surface height modulation is shown right above the main drawing. *Lower panel*: Original hand drawing by Lord Kelvin of Kelvin's ship-wave pattern [63]. The BCD wavefront belongs to the so-called transverse wave pattern. The so-called diverging waves connect the object at A to the dashed lines indicating the edges of the pattern

where ρ is the mass density of the fluid, g is the gravitational acceleration, and γ is the surface tension of the fluid-air interface.

In the simplest case of a *deep fluid*, the $\tanh kh$ factor can be approximated with 1 and the dispersion is characterized by two regions. For low wave-vectors $k \ll k_\gamma$, the dispersion follows the sub-linear square-root behaviour

$$\Omega(k) \simeq \pm\sqrt{gk}, \qquad (6.37)$$

of gravity waves, while for large $k \gg k_\gamma$ it is dominated by capillarity effects and has a super-linear growth as

$$\Omega(k) \simeq \pm\sqrt{\frac{\gamma}{\rho}k^{3/2}}. \qquad (6.38)$$

The characteristic wave-vector scale separating the two regions is fixed by the capillary wave-vector

$$k_\gamma = \sqrt{\frac{\rho g}{\gamma}}. \qquad (6.39)$$

For the specific case of water/air interface, $k_\gamma \simeq 370 \text{ m}^{-1}$, which corresponds to the value

$$\ell_\gamma = 1/k_\gamma = 2.7 \times 10^{-3} \text{ m} \qquad (6.40)$$

for the capillary length.

In fluids of finite depth, one can no longer approximate the tanh in (6.36) with 1. As a result, the dispersion in the low-wave-vector region recovers a sonic behaviour at low **k**'s

$$\Omega(k) \simeq \pm c_s k \qquad (6.41)$$

with a speed of sound $c_s = \sqrt{gh}$ proportional to the square root of the fluid depth. The sign of the first correction to the sonic behaviour (6.41),

$$\Omega(k)^2 \simeq ghk^2 + \left[\ell_\gamma^2 - \frac{h^2}{3}\right]c_s^2 k^4 \qquad (6.42)$$

critically depends on the depth of the fluid as compared to the capillary length (6.40). For relatively deep fluids such that $h > \sqrt{3}\ell_\gamma$, the dispersion has a sub-linear behaviour, while it recovers a super-linear behaviour analogue ous to the Bogoliubov dispersion (6.22) for very shallow fluids such that $h < \sqrt{3}\ell_\gamma$.

Independently of the fluid depth h, the super-linear behaviour of the dispersion $\Omega(k)$ at large k makes the locus Σ to be either empty or to consist of a closed curve. For very slow sub-sonic motions $v < v_{\min}$ (with $v_{\min}$ to be defined in the next subsection), the locus Σ is empty. For intermediate $c_s > v > v_{\min}$ (or infinitely deep fluids, $c_s = \infty$), the locus Σ shown in Fig. 6.7(e) consists of a smooth closed curve that does not encircle the origin point $\mathbf{k} = 0$. For super-sonic motions $v > c_s$, the locus Σ shown in Fig. 6.8(b, e) develops a conical Čerenkov singularity at $\mathbf{k} = 0$.

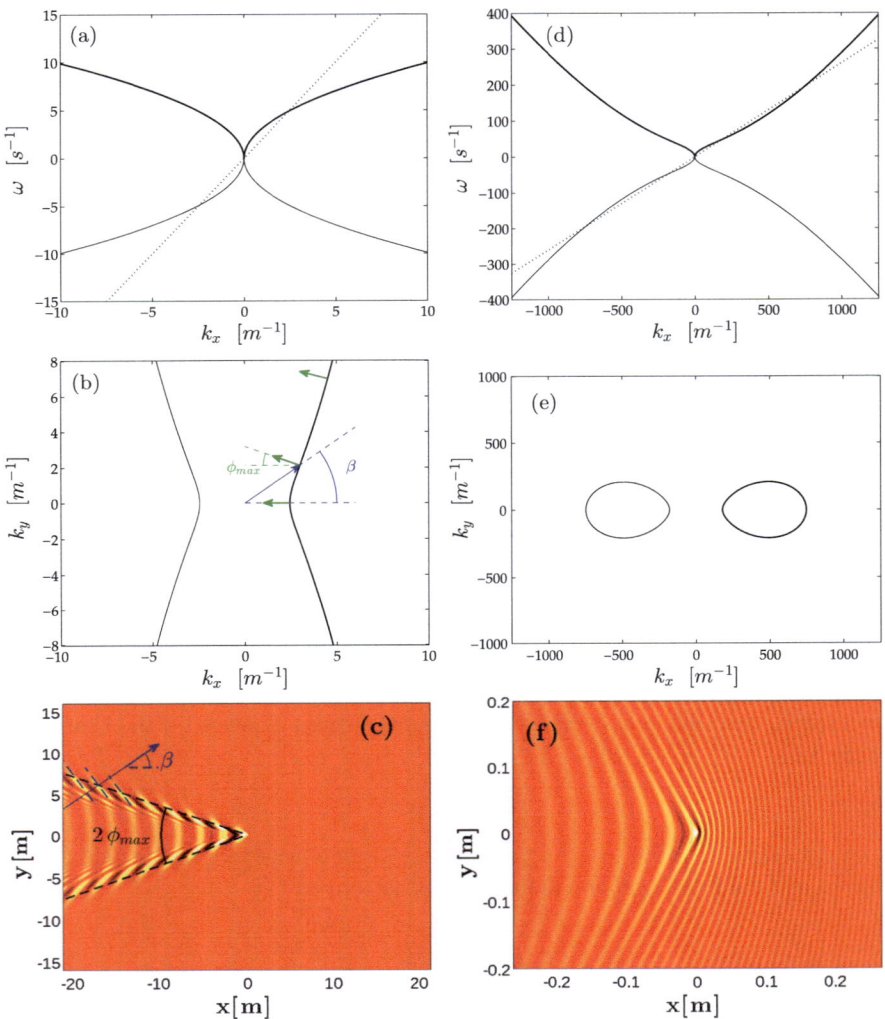

Fig. 6.7 *Top row*: dispersion of surface wave in the $h = \infty$ deep water limit. The *dashed line* indicates the $\Omega = \mathbf{k} \cdot \mathbf{v}$ plane for generic particle speeds $v = 2$ m/s ((**a**) panel), $v = 0.26$ m/s ((**d**) panel). *Middle row*: corresponding shapes of the **k**-space locus Σ of resonantly excited modes (panels (**b**, **e**)); the *green arrows* indicate the normal to the locus Σ, that is the direction of the relative group velocity $\mathbf{v}'_g$. *Bottom row*: real space patterns of the surface height modulation (panels (**c**, **f**)). These patterns are numerically obtained via a fast Fourier transform of the **k**-space perturbation (6.16) using the density and surface tension values of water. The (**c**) panel corresponds to the Kelvin's ship-wave pattern behind a duck swimming on a deep lake, a picture of which is shown in the upper left panel of Fig. 6.6. An original sketch by Lord Kelvin is shown in the lower panel of the same figure

6 The Čerenkov Effect Revisited: From Swimming Ducks to Zero Modes 131

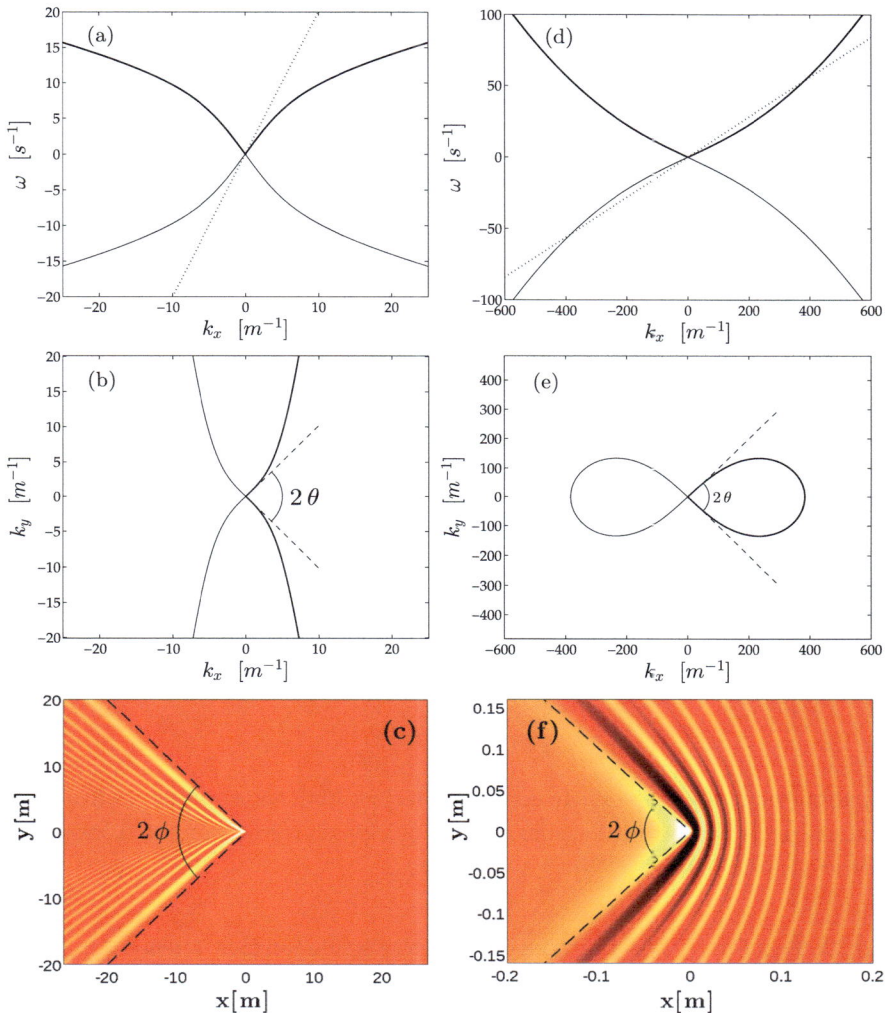

Fig. 6.8 *Top row*: dispersion of surface wave on shallow water of height $h = 0.2$ m (left (**a**) panel) and $h = 0.001$ m (right (**d**) panel). The *dashed line* indicates the $\Omega = \mathbf{k} \cdot \mathbf{v}$ plane for generic particle speeds $v = 2$ m/s (left (**a**) panel), $v = 0.14$ m/s (right (**d**) panel). *Middle row*: corresponding shapes of the **k**-space locus Σ of resonantly excited modes (panels (**b**, **e**)). The *dashed lines* indicate the Čerenkov cone in the low wave-vector region $k\xi \ll 1$. *Bottom row*: real space patterns of the surface height modulation (panels (**c**, **f**)). The *black dashed lines* indicate the Mach cone. These patterns are numerically obtained via a fast Fourier transform of the **k**-space perturbation (6.16) using the density and surface tension values of water. The *left panels* correspond to a case where $h > \sqrt{3}\ell_\gamma$ and the lowest-order correction to the sonic dispersion (6.42) is sub-linear. The *right panels* correspond to a case where $h < \sqrt{3}\ell_\gamma$ and the lowest-order correction has the same super-linear behaviour as the Bogoliubov dispersion (6.22) illustrated in Fig. 6.4

6.6.2 Deep Fluid

Let us start by investigating the deep fluid regime $h \to \infty$ for which the sonic speed $c_s \to \infty$. The structure of the locus Σ can be understood by looking at Fig. 6.7(d): for low speeds

$$v < v_{\min} = \left(\frac{4g\gamma}{\rho}\right)^{1/4}, \tag{6.43}$$

the locus Σ is empty and there is no emission. This critical speed depends on the surface tension of the fluid: for the case of a water/air interface it is equal to $v_{\min} \simeq 0.23$ m/s. The absence of emission corresponds to a vanishing *wave resistance* experienced by the slowly moving object which is able to slide with no friction on the surface of the fluid [12]. Still, the localized deformation of the surface around the object is responsible for a renormalization of the mass of the object [11].

An efficient emission of surface waves with the associated wave resistance [12] is suddenly recovered as soon as $v > v_{\min}$. In this regime, the locus Σ shown in Fig. 6.7(e) has a kind of oval shape [11], with two intersections with the k_x axis at respectively

$$k = k_x^{(1,2)} = k_\gamma \left[\frac{v^2}{v_{\min}^2} \pm \sqrt{\frac{v^4}{v_{\min}^4} - 1}\right]. \tag{6.44}$$

As expected, the two solutions merge to $k_x = k_\gamma$ for $v \gtrsim v_{\min}$, while at larger $v \gg v_{\min}$ they respectively tend to

$$k_x^{(1)} \simeq g/v^2, \tag{6.45}$$

$$k_x^{(2)} \simeq 2k_\gamma v^2/v_{\min}^2, \tag{6.46}$$

the former solution $k_x^{(1)}$ tends to zero in the large v limit and corresponds to almost pure gravity waves, while the latter one $k_x^{(2)}$ quickly diverges as v^2 and corresponds to almost pure capillary waves.

6.6.2.1 Fast Speed $v \gg v_{\min}$ (Deep Fluid, Negligible Surface Tension)

Within the $v \gg v_{\min}$ limit, we can start our discussion from the low wave vector region $k \ll k_\gamma$, where the waves have a mostly gravity nature, $\Omega(k) \simeq \sqrt{gk}$. Because of the fourth power of $v/v_{\min}$ that appears in (6.44), this limit is achieved already for moderate values of $v/v_{\min}$ of the order of a few unities. In this region, the locus Σ is approximately defined by the condition

$$k_y^2 = \frac{v^4 k_x^2}{g^2}\left(k_x^2 - \frac{g^2}{v^4}\right), \tag{6.47}$$

whose shape is plotted in the panel (b) of Fig. 6.7. The locus Σ extends in the $|k_x| \geq k_x^c = g/v^2$ regions: for $k_x \gtrsim k_x^c$, it has the form

$$k_y = \pm\sqrt{\frac{2g}{v^2}(k_x - k_x^c)}, \tag{6.48}$$

while for large $k_x \gg k_x^c$, one recovers an asymptotic behaviour

$$k_y = \pm \frac{v^2}{g} k_x^2. \qquad (6.49)$$

A most remarkable feature are the inflection points at $k_x^{\text{infl}} = \pm\sqrt{3/2}g/v^2$ where the slope dk_y/dk_x is minimum. At these points the normal to the locus Σ makes the maximum angle to the k_x axis, with a value $\phi_{\max}$ such that $\tan\phi_{\max} = 1/\sqrt{8} \simeq 19°28'$. This angle determines the aperture of the wake cone behind the moving object: remarkably, this value is universal and does not depend on the speed of the object. A picture of the *Kelvin's ship-wave pattern* behind a swimming duck on a quiet lake is shown in the upper left panel of Fig. 6.6; an original hand drawing by Lord Kelvin illustrating this physics is reproduced in the lower panel of the same figure.

Another, related feature that is worth noticing is that for each angle $|\phi| < \phi_{\max}$, there exist two points on the locus Σ such that the normal to Σ makes an angle ϕ with the negative k_x axis: according to the theory discussed in Sect. 6.2, these two solutions are responsible for the two inter-penetrating fringe patterns: the so-called *transverse* waves with a small k_y and the so-called *diverging* waves with large k_y.

The transverse waves are clearly visible as the long wavelength modulation right behind the object along the axis of motion: their wave-vector $k_x = k_x^c = g/v^2$ is determined by the intersection of the locus Σ with the k_x axis. Remarkably, the faster the object is moving, the smaller is the wave-vector k_x^c. On Kelvin's hand drawing of Fig. 6.6, the wavefront passing by point C belongs to the transverse wave pattern.

The diverging waves are easily identified in the hand drawing as the wavefronts with opposite curvature connecting the source at A with the edge of the wake pattern where the two patterns collapse onto each other. The direction of the peculiar fringe modulation of the edge of the wake [indicated by the blue dashed lines on Fig. 6.7(c)] is determined by the wave-vector $\mathbf{k}^{\text{infl}}$ of the inflection point of the **k**-space locus Σ: the orientation of $\mathbf{k}^{\text{infl}}$ fixes the angle β to a value such that $\tan\beta = k_y/k_x|_{\text{infl}} = 1/\sqrt{2}$, i.e. $\beta \simeq 35°$.

Of course, a complete treatment of the wake would require including the capillary waves at very high wave-vector $k \approx k_x^{(2)} \gg k_y$, i.e. the part of the locus Σ that closes the curve at large **k**'s outside the field of view of Fig. 6.7(b). However, the amplitude in these short-wavelength modes is quickly damped by viscous effects, so their contribution to the observable pattern turns out to be irrelevant in most practical cases.

6.6.2.2 Moderate Speed $v \gtrsim v_{\min}$ (Deep Fluid, Significant Surface Tension)

For moderate speeds $v \gtrsim v_{\min}$, surface tension effects are no longer negligible and all points of the **k**-space locus Σ contribute to the real-space pattern. In particular, the locus Σ is a closed curve that does not encircle the origin, as shown in Fig. 6.7(e): the two intersections with the k_x axis at $k_x = k_x^{(1,2)}$, corresponding to gravity and capillary waves propagate with relative group velocities directed in opposite directions from the fishing line of the celebrated experiment by Thomson. The pattern of long-wavelength gravity waves is located downstream of the fishing line, while the short-wavelength capillary waves are located in the upstream region, see Fig. 6.7(f) and the drawing by J.S. Russell reproduced in the middle panel of Fig. 6.6.

An approximate analytical understanding of this pattern can be obtained by approximating the locus Σ of Fig. 6.7(e) with a pair of circles analogue ously to the case of a parabolic dispersion discussed in Sect. 6.5 and shown in Fig. 6.5(g–i): within this approximation, the shape of the wavefronts consists a system of hyperbolas, with a closer spacing ahead of the object. The qualitative agreement of the hyperbolic wavefronts of Fig. 6.5(i) with the full calculations shown in Fig. 6.7(f) is manifest.

6.6.2.3 Effect of the Source *Structure Factor*

To complete the discussion, it is worth mentioning that the emission of waves can be hindered by the source *structure factor* $\tilde{S}(\mathbf{k})$ even at large $v > v_{\min}$. For example, the emission of surface waves will be strongly suppressed if the size ℓ of the source term (modelled as a Gaussian-shaped potential $S(\mathbf{r})$) is large enough to have $k\ell \gg 1$ for all points on Σ. For instance, for an object of typical size $\ell = 30$ cm, the argument such that $k_x^{(1)} \ell \leq 1$ imposes a lower critical speed $v \geq v_c^{\text{size}} = 3$ m/s to the emission of gravity waves. A similar argument for capillary waves was mentioned to explain the characteristic swimming speed of some floating insects [65, 66].

6.6.3 Shallow Fluid

When the wavelength of the perturbation is longer than the depth h of the fluid, the $\tanh(kh)$ term in the dispersion begins to be important and causes a radical change in the structure of the locus Σ. The left and right columns of Fig. 6.8 illustrate the two regimes $h > \sqrt{3}\ell_\gamma$ and $h < \sqrt{3}\ell_\gamma$ where the first correction to the sonic dispersion has either a sub-linear or a super-linear nature.

6.6.3.1 Small Surface Tension (Sub-luminal Dispersion)

We start here from the case where the surface tension is small enough to have $h > \sqrt{3}\ell_\gamma$. Depending on the speed v of the object, several regimes can be identified. For very low speeds $v < v_{\min}$, the locus Σ is empty and there is no perturbation to the fluid. For intermediate speeds $v_{\min} < v \ll \sqrt{gh}$, the shape of the locus Σ is determined by the high-$\mathbf{k}$ capillary region of the dispersion and is almost unaffected by the finite height h of the fluid: as in the deep fluid limit, Σ consists of closed, egg-shaped smooth curve and the real-space pattern again resembles a system of hyperbolas extending to the whole space, as illustrated in Fig. 6.7(d–f). For increasing, yet sub-sonic speeds $v < c_s = \sqrt{gh}$, gravity recovers an important role, while the finite height h keeps providing only a small correction to the deep water behaviour illustrated in Fig. 6.7(a–c).

The situation is completely different for supersonic speeds $v > c_s$ [Fig. 6.8(a–c)]: in this case, the locus Σ starts from the origin, where it exhibits a conical singularity as a result of the sonic dispersion. At larger **k**, the locus Σ recovers a shape similar to the infinitely deep fluid case: the sub-linear growth of the dispersion with k is responsible for the fast increase of k_y as a function of k_x. Of course, the supersonic dispersion of capillary waves at very large **k** [well outside the field of view of Fig. 6.8(a, b)] makes the locus Σ to close on itself. However, as already mentioned, these short-wavelength waves are quickly attenuated and hardly visible.

The singularity of Σ at the origin **k** = 0 is responsible for the Mach cone and the disappearance of the transverse wave pattern, as shown in Fig. 6.8(c). As usual for sonic dispersions, the aperture of the Mach cone [indicated by the dashed line on Fig. 6.8(c)] depends on the source speed v as $\sin\phi = c_s/v$: on the Σ locus shown in Fig. 6.8(b), this corresponds to the fact that the normal to Σ starts at a finite angle ϕ with the negative k_x axis for **k** = 0. For growing **k**'s, the angle monotonically decreases to 0 meaning that the perturbation is restricted to the spatial region *inside* the Mach cone. The absence of the transverse wave pattern is clearly visible in Fig. 6.8(c) as the absence of modulation along the negative x direction right behind the object.

6.6.3.2 Shallow One-Dimensional Channel

The restriction of this model to a one-dimensional geometry provides interesting insight on the physics of long wavelength surface waves in a spatially narrow channel of width W. Spatial confinement along the orthogonal direction (say y) makes the corresponding wave-vector to be quantized in discrete values determined by the boundary conditions at the edges of the channel, $k_y = \pi p/W$ with the integer $p = 0, 1, 2, \ldots$. For simplicity, we assume that the transverse shape of the source (e.g. a boat sailing along the channel) is broad enough to only excite the lowest mode at $k_y = 0$, corresponding to a transversally homogeneous wave.

Neglecting for simplicity also capillarity effects, a generalized Landau criterion for one-dimensional gravity waves anticipates that a uniformly moving object can emit $k_y = 0$ gravity waves only if its velocity is slower than a maximum velocity

$$v_{\max} = \max_{k_x}\left[\frac{\Omega(k_x, k_y = 0)}{k_x}\right] = c_s = \sqrt{gh}. \tag{6.50}$$

This feature is easily understood looking at the $k_y = 0$ cut of the dispersion shown in Fig. 6.8(a): for sub-sonic speed $v < c_s$, the $\Omega = k_x v$ straight line corresponding to the Čerenkov condition (6.9) has a non-trivial intersection with the dispersion law, while the intersection reduces to the irrelevant $k_x = 0$ point for supersonic speeds

$v > c_s$. In this case, no modes are any longer available for the emission, which reflects into a marked decrease of the *wave drag* friction experienced by the object. The observation of an effect of this kind when a ship travels at sufficiently fast speed long a channel was first reported by Scott Russel and often goes under the name of Houston paradox in the hydrodynamics and naval engineering literature [4, 5].

As compared to the standard Landau criterion for superfluidity, it is interesting to note that the condition on the object speed to have a (quasi-)frictionless flow is here reversed: friction is large at slow speeds and suddenly drops for $v > c_s$. This remarkable difference is due to the different sub-sonic rather than super-sonic shape of the gravity wave dispersion with respect to the Bogoliubov one. Of course, this suppression of friction is less dramatic when also higher $p > 0$ transverse modes of the channel can be excited and a richer phenomenology can be observed [67–69]. As we have previously discussed at length, in a transversally unlimited geometry the transition from sub-sonic to super-sonic speeds manifests itself as the disappearance of the transverse wave pattern from Kelvin's wake and a corresponding sudden but only partial decrease of the friction force.

6.6.3.3 Large Surface Tension (Super-luminal Dispersion)

In the opposite regime of large surface tension $h < \sqrt{3}\ell_\gamma$, the super-linear form of the dispersion (6.42) makes the physics to closely resemble the behaviour of impurities in a dilute superfluid discussed in Sect. 6.4. For a slowly moving object at $v < c_s$, the locus Σ is empty and the perturbation of the surface remains localized in the vicinity of the impurity. For a fast moving object at $v > c_s$, the locus Σ and the wake pattern closely resemble the corresponding ones for the case of a supersonically moving impurity in a superfluid shown in Fig. 6.4: a Mach cone of aperture $\sin\phi = c_s/v$ located behind the object [indicated by the dashed line on Fig. 6.8(f)] and a series of curved wavefronts ahead of the object. The most significant difference with the $h > \sqrt{3}\ell_\gamma$ case of Sect. 6.6.3.1 is the position of the modulation with respect to the Mach cone: in the sub-linear case of panel (c), it lies within (i.e. behind) the Mach cone, while in the super-linear case of panel (f), it stays outside (i.e. in front of) the Mach cone.

6.7 Čerenkov Processes and the Stability of Analogue Black/White Holes

The systems that were considered in the previous sections are presently among the most promising candidates for the realization of condensed matter analogues of gravitational black (or white) holes: the key idea of analogue models is to tailor the spatial structure of the flow in a way to obtain a horizon surface that waves can cross only in one direction. Upon quantization, a number of theoretical works have predicted that a condensed matter analogue of Hawking radiation should be emitted by the horizon. A complete review of this fascinating physics can be found in the other chapters of the book. In this last section, we shall review some consequences of Čerenkov processes that are most significant for the stability of analogue black and white holes based on either flowing superfluids or surface waves on flowing water.

The role of Čerenkov-like emission processes in the dynamics of the strong optical pulses that are used in optical analogue models based on nonlinear optics [70] is still in the course of being elucidated and interesting experimental observations in this direction have recently appeared [71]. Here, it is important to remind that, differently from the all-optical Čerenkov radiation experiments of [20, 35] where an effective moving dipole was generated by $\chi^{(2)}$ nonlinearity, the analogue models of [70] are based on the time- and space-dependent effective refractive index profile due to a $\chi^{(3)}$ optical nonlinearity: given the centro-symmetric nature of the medium under examination, no effective moving dipole can in fact appear unless the medium shows some material imperfection.

6.7.1 Superfluid-Based Analogue Models

Let us start from the simplest case of analogue black/white holes based on flowing superfluids for which a complete theoretical understanding is available [28, 72–76]. In a one dimensional geometry, the horizon consists of a point separating a region of sub-sonic flow from a region of super-sonic flow. In a black hole the sub-sonic region lies upstream of the horizon, while in a white hole the sub-sonic region lies downstream of the horizon. A sketch of both configurations is reproduced in Fig. 6.9 together with the Bogoliubov dispersion of excitations as observed from the laboratory frame: in the most common configurations, the flow has a non-trivial structure only in a small region around the horizon and recovers a homogeneous shape with space-independent density and speed farther away from the horizon. In the laboratory reference frame (corresponding to the rest frame of the impurity), the Čerenkov emission occurs in the zero-frequency Bogoliubov modes, the so-called *zero modes*.

As we have discussed in detail in Sect. 6.4, the super-linear nature of Bogoliubov dispersion restricts Čerenkov emission processes to super-sonic flows, where they generate waves that propagate in the upstream direction. In the geometry under consideration here, the flow is everywhere smooth exception made for the horizon region. Combining these requirements immediately rules out the possibility of Čerenkov emission in black hole configurations: the group velocity of the zero mode waves emitted at the horizon points in the direction of the sub-sonic region, where it can no longer be supported. This simple kinematic argument contributes to explaining the remarkable dynamical stability of acoustic black hole configurations, as observed in numerical simulations of their formation starting from a uniformly moving fluid hitting a localized potential barrier [77].

In contrast, white hole configurations are much more sensitive to the dissipation of energy via Čerenkov processes: Bogoliubov excitations can appear in the supersonic region upstream of the horizon and give rise to significant modulations of

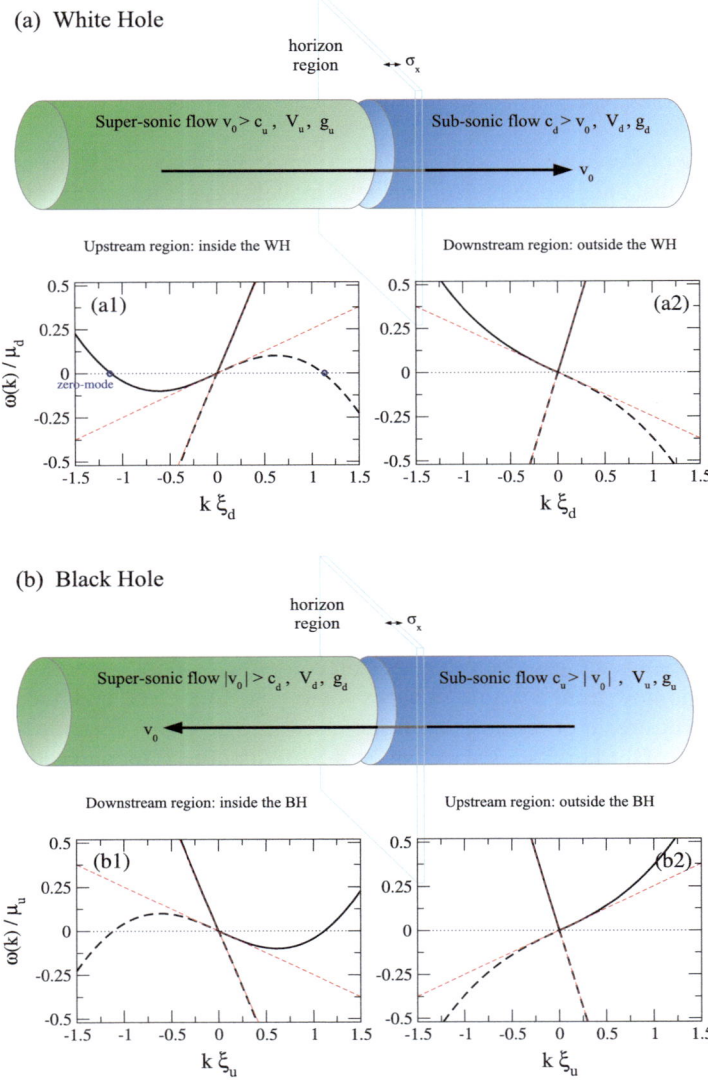

Fig. 6.9 *Main panels*: sketch of the flow geometry for white (**a**) and black (**b**) hole configurations based on superfluids. *Smaller panels* (**a1, a2, b1, b2**): dispersion of Bogoliubov excitations in the asymptotic regions far from the horizon. Čerenkov emission is only possible for white hole configurations: the corresponding *zero mode* is indicated in blue in (**a1**). Figure from [28]

the density and flow speed, the so-called *undulation* patterns. Several reasons make such processes to be potentially harmful to the study of quantum features of the white hole radiation. To the best of our knowledge, the only known realistic scheme to generate a white hole configuration in a flowing superfluid is the one of [28] using a simultaneous spatial and temporal modulation of both the atom-atom interaction strength and the external confining potential. The main difficulty of this configuration is that it requires a very precise tuning of the system parameters to eliminate unwanted Čerenkov emission processes that may mask the quantum vacuum radiation.

Even if a perfect preparation of the white hole is assumed, Čerenkov emission processes may still be triggered by nonlinear effects in the horizon region. As it was shown in [28], an incident classical Bogoliubov wavepacket is able to induce a distortion of the horizon proportional to the square of its amplitude, which then results in the onset of a continuous wave Čerenkov emission from the horizon and the appearance of a spatially oscillating modulation in the density profile upstream of the horizon. Of course, a similar mechanism is expected to be initiated by quantum fluctuations when back-reaction effects are included in the model, i.e. the non-linear interaction of quantum fluctuations with the underlying flow. A third, more subtle mechanism of instability of a white hole configuration was unveiled in [28]: the $1/\sqrt{\omega}$ divergence of the matrix elements of the **S**-matrix for low-frequency outgoing modes in the neighbourhood of the finite wave-vector zero mode is responsible for a steady growth of the density fluctuation amplitude in time since the formation of the white hole. Even if the temporal growth of fluctuations follows a slow logarithmic (linear at a finite initial temperature $T > 0$) law, still it is expected to strongly affect the properties of the horizon at long times.

The situation is expected to be different if fully three dimensional systems without transverse confinement are considered. In this case, Čerenkov emission can take place also in a black hole configuration: a distortion of the horizon by classical or quantum fluctuations with a non-trivial transverse structure is in fact able to excite Bogoliubov modes with a finite transverse component $k_y \neq 0$. As we have seen in Fig. 6.4, there exist such modes that can propagate in the downstream direction into the supersonic region inside the black hole.

6.7.2 Analogue Models Based on Surface Waves

The different dispersion of surface waves is responsible for dramatic qualitative differences in the wave propagation from the horizon of analogue black and white holes configurations. As it is sketched in Fig. 6.10(a, b), the trans-sonic interface is generally created in these systems by means of a spatial variation of the fluid depth h [78–80]. Depending on the detailed shape of the transition region and/or on the presence of fluctuations, Čerenkov emission by the horizon can occur into the zero modes of zero energy in the laboratory frame, as observed e.g. in [81]: a quantitative estimate of the amplitude of the resulting stationary undulation pattern for specific configurations of actual experimental interest requires however a complete solution of the hydrodynamic equations, which goes beyond the scope of the present work.

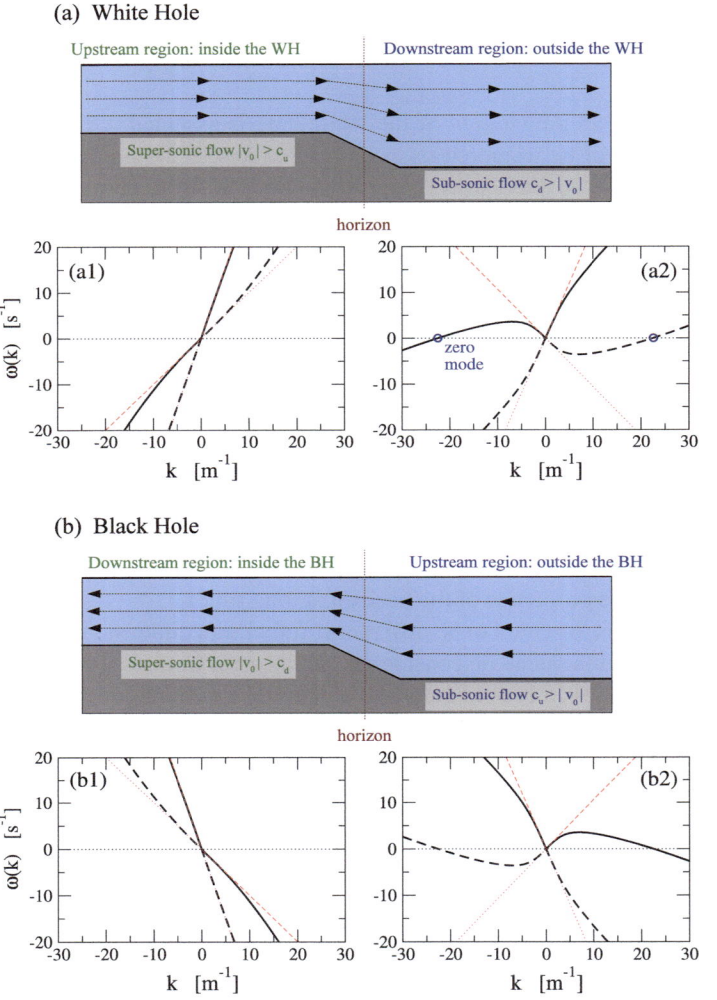

Fig. 6.10 *Main panels*: sketch of the flow geometry for white (**a**) and black (**b**) hole configurations based on surface waves on a fluid. *Smaller panels* (**a1**, **a2**, **b1**, **b2**): surface wave dispersion in the asymptotic regions far from the horizon. Čerenkov emission is only possible for white hole configurations: the corresponding *zero mode* is indicated in blue in (**a2**). Differently from the superfluid case of Fig. 6.9, the zero mode now has a group velocity in the downstream direction into the subsonic region. Parameters: flow speed $v = 2$ m/s, fluid depth $h = 0.1$ m [white hole, upstream inner region, panel (**a1**)], $v = 0.666$ m/s, fluid depth $h = 0.3$ m [white hole, downstream outer region, panel (**a2**)]; flow speed $v = -2$ m/s, fluid depth $h = 0.1$ m [black hole, downstream inner region, panel (**b1**)], $v = -0.666$ m/s, fluid depth $h = 0.3$ m [black hole, upstream outer region, panel (**b2**)]. For the chosen parameters, the effect of surface tension at the water/air interface is negligible

With an eye to the experiments of [78–80], we can restrict our attention to low-wave-vector gravity waves and neglect surface tension effects in a simplest one-dimensional geometry. In this case, Čerenkov emission processes only occur for flow speeds lower than the sonic speed $c_s = \sqrt{gh}$ and result in an emission in the downstream direction. From the surface wave dispersions shown in Fig. 6.10(a1, a2, b1, b2), one can easily see that Čerenkov emission can again only occur in white hole configurations, which are then expected to be again less stable than black hole ones. As in the superfluid case, Čerenkov emission of surface waves with a finite $k_y \neq 0$ become possible also in the black hole case as soon as a two dimensional geometry is considered and Kelvin's diverging waves are allowed. In spite of its importance in view of on-going experiments, we are not aware of any comprehensive work having studied in full detail the dynamical stability of surface wave-based analogue white/black holes as in the case of superfluid-based ones [28].

6.8 Conclusions

In this chapter we have presented a review on some most significant aspects of the Čerenkov effect from a modern and interdisciplinary point of view. In our perspective the same basic process of generalized Čerenkov emission encompasses all those emission processes that take place when a uniformly moving source is coupled to some excitation field: as soon as the source velocity exceeds the phase velocity of some mode of the field, this gets continuously excited. Simple geometrical arguments are presented that allow to extract the shape of the emission pattern in real and **k** space from the dispersion law $\Omega(\mathbf{k})$ of the field. Application of the general concepts to some most illustrative cases is discussed, from the standard Čerenkov emission of relativistically moving charged particles in non-dispersive media, to the Mach cone behind a supersonically moving impurity in a superfluid, to the wake of gravity and capillary waves behind a duck swimming on the surface of a quiet lake. The impact of Čerenkov emission processes on condensed-matter analogue models of gravitational physics is finally discussed. Open questions in this direction are reviewed.

Acknowledgements I.C. is grateful to G. C. La Rocca and M. Artoni for triggering his interest in the rich physics of the Čerenkov effect, to C. Ciuti, M. Wouters, A. Amo, A. Bramati, E. Giacobino, and A. Smerzi for a long-lasting collaboration on (among other) the Bogoliubov-Čerenkov emission in atomic and polaritonic superfluids, and to R. Balbinot, A. Fabbri, C. Mayoral, R. Parentani, and A. Recati for fruitful collaboration on analogue models. I.C. acknowledges financial support from ERC through the QGBE grant. G.R. is grateful to Conseil Général 06 and région PACA (HYDRO Project) for financial support.

References

1. Whitham, G.B.: Linear and Nonlinear Waves. Wiley-Interscience, New York (1974)

2. Lighthill, J.: Waves in Fluids. Cambridge University Press, Cambridge (1979)
3. Mei, C.C., Stiassnie, M., Yue, D.K.P.: Theory and Applications of Ocean Surface Waves: Part I, Linear Aspects; Part II, Nonlinear Aspects. World Scientific, Singapore (2005)
4. Darrigol, O.: Worlds of Flow: A History of Hydrodynamics from the Bernoulli to Prandtl. Oxford University Press, Oxford (2005)
5. Darrigol, O.: Arch. Hist. Exact Sci. **58**, 21–95 (2003)
6. Sorensen, R.M.: Adv. Hydrosci. **9**, 49–83 (1973)
7. Yih, C.-S., Zhu, S.: Q. Appl. Math. **47**, 17–33 (1989)
8. Yih, C.-S., Zhu, S.: Q. Appl. Math. **47**, 35–44 (1989)
9. Torsvik, T.: In: Quak, E., Soomere, T. (eds.) Applied Wave Mathematics. Springer, Berlin (2009)
10. Soomere, T.: In: Quak, E., Soomere, T. (eds.) Applied Wave Mathematics. Springer, Berlin (2009)
11. Raphaël, É., de Gennes, P.-G.: Phys. Rev. E **53**, 3448 (1996)
12. Le Merrer, M., Clanet, C., Quéré, D., Raphaël, É., Chevy, F.: Proc. Natl. Acad. Sci. USA **108**, 15064 (2011)
13. Jelley, J.V.: Cherenkov Radiation and Its Applications. Pergamon Press, London (1958)
14. Carusotto, I., Ciuti, C.: Phys. Rev. Lett. **93**, 166401 (2004)
15. Ciuti, C., Carusotto, I.: Phys. Status Solidi B **242**, 2224 (2005)
16. Amo, A., Lefrère, J., Pigeon, S., Adrados, C., Ciuti, C., Carusotto, I., Houdré, R., Giacobino, E., Bramati, A.: Nat. Phys. **5**, 805 (2009)
17. Cornell, E.: Talk at the KITP Conference on Quantum Gases (2004). http://online.itp.ucsb.edu/online/gases_c04/cornell/
18. Carusotto, I., Hu, S.X., Collins, L.A., Smerzi, A.: Phys. Rev. Lett. **97**, 260403 (2006)
19. Afanasiev, G., Kartavenko, V.G., Magar, E.N.: Physica B **269**, 95 (1999)
20. Stevens, T.E., Wahlstrand, J.K., Kuhl, J., Merlin, R.: Science **291**, 627 (2001)
21. Carusotto, I., Artoni, M., La Rocca, G.C., Bassani, F.: Phys. Rev. Lett. **87**, 064801 (2001)
22. Luo, C., Ibanescu, M., Johnson, S.G., Joannopoulos, J.D.: Science **299**, 368 (2003)
23. Xi, S., Chen, H., Jiang, T., Ran, L., Huangfu, J., Wu, B.-I., Kong, J.A., Chen, M.: Phys. Rev. Lett. **103**, 194801 (2009)
24. Pitaevskii, L., Stringari, S.: Bose-Einstein Condensation. Oxford University Press, Oxford (2003)
25. Pines, D., Nozieres, P.: The Theory of Quantum Liquids, vols. 1 and 2. Addison-Wesley, Redwood City (1966)
26. Leggett, A.J.: Rev. Mod. Phys. **71**, S318 (1999)
27. Landau, L.D., Lifshitz, E.M., Pitaevskii, L.P.: Statistical Physics, vol. 2. Pergamon, Oxford (1980)
28. Mayoral, C., Recati, A., Fabbri, A., Parentani, R., Balbinot, R., Carusotto, I.: New J. Phys. **13**, 025007 (2011)
29. Lock, E.H.: Phys. Usp. **51**, 375–393 (2008)
30. Pomeau, Y., Rica, S.: Phys. Rev. Lett. **71**, 247 (1993)
31. Astrakharchik, G.E., Pitaevskii, L.P.: Phys. Rev. A **70**, 013608 (2004)
32. Vavilov, S.: Dokl. Akad. Nauk SSSR **2**, 457 (1934)
33. Cherenkov, P.A.: Dokl. Akad. Nauk SSSR **2**, 451 (1934)
34. Frank, I., Tamm, I.: Dokl. Akad. Nauk SSSR **14**, 107 (1937)
35. Auston, D.H., Cheung, K.P., Valdmanis, J.A., Kleinman, D.A.: Phys. Rev. Lett. **53**, 1555 (1984)
36. Tamm, I.: J. Phys. (Mosc.) **1**, 439 (1939)
37. Frank, I.M.: Nucl. Instrum. Methods Phys. Res., Sect. A **248**, 7 (1986)
38. Hau, L.V., Harris, S.E., Dutton, Z., Behroozi, C.H.: Nature **397**, 594 (1999)
39. Kash, M.M., et al.: Phys. Rev. Lett. **82**, 5229 (1999)
40. Inouye, S., et al.: Phys. Rev. Lett. **85**, 4225 (2000)
41. Budker, D., Kimball, D.F., Rochester, S.M., Yashchuk, V.V.: Phys. Rev. Lett. **83**, 1767 (1999)
42. Frisch, T., Pomeau, Y., Rica, S.: Phys. Rev. Lett. **69**, 1644 (1992)

43. Neely, T., Samson, E., Bradley, A., Davis, M., Anderson, B.: Phys. Rev. Lett. **104**, 160401 (2010)
44. Amo, A., Pigeon, S., Sanvitto, D., Sala, V.G., Hivet, R., Carusotto, I., Pisanello, F., Lemenager, G., Houdre, R., Giacobino, E., Ciuti, C., Bramati, A.: Science **332**, 1167 (2011)
45. Nardin, G., Grosso, G., Léger, Y., Pietka, B., Morier-Genoud, F., Deveaud-Plédran, B.: Nat. Phys. **7**, 635 (2011)
46. Sanvitto, D., Pigeon, S., Amo, A., Ballarini, D., De Giorgi, M., Carusotto, I., Hivet, R., Pisanello, F., Sala, V.G., Soares-Guimaraes, P.S., Houdr, R., Giacobino, E., Ciuti, C., Bramati, A., Gigli, G.: Nat. Photonics **5**, 610 (2011)
47. Roberts, D.C., Pomeau, Y.: Phys. Rev. Lett. **95**, 145303 (2005)
48. Rayfield, G.W.: Phys. Rev. Lett. **16**, 934 (1966)
49. Phillips, A., McClintock, P.V.E.: Phys. Rev. Lett. **33**, 1468 (1974)
50. Ketterle, W., Inouye, S.: Lecture notes from Summer School on Bose-Einstein condensates and atom lasers, Cargèse, France (2000). Available as preprint arXiv:cond-mat/0101424
51. Leboeuf, P., Pavloff, N.: Phys. Rev. A **64**, 033602 (2001)
52. Pavloff, N.: Phys. Rev. A **66**, 013610 (2002)
53. Gladush, Yu.G., El, G.A., Gammal, A., Kamchatnov, A.M.: Phys. Rev. A **75**, 033619 (2007)
54. Freixanet, T., Sermage, B., Bloch, J., Marzin, J.Y., Planel, R.: Phys. Rev. B **60**, R8509 (1999)
55. Houdré, R., Weisbuch, C., Stanley, R.P., Oesterle, U., Ilegems, M.: Phys. Rev. B **61**, 13333R (2000)
56. Langbein, W., Hvam, J.M.: Phys. Rev. Lett. **88**, 047401 (2002)
57. Bazaliy, Y.B., Jones, B.A., Zhang, S.C.: Phys. Rev. B **57**, R3213 (1998)
58. Fernández-Rossier, J., Braun, M., Núñez, A.S., MacDonald, A.H.: Phys. Rev. B **69**, 174412 (2004)
59. Vlaminck, V., Bailleul, M.: Science **322**, 410 (2008)
60. Cohen-Tannoudji, C., Diu, D., Laloë, F.: Quantum Mechanics. Wiley, New York (1978)
61. Jannes, G., Piquet, R., Maïssa, P., Mathis, C., Rousseaux, G.: Phys. Rev. E **83**, 056312 (2011)
62. Russell, J.S.: Report on waves. British Association for the Advancement of Science, Annual Report (1844). Figure reproduced from [4, 5]
63. Thomson, W. (Lord Kelvin): On ship waves. Lecture delivered at the "Conversazione" in the Science and Art Museum, Edinburgh, on 3 August 1887. Published in: Institution of Mechanical Engineers, Minutes of proceedings, 409–434 (1887). Figure reproduced from [4, 5]
64. Chepelianskii, A., Chevy, F., Raphaël, É.: Phys. Rev. Lett. **100**, 074504 (2008)
65. Bush, J.W.M., Hu, D.L.: Annu. Rev. Fluid Mech. **38**, 339 (2006)
66. Voise, J., Casas, J.: J. R. Soc. Interface **7**, 343 (2010)
67. Chen, X.-N., Sharma, S.-D.: J. Fluid Mech. **335**, 305–321 (1997)
68. Chen, X.-N., Sharma, S.-D., Stuntz, N.: J. Fluid Mech. **478**, 111–124 (2003)
69. Chen, X.-N., Sharma, S.-D., Stuntz, N.: J. Ship Res. **47**, 1–10 (2003)
70. Belgiorno, F., Cacciatori, S.L., Ortenzi, G., Sala, V.G., Faccio, D.: Phys. Rev. Lett. **104**, 140403 (2010)
71. Rubino, E., McLenaghan, J., Kehr, S.C., Belgiorno, F., Townsend, D., Rohr, S., Kuklewicz, C.E., Leonhardt, U., König, F., Faccio, D.: Preprint. arXiv:1201.2689
72. Balbinot, R., Fabbri, A., Fagnocchi, S., Recati, A., Carusotto, I.: Phys. Rev. A **78**, 021603 (2008)
73. Carusotto, I., Fagnocchi, S., Recati, A., Balbinot, R., Fabbri, A.: New J. Phys. **10**, 103001 (2008)
74. Recati, A., Pavloff, N., Carusotto, I.: Phys. Rev. A **80**, 043603 (2009)
75. Macher, J., Parentani, R.: Phys. Rev. A **80**, 043601 (2009)
76. Macher, J., Parentani, R.: Phys. Rev. D **79**, 124008 (2009)
77. Kamchatnov, A.M., Pavloff, N.: Preprint. arXiv:1111.5134
78. Rousseaux, G., Mathis, C., Maïssa, P., Philbin, T.G., Leonhardt, U.: New J. Phys. **10**, 053015 (2008)

79. Rousseaux, G., Maïssa, P., Mathis, C., Coullet, P., Philbin, T.G., Leonhardt, U.: New J. Phys. **12**, 095018 (2010)
80. Weinfurtner, S., Tedford, E.W., Prentice, M.C.J., Unruh, W.G., Lawrence, G.A.: Phys. Rev. Lett. **106**, 021302 (2010)
81. Unruh, W.G.: Philos. Trans. R. Soc. Lond. **366**, 2905–2913 (2008)

Chapter 7
Some Aspects of Dispersive Horizons: Lessons from Surface Waves

Jennifer Chaline, Gil Jannes, Philppe Maïssa, and Germain Rousseaux

Abstract Hydrodynamic surface waves propagating on a moving background flow experience an effective curved spacetime. We discuss experiments with gravity waves and capillary-gravity waves in which we study hydrodynamic black/white-hole horizons and the possibility of penetrating across them. Such possibility of penetration is due to the interaction with an additional "blue" horizon, which results from the inclusion of surface tension in the low-frequency gravity-wave theory. This interaction leads to a dispersive cusp beyond which both horizons completely disappear. We speculate the appearance of high-frequency "superluminal" corrections to be a universal characteristic of analogue gravity systems, and discuss their relevance for the trans-Planckian problem. We also discuss the role of Airy interference in hybridising the incoming waves with the flowing background (the effective spacetime) and blurring the position of the black/white-hole horizon.

7.1 Introduction

Several physical systems reproduce certain properties of astrophysical objects like black holes: they exhibit an effective curved spacetime when a wave propagates in a "moving" medium [1, 2]. Examples can be found in acoustics, dielectrics, optical fibres, micro-wave guides, Bose-Einstein condensates, superfluids, ion traps... Starting with the seminal works of White [3], Anderson & Spiegel [4], Moncrief [5] and Unruh [6], there has been a growing interest for such analogue models of gravity in order to simulate and understand the physics of wave propagation on a curved spacetime. The case of interface and surface waves was initiated by Schützhold & Unruh, who derived the equation of propagation of long gravity waves moving on a background flow in terms of a general relativistic metric [7]. A black hole, or its time-inverse: a white hole, can indeed be mimicked by the interaction between interfacial waves and a liquid current [7–11]. Just like the expected behaviour of light near the event horizon of a black hole, gravity waves cannot escape a hydrodynamic

---

J. Chaline · G. Jannes · P. Maïssa · G. Rousseaux (✉)
Laboratoire J.-A. Dieudonné, UMR CNRS-UNS 6621, Université de Nice-Sophia Antipolis, Parc Valrose, 06108 Nice Cedex 02, France
e-mail: germain.rousseaux@unice.fr

black hole featuring a trapping line caused by the velocity gradient of a sufficiently strong background flow. In hydrodynamic experiments, it is usually more convenient to simulate a white hole. Then, gravity waves cannot enter a hydrodynamic white hole featuring a blocking line.

In a previous series of experiments, such an artificial white hole was created and observed in a laboratory, using water waves in the presence of a counter-flow inside a wave-tank with varying bottom profile [8]. Here, we provide a status report on a new series of ongoing experiments, in which we study the behaviour of surface waves of various frequencies in the gravity and gravity-capillary regime in the presence of such white-hole horizons. We encounter a good qualitative agreement with the theory developed in [9, 10], and interpret the interesting quantitative differences. We are in particular interested in the following key points of the theory of gravity-capillary surface waves. When including surface tension, a second and third horizon appear on top of the white-hole horizon for gravity waves: a "blue" horizon associated with mode-converted blue-shifted waves, and a "negative" horizon. This negative horizon is associated with the appearance of negative-energy waves [8] (recently observed in a similar setup [11]), an essential feature of Hawking radiation (the quantum glow of super-massive objects). The blue horizon can interact and even merge with the white horizon providing two scenarios in order to escape an artificial black hole. One scenario, consisting of a double bounce with mode conversion, was experimentally discovered nearly three decades ago by Badulin et al. [12], and interpreted in terms of the black/white-hole analogy in [10]. The other scenario is novel and consists of a direct dispersive penetration across the white-hole horizon. We have observed strong indications to validate this second scenario, and hope to confirm these quantitatively in the near future. This would establish that one can enter into the (normally forbidden) white-hole region by sending high-frequency capillary waves. These are not blocked at the primary white-hole horizon, contrarily to the low-frequency gravity waves, nor even at higher counter-flow velocities.[1] Indeed, the horizon completely disappears above a certain frequency due to the merging and consequent disappearance of the blue and white horizons.

These scenarios are not limited to hydrodynamic models, but are generic consequences of the dispersive properties beyond the relativistic regime. A double-bouncing scenario is possible in any system where "subluminal" dispersion (group velocity c_g decreasing with the wave number k) at intermediate wavenumbers k gives place to "superluminal" dispersion (c_g increasing with k) at higher k. This latter condition (superluminal high-k dispersion, irrespective of the behaviour at intermediate k) alone is sufficient for direct dispersive penetration.

---

[1] In the presence of an ever-increasing counter-flow velocity, the capillary waves will continuously blueshift and ultimately vanish through viscous damping. We will come back to this point in Sect. 7.6.

7.2 Preliminaries

Water waves in the presence of a uniform current are described by the dispersion relation [12–15]

$$(\omega - \mathbf{U}\cdot\mathbf{k})^2 = \left(gk + \frac{\gamma}{\rho}k^3\right)\tanh(kh), \tag{7.1}$$

where $\omega/2\pi$ is the frequency of the wave in the rest frame and k the wavenumber; g denotes the gravitational acceleration at the water surface, ρ the fluid density, γ the surface tension, $U < 0$ the constant velocity of the background flow and h the water depth. For water, $\rho = 1000 \text{ kg m}^{-3}$ and $\gamma = 0.073 \text{ N m}^{-1}$. The flow induces a Doppler shift of the pulsation ω.

The relativistic regime with its Schwarzschild-like metric corresponds to the shallow-water limit ($kh \ll 1$) of gravity-wave propagation [7]: $(\omega - \mathbf{U}\cdot\mathbf{k})^2 \simeq ghk^2$, with the relativistic "invariant" speed $c = \sqrt{gh}$. An analogue black/white-hole horizon will then appear when $|U| = \sqrt{gh}$. But the concept of horizon can easily be generalized to any location where the group velocity $c_g \equiv d\omega/dk$ vanishes. The shallow-water limit $kh \ll 1$ can be realized for example in the circular hydraulic jump, which spontaneously forms a hydrodynamic white hole [16]. In wave-channel experiments, both the shallow-water and the deep-water limits can be probed, depending on the setup. Interaction with a counter-current will lead to a blueshifting of the incident waves: k increases. But variations of the counter-current flow rate are typically achieved by a variation of h (e.g. through the immersion of a bump). These two effects compete and various kh regimes can in principle be realised. At sufficiently high wavenumbers k, small-scale dispersive corrections appear due to capillarity, which will turn out to be crucial to cross a horizon. Since this is our main object of study, we will from now on focus on the deep-water limit ($kh \gg 1$), and start again from long-wavelength gravity waves, proceeding step by step towards higher k.

When a gravity wave meets a counter-current, the incident wavelength diminishes and the wave height increases [12–14, 17–26]. According to the ray theory, the wave amplitude would diverge when blocking occurs. However, such caustic for the energy is avoided by a regularization process. Due to the velocity gradient, the incoming waves are somewhat diffracted before being stopped at the blocking point, where c_g changes sign and the waves are reflected. Blue-shifted modes are therefore created through a process of mode conversion at the blocking point. Since the incident and blue-shifted waves have the same wavenumber at the blocking point, they interfere and a spatial resonance appears. The diffraction implies that the figure of interference is not a simple standing wave. In the deep-water limit valid for our experiments, an Airy interferences pattern appears [9, 19–21]. The blocking point itself is a saddle-node or tangent bifurcation [9]: it marks the point where the two real solutions disappear.

When including surface tension [10], the deep-water ($kh \gg 1$) dispersion relation becomes

$$(\omega - \mathbf{U}\mathbf{k})^2 \simeq gk + \frac{\gamma}{\rho}k^3, \tag{7.2}$$

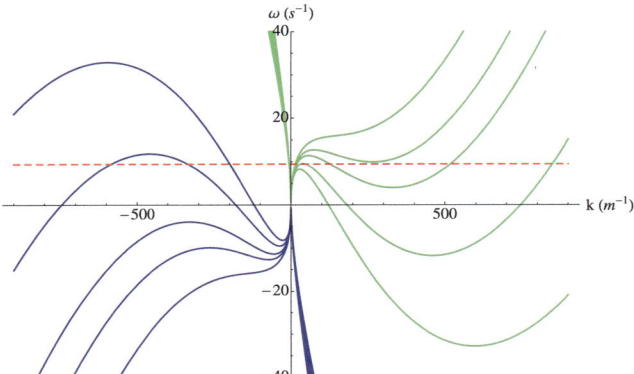

Fig. 7.1 Dispersion relation (7.2) for varying counter-flow U (the curves rotate clockwise with increasing $|U|$), plotted in the form $\omega = \mathbf{U}\mathbf{k} \pm \sqrt{(gk + \frac{\gamma}{\rho}k^3)\tanh(kh)}$. The group velocity $c_g = \frac{d\omega}{dk}$ corresponds to the slope of the *green/blue curves* (*green*: positive co-moving frequency $\omega' = \omega - \mathbf{U}.\mathbf{k}$, *blue*: negative ω'), and horizons are characterized by local minima/maxima. The double-bouncing observed by Badulin [12] corresponds to the *dashed red line* (see also [10])

or $(\omega - \mathbf{U}\mathbf{k})^2 \simeq gk(1 + l_c^2 k^2)$, where $l_c \equiv \sqrt{\frac{\gamma}{\rho g}}$ is the capillary length ($l_c \simeq 2.7$ mm for water). The gravity waves are still blocked, with the blocking velocity for pure gravity waves ($\gamma = 0$) given by $|U_g| = gT/8\pi$, with $T = 2\pi/\omega$ the period. Moreover, the blue-shifted waves are also stopped at a new blocking point on their backward drift (group velocity $c_g < 0$, along with the background flow U, although the phase velocity $c_\phi \equiv \omega/k > 0$). The capillary asymptotic limit for the blocking velocity of these blue-shifted waves is $U^*_{T\to\infty} = U_\gamma = -\sqrt{2}(\frac{\gamma g}{\rho})^{1/4}$ [10]. At this second blocking point, the blue-shifted wave merges with a new capillary solution, which appears (again through mode conversion, see Fig. 7.1) at this secondary turning point. The capillary waves propagate in the same direction as the original incident gravity waves. These newly created capillary waves are not blocked by the primary saddle-node line but go through the gravity horizon [10]. Badulin et al. observed experimentally that gravity waves can undergo such double bouncing behaviour followed by conversion to capillary waves which propagate into the forbidden region, and finally vanish by viscous damping [12]. Figure 7.1 illustrates the process. In the context of the black hole analogy, when time-reversing these observations, one concludes that incident long-wavelength gravity waves cannot escape from a trapping region (black hole) unless they are converted into the capillary range.

The second escape route consists of creating incident waves from the start in the capillary range. The horizon completely disappears below some critical period T_c, determined by the cusp formed through the interaction of the white and blue horizons. Waves with $T < T_c$ can then propagate straight ahead, avoiding any horizons, and enter the "forbidden" region (or escape from the trapping region). The cusp can

7 Some Aspects of Dispersive Horizons: Lessons from Surface Waves

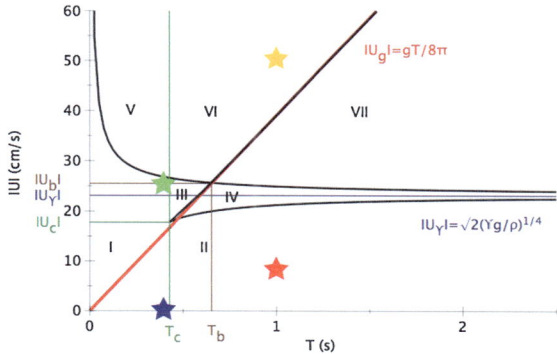

Fig. 7.2 Phase space: background counter-flow velocities U versus wave period T. The blocking curves U^* are marked in *thick black lines*. The white-hole horizon (corresponding to the gravity-wave blocking curve U_g—*thick red line*—when $\gamma = 0$) intersects the blue horizon or blocking curve for the blueshifted waves (the *lowest thick black line*, which asymptotes to U_γ from below) and creates a cusp (T_c, U_c) below which both horizons disappear. Note that there also exists a blocking line for the negative-frequency waves, which intersects U_g at (T_b, U_b) and also asymptotes to U_γ (from above). See also [10]. The four coloured stars correspond to the experimental results reported in Sect. 7.4

clearly be identified graphically from the (U^* vs T) phase diagram for deep-water waves ($kh \gg 1$), see Fig. 7.2, where U^* represents any critical or blocking velocity.

The theoretical phase diagram Fig. 7.2 was derived in [10]. We briefly recall the main steps in its derivation. We start from the cubic dispersion relation (7.2) for water waves: $(\omega - Uk)^2 \simeq gk + \frac{\gamma}{\rho}k^3$, where we have taken $k > 0$. A double root k_2 of this cubic equation, characteristic of a saddle-node point (and hence a horizon or turning point), is such that $(k - k_1)(k - k_2)^2 = 0$, where k_1 is the remaining simple root.

After some straightforward but tedious algebra, this constraint leads to a quintic equation for the critical velocity U^*, corresponding to all possible resonances (for $k > 0$):

$$4\rho^2 g\omega \left[U^5 + \frac{1}{4}\frac{g}{\omega}U^4 + \frac{\gamma\omega^2}{\rho g}U^3 - \frac{15}{2}\frac{\gamma\omega}{\rho}U^2 - 6\frac{g\gamma}{\rho}U - \left(\frac{\gamma g^2}{\rho\omega} + \frac{27}{4}\frac{\gamma^2\omega^3}{\rho^2 g}\right) \right] = 0. \quad (7.3)$$

A dual quintic is obtained for $k < 0$ by reversing the velocity $U \to -U$.

Using this quintic, one can numerically compute and plot the velocity for saddle-node bifurcations or blocking points as a function of the period of the incident waves, as in Fig. 7.2. Here we are mainly concerned with the gravity-wave blocking speed U_g and the blocking speed for blue-shifted waves (of which U_γ is the asymptotic limit as $T \to \infty$), but we note that there also exists a blocking line for the negative-frequency waves [10].

The well-known result $U_g = -g/4\omega$ in the gravity-wave limit immediately follows from Eq. (7.3) when setting $\gamma = 0$ (or $|U| \to \infty$). The asymptotic capillary limit $U_\gamma = -\sqrt{2}(\frac{\gamma g}{\rho})^{1/4}$ is likewise obtained for $T = \frac{2\pi}{\omega} \to \infty$. From Eq. (7.3), several analytic approximations can also be obtained. For example, to find an approximate expression for the blue horizon (the blocking of blue-shifted waves), we keep the terms up to first order in ω and introduce the perturbative development

$$U^* \simeq U_\gamma \left(1 + \frac{\epsilon}{U_\gamma}\right), \tag{7.4}$$

in order to write:

$$4\rho\omega U_\gamma^5 \left(1 + \frac{5\epsilon}{U_\gamma}\right) + g\rho U_\gamma^4 \left(1 + \frac{4\epsilon}{U_\gamma}\right) - 24g\omega\gamma U_\gamma \left(1 + \frac{\epsilon}{U_\gamma}\right) - 4g^2\gamma \simeq 0. \tag{7.5}$$

Solving for ϵ, using $\omega \to 0$, we obtain

$$\epsilon \simeq \omega \left(\frac{6\gamma}{\rho U_\gamma^2} - \frac{U_\gamma^2}{g}\right), \tag{7.6}$$

or

$$U^* \simeq U_\gamma + \frac{2\pi}{T} \sqrt{\frac{\gamma}{g\rho}} = -\sqrt{2} \left(\frac{g\gamma}{\rho}\right)^{1/4} + 2\pi \frac{l_c}{T}. \tag{7.7}$$

Note that an identical computation for the negative quintic leads to a simple change of sign in the second term: to first order, the negative-frequency blocking line has the same departure (but in opposite direction) from the capillary asymptotic limit as the blue horizon (for $T \geq T_c$).

The cusp $(T_c, U_c) = (0.425 \text{ s}, -0.178 \text{ m s}^{-1})$ corresponds to the intersection of two saddle-node lines (the white and blue horizons), and is associated with an inflection point of the dispersion relation (7.2). In dynamical-systems theory, we anticipate a so-called pitchfork bifurcation [27]: the merging of two saddle-node bifurcations is equivalent to the appearance of a fictive symmetry in the representation space of the cubic-in-k Eq. (7.2). The resulting pitchfork bifurcation is then associated with the breaking of this symmetry, which is absent from the original system. In other words, below T_c, the blocking lines or horizons corresponding to both saddle-node bifurcations completely disappear.

Note that the cusp is analogous to a critical point (second-order phase transition) in a thermodynamical phase diagram (Fig. 7.2). The saddle-node lines (first-order phase transitions) separate the analogues of thermodynamical phases [10]. One can distinguish seven regions of interest, marked by Roman numerals, which can be grouped into four phases: $A = I + II + III$, $B = IV$, $C = V + VI$ and $D = VII$.

The vertical line $T = T_c$ roughly separates the capillary ($T < T_c$) and the gravity ($T > T_c$) regimes. The A phase corresponds to a simple root of the cubic dispersion relation, where the incident wave can be of capillary (I) or gravity (II or III) type. The saddle-node line ending on U_c (critical point) and U_γ (tricritical point at infinity) corresponds to the threshold for the simultaneous appearance of blue-shifted waves and capillary waves propagating in the same direction as the incident ones (B). The C and D phases are characterized by the presence of negative energy

7 Some Aspects of Dispersive Horizons: Lessons from Surface Waves   151

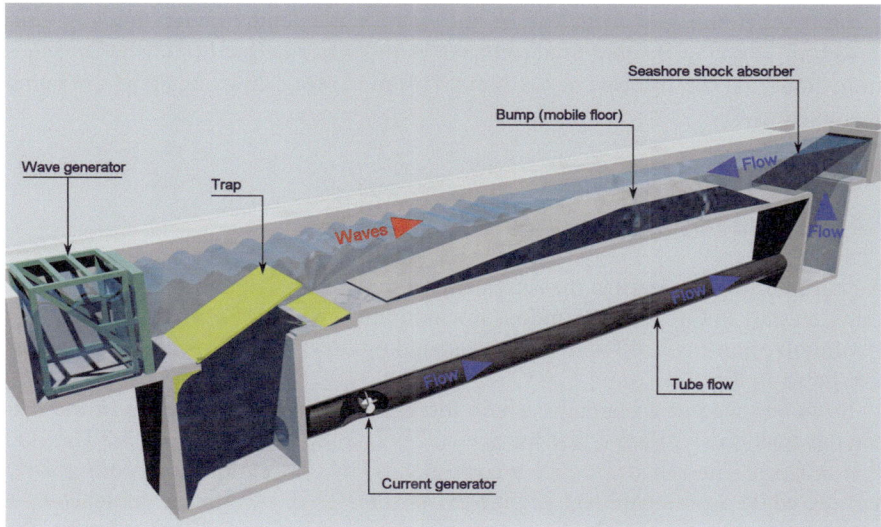

Fig. 7.3 Experimental setup. Characteristics of the bump (from left to right): linear slope with angle $\alpha_1 = 7.5°$ and length $l_1 = 8$ m, flat part $l_2 = 4.80$ m, linear slope $\alpha_3 = -18.5°$ and $l_3 = 3.30$ m. Water depth (min–max): 50 cm–160 cm

waves. In the D phase, these negative energy waves are of the gravity and capillary type and propagate in the same direction as the counter-flow. The C phase is the forbidden region for gravity waves coming from A across B or D. Only gravity waves from the A phase (after mode conversion) or directly capillary waves are allowed to go into the C phase.

7.3 Experimental Setup

We performed laboratory experiments to corroborate various aspects of the phase diagram in Fig. 7.2. These experiments were performed at ACRI, a private research company working on environmental fluid mechanics such as coastal engineering. The experiment features a wave-tank 30 m long, 1 m 80 large and 1 m 80 deep, see Fig. 7.3. The piston-type wave-maker can generate waves with periods $T = 0.35$–3 s and typical wave heights of 0.5–30 cm. A current can be created along or opposite to the direction of wave propagation with a maximum flow rate around 1.2 m^3 s^{-1}. The waves themselves are recorded using several video cameras and the videos are digitalized and calibrated.

In order to generate a gravity-wave horizon, a bump is immersed into the channel. The bump has a positive and a negative slope separated by a flat section. We send a train of progressive water waves onto the bump, hindered by a reverse fluid flow produced by a pump.

The background flow velocity depends on the water depth through flow rate conservation. The counter-current accelerates as the water height diminishes, reaches a maximum on the flat part of the bump before slowing down again as the bump height decreases.

7.4 Experimental Results

We have performed detailed measurements at the (U, T) values marked by the four coloured stars in Fig. 7.2. Continuous low-amplitude wave-trains were used in order to minimize non-linear effects. The corresponding experimental spacetime diagrams are shown in Fig. 7.4.

In the absence of a counter-current, the gravity and capillary terms in the dispersion relation (7.2) are equal for $\omega = \sqrt{2g/l_c} = 86$ rad s^{-1} ($f = 13.7$ Hz), i.e. $T = 0.073$ s. The range $T > 0.1$ s corresponds (by convention) to a pure gravity regime, while a pure capillary regime exists for $T < 0.04$ s. In the presence of a counter-current, one must look at the phase diagram in Fig. 7.2 to distinguish the gravity and capillary influence. The upper diagrams in Fig. 7.4 correspond to waves with a period of $T = 1$ s. These are therefore pure gravity waves, and we expect them to be blocked near $|U_g| = gT/8\pi = 0.39$ m s^{-1}. The lower diagrams show the propagation of waves with a period $T = 0.4$ s, approximately the lowest period allowed by the wave-maker. For weak counter-currents, these behave as pure gravity waves, but they should suffer a strong blueshifting towards the pure capillary regime as the counter-current increases, and penetrate through the gravity-wave white-hole horizon. The four diagrams in Fig. 7.4 thus correspond to the following cases.

(a) First (red star), we recorded the normal propagation of a gravity wave of $T = 1$ s and amplitude $A = 3$ cm against a moderate counter-current ($|U|$ increases from 0.074 to 0.087 m s^{-1} from left to right, well below $|U_g| = gT/8\pi = 0.39$ m s^{-1}), see Fig. 7.4 (top left).

(b) Second (yellow star), still for gravity waves of $T = 1$ s and $A = 3$ cm, we depict the range $|U| = 0.45$–0.55 m s^{-1} for which we recover the existence of a white hole marking a forbidden region into which gravity waves cannot enter ($|U_g^{\text{exp}}| \approx 0.53$ m s^{-1}), see Fig. 7.4 (top right), in agreement with the value measured in [22, 28]. We will come back to the apparent mismatch with the theoretical prediction $|U_g^{\text{th}}| = 0.39$ m s^{-1} in the following section. Note that the white horizon or blocking line is actually blurred into a "blocking region". Also notice the clear blue-shifting due to the increasing counter-current: the slope of the world-lines increases from left to right, in full analogy with the behavior of light close to a gravitational fountain. Indeed, the slope is the inverse of the phase velocity and therefore proportional to the wavelength. The slope of incoming rays grows until the rays disappear at the horizon. We point out forcefully that the slope does not increase to infinity. This is due to the dispersive effect close to the horizon, which leads to an Airy regularization mechanism, which we will discuss below. A "trans-Planckian" problem is thus avoided for

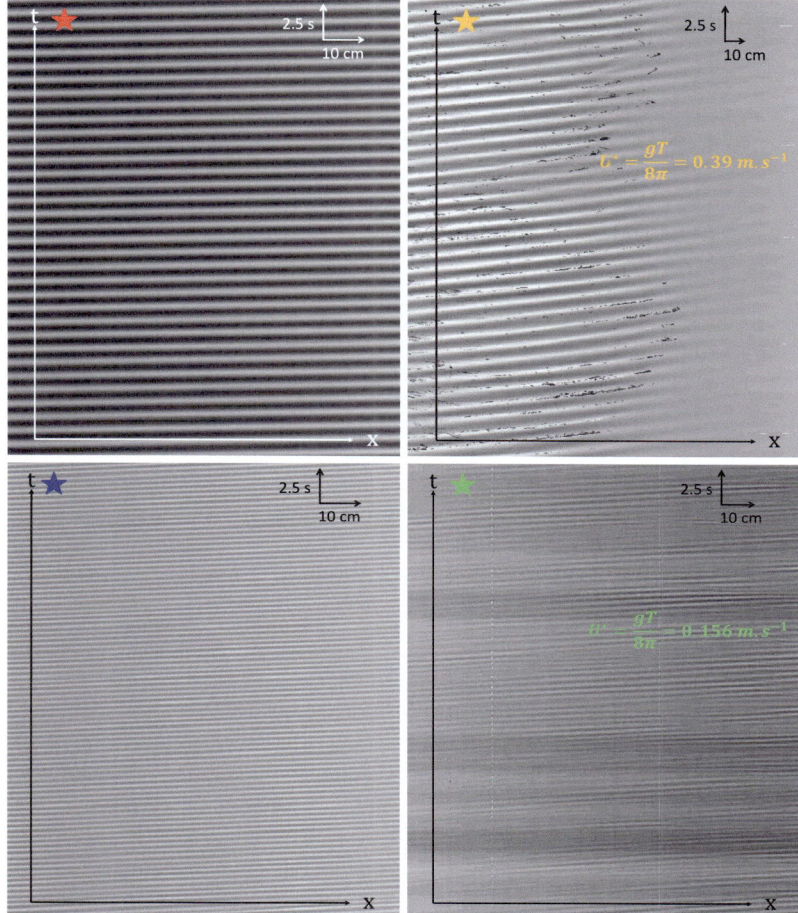

Fig. 7.4 Spatio-temporal diagrams for the four values of (U, T) marked with coloured stars in Fig. 7.2. The *light* and *dark* lines represent the world-lines of crests and troughs, respectively. The *diagrams* show: the normal propagation of a gravity wave against a moderate counter-current (*top left*); the blocking of a gravity wave at a blocking line or white-hole horizon due to a strong counter-current (*top right*); the normal propagation of a capillary-gravity wave in the absence of a counter-current (*bottom left*); and the propagation of a capillary-gravity wave across a region with a counter-flow velocity well above the gravity-wave blocking value (*bottom right*). The width of the images along the x-coordinate is 78.1 cm (*top left*), 191.7 cm (*top right*), 78.1 cm (*bottom left*) and 113.92 cm (*bottom right*). Other parameters: see main text

the incident wave, since the blue-shifting does not become infinite, as in the purely relativistic case. However, as we will also discuss below (see Fig. 7.9), there is still a problem for the mode-converted blue-shifted waves (as well as for the negative waves). In the pure gravity case, these would have $k \to \infty$ as

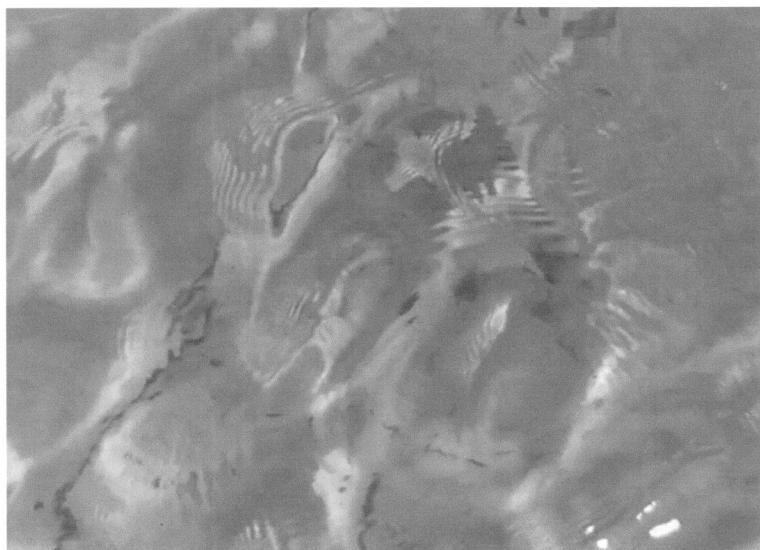

Fig. 7.5 Appearance of capillary waves for $T = 0.4$ s in the presence of a strong counter-current U

$U \to 0$: if capillarity did not come into play, then the trans-Planckian problem would, in a sense, simply be displaced to flat spacetime.

(c) Third (blue star), we show the normal propagation of a capillary-gravity wave ($T = 0.4$ s and $A = 1$–2 cm) in the absence of a counter-current ($U = 0$), see Fig. 7.4 (bottom left). Note that there is an excellent agreement between theory and experiment about the wavelengths in this case: $\lambda_{\text{exp}} = 0.25$ m versus $\lambda_{\text{th}} = 0.249$ m in the whole range $h = 0.50$–1.60 m.

(d) Finally (green star), we show the propagation of a capillary-gravity wave ($T = 0.4$ s $< T_c$, $A = 1$ cm) against a counter-current well above the blocking value: $|U| = 0.232$–0.275 m s^{-1} versus $|U_g| = gT/8\pi = 0.156$ m s^{-1}, see Fig. 7.4 (bottom right). From the diagram, we conclude that there is a complete penetration across this part of the white-hole region (forbidden for gravity waves), and barely any noticeable blue-shifting, contrarily to the case (b) of the blocked gravity waves. This last diagram therefore shows a double discrepancy with the theoretical expectation.

First, the lack of any strong blueshifting with respect to the previous case means that the transition to the capillary regime has not been fully completed for these values of U. Indeed, for the counter-currents in the range of this camera position ($|U| = 0.233$–0.275 m s^{-1}), $\lambda_{\text{th}} = 8.95$–$6.13 \times 10^{-3}$ m whereas the measured λ_{exp} is of the same order of magnitude as the incident wavelength. The capillary conversion seems to have taken place at a much higher value of the counter-current than expected. Indeed, further upstream (at higher values of the counter-current $|U|$), the appearance of capillary waves can be observed with the naked eye, see Fig. 7.5. We believe that the mismatch is due to the appearance of a transversal instability, see

Fig. 7.6 Development of transversal instability for $T = 0.4$ s: near the wave-maker (at the furthest end of the channel), the wavefronts are nearly perfectly perpendicular to the channel's edges. Towards the bottom of the picture, the wavefronts start to deform under the effect of a transversal instability, and the free water surface acquires a 'fish-scale' pattern. The capillary waves in Fig. 7.5 appear on the fronts of these fish scales

Fig. 7.6, which blurs the capillary conversion. We plan experiments with a narrower wave channel in the near future in order to reduce this transversal instability and study the conversion to the capillary regime in more detail.

Second, in the absence of such a full transition to the capillary regime, the waves should have been blocked at (or near) the gravity-wave blocking velocity $|U_g| = gT/8\pi = 0.156 \text{ m s}^{-1}$. This blocking has not taken place either. This second discrepancy is in the line of the mismatch mentioned above in the pure gravity case, and we will come back to it in the next section.

7.5 Airy Interference and Gravity-Wave Blocking

Airy interference provides the crucial mechanism through which a divergence of the amplitude is avoided [9] in the regime $kh \gg 1$, see Fig. 7.7. An explicit expression for the stopping length L_s associated with the Airy interference in the case of pure gravity waves is

$$L_s = \frac{1}{16(2\pi^5)^{1/3}} g T^{5/3} \left(\frac{dU}{dx}\right)_{x=x_*}^{-1/3}, \tag{7.8}$$

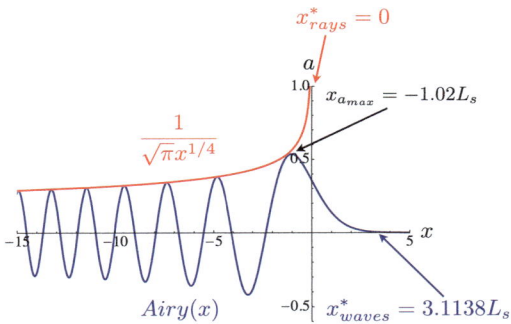

Fig. 7.7 Blocking of rays (in *red*) versus waves (in *blue*). In the ray approximation, the amplitude theoretically diverges towards the blocking point x^*_{rays}, beyond which it vanishes discontinuously. This divergence is regularized in the wave picture through Airy interference. The wave blocking point x^*_{waves} lies several Airy stopping lengths L_s further than x^*_{rays}

with x_* the horizontal blocking position. A simple derivation of this expression can be found in the Appendix. Because of the geometry of our experiment, the background surface velocity evolves linearly: $|U(x)| = 0.51 - 0.05x$, where $x = 0$ corresponds to the kink in the mobile floor (note that the x-axis is oriented along the background flow, i.e. from right to left in Fig. 7.3). From Fig. 7.7, the apparent mismatch between the well-known theoretical prediction $U_g^{\text{th}} = -\frac{gT}{8\pi}$ for the counter-flow velocity at blocking ($|U_g^{\text{th}}| = 0.39 \text{ m s}^{-1}$ for $T = 1$ s), and the measured value ($|U_g^{\text{exp}}| = 0.53 \text{ m s}^{-1}$) can also partially be understood. U_g^{th} is obtained in the ray approximation. The waves will actually be blocked a certain distance $\Delta x^*_{\text{waves}}$ further due to the Airy interference. The experimental blocking position $x_*^{\text{exp}} = -0.36$ m corresponds to the wave-blocking position. Taking the conservative assumption that a wave is considered "blocked" (i.e., it is no longer detected on camera) when its amplitude has decreased below 1 % of its maximum, one obtains $\Delta x^*_{\text{waves}} = 3.11 L_s$, i.e. (for $T = 1$ s) $\Delta x^*_{\text{waves}} = 0.61$ m. The ray-divergence position would then be at $x_*^{\text{exp}} + \Delta x^*_{\text{waves}} = 0.25$ m, corresponding to a velocity $U_g^{\text{exp-ray}} = 0.50 \text{ m s}^{-1}$. The theoretical blocking position corresponding to $|U_g^{\text{th}}| = 0.39 \text{ m s}^{-1}$ is $x_*^{\text{th}} = +2.42$ m. In other words, Airy interference explains roughly 20–25 % of the difference between the theoretical prediction and the experimental measure.

A second element which contributes to the difference between U_g^{th} and U_g^{exp} lies in the decrease with depth of the real velocity profile. The theoretical prediction for the value at the surface should, in a real experiment, be considered as an averaged (integrated) value over some depth, necessarily leading to a slightly higher value at the surface. The vertical velocity profile is approximately of the so-called plug type on the flat part of the bump and acquires a parabolic form after the flow has decelerated on the descending slope of the bump, see Fig. 7.8. Comparison of the surface values with the average, vertically integrated values gives differences of 5–20 %. Interpolating between the cases represented in Fig. 7.8, we obtain an estimated difference of ~ 10 % between the theoretically predicted value and the value measured experimentally at the surface near the blocking point ($x_*^{\text{exp}} = -0.36$ m for $T = 1$ s).

7 Some Aspects of Dispersive Horizons: Lessons from Surface Waves

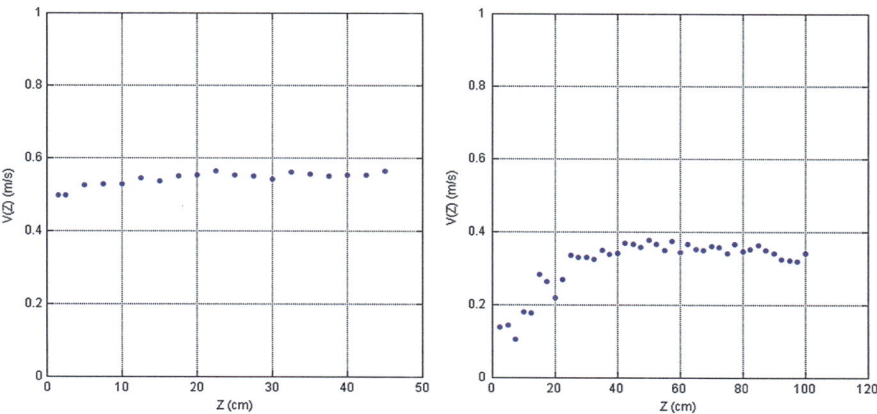

Fig. 7.8 Vertical background-flow velocity profiles: nearly plug-type profile on the bump at $x = -2.43$ m *(left)* and parabolic profile towards the wave-maker at $x = 4.20$ m *(right)*, where $x = 0$ corresponds to the kink in the bump, see Fig. 7.3

A third element which might be thought to be important is that the blocking velocity $U_g^{th} = -\frac{gT}{8\pi}$ is obtained in the pure gravity-wave limit, whereas even for $T = 1$ s, a small capillary influence persists and slightly increases the real blocking velocity. However, this difference is negligible, as can be seen from Fig. 7.4, where it corresponds to the departure between the red line U_g and the corresponding black line U^* obtained numerically from the full Eq. (7.3). It can be estimated quantitatively as follows. In the regime $kl_c \ll 1$, the dispersion relation (7.2) can be approximated by

$$\omega \simeq Uk + \sqrt{gk}\left(1 + \frac{(kl_c)^2}{2}\right). \tag{7.9}$$

The condition $\frac{d\omega}{dk} = 0$ for wave blocking becomes

$$U^* = -\sqrt{\frac{g}{4k}}\left(1 + \frac{5}{2}l_c^2 k^2\right). \tag{7.10}$$

In the limit $l_c = 0$, the blocking wavenumber k_g (corresponding to $U_g = -g/4\omega$) gives $k_g = \frac{4\omega^2}{g}$. Inserting this value in the previous expression leads to

$$U^* \simeq U_g + \Delta U^* = -\frac{gT}{8\pi} - 80\pi^3 \frac{l_c^2}{gT^3}. \tag{7.11}$$

For $T = 1$ s, $\Delta U^* \simeq -0.002$ m s$^{-1} \ll U_g$.

Non-linear effects could also play an important role, in spite of our attempts to limit them by working with small amplitudes. We limit ourselves to two comments. First, the finite wave amplitude A increases the blocking velocity. This well-known (but poorly understood) phenomenon [22, 24] can to a first approximation be modelled as an effective surface tension (see e.g. [29]):

$$\omega^2 = gk(1 + A^2 k^2) \quad (7.12)$$

(for pure gravity waves in the absence of a counter-current). Since $A/l_c \sim 10$ in our experiments, it is clear that the finite amplitude has a much stronger influence on the blocking velocity than the intrinsic surface tension, and is perhaps the main contributor to the difference between U_g^{th} and U_g^{exp}. This illustrates the importance of generating waves with a low factor Ak ($Ak \sim 0.1$ at $T = 1$ s in our current setup). A second non-linear effect which plays an important role in certain other wave-blocking experiments [23, 28] is the Benjamin-Feir instability, which leads to the appearance of so-called side-bands (excitations at frequencies slightly different from the fundamental one). However, we have verified conservation of period in our experiments, thereby nearly excluding this possibility.

We believe the above elements to provide a reasonable explanation for the apparent mismatch between $|U_g^{\text{exp}}| \approx 0.53$ m s^{-1} and $|U_g^{\text{th}}| = 0.39$ m s^{-1}. It should be noted that this mismatch is well known in the fluid-mechanics community, but not well understood. For example, [22] and [28] also observe the blocking of $T = 1$s waves at a countercurrent velocity $|U_g^{\text{exp}}| \approx 0.53$ m s^{-1}, but do not attempt to interpret this discrepancy with the theoretical prediction.

Also note that similar arguments would apply to the case of the gravity-capillary waves at $T = 0.4$ s. There, however, the additional appearance of a transversal instability mentioned above further complicates matters and further experiments are required to clarify the situation.

It is remarkable that the Airy interference hybridizes the character of the incoming wave. A hybrid is created between the original wave (through the period T, in the expression (7.8) of the stopping length L_s) and the background flow (through $\frac{dU}{dx}$). In our experiment, we are sending continuous wave-trains. If one were to send wave-packets ("particles", i.e.: superpositions of waves), then these would be deformed into superpositions of wave-flow hybrids, or "hybridons". As a matter of fact, we can take this observation further. The Airy stopping length obeys $L_s \propto \lambda^* \mathscr{U}_n^{1/3}$, where λ^* is the wavelength at blocking, and $\mathscr{U}_n$ the dimensionless "Unruh" number $\mathscr{U}_n = \omega (\frac{dU}{dx})_{x_*}^{-1}$ obtained from the two characteristic "frequencies" at blocking: the wave frequency ω and the flow gradient $\frac{dU}{dx}$. This leads to the following interpretation. Dispersion has a double role in the near-horizon physics. It keeps the wavenumber finite (i.e., it solves the trans-Planckian problem—see next section), thereby avoiding the first relativistic ray-theory pathology $\lambda^* \to 0$. Second, it creates an interference mechanism which hybridizes this wavelength with the background flow by modulating it through $\mathscr{U}_n$. Dispersion thus replaces the wavelength by a characteristic interference length L_s. This mechanism of interferences allows to solve the second pathology associated with the ray theory: the infinite amplitude at the blocking point. Indeed, when $\mathscr{U}_n \gg 1$, i.e. when the frequency of the wave is large compared to the spatial variation of the background flow velocity, then the WKB-approximation is valid. Note that $\mathscr{U}_n \gg 1$ also leads to $\exp(\omega/(\frac{dU}{dx})_{x_*}) - 1 \gg 1$ and therefore to negligible Hawking radiation. Near the blocking point, though, one always has $\mathscr{U}_n \sim 1$. The WKB approximation then breaks down, and two resonance mechanisms come into play. The first one (Airy interference) is an adiabatic process and the second one (Hawking radiation) is a non-adiabatic process, see the Chapter on "The Basics of Water Waves Theory for Analogue Gravity" elsewhere in this Volume.

As a final note, we should point out that we have neglected the presence of a zero mode in our considerations on the Airy mechanism. Such a zero mode (an $\omega = 0$ solution to the dispersion relation, or superposition of various such solutions) would deform the free surface and complicate the interference pattern, see [11, 30] and the discussion in the Chapter "The Cerenkov effect revisited: from swimming ducks to zero modes in gravitational analogs" elsewhere in this Volume. This omission is justified since in the regime $kh \gg 1$ one can minimize the amplitude of the zero mode by working at low velocities and limiting the slope of the bump. Note that $\gamma \neq 0$ implies the existence of a threshold $|U| \geq |U_\gamma|$ for the appearance of a zero mode, contrarily to the pure gravity case.

7.6 The Trans-Planckian Problem

Our experimental results to corroborate the theory developed in [10] have been slightly marred by the appearance of a transversal instability, which we hope to remedy using a narrower wave-channel. Nevertheless, there is little doubt that capillary waves can penetrate through a gravity-wave blocking line. The full strength of this statement becomes clear in the context of the gravitational analogy. The surface tension constitutes a high-k dispersive correction to the low-k gravity-wave theory. Such dispersive corrections can therefore completely alter the properties of a horizon: dispersive horizons are no longer one-way membranes, and the infinite blueshifting associated with strictly relativistic horizons disappears. The example of surface waves shows that this statement can be true even if the dispersion is (initially) subluminal.

The idea that dispersive corrections could solve the trans-Planckian problem of gravity has from the start been one of the cornerstones of the analogue gravity programme [6]. Most work has historically focused on the study of subluminal ("normal") dispersion. This seems curious in the light of the following observations. In the pure gravity-wave (subluminal) regime, the trans-Planckian problem is indeed avoided for the incident wave: it is mode-converted into a blue-shifted wave and the wave bounces away from the horizon (in terms of the group velocity; the phase velocity is still directed towards the white hole). However, as this blue-shifted wave approaches "flat spacetime" (i.e., as the counter-current velocity $|U| \to 0$), a new, secondary trans-Planckian problem arises: the wavenumber of the blue-shifted wave $k_B \to \infty$, see Fig. 7.9. This was observed earlier [31] with respect to Unruh's original subluminal model [32], and related problems with other subluminal models were also discussed in [33] and [34]. The same secondary trans-Planckian problem occurs

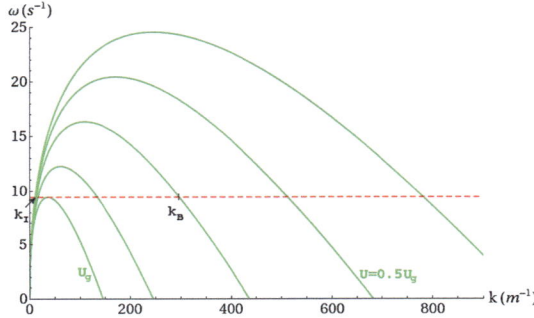

Fig. 7.9 Gravity waves and the trans-Planckian problem. For a given frequency ω (*dashed red line*), a blue-shifted wave k_B is created from the incident wave k_I through mode conversion at the blocking line $U = U_g$. Since k_B has $c_g < 0$, it moves away from the horizon towards lower $|U|$. As $|U| \to 0$, $k_B \to \infty$, leading to a secondary trans-Planckian problem in flat spacetime

for the negative-frequency waves associated with a Hawking-like process. Actually, even without invoking any horizon effects, a similar problem arises: Any countercurrent, no matter how small, would allow for the existence of both blue-shifted and negative-energy waves with infinitely small wavelengths. Purely subluminal dispersion would then solve the primary trans-Planckian problem associated with the horizon, at the cost of creating a new one in flat spacetime.

More complicated dispersion relations, e.g. as in Helium-II (superfluid ^{4}He), which has a "roton" minimum in the ω'–vs–k diagram, could overcome this problem by letting the outgoing blue-shifted wave decay at the end of the quasiparticle spectrum into two rotons, with the same total energy and momentum (i.e., $\omega \to \omega/2 + \omega/2$; $k \to k/2 + k/2$) [31]. The secondary trans-Planckian problem is then avoided because the dispersion curve ends in such a two-roton decay channel.[2]

Another obvious way of resolving the secondary trans-Planckian problem with subluminal dispersion is through dissipation, for example due to viscosity. Note that viscosity, apart from leading to dissipation, necessarily also introduces dispersion [37]. Dissipation in media is quite generic, often unavoidable (as in optical media, where dissipation and dispersion are coupled through the Kramers-Kronig relations), and might be relevant for our "fundamental" spacetime as well. Actually,

---

[2]Actually, this is not the end of the story: these rotons are still subject to the background flow, and will therefore also start blueshifting, just like the original mode, and again split into two rotons each at the end of the quasiparticle dispersion curve, etcetera, leading to an apparently endless creation of rotons. This process is limited because the roton creation will deplete the superfluid component, and ultimately destabilizes the white-hole configuration. Also note that the rotons will relax after some time due to interaction with the environment and condense into a roton BEC [35]. If one takes such a ^{4}He-like dispersion model seriously for true gravity, this could lead to the creation of a photon condensate near a white hole, or vice versa: outgoing particles from a black hole might originate from a condensate in curved spacetime. Although such a scenario might not be as crazy as it sounds [36], we will not pursue this exotic line of thought further here, and stick to "simpler" solutions of the trans-Planckian problem.

from a theoretical QFT point of view, Lorentz symmetry violation is automatically accompanied by dissipation under quite general assumptions, and dissipative effects should therefore in principle be treated together with dispersive ones [38]. However, dissipation is formally much harder to treat than dispersion, and it has received little attention in the context of (analogue) gravity. Also, the relevance of dissipation obviously depends on its characteristic scale. Here, we mainly wish to stress that the secondary trans-Planckian problem in surface waves is solved through dispersion before dissipation becomes relevant.

These observations suggest a more general interpretation for our results with water waves. The mesoscopic scale of surface tension "saves" the fluid continuum approximation from breaking down in the presence of a counter-current: The capillary behaviour at high k is essential in order to avoid a trans-Planckian pathology. We therefore expect that any system displaying analogue gravity behaviour at low k through propagation on a moving background medium will necessarily have superluminal dispersion at sufficiently high k's, unless dissipation kills the whole phenomenon before such scales are actually reached. The exotic case of Helium-II mentioned above might be the exception that confirms the rule, since even then, the dispersion relation is superluminal for a certain range of high wavenumbers beyond the roton minimum.

Moreover, we can establish the following general rules. If the first corrections at intermediate k are subluminal, followed by superluminal corrections at high k, as in the case of deep-water waves, then there will always be a white horizon and a blue horizon, leading to the possibility of a Badulin-type double-bouncing scenario. There will then also always exist some counter-current velocity U_c for which these white and blue saddle-node points merge into a pitchfork bifurcation and both horizons disappear, allowing for direct dispersive penetration. Such direct dispersive penetration is actually even more universal: it suffices to have a superluminal correction at high k to a relativistic low-k behaviour, irrespective of the intermediate regime. This is true, e.g., for phonons in BECs [39], just like for capillary surface waves: sufficiently blueshifted ("superblueshifted") modes will always be able to penetrate through any counter-flow barrier, unless dissipation prevents such superblueshifting. The case of surface waves is in a sense richer than that of BEC-phonons, in that there is a true blocking line for low-frequency gravity waves, which cannot directly penetrate the horizon. For BECs, the absence of an intermediate subluminal correction implies that even low-frequency phonons can in principle superblueshift and cross the horizon directly. Finally, in spite of our several comments regarding dissipation, it seems that both horizon-crossing scenarios can indeed be fully realized for surface waves before being dissipated.

To sum up, to enter a white hole—or, by time-inversion: to escape a black hole—one has to either tune the period to be subcritical, or bounce on two horizons.

The bottom line is of course what this teaches us for real gravity. Here the issue is more complicated, because extrapolation from the current state of observations has so far not given any evidence for dispersion (or dissipation) even at the Planck scale. Other complications might also arise which are peculiar to real gravity. For example, in [40] it was shown using a simple toy model that superluminal dispersion would render gravitational black holes strongly unstable, due to the leaking of resonant modes. In any case, we may conclude that, if dispersion is indeed relevant in gravity (possibly even far beyond the Planck scale), then subluminal dispersion alone would most certainly not be sufficient to solve the trans-Planckian problem.

Acknowledgements The authors thank C. Barceló, D. Faccio, L.J. Garay, Th.G. Philbin and G.E. Volovik for useful discussions and comments. GR is grateful to Conseil Général 06 and région PACA (HYDRO Project) for financial support.

Appendix: Airy Stopping Length

Smith was the first to derive the Airy equation in 1975 by performing an asymptotic expansion of the water-wave equations (Euler equations + continuity equation + boundary conditions) close to the caustic [19]. He inferred the so-called amplitude equation which is a nonlinear Schrödinger equation with a term proportional to the distance to the caustic. When the cubic term is negligible, the amplitude equation reduces to the Airy equation.

Following Smith and after tedious algebra (matched asymptotics and WKB solutions), Trulsen and Mei computed the following stopping length (for $\gamma = 0$) in 1993 [41]:

$$L_s = \left(\frac{U_*^2}{2k_*\omega(\frac{dU}{dx})_{x_*}} \right)^{1/3}. \tag{7.13}$$

In 1977, Basovich & Talanov [21] provided another derivation of the Airy equation by noticing that $\frac{dU}{dk} = 0$ at the blocking point. Taylor-expanding the function $U(k)$ close to its parabolic minimum and the function $U(x)$ close to the stopping point x_* and taking the inverse Fourier transformation, they deduced the Airy function and the associated stopping length:

$$L_s = \left(\frac{\omega}{4k_*^3(\frac{dU}{dx})_{x_*}} \right)^{1/3}. \tag{7.14}$$

In 1979, Peregrine & Smith [20] used an operator expansion method: the idea is to inverse Fourier-transform a truncated series expansion of the dispersion relation written in the form $G(\omega, k, x) = 0$. Again, the cubic Schrödinger equation with a spatial term was derived with another expression for the stopping length:

$$L_s = \left(\frac{G_{kk}}{2G_x} \right)^{1/3}, \tag{7.15}$$

where the subscripts mean partial derivative and the derivatives are taken at the blocking line. The method was generalized in 2004 by Suastika [24, 25] to include viscous dissipation and wave breaking. In 2003, Lavrenov [17] applied a saddle-point method to the Maslov integral representation of the uniform wave field asymptotics in the vicinity of the blocking line. He found:

$$L_s = \left(\frac{\Omega_{kk}}{2\Omega_x}\right)^{1/3}, \tag{7.16}$$

where $\omega = \Omega(k,x)$ is the dispersion relation function and the derivatives are taken at the caustic.

We will show that it is possible to derive the stopping length in a very simple fashion, inspired by the method of Basovich & Talanov, and derive a previously unnoticed scaling law for L_s.

We write the background flow velocity near the critical value $U_* = -\frac{gT}{8\pi}$ as a function of x and k, and develop both to lowest non-zero order around the stopping length:

$$U(k) \simeq U(k_*) + \frac{U''(k_*)}{2}(k-k_*)^2 = U_* - \frac{U_*^3}{4\omega^2}(k-k_*)^2, \tag{7.17}$$

$$U(x) \simeq U_* + \left(\frac{dU}{dx}\right)_{x_*}(x-x_*). \tag{7.18}$$

Equating both into

$$\frac{U_*^3}{4\omega^2}(k-k_*)^2 + \left(\frac{dU}{dx}\right)_{x_*}(x-x_*) \simeq 0, \tag{7.19}$$

and making the substitution $H(x) \simeq e^{i(k-k_*)x}$ immediately leads to the Airy differential equation

$$\frac{d^2 H}{dX^2} - XH = 0, \tag{7.20}$$

where $X = \frac{x-x_*}{L_s}$, with $L_s = \frac{|U_*|}{(4\omega^2(\frac{dU}{dx})_{x_*})^{1/3}}$. Thus, $H(x)$ is an Airy function

$$H(x) \simeq \text{Ai}\left(\frac{x-x_*}{L_s}\right) = \frac{1}{\pi}\int_0^\infty \cos\left(\frac{1}{3}t^3 + \frac{x-x_*}{L_s}t\right)dt, \tag{7.21}$$

and

$$L_s = \frac{1}{16(2\pi^5)^{1/3}}gT^{5/3}\left(\frac{dU}{dx}\right)_{x=x_*}^{-1/3} \tag{7.22}$$

is the Airy stopping length, which depends both on the incident wave and the background flow: it scales with the period T of the incident wave as $L_s \propto T^{5/3}$ and with the background flow acceleration as $L_s \propto (\frac{dU}{dx})_{x=x_*}^{-1/3}$. A straightforward dimensional analysis ($L_s \simeq g^\alpha T^\beta (\frac{dU}{dx})_{x=x_*}^\gamma$) would only lead to $\alpha = 1$ and $\beta - \gamma = 2$.

Note that Airy interference requires the flow gradient to remain approximately constant over the characteristic length of the interference process. This can easily be seen in our derivation: Eq. (7.18) is only a good approximation if $\frac{d^2U}{dx^2} \approx 0$ for $x - x_* = \mathcal{O}(L_s)$. In our experiments, $\frac{dU}{dx}$ is a constant on the whole linear slope of the bump where the horizon x_* is located, so we do not need to worry about this issue.

References

1. Barceló, C., Liberati, S., Visser, M.: Analogue gravity. Living Rev. Relativ. **8**, 12 (2005)
2. Schützhold, R., Unruh, W.G. (eds.): Quantum Analogues: From Phase Transitions to Black Holes and Cosmology. Springer, Berlin (2007)
3. White, R.W.: Acoustic ray tracing in moving inhomogeneous fluids. J. Acoust. Soc. Am. **53**, 1700–1704 (1973)
4. Anderson, J.L., Spiegel, E.A.: Radiative transfer through a flowing refractive medium. Astrophys. J. **202**, 454–464 (1975)
5. Moncrief, V.: Stability of stationary, spherical accretion onto a Schwarzschild black hole. Astrophys. J. **235**, 1038–1046 (1980)
6. Unruh, W.G.: Experimental black hole evaporation? Phys. Rev. Lett. **46**, 1351–1353 (1981)
7. Schützhold, R., Unruh, W.G.: Gravity wave analogues of black holes. Phys. Rev. D **66**, 044019 (2002)
8. Rousseaux, G., Mathis, C., Maïssa, P., Philbin, T.G., Leonhardt, U.: Observation of negative-frequency waves in a water tank: a classical analogue to the Hawking effect? New J. Phys. **10**, 053015 (2008)
9. Nardin, J.-C., Rousseaux, G., Coullet, P.: Wave-current interaction as a spatial dynamical system: analogies with rainbow and black hole physics. Phys. Rev. Lett. **102**, 124504 (2009)
10. Rousseaux, G., Maïssa, P., Mathis, C., Coullet, P., Philbin, T.G., Leonhardt, U.: Horizon effects with surface waves on moving water. New J. Phys. **12**, 095018 (2010)
11. Weinfurtner, S., Tedford, E.W., Penrice, M.C.J., Unruh, W.G., Lawrence, G.A.: Measurement of stimulated Hawking emission in an analogue system. Phys. Rev. Lett. **106**, 021302 (2011)
12. Badulin, S.I., Pokazeev, K.V., Rozenberg, A.D.: Laboratory study of the transformation of regular gravity-capillary waves in inhomogeneous currents. Izv. Akad. Nauk SSSR, Fiz. Atmos. Okeana **19**, 1035–1041 (1983)
13. Fabrikant, A.L., Stepanyants, Y.A.: Propagation of Waves in Shear Flows. World Scientific, Singapore (1998)
14. Dingemans, M.W.: Water Wave Propagation over Uneven Bottoms. World Scientific, Singapore (1997)
15. Huang, H.: Linear surface capillary-gravity short-crested waves on a current. Chin. Sci. Bull. **53**, 3267–3271 (2008)
16. Jannes, G., Piquet, R., Maissa, P., Mathis, C., Rousseaux, G.: Experimental demonstration of the supersonic-subsonic bifurcation in the circular jump: a hydrodynamic white hole. Phys. Rev. E **83**, 056312 (2011)
17. Lavrenov, I.: Wind-Waves in Oceans. Springer, Berlin (2003)
18. Peregrine, D.H.: Interactions of water waves and currents. Adv. Appl. Mech. **16**, 9–117 (1976)
19. Smith, R.: Giant waves. J. Fluid Mech. **77**, 417–431 (1976)
20. Peregrine, D.H., Smith, R.: Nonlinear effects upon waves near caustics. Philos. Trans. R. Soc. Lond. A **292**, 341–370 (1979)
21. Basovich, A.Y., Talanov, V.I.: On the short-period surface wave transformation in heterogeneous flows. Izv. Akad. Nauk SSSR, Fiz. Atmos. Okeana **13**, 766–773 (1977)

22. Chawla, A., Kirby, J.T.: Experimental study of wave breaking and blocking on opposing currents. In: Proc. 26th Int. Conf. on Coastal Engineering, Copenhagen, Denmark, ASCE, pp. 759–772 (1998)
23. Chawla, A., Kirby, J.T.: Monochromatic and random wave breaking at blocking points. J. Geophys. Res. **107**, 3067 (2002)
24. Suastika, I.K.: Wave blocking. PhD thesis, Technische Universiteit Delft, The Netherlands (2004)
25. Suastika, I.K., Battjes, J.A.: A model for blocking of periodic waves. Coast. Eng. **51**(2), 81–99 (2009)
26. Baschek, B.: Wave-current interaction in tidal fronts. In: 14th Aha Hulikoa Winter Workshop: Rogue Waves, Honolulu, Hawaii (2005)
27. Poston, T., Stewart, I.: Catastrophe Theory and Its Applications. Dover, New York (1998)
28. Ma, Y., Dong, G., Perlin, M., Ma, X., Wan, G., Xu, J.: Laboratory observations of wave evolution, modulation and blocking due to spatially varying opposing currents. J. Fluid Mech. **661**, 108–129 (2010)
29. Lamb, H.: Hydrodynamics, 6th edn. Cambridge University Press, Cambridge (1975)
30. Unruh, W.G.: Dumb holes: analogues for black holes. Philos. Trans. R. Soc. Lond. A **366**, 2905 (2008)
31. Jacobson, T.: On the origin of the outgoing black hole modes. Phys. Rev. D **53**, 7082–7088 (1996)
32. Unruh, W.G.: Sonic analog of black holes and the effects of high frequencies on black hole evaporation. Phys. Rev. D **51**, 2827–2838 (1995)
33. Corley, S., Jacobson, T.: Hawking spectrum and high frequency dispersion. Phys. Rev. D **54**, 1568–1586 (1996)
34. Jacobson, T.: Trans-Planckian redshifts and the substance of the space-time river. Prog. Theor. Phys. Suppl. **136**, 1–17 (1999)
35. Iordanskii, S.V., Pitaevskii, L.P.: Bose condensation of moving rotons. Sov. Phys. Usp. **23**, 317 (1980)
36. Hu, B.L.: Can spacetime be a condensate? Int. J. Theor. Phys. **44**, 1785 (2005)
37. Visser, M.: Acoustic black holes: horizons, ergospheres, and Hawking radiation. Class. Quantum Gravity **15**, 1767–1791 (1998)
38. Parentani, R.: Constructing QFT's wherein Lorentz invariance is broken by dissipative effects in the UV. PoS **QG-PH**, 031 (2007)
39. Pitaevskii, L., Stringari, S.: Bose-Einstein Condensation. Oxford Science Publications, Oxford (2003)
40. Barbado, L.C., Barceló, C., Garay, L.J., Jannes, G.: The trans-Planckian problem as a guiding principle. J. High Energy Phys. **11**, 112 (2011)
41. Trulsen, K., Mei, C.C.: Double reflection of capillary-gravity waves on a variable current. J. Fluid Mech. **251**, 239–271 (1993)

Chapter 8
Classical Aspects of Hawking Radiation Verified in Analogue Gravity Experiment

Silke Weinfurtner, Edmund W. Tedford, Matthew C.J. Penrice,
William G. Unruh, and Gregory A. Lawrence

Abstract There is an analogy between the propagation of fields on a curved spacetime and shallow water waves in an open channel flow. By placing a streamlined obstacle into an open channel flow we create a region of high velocity over the obstacle that can include wave horizons. Long (shallow water) waves propagating upstream towards this region are blocked and converted into short (deep water) waves. This is the analogue of the stimulated Hawking emission by a white hole (the time inverse of a black hole). The measurements of amplitudes of the converted waves demonstrate that they appear in pairs and are classically correlated; the spectra of the conversion process is described by a Boltzmann-distribution; and the Boltzmann-distribution is determined by the change in flow across the white hole horizon.

8.1 Motivation

There is a broad class of systems where perturbations propagate on an effective $(d+1)$ dimensional spacetime geometry. In the literature this phenomenon

---

S. Weinfurtner (✉)
SISSA—International School for Advanced Studies, Via Bonomea 265, 34136 Trieste, Italy
e-mail: silkiest@gmail.com

E.W. Tedford · G.A. Lawrence
Department of Civil Engineering, University of British Columbia, 6250 Applied Science Lane, Vancouver V6T 1Z4, Canada

G.A. Lawrence
e-mail: lawrence@civil.ubc.ca

M.C.J. Penrice
Department of Physics and Astronomy, University of Victoria, Victoria V8W 3P6, Canada
e-mail: mattpen@uvic.ca

W.G. Unruh
Department of Physics and Astronomy, University of British Columbia, Vancouver V6T 1Z1, Canada
e-mail: unruh@physics.ubc.ca

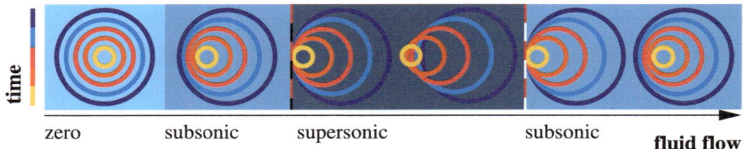

Fig. 8.1 Acoustic horizons. The propagation of sound waves in a convergent fluid flow exhibiting sub- and supersonic flow regions are depicted. The *dashed (red) black/white lines*, separating the sub- and supersonic regions, indicate the location of the acoustic black/white horizon. From the left to the right the flow velocity is speeding up and slowing down again

is referred to as an *analogue model*. The first modern paper on analogue spacetime geometry was published in 1981 by W.G. Unruh [22], followed by Matt Visser [27] in 1993. It was demonstrated that *sound waves* in a fluid flow propagate along geodesics of an *acoustic* spacetime metric. More generally, for a single scalar field ϕ whose dynamics is governed by some generic Lagrangian $\mathscr{L}(\partial_a \phi, \phi)$, the kinematics of small perturbations around some background solution, $\phi(t, \mathbf{x}) = \phi_0(t, \mathbf{x}) + \varepsilon \phi_1(t, \mathbf{x}) + \frac{\varepsilon^2}{2} \phi_2(t, \mathbf{x}) + \cdots$, can be described by a minimally coupled free scalar field, $(\Delta_{g(\phi_0)} - V(\phi_0))\phi_1 = 0$, where $\Delta_{g(\phi_0)}$, a d'Alembertian operator with metric tensor

$$g_{ab}(\phi_0) = \left[-\det\left(\frac{\partial^2 \mathscr{L}}{\partial(\partial_a \phi)\partial(\partial_b \phi)}\right)\right]^{\frac{1}{d-1}}\bigg|_{\phi_0} \left(\frac{\partial^2 \mathscr{L}}{\partial(\partial_a \phi)\partial(\partial_b \phi)}\right)^{-1}\bigg|_{\phi_0}, \quad (8.1)$$

an effective curved spacetime geometry [3]. Over the last 25 years the basic concept of analogue models has been transferred to many different media, and by now we know of a broad class of systems that possess an effective spacetime metric tensor as seen by linear excitations. Detailed background information and current developments can be found in [4].

Analogue models of gravity provide not only a theoretical but also an experimental framework in which to verify predictions of classical and quantum field theory in curved spacetimes. For example, the first model, proposed by W.G. Unruh in 1981, is based on the fact that sound waves propagating on an inviscid and irrotational fluid flow satisfy the Klein–Gordon equation in an effective curved background [22]. If the velocity of the fluid exceeds the velocity of sound at some closed surface, a dumb hole, i.e. an analogue of a black hole, forms, see Fig. 8.1. The presence of effective horizons opens up new possibilities to experimentally explore the black hole evaporation/Hawking radiation process.

There are several hinderances that one has to overcome before testing analogue gravity systems in a laboratory experiment. Any experimental setup has to fall within the approximations made when deriving the analogy. For example, the main difficulty in implementing acoustic black hole (dumb hole) horizons, is to ensure that the waves obey the linear approximation throughout. Shock waves (sonic booms) occur far too readily at transitions between sub- and supersonic flows. In fact, we are all familiar with the sonic boom related to the shock wave generated by supersonic aircrafts.

In 2002 it was argued that surface waves in an open channel flow with varying depth are an ideal toy model for black hole experiments [19]. Unruh's 1981 paper raised the possibility of doing experiments with these analogues. One issue with Hawking's derivation is its apparent reliance on arbitrarily high frequencies, this phenomenon is commonly referred to as the trans-Planckian problem.[1] The dispersion relation of gravity waves creates a natural physical short wavelength cutoff, which obviates this difficulty. Thus the dependence of the Hawking effect on the high-frequency behavior of the theory can be tested in such analogue experiments [7, 23, 24]. While numerical studies indicate that the effect is independent of short-wavelength physics, experimental verification of this would strengthen our faith in the process. The presence of this effect in our physical system, which exhibits turbulence, viscosity, and non-linearities, would indicate the generic nature of the Hawking thermal process. Below we present all the necessary steps to understand and carry out such an experiment.

8.2 Black & White Hole Evaporation Process

One of the most striking findings of general relativity is the prediction of black holes, accessible regions of no escape surrounded by an event horizon. In the early 70s, Hawking suggested that black holes evaporate via a quantum instability [11, 21]. The study of classical and quantum fields around black holes shows that small classical as well as quantum field excitations are being amplified. In particular, a pair of field excitations at temporal frequency f are created, with amplitudes α_f, β_f (Bogoliubov coefficients) related by,

$$\frac{|\beta_f|^2}{|\alpha_f|^2} = \exp\left(\frac{-4\pi^2 f}{g_H}\right) \tag{8.2}$$

where g_H is the surface gravity of the black hole, and α_f and β_f are positive and negative norm components [11, 21]. Positive norm modes are emitted, while negative ones are absorbed by the black hole, effectively reducing its mass. The surface gravity for a non-rotating black hole with a mass M is given by $g_H = 1.0 \times 10^{35}/M$ [kg/s]. Equation (8.2) is applicable for both stimulated and spontaneous emission, and at regimes where the quantum physics is dominant. A comparison of (8.2) with the Boltzmann-distribution allows one to associate a temperature T with the black hole,

$$T = \frac{\hbar g_H}{2\pi k_B} = 1.2 \times 10^{-12} \cdot g_H \, [\text{sK}] = 6.03 \times 10^{-8} \frac{M_\circ}{M} \, [\text{K}]. \tag{8.3}$$

Here $M_\circ$ is a solar mass, and the smallest observed black holes are of this order. Thus black hole evaporation is clearly difficult to observe directly [6].

---

[1]The original derivation by Hawking radiation predicts that the quantum field excitations in the initial state—which are responsible for the late time radiation—have frequencies exponentially higher than the frequency associated with the Planck scale [5, 12].

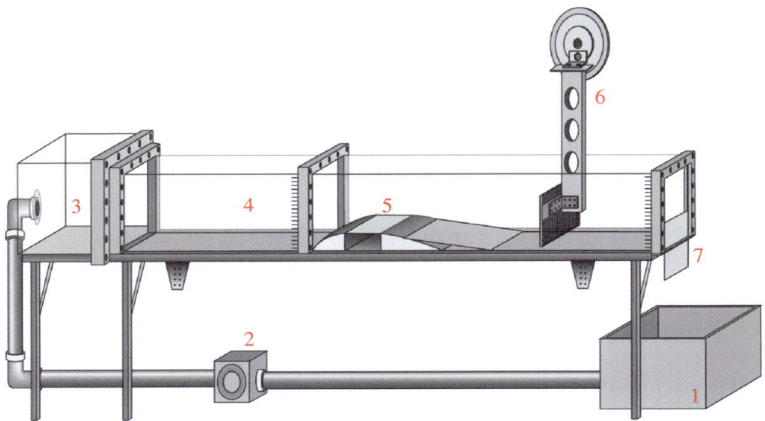

Fig. 8.2 Experimental apparatus. The experimental apparatus used in our experiments: (*1*) holding reservoir, (*2*) pump and pump valve, (*3*) intake reservoir, (*4*) flume, (*5*) obstacle, (*6*) wave generator, and (*7*) adjustable weir

The situation is not as challenging in an analogue gravity experiment, where one is dealing with table-top experiments that are under much better control and significantly easier to access. The question then arises as to how to collect conclusive experimental evidence to be assured one is dealing with analogue black hole evaporation, and not with some other classical or quantum process. As we will demonstrate below, the Hawking process exhibits in principle the following measurable characteristics: (i) the emission of field excitations is correlated; (ii) the spectra of the emission process is described by a Boltzmann-distribution; (iii) the Boltzmann-distribution is determined by the surface gravity at the effective horizon; and (iv) the emitted field excitations are stronger-than-classically/quantum correlated. In the following we will present an analogue gravity experiment in which we observe all classical features of the Hawking process, i.e. (i)–(iii). We will later argue that it is not practical to look for (iv) due to the particular analogue system we are using.

8.3 Experimental Setup

Our experiments were performed in a 6.2 m long, 0.15 m wide and 0.48 m deep flume (Fig. 8.2), and were partly motivated by experiments in similar flumes [2, 15, 18]. We set up a spatially varying background flow by placing a 1.55 m long and 0.106 m high obstacle in the flume.

Particular care was taken to design an obstacle to minimize, or avoid, flow separation. Especially downstream of the obstacle as the flow slows down it has the tendency to separate and create a recirculating flow, see Fig. 8.3. Initially our obsta-

Fig. 8.3 Image of flow separation. The image visualize the flow behavior at the lee side of an obstacle with a trapezoidal profile. The visualization technique utilizes emerged neutrally buoyant particles. The motion of the particles during the exposure time causes streak lines indicating approximately the velocity field of the flow

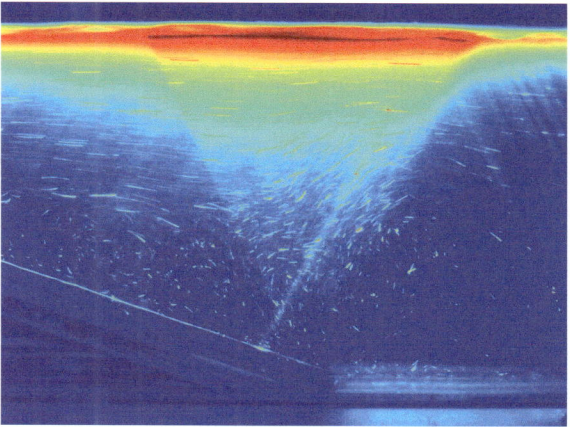

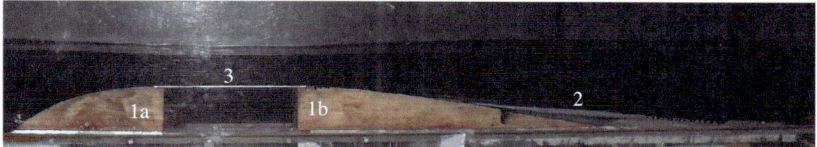

Fig. 8.4 Obstacle used for experiments: (*1a*) and (*1b*) curved parts motivated by airplane wing; (2) flat aluminum plate to further reduce flow separation; and (3) flat top aluminum plate to reduce wave tunneling effects

cle was modeled after an airplane wing with a flat top and a maximum downstream slope of 5.2 degrees designed to prevent flow separation, with a profile given by

$$H(x) = 2a\bigl(1 - x - \exp(-bx)\bigr), \tag{8.4}$$

where $a = 0.094$ of a meter and $b = 5.94$ per meter. However the gradual change in slope along the down stream side of the profile, as well as the absence of any sharp transitions, were not sufficient to fully prevent flow separation. To address this issue we added a constant slope along the backside of the obstacle. The plate is 0.81 meters long. It tapers at a 4.5 degree angle on each end so as to create a smooth transition from obstacle to plate and then from the plate to the bottom of the flume. The gradual slope eliminated apparent flow separation. Maximum flow velocity occurs at the crest of the obstacle. In order to reduce wave tunneling effects between the effective black and white hole horizons, the crest of the obstacle was extended. This was done by cutting the obstacle at the crest and adding a plate (15 cm in length) to join the sections. The extended flat section at the crest of the obstacle resulted in a region of relatively uniform maximum flow velocity. The final obstacle is displayed in Fig. 8.4.

We used particle imaging velocimetry [1] to determine the flow rate q, and to verify the suppression of flow separation. In this technique, small neutrally buoyant, tracer particles are added to the fluid. A short light pulse from each of two lasers (with different colors, red and green, say) is focussed into a narrow sheet within the fluid, the two pulses being separated in time by a few milliseconds. The light scattered in a direction normal to the sheet by the tracer particles within the sheet is focussed so that it forms an image of the particles in a monochrome CCD camera. Each particle produces two images, separated by a distance that is a measure of the component of the velocity in the plane of the light sheet with which the particle is moving; the distance between the two images is therefore a measure of the component of the local fluid velocity. Analysis requires that the pair of images belonging to a given particle be identified, and this is achieved by a cross-correlation technique based on the assumption that particles within a small interrogation area are moving with approximately the same velocity. As a result one obtains the flow velocity as a function of height $v(h)$, and the flow rate is given by $q = \int_0^{h_0} v(h)dh$.

Shallow water waves of approximately 2 mm amplitude were generated 2 m downstream of the obstacle, by a vertically oscillating mesh, which partially blocked the flow as it moved in and out of the water. The intake reservoir had flow straighteners and conditioners to dissipate turbulence, inhomogeneous flow, and surface waves caused by the inflow from the pump. The flume was transparent to allow photography through the walls, and the experimental area was covered to exclude exterior light.

We are interested in the excitations propagating on the background flow in our setup. We measured and analyzed the variations in water surface height using essentially the same techniques as in [10]. The water surface was illuminated using laser-induced fluorescence, and photographed with a high-resolution (1080p) monochrome camera. The fluoresces served to scatter the light to the sides where it could be photographed, to sharply delineate the surface, since the mean free path of the laser in the dyed water was less than 1 mm, and to suppressed the speckle which bedevils all laser illuminated objects.

The camera was set up such that the pixel size was 1.3 mm, the imaged area was 2 m wide and 0.3 m high, and the sampling rate was 20 Hz. The green (532 nm) 0.5 W laser light passed through a Powell lens to create a thin (∼2 mm) light sheet (Fig. 8.5). Rhodamine-WT dye was dissolved in the water, which fluoresced to create a sharp (<0.2 mm) surface maximum in the light intensity. We interpolated the intensity of light between neighboring pixels to determine the height of the water surface to subpixel accuracy.

8.4 Quasi-Particle Excitations

The excitation spectrum of gravity waves on a slowly varying background flow is well understood and has a dispersion relation given by,

$$f^2 = \left(\frac{gk}{2\pi}\right) \cdot \tanh(2\pi kh), \tag{8.5}$$

with the frequency, $f = 1/\omega$, where is the wave period; the wavenumber, where $k = 1/\lambda$ is the wavelength; g is the gravitational acceleration, and h the depth of the

Fig. 8.5 Surface wave detection. Diagram of light-sheet projection for surface wave detection: (*1*) water with dye, (*2*) Powell lens, (*3*) light sheet, and (*4*) fluorescing water surface

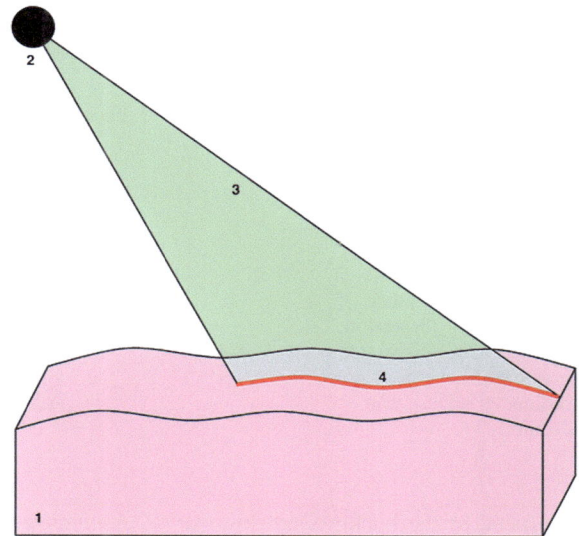

fluid. We neglect surface tension and viscosity. We classify waves according to the value of $2\pi kh$. Our waves all had wavelengths longer than about 2.1 m (still water wavelengths), and surface tension would only play a role for waves with wavelengths less than about 1 cm.

For $2\pi kh < 1$ the dispersion relation can be approximated by $f = \sqrt{gh}k$. These shallow water waves (called that because their wavelength $1/k$ is much longer than the depth of the water h) have both group and phase speed approximately equal to $\sqrt{gh}$. For $2\pi kh > 1$, the dispersion relation is approximated by $f = (gk/2\pi)^{1/2}$. The group speed of these deep water waves is approximately half the phase speed, and both vary as the square root of the wavelength. For a given water depth, both the group and phase speeds of deep water waves are less than the group and phase speeds of shallow water waves.

To determine the ambient wave noise in our facility, and to check the effectiveness of our procedures, we conducted an experiment without the obstacle in place and with no wave generation. The space and time Fourier transform of the noise match the dispersion relation for this flow ($q = 0.039$ m^2/s and $h = 0.24$ m) extremely well (Fig. 8.6). In general, the amplitude of the Fourier components has a noise level of less than 0.2 mm away from the dispersion curves. The apparently elevated noise energy crossing the k axis at $f = \pm 3.1$ Hz is due to the second transverse mode branch of the dispersion relation (the first transverse mode has a node at the location of the light sheet).

In [19] Schützhold and Unruh argued that the equation of motion of shallow water waves can be cast into a wave equation on a curved spacetime background if

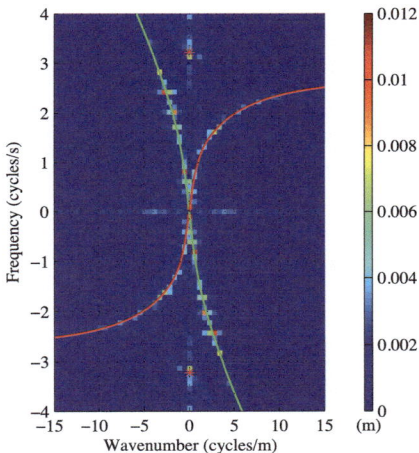

Fig. 8.6 Background noise. Fourier transform of water surface in flat bottom flume without waves; $q = 0.039 \, \text{m}^2/\text{s}$ and $h = 0.24$ m. Fluctuations lie on upstream (*red line*) and downstream (*green line*) branches of the dispersion relation. Just visible at $f = \pm 3.1$, $k = 0$ are the second transverse mode branches of the dispersion relation. Off the dispersion curves, the background noise amplitudes are less than 0.1 mm

the speed of the background flow varies. Assuming a steady, incompressible flow the velocity

$$v(x) = \frac{q}{h(x)}. \quad (8.6)$$

Here the two-dimensional flow rate q is fixed. The dispersion relation in the presence of a non-zero background velocity becomes,

$$(f + vk)^2 = \left(\frac{gk}{2\pi}\right) \cdot \tanh(2\pi k h). \quad (8.7)$$

In Fig. 8.7, the dispersion relation is plotted for a flow typical of our experiments. Only the branch corresponding to waves propagating against the flow is plotted. For low frequencies, there are three possible waves, which we denote according to wavenumber. The first, k_{in}^+, is a shallow water wave with both positive phase and group velocities, and corresponds to the wave that we generate in our experiments. The second, k_{out}^+, has positive phase velocity, but negative group velocity. Both waves, k_{in}^+ and k_{out}^+, are on the positive norm branch of the dispersion relation. The third, k-out, has both negative phase and group velocities, and it lies on the negative norm branch. In our experiment, generated shallow water waves move into a region where they are blocked by a counter-current, and converted into the other two waves. The goal of our experiment was the measurement of the relative amplitudes of the outgoing positive and negative norm modes to test the validity of (8.2). (Further conversion from deep-water waves to capillary waves [2, 17] are also possible but are not studied here.) The conversion from shallow water to deep water waves occurs where a counter-current become sufficiently strong to block the upstream propagation of shallow water waves [17, 20, 25]. It is this that creates the analogy with the white hole horizon in general relativity. That is, there is a region that the shallow water waves cannot access, just as light cannot enter a white hole horizon. Note that while our experiment is on white hole horizon analogues, it is

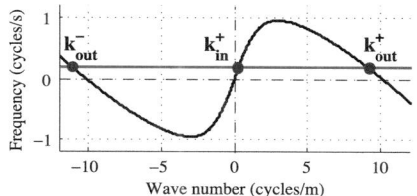

Fig. 8.7 Conversion process. Dispersion relation for waves propagating against a flow typical of our experiments. A shallow water wave, k_{in}, sent upstream, is blocked by the flow and converted to a pair of deep water waves (k_{out}^+ and k_{out}^-) that are swept downstream

because they are equivalent to the time inverse of black hole analogues that we can apply our results to the black hole situation.

8.5 Experimental Procedure

We are interested in the physics around white hole, not black hole horizon. In our particular analogue gravity system the two "outgoing" modes are now not on either side of the horizon but both come out downstream of the white hole horizon. In order to measure the effect of the horizon on incident waves, we sent shallow water waves toward the effective white hole horizon, which sits on the lee side of the obstacle. We conducted a series of experiments, with $q = 0.045$ m^2/s and $h = 0.194$ m, and examined 9 different ingoing frequencies between 0.02 and 0.67 Hz, with corresponding still water wavelengths between 2.1 and 69 meters, corresponding to 0.67 to 0.02 Hz frequencies. This surface was imaged at 20 frames per second, for about 200 s. In all cases we analyzed a period of time which was an exact multiple of the period of the ingoing wave, allowing us to carry out sharp temporal frequency filtering of the signals (i.e., eliminating spectral leakage).

The analysis of the surface wave data was facilitated by introducing the convective derivative operator $\partial_t + v(x)\partial_x$. We redefine the spatial coordinate using,

$$\xi = \int_{x=0} \frac{dx}{v(x)} dx, \tag{8.8}$$

where x is the distance downstream from the right hand edge of the flat portion of the obstacle. The coordinate has dimensions of time, and its associated wave number has units of Hz. The convective derivative becomes $\partial_t + \partial_\xi$, or, in Fourier transform space, $f + \kappa$. This is the term that enters the conserved norm. From Eqs. (35), (36) and (87) of Ref. [19] we find that the conserved norm has the form

$$\int \frac{|A(f,\kappa)|^2}{(f+\kappa)} d\kappa, \tag{8.9}$$

where $A(f,\kappa)$ is the t, ξ Fourier transform of the vertical displacement of the wave. In using this coordinate system the outgoing waves have an almost uniform wavelength even over the obstacle slope.

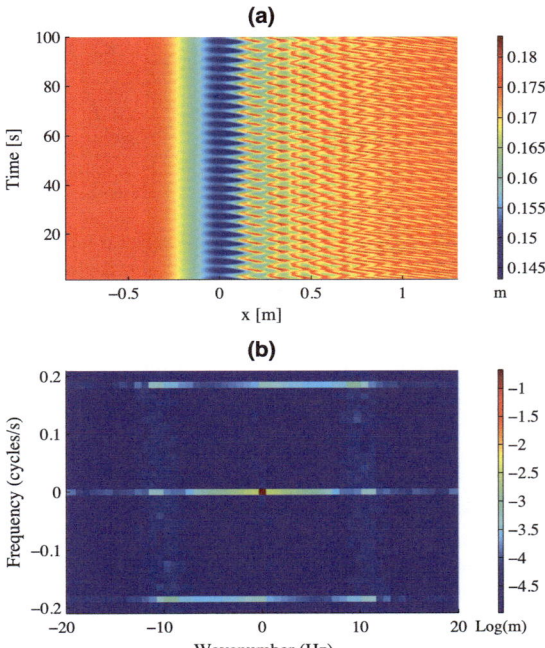

Fig. 8.8 The wave characteristics (**a**) shows the changes in the free surface. Notice the colors indicate the water level relative to the bottom of the tank, not that actual water heights. The double Fourier transformation (**b**) of the wave characteristics shows three excited frequency bands. The one at $\omega = 0$ represents the background (at $k = 0$) and a standing wave, refereed to as the undulation (to peaks at $k \sim \pm 10$ Hz). The other two excited frequency bands at ± 0.185 cycles/s correspond to the stimulated frequency bands

8.6 Data Analysis and Results

We will illustrate the pair-wave creation process by presenting the results for $f_{in} = 0.185$ Hz. In this case we analyzed images from exactly 18 cycles, measuring the free surface along approximately 2 m of the flow including the obstacle. We calculated from the wave characteristics, and after converting to ξ-coordinates (8.8), the two-dimensional Fourier transformation as displayed in Figs. 8.8(a) and (b). Note that the amplitudes of the Fourier transform at frequencies above and below ± 0.185 Hz are very small, indicating that the noise level is small.

As expected, there are three peaks, one corresponding to the ingoing shallow water wavelength around $\kappa = 0$, and the other two corresponding to converted deep water waves peaked near $\kappa_{out}^+ = 9.7$ Hz and $\kappa_{out}^- = -10.5$ Hz. The former is a positive norm and the latter a negative norm outgoing wave, see Eq. (8.9).

In Fig. 8.9 we plot the wave characteristics (amplitude as function of t and ξ) filtered to give only the temporal 0.185 Hz band. Figures 8.9(b) and (c) are the characteristic plots where we further filter to include only $\kappa < -1$ Hz and $\kappa > 1$ Hz respectively. These are the negative and positive norm outgoing components without the central peak of the ingoing wave (because of their very long wavelengths and the rapid change in wavelength as they ascend the slope, the incoming waves have a very broad Fourier transform). Recall, because we are only interested in counter-propagating waves, we defined positive phase and group speeds as pointing to the left. As expected from the dispersion relationship, see Fig. 8.7, the negative norm

Fig. 8.9 Demonstration of pair-wave conversion of an ingoing frequency of 0.185 cycles/s: (**a**) Fourier transform of unfiltered wave characteristic. (**b**) Filtered wave characteristic containing only the ingoing frequency band. (**c**) and (**d**) Wave characteristics for filtered negative and positive norm modes (The *colours* represent the amplitudes of the waves, see *color bars*)

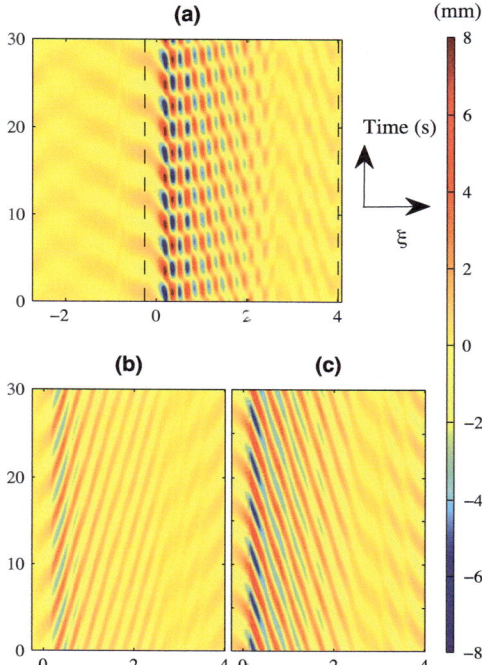

waves have negative phase velocity, while the positive norm waves have positive phase velocity. The complex structure in the characteristics of Fig. 8.9(a) arises because of the interference between the three components, the original ingoing wave, and the positive and negative norm outgoing waves. In Fig. 8.9(b), we see that the ingoing wave is blocked around $\xi = 0$, with only a small component penetrating into the region over the top of the obstacle $\xi < 0$.

Our key results are presented in Fig. 8.10. Figure 8.10(a) shows the amplitude of the spatial Fourier transform at three selected ingoing frequencies. As the frequency increases, the ratio of the negative norm peak to positive norm peak decreases. Furthermore, the location of the positive norm peak moves slightly toward zero as the frequency increases, while the negative norm peak moves away from zero. This is to be expected from the location of the allowed spatial wavenumber from the dispersion plot, see Fig. 8.7. The red-dashed curve in Fig. 8.10(a) shows the Fourier transform in the adjacent temporal frequency bands for the sample case of 0.185 Hz. This is a representation of the noise, and is a factor of at least 10 lower than the signal in the 0.185 frequency band.

To test whether or not the negative norm wave creation was due to non-linearities we repeated the runs at all frequencies with 50 % larger amplitudes. The converted wave amplitudes did, in fact, scale linearly.

The crucial question is: Does the ratio of the negative to positive norm outgoing waves scale as predicted by the thermal hypothesis of Eq. (8.2)? This is shown to be the case in Fig. 8.10(b), where the norm ratios are plotted as a function

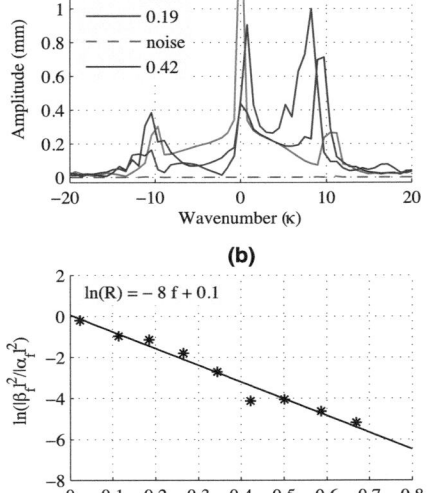

Fig. 8.10 Amplitudes and thermal spectrum. (**a**) Absolute value of three different ingoing frequency bands, and typical noise level (*red line*). (**b**) Ratio between negative and positive norm components in between 0.02 and 0.67 cycles/s (*red stars*), and linear least-squares fit (*red line*)

of ingoing frequency. To calculate the norm of the outgoing waves we integrate $\int |A(f,\kappa)|^2/(f+\kappa)d\kappa$ over the peaks. In Fig. 8.10(b) the points represent the log of the ratios of these areas for each of the input frequencies we tested. The thermal hypothesis is strongly supported, with linear regression giving an inverse slope of 0.12 Hz and an offset close to zero. The slope corresponds to a temperature of $T = 6 \times 10^{-12}$ K, and the offset is zero within our error bounds.

We see from Figs. 8.9(b), (c) that the region of "wave blocking" where the ingoing wave is converted to a pair of outgoing waves, is not a phase velocity horizon (where the phase velocity in the laboratory frame goes to zero). This is true even for the very lowest frequencies. The usual derivation of the temperature from the surface gravity relies on this conversion occurring at a phase velocity horizon. This makes the calculation of the surface gravity, and thus the predicted temperature uncertain. In our case estimates of the surface gravity give a predicted temperature of the same order as the measured temperature. What is important is that the conversion process does exhibit the thermal form predicted for the Hawking process.

This, together with the loss of irrotational flow near the horizon, and absence of a dependable theory of surface waves over an uneven bottom make prediction of the temperature from the fluid flow difficult. Our estimates—using the background flow parameters (i.e. flow rate and water height) to calculate $g_H = 1/2\partial(c^2 - v^2)/\partial x$—give us a value somewhere between about 0.08 and 0.18 Hz. What is important is that the conversion process does exhibit the thermal form predicted for the Hawking process.

8.7 Conclusions and Outlook

We have conducted a series of experiments to verify the stimulated Hawking process at a white hole horizon in a fluid analogue gravity system. These experiments demonstrate that the pair-wave creation is described by a Boltzmann-distribution, indicating that the thermal emission process is a generic phenomenon. It survives fluid-dynamical properties, such as turbulence and viscosity that, while present in our system, are not included when deriving the analogy. The ratio is thermal despite the different dispersion relation from that used by Hawking in his black hole derivation. This increases our trust in the ultraviolet independence of the effect, and our belief that the effect depends only on the low frequency, long wavelength aspects of the physics. When the thermal emission was originally discovered by Hawking, it was believed to be a feature peculiar to black holes. Our experiments, and prior numerical work [7, 22], demonstrate that this phenomenon seems to be ubiquitous, and not something that relies on quantum gravity or Planck-scale physics.

Black holes are linear phase-insensitive field amplifiers of a very peculiar kind [16, 26]. As mentioned in the introduction, the energy of the modes suffers an extremely severe de-amplification, going from frequencies and wave numbers far far higher than the Planck scale, to ones in the kHz regime for solar mass black holes. Nevertheless, when looking at the norms of the modes, they act just like any other amplifier. The Hawking effect is the quantum noise which must accompany any amplifier, but the characteristics of that noise are entirely determined by the amplification properties of the amplifier, which can of course be measured in the classical regime. This relation between the classical and quantum behavior was first pointed out by Einstein in his relation between stimulated and spontaneous emission, by Haus and Mullen in their characterization of quantum noise in a linear amplifier, and by many others [8, 9]. In our case, the direct measurement of the quantum noise, with a characteristic temperature of the order of $T = 6 \times 10^{-12}$ K, is of course impossible. A possible step forward is to study the behavior of quantum noise in analogue gravity systems in Bose–Einstein condensates, as described e.g. in Refs. [13, 14] and in the following chapter of this book.

References

1. Adrian, R.J.: Twenty years of particle image velocimetry. Exp. Fluids **39**, 159–169 (2005)
2. Pokazeyev, K.V., Rozenberg, A.D., Badulin, S.I.: A laboratory study of the transformation of regular gravity-capillary waves on inhomogeneous flows. Izv. Atmos. Ocean. Phys. **19**(10) (1983)
3. Barceló, C., Liberati, S., Visser, M.: Refringence, field theory, and normal modes. Class. Quantum Gravity **19**, 2961–2982 (2002)
4. Barceló, C., Liberati, S., Visser, M.: Analogue gravity. Living Rev. Relativ. **8**, 12 (2005)
5. Brout, R., Massar, S., Parentani, R., Spindel, P.: Hawking radiation without trans-Planckian frequencies. Phys. Rev. D **52**, 4559–4568 (1995)
6. Carr, S.B., Giddings, B.J.: Quantum black holes. Sci. Am. **292**(5), 48–55 (2005)
7. Corley, S., Jacobson, T.: Hawking spectrum and high frequency dispersion. Phys. Rev. D **54**, 1568–1586 (1996)

8. Einsten, A.: Zur Quantentheorie der Strahlung. Phys. Z. **XVIII**, 121–128 (1917)
9. Haus, H.A., Mullen, J.A.: Phys. Rev. **128**, 2407–2413 (1962)
10. Tedford, E.W., Pieters, R., Lawrence, G.A.: Symmetric Holmboe instabilities in a laboratory exchange flow. J. Fluid Mech. **636**, 137–153 (2009)
11. Hawking, S.W.: Black hole explosions. Nature **248**, 30–31 (1974)
12. Jacobson, T.: Trans-Planckian redshifts and the substance of the space-time river. Prog. Theor. Phys. Suppl. **136**, 1–17 (1999)
13. Bonneau, M., Lopes, R., Ruaudel, J., Boiron, D., Westbrook, C.I., Jaskula, J.-C., Partridge, G.B.: An acoustic analog to the dynamical Casimir effect in a Bose–Einstein condensate (2012). arXiv:1207.1338v1
14. Kheruntsyan, K.V., Jaskula, J.-C., Deuar, P., Bonneau, M., Partridge, G.B., Ruaudel, J., Lopes, R., Boiron, D., Westbrook, C.I.: Violation of the Cauchy-Schwarz inequality with matter waves. Phys. Rev. Lett. **108**, 260401 (2012)
15. Lawrence, G.A.: Steady flow over an obstacle. J. Hydraul. Eng. **113**(8), 981–991 (1987)
16. Richartz, M., Prain, A., Weinfurtner, S., Liberati, S.: Superradiant scattering of dispersive fields (2012)
17. Rousseaux, G., Maissa, P., Mathis, C., Coullet, P., Philbin, T.G., et al.: Horizon effects with surface waves on moving water. New J. Phys. **12**, 095018 (2010)
18. Rousseaux, G., Mathis, C., Maissa, P., Philbin, T.G., Leonhardt, U.: Observation of negative-frequency waves in a water tank: a classical analogue to the hawking effect? New J. Phys. **10**(5), 053015 (2008)
19. Schützhold, R., Unruh, W.G.: Gravity wave analogs of black holes. Phys. Rev. D **66**, 044019 (2002)
20. Suastika, I.K.: Wave blocking. Ph.D. thesis, Technische Universiteit Delft, The Netherlands (2004). http://repository.tudelft.nl/file/275166/201607
21. Unruh, W.G.: Notes on black hole evaporation. Phys. Rev. D **14**, 870 (1976)
22. Unruh, W.G.: Experimental black hole evaporation. Phys. Rev. Lett. **46**, 1351–1353 (1981)
23. Unruh, W.G.: Dumb holes and the effects of high frequencies on black hole evaporation. Phys. Rev. D **51**(6), 2827–2838 (1995)
24. Unruh, W.G.: Sonic analogue of black holes and the effects of high frequencies on black hole evaporation. Phys. Rev. D **51**(6), 2827–2838 (1995)
25. Unruh, W.G.: Dumb holes: analogues for black holes. Philos. Trans. R. Soc. Lond. A **366**, 2905–2913 (2008)
26. Unruh, W.G.: Quantum noise in amplifiers and Hawking/Dumb-hole radiation as amplifier noise (2011)
27. Visser, M.: Acoustic propagation in fluids: an unexpected example of Lorentzian geometry. gr-qc/9311028 (1993)

Chapter 9
Understanding Hawking Radiation from Simple Models of Atomic Bose-Einstein Condensates

Roberto Balbinot, Iacopo Carusotto, Alessandro Fabbri, Carlos Mayoral, and Alessio Recati

Abstract This chapter is an introduction to the Bogoliubov theory of dilute Bose condensates as applied to the study of the spontaneous emission of phonons in a stationary condensate flowing at supersonic speeds. This emission process is a condensed-matter analog of Hawking radiation from astrophysical black holes but is derived here from a microscopic quantum theory of the condensate without any use of the analogy with gravitational systems. To facilitate physical understanding of the basic concepts, a simple one-dimensional geometry with a stepwise homogenous flow is considered which allows for a fully analytical treatment.

9.1 Introduction

One of the most spectacular predictions of Einstein's General Relativity is the existence of Black Holes (BHs), mysterious objects whose gravitational field is so

---

R. Balbinot (✉)
Dipartimento di Fisica, Università di Bologna and INFN sezione di Bologna, Via Irnerio 46, 40126 Bologna, Italy
e-mail: balbinot@bo.infn.it

I. Carusotto · A. Recati
INO-CNR BEC Center and Dipartimento di Fisica, Università di Trento, via Sommarive 14, 38123 Povo, Trento, Italy

I. Carusotto
e-mail: carusott@science.unitn.it

A. Recati
e-mail: recati@science.unitn.it

A. Fabbri · C. Mayoral
Departamento de Física Teórica and IFIC, Universidad de Valencia-CSIC, C. Dr. Moliner 50, 46100 Burjassot, Spain

A. Fabbri
e-mail: afabbri@ific.uv.es

C. Mayoral
e-mail: carlosmsaenz@gmail.com

strong that not even light can escape from them but remains trapped inside a horizon. According to the standard view, BHs are formed by the collapse of massive stars ($M > 3M_{Sun}$) at the end of their thermonuclear evolution when the internal pressure is no longer able to balance the gravitational self attraction of the star. Furthermore supermassive BHs ($M > 10M_{Sun}$) are supposed to constitute the inner core of active galaxies.

As no light can escape from them, BHs are expected to be really "black" objects. In particular, their observational evidence can only be indirect: typically, the presence of a black hole is deduced by observing the behavior of matter (typically hot gas) orbiting outside the horizon. A hypothetical isolated BH (i.e. a BH surrounded by vacuum) would not manifest its presence except for its gravitational field, which after a short time becomes stationary (even static if there is no angular momentum).

In 1974 Hawking showed [1] that this common belief is incorrect. If one takes into account Quantum Mechanics, static and stationary BHs are no longer "black", but rather emit a steady radiation flux with a thermal spectrum at a temperature given, simply speaking, by the gradient of the gravitational potential at the horizon. This intrinsically quantum mechanical process is triggered by the formation of the horizon and proceeds via the conversion of off-shell vacuum fluctuations into on-shell particles. This effect is a universal feature of BHs, completely independent of the details of the BH formation.

In spite of the interest that this fascinating effect has raised in a wide audience, no experimental evidence is yet available to support this amazing theoretical prediction. Since the emission temperature scales as the inverse of the BH mass ($T \sim 10^{-7}$ K for a solar mass BH), the expected Hawking signal is in fact many order of magnitudes below the 2.7 K cosmic microwave background. As a result, the Hawking radiation by BHs appears to be a completely irrelevant process in any realistic astrophysical situation, with no hope to be detected in the sky. This situation is rather frustrating, since the conceptual relevance of Hawking discovery is extremely profound: the existence of Hawking radiation allows such a beautiful synthesis between gravity and thermodynamics that it cannot be just an accident; many people indeed regard Hawking result as a milestone in the yet undiscovered quantum theory of gravity.

After almost 40 years of research into BHs, the attitude nowadays appears a bit different and more promising on the experimental side. In particular, it was realized that the Hawking emission process is not exclusively bound to gravitational physics: its "kinematical" rather than "dynamical" nature makes it manifest itself in different physical contexts. This way of looking at the Hawking effect has its origin in a paper by Unruh in 1981 [2] where a steady emission of thermal phonons was predicted to appear in any fluid with a transition from stationary to supersonic flow: the basic process underlying this phonon emission is completely identical to the one discussed by Hawking for the gravitational BH, in the sense that the mathematical equations describing it are exactly the same as the ones describing Hawking radiation from gravitational BHs. The reason for this amazing and unexpected "analogy" is that the equation describing the propagation of long wavelength sound waves in a moving fluid can be recast in terms of a massless scalar field propagating in a curved spacetime with a suitably chosen "acoustic metric". In particular, the point where a sub-sonic flow turns supersonic plays the role of an "acoustic horizon", since sound waves in the supersonic region are no longer able to propagate upstream. Similarly to light inside a BH, sound waves are trapped inside the sonic horizon of the "acoustic black hole": upon quantization, Hawking radiation by the horizon is predicted. Nowadays, we know that this analogy with gravitational systems is not limited to fluids but can be developed for many other condensed matter and optical systems [3]. Unlike gravitational BHs, these analog condensed matter models often possess a well understood quantum mechanical description at the microscopic level, which allows for a complete control of their physics. This is the case, in particular, for atomic Bose-Einstein condensates which are the subject of the present chapter.

The relevance of the analogy is therefore twofold. On the one hand, one can concretely consider investigating the actual existence of Hawking radiation using table top experiments with complete control over the physical system. On the other hand, the detailed knowledge of the microscopic quantum theory that underpins these systems allows us to address a very delicate point in the theory of Hawking radiation and possibly to eliminate some intrinsic inconsistencies in its standard derivation.

In the absence of a complete and self-consistent quantum theory of gravity, one typically adopts a semi-classical framework where gravity is treated classically according to General Relativity, whereas light and matter fields propagating in the curved spacetime are quantized. This is the so called Quantum Field Theory in Curved Space [4]. One expects this scheme to provide a sufficiently accurate description of the gravity-matter systems for scales which are sufficiently large when compared to the fundamental quantum scale for gravity, the so-called Planck scale equal to 10^{-33} cm or 10^{19} GeV. Approaching this Planck scale, one can reasonably expect that this semiclassical description becomes inaccurate and has to be replaced by a (yet to be discovered) complete theory of quantum gravity.

Now because of the infinite (exponential) redshift suffered by the Hawking phonons in their journey from near the horizon to infinity, a given mode of Hawking radiation measured at time t with frequency ν far from the BH appears to have had a frequency $\nu' = \nu e^{ct/2R}$ near the BH horizon (R is the radius), which rapidly exceeds the Planck energy. This feature makes the derivation clearly inconsistent and casts serious doubts over the very existence of Hawking BH radiation. This is the so called transplanckian problem [5].

The same kind of argument can also be repeated for Hawking-like radiation in condensed matter systems: because of the infinite Doppler shift at the sonic horizon, the modes responsible for Hawking-like radiation oscillate near the horizon at a wavelength much smaller than the intermolecular or interatomic spacing, which makes the hydrodynamical long-wavelength approximation inconsistent. On this basis, it would therefore be difficult to rule out the possibility that Hawking radiation is an artifact, illegitimately extrapolated from of the long-wavelength approximation, i.e. a spurious outcome without any physical reality.

From this perspective, analogue condensed matter systems provide a new angle from which the transplanckian problem may be attacked: as they possess a detailed and well understood microscopic quantum description, the question of the existence of Hawking radiation can be investigated from first principles, without using the hydrodynamical approximation and hence any of the concepts borrowed from the gravitational analogy such as the effective metric, horizon, etc. So far, most of the work in this direction has been performed using atomic BECs, but the entirely positive results originating from these studies appears to hold under very general assumptions: Hawking radiation is indeed a real physical phenomenon!

A closer look at the spectral and coherence properties of the predicted Hawking radiation match the original expectation that, if the transition is sufficiently smooth with respect to the microscopic scales of the fluid, the Hawking emission of Bogoliubov phonons is thermal with a temperature proportional to the gradient of the flow potential at the horizon [6–8]. In addition, several novel interesting features have pointed out in regimes beyond the hydrodynamical approximation as well as in different configurations, e.g. white holes (the time-reversed black hole) [9] and the so-called black-hole lasers (a pair of adjacent black and white hole horizons) [44].

In parallel to these theoretical and conceptual advances, a great effort has been devoted in the last number of years to the identification of the most promising physical systems for experimental investigation of analogue Hawking radiation. Having established that the Hawking effect exists, one can start to think at the best experimental setting to reveal it. There are many systems proposed at this end, like ultracold atoms, optical systems, water tank experiments and others. At the moment, experiments with water tanks [10] have detected the classical counterpart of Hawking emission in flows exhibiting white hole horizons: stimulated emission by the Hawking mechanism is probed by sending a classical incident wavepacket of surface waves against the horizon. Unfortunately, these room-temperature experiments do not appear suitable for investigating the quantum-mechanical nature of Hawking radiation, that is the conversion of zero-point fluctuations into observable quanta by the horizon. An observation of Hawking radiation from laser pulses propagating in nonlinear optical media has been recently reported [11], but this result is still object of intense discussion in the community [12–16].

The main experimental difficulty in the quest for analog Hawking radiation in condensed matter systems is the extremely weak intensity of the signal in realistic systems, which is therefore easily obscured by competing effects such as thermal emission due to the non zero temperature of the systems as well as quantum noise. In this respect, atomic gases appear to be the most promising candidate system [19, 20], as they combine a variety of tools for the manipulation of the state of a system on a microscopic level with the possibility of cooling the system to very low temperatures where zero point quantum fluctuations start playing an important role. Still, even in these systems temperatures below the expected Hawking temperature—of the order of 10 nK—are hardly reached, and further difficulty comes from the detection of the Hawking phonons emitted from the horizon.

A major breakthrough that appears to bypass both these problems was proposed by us in 2008 [17] and is based on the use of density correlations, a modern powerful tool to investigate microscopic properties of strongly correlated atomic gases and in particular of their elementary excitations. Taking advantage of the fact that the Hawking radiation consists of correlated pairs of quanta emitted in opposite directions from the horizon, a characteristic signal will appear in the density-density correlation function for points situated on opposite sides with respect to the horizon. Quantitative analysis of this unique signature was made using methods from gravitational physics and then numerically confirmed by *ab initio* simulations of the condensate dynamics based on a microscopic description of their collective properties [18]. As a result, it appears to be an ideal tool to isolate the Hawking radiation signal from the background due to competing processes and experimental noise even at non-zero temperatures. Of course, a similar strategy would be clearly impossible in astrophysical black holes as no access is possible to the region beyond the horizon.

In this paper we shall use standard tools of the theory of a dilute Bose gas to show in a rather pedagogical way how Hawking radiation emerges in an atomic BEC and to explain its features on a simple and analytically tractable toy model. Our treatment, as we shall see, closely resembles a model used to teach elementary Quantum Mechanics—a one dimensional Schrödinger equation with square potential. Most of the material presented here was originally published in Refs. [21, 22].

9.2 The Theory of Dilute Bose-Einstein Condensates in a Nutshell

In this section we give a brief and rapid introduction to the theory of BECs. In particular, we shall review the Gross-Pitaevskii equation describing the dynamics of the condensate at the mean field level and the Bogoliubov description of quantum fluctuations on it. More details can be found in textbooks [45] and in dedicated reviews [46].

Bose-Einstein condensation is characterized by the accumulation of a macroscopic fraction of the particles into a single quantum state. To achieve such a quantum degeneracy very low temperatures are required (on the order of $T = 100$ nK for the typical densities of ultracold atomic gases in magnetic or optical traps), where particles are no longer distinguishable and their Bose statistics start to become relevant.

9.2.1 The Gross-Pitaevskii Equation and the Bogoliubov Theory

The model Hamiltonian describing a many-body system composed of N interacting bosons confined in an external potential $V_{ext}(\mathbf{x})$ can be written in a second quantized formalism as:

$$\hat{H} = \int d^3x \left[\hat{\Psi}^\dagger \left(-\frac{\hbar^2}{2m}\nabla^2 + V_{ext} \right) \hat{\Psi} + \frac{g}{2} \hat{\Psi}^\dagger \hat{\Psi}^\dagger \hat{\Psi} \hat{\Psi} \right] \quad (9.1)$$

where $\hat{\Psi}(t, \mathbf{x})$ is the field operator which annihilates an atom at position $\mathbf{x}$ and obeys standard bosonic equal time commutation rules

$$[\hat{\Psi}(\mathbf{x}), \hat{\Psi}^\dagger(\mathbf{x}')] = \delta^3(\mathbf{x} - \mathbf{x}'). \qquad (9.2)$$

The model Hamiltonian (9.1) is generally used within the dilute gas approximation where the two body interatomic potential can be approximated by a local term $V(x - x') = g\delta^3(\mathbf{x} - \mathbf{x}')$ with an effective coupling constant g related to the atom-atom scattering length a by $g = 4\pi \hbar^2 a/m$.

At sufficiently low temperatures well below the Bose-Einstein condensation temperature, a macroscopic fraction of the atoms are accumulated into the single particle, lowest energy state, described by the macroscopic wavefunction $\Psi_0(\mathbf{x})$. The time evolution of the macroscopic wavefunction in response of some excitation (e.g. a temporal variation of the confining potential V_{ext}) is described by the *Gross-Pitaevski equation*

$$i\hbar \frac{\partial \Psi}{\partial t} = \left(-\frac{\hbar^2}{2m}\nabla^2 + V_{ext} + g|\Psi|^2\right)\Psi: \qquad (9.3)$$

whose form can be heuristically derived by performing a mean-field approximation $\hat{\Psi} \to \Psi_0$ in the Heisenberg equation

$$i\hbar \frac{\partial \hat{\Psi}(t, \mathbf{x})}{\partial t} = [\hat{\Psi}(t, \mathbf{x}), \hat{H}] \qquad (9.4)$$

for the time-evolution of the atomic quantum field operator $\hat{\Psi}$. The ground state wavefunction naturally emerges as the lowest-energy steady-state $\Psi_0(\mathbf{x})$ of the Gross-Pitaevskii equation and oscillates at a frequency $\mu/\hbar$.

Small fluctuations around the mean-field can be studied within the so-called Bogoliubov approximation, where the bosonic field operator $\hat{\Psi}$ is written as the sum of a classical mean-field plus quantum fluctuations. In its usual formulation to describe weakly excited condensates, one takes a steady state Ψ_0 as the mean-field,

$$\hat{\Psi}(t, \mathbf{x}) = \Psi_0(\mathbf{x})\left[1 + \hat{\phi}(t, \mathbf{x})\right]e^{-i\mu t/\hbar}. \qquad (9.5)$$

The field operator $\hat{\phi}$ describing fluctuations then satisfies the Bogoliubov-de Gennes (BdG) equation

$$i\hbar \frac{\partial \hat{\phi}}{dt} = -\left(\frac{\hbar^2 \nabla^2}{2m} + \frac{\hbar^2}{m}\frac{\nabla \Psi_0}{\Psi_0}\nabla\right)\hat{\phi} + ng(\hat{\phi} + \hat{\phi}^\dagger), \qquad (9.6)$$

where $n = |\Psi_0|^2$. The next subsections will be devoted re-expressing the BdG equation in terms of a curved spacetime with an effective metric determined by the spatial profiles of the local speed of sound $c = \sqrt{ng/m}$ and of the local flow velocity $\mathbf{v}_0$.

9.2.2 Analogue Gravity in Atomic BECs

Before continuing the formal development of BEC theory, we will show that by parameterizing the field operator in a different way leads to a reinterpretation of the above equations in a hydrodynamical language and then to the gravitational analogy [3].

Using the so called density-phase representation of the condensate wavefunction $\Psi_0 = \sqrt{n} e^{i\theta}$, the Gross-Pitaevskii equation (9.3) can be rewritten as a pair of real equations,

$$\partial_t n + \nabla (n\mathbf{v}) = 0, \tag{9.7}$$

$$\hbar \partial_t \theta = -\frac{\hbar^2}{2m}(\nabla \theta)^2 - gn - V_{ext} - V_q: \tag{9.8}$$

the former equation Eq. (9.7) is the continuity equation with an irrotational[1] condensate velocity $\mathbf{v}_0 = \hbar \nabla \theta / m$. The latter is analogous to Euler equation for an irrotational inviscid fluid, with an additional "quantum pressure" term $V_q(\mathbf{x})$

$$V_q \equiv -\frac{\hbar^2}{2m} \frac{\nabla^2 \sqrt{n}}{\sqrt{n}} \tag{9.9}$$

describing a kind of stiffness of the macroscopic wavefunction.

In this density-phase representation, the Bogoliubov expression (9.5) of the field operator is rewritten as

$$\hat{\Psi} = \sqrt{n + \hat{n}_1} e^{i(\theta + \hat{\theta}_1)} \simeq \Psi_0 \left(1 + \frac{\hat{n}_1}{2n} + i\hat{\theta}_1\right) \tag{9.10}$$

and the Bogoliubov equation (9.6) reduce to a pair of equations of motion for the fluctuations in the density $\hat{n}_1$ and in the phase $(\hat{\theta}_1)$ in the form

$$\hbar \partial_t \hat{\theta}_1 = -\hbar \mathbf{v}_0 \nabla \hat{\theta}_1 - \frac{mc^2}{n} \hat{n}_1 + \frac{mc^2}{4n} \xi^2 \nabla \left[n \nabla \left(\frac{\hat{n}_1}{n}\right)\right] = 0, \tag{9.11}$$

$$\partial_t \hat{n}_1 = -\nabla \left(\mathbf{v}_0 \hat{n}_1 + \frac{\hbar n}{m} \nabla \hat{\theta}_1\right). \tag{9.12}$$

Here, a fundamental length scale is set by the so-called *healing length* defined as $\xi \equiv \hbar/mc$ in terms of the local speed of sound $c = \sqrt{ng/m}$.

If one is probing the system on length scales much larger than ξ (the so-called *hydrodynamic approximation*), the last term in Eq. (9.11) can be neglected. As a result, the density fluctuations can be decoupled as

$$\hat{n}_1 = -\frac{\hbar n}{mc^2} [\mathbf{v}_0 \nabla \hat{\theta}_1 + \partial_t \hat{\theta}_1]. \tag{9.13}$$

---

[1] From the definition of the velocity field $\mathbf{v}_0$, it is immediate to see that the vorticity in the condensate can only appear at points where the density vanishes.

When this form is inserted in Eq. (9.12), the equation of motion for the phase perturbation

$$-(\partial_t + \nabla \mathbf{v}_0)\frac{n}{mc^2}(\partial_t + \mathbf{v}_0\nabla)\theta_1 + \nabla \frac{n}{m}\nabla\theta_1 = 0 \qquad (9.14)$$

can be rewritten in a matrix form

$$\partial_\mu \left(f^{\mu\nu} \partial_\nu \hat{\theta}_1 \right) = 0 \qquad (9.15)$$

where the matrix elements $f^{\mu\nu}$ are defined as

$$f^{00} = -\frac{n}{c^2}, \qquad f^{0i} = f^{i0} = -\frac{n}{c^2} v_0^i, \qquad f^{ij} = \frac{n}{c^2}\left(c^2 \delta^{ij} - v_0^i v_0^j\right) \qquad (9.16)$$

in terms of the condensate density n and local velocity $\mathbf{v}_0$. Greek indices $\mu, \nu = 0, 1, 2, 3$ indicate 4-dimensional objects, while Latin ones $i = 1, 2, 3$ indicate the space coordinates.

Now in any Lorentzian manifold the curved space scalar d'Alembertian operator can be written as

$$\Box = \frac{1}{\sqrt{-g}} \partial_\mu \left(\sqrt{-g} g^{\mu\nu} \partial_\nu \right) \qquad (9.17)$$

where g is the metric, $g^{\mu\nu}$ its inverse and $g = \det(g_{\mu\nu})$. Keeping this in mind, Eq. (9.14) for the condensate phase dynamics can be rewritten in the form of a curved space wave equation

$$\Box \theta_1 = 0, \qquad (9.18)$$

provided one identifies

$$\sqrt{-g} g^{\mu\nu} \equiv f^{\mu\nu}, \qquad (9.19)$$

which can be inverted leading to the effective metric

$$g_{\mu\nu} = \frac{n}{mc}\begin{pmatrix} -(c^2 - \mathbf{v}_0^2) & -v_0^i \\ -v_0^j & \delta_{ij} \end{pmatrix}. \qquad (9.20)$$

To summarise, we have shown that under the hydrodynamical approximation, the equation of motion for the phase fluctuation in a BEC can be rewritten in terms of a Klein-Gordon equation for a massless scalar field propagating in a fictitious spacetime described by the metric $g_{\mu\nu}$ defined by Eq. (9.20). This is the core of the gravitational analogy.

It should be stressed that this Lorentzian spacetime has nothing to do with the real spacetime in which our BEC lives. Note also that the invariance of Eq. (9.18) under general coordinate transformation is fake. The underlying BEC theory is not even (special) relativistic, but Newtonian, with an absolute time, the laboratory time, with respect to which the equal time commutators (Eq. (9.2)) are given.

However, having said this, taking a closer look at the metric $g_{\mu\nu}$ given by Eq. (9.20): a particularly interesting situation arises in the fluid where there is a transition from sub to supersonic flow (i.e. $|\mathbf{v}_0| > c$). The gravitational analogy of such a configuration corresponds to a black hole as described in the so-called Painlevé—Gullstrand coordinate system and is therefore called a "sonic black hole"—since sound waves travel at a velocity c which is lower than the fluid velocity $\mathbf{v}_0$, they are not able to propagate back and therefore become trapped inside the supersonic region beyond the "sonic horizon", i.e. the locus where $|\mathbf{v}_0| = c$.

In such a setting Eq. (9.18) describes a massless scalar field propagating in a black hole spacetime. But this is exactly the system considered by Hawking to obtain his famous result. One can therefore repeat, step by step, Hawking's derivation of black hole radiation. First of all, one has to perform a modal expansion of the field and then focus on the upstream propagating modes which are barely able to avoid being trapped by the horizon and escape into the subsonic region. Upon quantization, the a comparison of the 'in' and 'out' vacuum states shows that they are inequivalent since the corresponding annihilation and creation operators are related by a Bogoliubov transformation that mixes them in a non-trivial way. As a result, one can expect that the emission of Bogoliubov phonons by the horizon, appearing in sub-sonic region and thermally distributed at a temperature given by the surface gravity κ of the sonic horizon defined as

$$\text{with } \kappa = \frac{1}{2c} \frac{d(c^2 - \mathbf{v}_0^2)}{dn}\bigg|_{hor}, \qquad (9.21)$$

where n is the spatial coordinate normal to the horizon.

It is however crucial to note that this conclusion is based on a very strong assumption, namely the large wavelength approximation, used to neglect the last term in Eq. (9.11) and therefore rewrite this equation as $\Box^2 \theta_1 = 0$, and introduce the gravitational analogy. As explained in the introduction, the modes of the field responsible for the Hawking emission experience an infinite Doppler shift when leaving the near-horizon region in the upstream direction. As their wavelength in this region is many order of magnitude smaller than the healing length of the atomic gas, all the derivation of Hawking radiation in atomic BEC, outlined above, is at least questionable.

For this reason we need to go back to the original microscopic BEC theory of Sect. 9.2 and try to derive Hawking radiation without making any hydrodynamical approximations and without any reference to the gravitational analog: the emission of Hawking radiation in BEC supersonic configurations will then appear as a natural outcome of the underlying quantum theory.

9.3 Stepwise Homogeneous Condensates

A simple analytical treatment can be developed to show the occurrence of Bogoliubov phonon creation "à la Hawking" in an atomic BEC undergoing supersonic motion in a very idealized setting consisting of two semi-infinite stationary homogeneous one dimensional condensates (left and right sector) connected by a step-like discontinuity [22]. As the purpose of this article is mostly a pedagogical one, we will not discuss the actual experimental feasibility of such a configuration, but instead refer to the most recent research literature [47, 48].

In particular, we will assume the condensate to have uniform density n in both sections as well as a spatially uniform flow velocity v along the negative x axis. The external potential V_{ext} and the repulsive atom-atom interaction coupling g are taken as constants within each sector, but to have different values in each sector, satisfying

$$V_{ext}^l + g^l n = V_{ext}^r + g^r n. \tag{9.22}$$

Here, the superscripts "l" and "r" refer to left ($x < 0$) sector and right ($x > 0$) sector respectively with the discontinuity located at $x = 0$. The change in the interaction constant g can be obtained either via the dependence of the atom-atom scattering length on the value of a static external magnetic field, or by modulating the transverse confinement orthogonal to the x direction. Such a change in g directly translates into a change in the local sound speed, which therefore has a different value c^l and c^r in the two sections, defined as usual by $m(c^{l,r})^2 = n g^{l,r}$. Due to condition (9.22), the plane wave form

$$\Psi_0(t, x) = \sqrt{n} e^{i k_0 x - i \omega_0 t} \tag{9.23}$$

of the condensate wavefunction is a solution of the Gross-Pitaevski equation (9.3) at all times t and positions x. The wavevector k_0 and the frequency ω_0 are related to the flow velocity v_0 by $v_0 = \hbar k_0 / m$ and $\hbar \omega_0 = \hbar^2 k_0^2 / (2m) + gn$.

Let us look now at the solutions of the BdG equation Eq. (9.6) for the fluctuation field $\hat{\phi}$ within each sector. Exploiting the stationarity of the configuration, it is convenient to separate $\hat{\phi}$ into its "particle" and "antiparticle" components

$$\hat{\phi}(t, x) = \sum_j [\hat{a}_j \phi_j(t, x) + \hat{a}_j^\dagger \varphi_j^*(t, x)], \tag{9.24}$$

where $\hat{a}_j$ and $\hat{a}_j^\dagger$ are phonons annihilation and creation operators, satisfying the usual bosonic commutations rules $[\hat{a}_i, \hat{a}_j^\dagger] = \delta_{ij}$. The mode functions ϕ_j and φ_j satisfy the equations of motion

$$\begin{aligned} \left[i(\partial_t + v_0 \partial_x) + \frac{\xi c}{2} \partial_x^2 - \frac{c}{\xi} \right] \phi_j &= \frac{c}{\xi} \varphi_j, \\ \left[-i(\partial_t + v_0 \partial_x) + \frac{\xi c}{2} \partial_x^2 - \frac{c}{\xi} \right] \varphi_j &= \frac{c}{\xi} \phi_j \end{aligned} \tag{9.25}$$

which follow from Eq. (9.6) and its conjugate and are to be chosen as oscillating at a frequency ω_j. Imposing that the equal time commutators satisfy

$$[\hat{\phi}(t, x), \hat{\phi}^\dagger(t, x')] = \frac{1}{n} \delta(x - x'), \tag{9.26}$$

allows the modes to be normalised:

$$\int dx [\phi_j \phi_{j'}^* - \varphi_j^* \varphi_{j'}] = \pm \frac{\delta_{jj'}}{\hbar n}. \tag{9.27}$$

The sum over j in (9.24) only involves positive norm modes for which the sign in (9.27) is positive.

Within each of the two $x < 0$ and $x > 0$ spatially uniform regions, the mode functions have a plane wave form

$$\phi_\omega = D(\omega) e^{-i\omega t + ik(\omega)x}, \qquad \varphi_\omega = E(\omega) e^{-i\omega t + ik(\omega)x}, \tag{9.28}$$

where $D(\omega)$ and $E(\omega)$ are normalization factors to be determined using Eq. (9.27). Inserting Eqs. (9.28) into (9.25) yields

$$\left[(\omega - v_0 k) - \frac{\xi c k^2}{2} - \frac{c}{\xi} \right] D(\omega) = \frac{c}{\xi} E(\omega),$$
$$\left[-(\omega - v_0 k) - \frac{\xi c k^2}{2} - \frac{c}{\xi} \right] E(\omega) = \frac{c}{\xi} D(\omega): \tag{9.29}$$

the existence of nontrivial solutions requires that the determinant associated with the above homogeneous system vanishes,

$$(\omega - v_0 k)^2 = c^2 \left(k^2 + \frac{\xi^2 k^4}{4} \right). \tag{9.30}$$

Solving this implicit equation provides the so-called Bogoliubov dispersion of weak excitations on top of a spatially uniform condensate,

$$\omega - v_0 k = \pm c \sqrt{k^2 + \frac{\xi^2 k^4}{4}} \equiv \Omega_\pm(k): \tag{9.31}$$

here, $\Omega_\pm$ is the excitation frequency as measured in the frame co-moving with the fluid. The $+$ ($-$) sign refers to the positive (negative) norm branch. As expected, for small k such that $k\xi \ll 1$, the dispersion relation is linear

$$\omega - v_0 k = \pm c k, \tag{9.32}$$

this is the hydrodynamical regime to which the gravitational analogy is strictly speaking restricted. At higher k, the corrections to the linear dispersion are positive and the modes propagate supersonically. For large k such that $k\xi \gg 1$, the relation tends to the quadratic dispersion of single particles.

The normalization condition gives

$$|D(\omega)|^2 - |E(\omega)|^2 = \pm \frac{1}{2\pi \hbar n} \left| \frac{dk}{d\omega} \right|. \tag{9.33}$$

which using Eqs. (9.29) yields the normalization factors

$$D(\omega) = \frac{\omega - v_0 k + \frac{c\xi k^2}{2}}{\sqrt{4\pi\hbar nc\xi k^2|(\omega - v_0 k)(\frac{dk}{d\omega})^{-1}|}},$$

$$E(\omega) = -\frac{\omega - v_0 k - \frac{c\xi k^2}{2}}{\sqrt{4\pi\hbar nc\xi k^2|(\omega - v_0 k)(\frac{dk}{d\omega})^{-1}|}}:$$

(9.34)

as expected, positive (negative) norm states correspond to the branch of the dispersion relation at a positive (negative) comoving frequency. Remarkably, for any positive norm branch of frequency ω and wavevector k, there exists a negative norm branch of opposite frequency $-\omega$ and wavevector $-k$. Taking advantage of this duality, one can use both positive and negative norm states, replacing the sum over j in (9.24) with an integral over ω and restrict to the positive frequency ones.

Let us go back to the dispersion relation Eq. (9.30). At fixed ω (>0) this is a fourth order equation in k. It has four solutions $k_\omega^{(i)}$ and in general ϕ_ω is a linear combination of four plane waves constructed from these solutions

$$\phi_\omega(x,t) = e^{-i\omega t} \sum_{i=1}^{4} A_i^{(\omega)} D_i(\omega) e^{ik_\omega^{(i)} x}$$

(9.35)

where the $A_i(\omega)$ are the amplitudes of the modes, not to be confused with the normalization coefficients $D(\omega)$. Similarly for $\varphi_\omega(x,t)$.

As said before, our systems consist of two semi-infinite homogeneous condensates joined at $x = 0$ where there is a step-like discontinuity in the speed of sound. Looking at the modes equations (9.25), one has to require that the solutions in the left region and the ones in the right region satisfy the following matching conditions at $x = 0$

$$[\phi] = 0, \quad [\phi'] = 0, \quad [\varphi] = 0, \quad [\varphi'] = 0,$$

(9.36)

where $[f(x)] = \lim_{\epsilon \to 0}[f(x+\epsilon) - f(x-\epsilon)]$ and $'$ means $\frac{d}{dx}$. These four conditions allow to establish a linear relation between the left and right amplitudes

$$A_i^l = M_{ij} A_j^r$$

(9.37)

where M is a 4×4 matrix called the matching matrix, not to be confused with the scattering matrix S we will introduce later, whose dimensionality may vary.

To proceed further in the analysis and explicitly solve the dispersion relation to get the four roots $k_\omega^{(i)}$, one has to specify the flow configuration under investigation, as the position of the roots in the complex plane varies according to the subsonic or supersonic character of the flow. In the next sections we shall separately consider the different cases.

9 Simple Models of Hawking Radiation in Atomic BECs

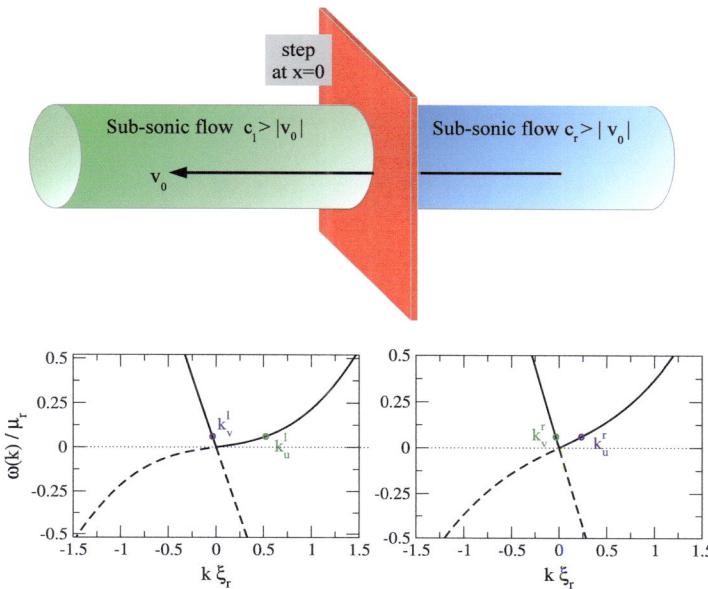

Fig. 9.1 *Upper panel*: sketch of the subsonic-subsonic flow configuration. *Low panels*: dispersion relation of Bogoliubov modes in the asymptotic flat regions away from the transition region

9.4 Subsonic-Subsonic Configuration

We start by considering a uniformly sub-sonic flow, that is with a flow speed v_0 smaller in magnitude than both c_l and c_r, that is $|v_0| < c_{r,l}$. A sketch of the configuration under investigation is given in the upper panel of Fig. 9.1.

9.4.1 The Bogoliubov Modes and the Matching Matrix

The Bogoliubov dispersion in a subsonic flow is graphically displayed in the two lower panels of Fig. 9.1 for two different relative values of the speed of sound c^l (left) and $c^r > c^l$ (right). The positive (negative) norm branches are plotted as solid (dashed) lines. For any given $\omega > 0$, two real solutions belonging to the positive norm branch exist within each l, r region: one, k_u, has a positive group velocity $v_g = \frac{d\omega}{dk}$ and propagates in the rightward, upstream direction; the other, k_v has a negative group velocity and propagates in the leftward, downstream direction. The u, v labels used to indicate these solutions are the conventional ones in General

Relativity. These two real roots admit a perturbative expansion

$$k_v = \frac{\omega}{v_0 - c}\left(1 + \frac{c^3 z^2}{8(v_0 - c)^3} + O(z^4)\right),$$
$$k_u = \frac{\omega}{v_0 + c}\left(1 - \frac{c^3 z^2}{8(v_0 + c)^3} + O(z^4)\right) \quad (9.38)$$

where the dimensionless expansion parameter is $z \equiv \xi\omega/c$. To zeroth order in z, one recovers the well known hydrodynamical results $k_v = \omega/(v_0 - c)$ and $k_u = \omega/(v_0 + c)$. In the following, we shall indicate as $k_{u,v}$ the value of these roots in each of the two homogeneous sections on either side of the interface.

The other two solutions of the dispersion relation are a pair of complex conjugate roots. Within the right sector at $x > 0$, we call k_+^r the root with positive imaginary part, which represents a decaying mode when one goes away from the horizon in the positive x direction. The other solution with a negative imaginary part k_-^r corresponds instead to a growing (and therefore non-normalizable) mode. The opposite holds in the left sector at $x < 0$; the k_+^l root with a positive imaginary part represents a growing mode away from the horizon, while the other root k_-^l with a negative imaginary part represents the decaying mode. Within each l, r region, the wavevector of these modes can be expanded in powers of $z = \xi\omega/c$ as

$$k_\pm = \frac{\omega v_0}{c^2 - v_0^2}\left[1 - \frac{(c^2 + v_0^2)c^4 z^2}{4(c^2 - v_0^2)^3} + O(z^4)\right]$$
$$\pm \frac{2i\sqrt{c^2 - v_0^2}}{c\xi}\left[1 + \frac{(c^2 + 2v_0^2)c^4 z^2}{8(c^2 - v_0^2)^3} + O(z^4)\right]. \quad (9.39)$$

To summarise, the decomposition of ϕ_ω and φ_ω in the left (right) regions reads

$$\phi_\omega^{l(r)} = e^{-i\omega t}\left[A_v^{l(r)} D_v^{l(r)} e^{ik_v^{l(r)}x} + A_u^{l(r)} D_u^{l(r)} e^{ik_u^{l(r)}x}\right.$$
$$\left. + A_+^{l(r)} D_+^{l(r)} e^{ik_+^{l(r)}x} + A_-^{l(r)} D_-^{l(r)} e^{ik_-^{l(r)}x}\right], \quad (9.40)$$

$$\varphi_\omega^{l(r)} = e^{-i\omega t}\left[A_v^{l(r)} E_v^{l(r)} e^{ik_v^{l(r)}x} + A_u^{l(r)} E_u^{l(r)} e^{ik_u^{l(r)}x}\right.$$
$$\left. + A_+^{l(r)} E_+^{l(r)} e^{ik_+^{l(r)}x} + A_-^{l(r)} E_-^{l(r)} e^{ik_-^{l(r)}x}\right]. \quad (9.41)$$

We stress again the fact that the coefficients $A_{u,v,\pm}^{l(r)}$ are the amplitudes of the different modes, not to be confused with the normalization coefficients, $D_{u,v,\pm}^{l(r)}$ for ϕ_ω and $E_{u,v,\pm}^{l(r)}$ for φ_ω: these latter coefficients are uniquely fixed by the commutator relations and the equation of motion, while the amplitudes depend on the choice of basis for the scattering states as we shall see in Sect. 9.4.2. Note that the amplitudes $A_{u,v\pm}^{l(r)}$ are the same for ϕ_ω and φ_ω as required by the equation of motion.

The matching conditions at $x=0$, $[\phi]=0$, $[\varphi]=0$, $[\phi']=0$, $[\varphi']=0$ impose a linear relation between the four left amplitudes $A^l_{u,v,\pm}$ and the right ones $A^r_{u,v,\pm}$

$$W_l \begin{pmatrix} A^l_v \\ A^l_u \\ A^l_+ \\ A^l_- \end{pmatrix} = W_r \begin{pmatrix} A^r_v \\ A^r_u \\ A^r_+ \\ A^r_- \end{pmatrix}, \tag{9.42}$$

where the 4×4 matrices $W_{l(r)}$ are

$$W_{l(r)} = \begin{pmatrix} D^{l(r)}_v & D^{l(r)}_u & D^{l(r)}_+ & D^{l(r)}_- \\ ik^{l(r)}_v D^{l(r)}_v & ik^{l(r)}_u D^{l(r)}_u & ik^{l(r)}_+ D^{l(r)}_+ & ik^{l(r)}_- D^{l(r)}_- \\ E^{l(r)}_v & E^{l(r)}_u & E^{l(r)}_+ & E^{l(r)}_- \\ ik^{l(r)}_v E^{l(r)}_v & ik^{l(r)}_u E^{l(r)}_u & ik^l_+ D^{l(r)}_+ & ik^{l(r)}_- D^{l(r)}_- \end{pmatrix}. \tag{9.43}$$

Multiplying both sides by W_l^{-1} one finally gets

$$\begin{pmatrix} A^l_v \\ A^l_u \\ A^l_+ \\ A^l_- \end{pmatrix} = M \begin{pmatrix} A^r_v \\ A^r_u \\ A^r_+ \\ A^r_- \end{pmatrix}, \tag{9.44}$$

in terms of the matching matrix $M = W_l^{-1} W_r$ whose explicit form is rather involved and is not given here.

9.4.2 The "In" and "Out" Basis

We now proceed to construct a complete and orthonormal (with respect the scalar product Eq. (9.27)) basis for the scattering states of our operator. This can be done in two ways: either choosing a "in" basis constructed with incoming scattering states (i.e. states that propagate from the asymptotic regions $x = \pm\infty$ towards the discontinuity at $x = 0$) or an "out" basis constructed with outgoing scattering states (i.e. states that propagate away from the discontinuity to $x = \pm\infty$).

Let us start with the "in" basis, whose construction is sketched in Fig. 9.2. We define the in v-mode $\phi^{v,in}_\omega$ as a left-moving scattering state with a unit initial amplitude incident on the discontinuity from the right ($x = +\infty$), i.e. $D^r_v e^{-i\omega t + ik^r_v x}$. The incident wave is scattered by the discontinuity at $x = 0$ into a transmitted v-mode in the left region with amplitude A^l_v (i.e. $A^l_v D^l_v e^{-i\omega t + ik^l_v x}$) and partially reflected in the right region with amplitude A^r_u (i.e. $A^r_u D^r_u e^{-i\omega t + ik^r_u x}$). In order to complete the construction, we have to include, in both regions, the complex decaying modes as

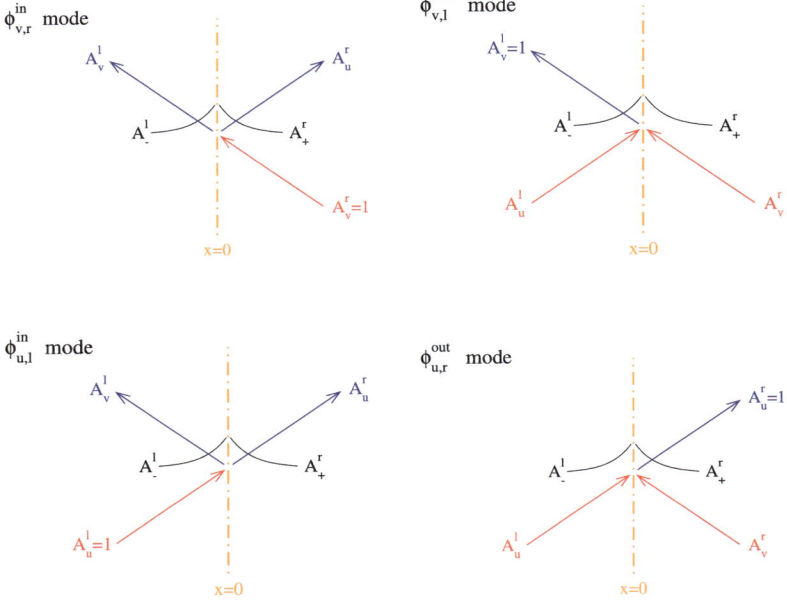

Fig. 9.2 Sketch of the Bogoliubov modes involved in the "in" (*left panels*) and "out" (*right panels*) basis. The mode labels refer to the dispersion shown in the lower panels of Fig. 9.1

well: $A_+^r D_+^r e^{-i\omega t + i k_+^r x}$ and $A_-^l D_-^l e^{-i\omega t + i k_-^l x}$. Growing modes are not included as they diverge at infinity.

The general matching equation (9.44) becomes in this case

$$\begin{pmatrix} A_v^l \\ 0 \\ 0 \\ A_-^l \end{pmatrix} = M \begin{pmatrix} 1 \\ A_u^r \\ A_+^r \\ 0 \end{pmatrix}. \tag{9.45}$$

Treating M perturbatively in $z_l \equiv \frac{\omega \xi_l}{c_l}$ we obtain (for the simplest case $v_0 = 0$; for the general subsonic $v_0 \neq 0$ case the amplitudes are given in the appendix of [21])

$$A_v^l \equiv T = \frac{2\sqrt{c_l c_r}}{c_l + c_r} - \frac{i\sqrt{c_l}(c_l - c_r)^2 z_l}{c_r^{3/2}(c_l + c_r)} + \frac{c_l(c_l - c_r)^2(c_l^2 + c_r^2)z_l^2}{2c_r^3(c_l + c_r)^2}, \tag{9.46}$$

$$A_u^r \equiv R = \frac{c_l - c_r}{c_l + c_r} - \frac{ic_l(c_l - c_r)^2 z_l}{c_r^2(c_l + c_r)} - \frac{c_l(c_l - c_r)(2c_l^3 - 3c_l^2 c_r + 2c_l c_r^2 + c_r^3)z_l^2}{4c_r^4(c_l + c_r)}, \tag{9.47}$$

$$A_-^l = \frac{(c_l - c_r)\sqrt{z_l}}{D_-^l \sqrt{c_r}(c_l + c_r)} - \frac{(c_l - c_r)z_l^2}{2D_-^l c_r^{5/2}(c_l + c_r)}[c_r^2 + i(c_l^2 + c_r^2 - c_r c_l)], \tag{9.48}$$

$$A_+^r = \frac{c_l(-c_l + c_r)\sqrt{z_l}}{D_+^r c_r^{3/2}(c_l + c_r)} + \frac{c_l^2(c_l - c_r)z_l^2}{2D_+^r c_r^{7/2}(c_l + c_r)}[c_l + i(c_l - 2c_r)]. \tag{9.49}$$

Note that these combine in such a way that the unitarity relation $|R|^2 + |T|^2 = 1$ is satisfied. Note also that even though they do not enter the unitarity relation, the amplitudes of the decaying modes are part of the full mode and their presence will show up as an explicit contribution to the density-density correlation.

In a similar way we can construct the $\phi_\omega^{u,in}$ as a scattering state with a unit initial amplitude moving to the right and incident on the discontinuity from the left ($x = -\infty$), which is partially reflected back and partially transmitted, as shown in Fig. 9.2. Here too, we have to include the decaying modes.

The matching relation now reads

$$\begin{pmatrix} A_v^l \\ 1 \\ 0 \\ A_-^l \end{pmatrix} = M \begin{pmatrix} 0 \\ A_u^r \\ A_+^r \\ 0 \end{pmatrix} \tag{9.50}$$

yielding

$$A_v^l \equiv R' = \frac{c_r - c_l}{c_l + c_r} - \frac{i(c_l - c_r)^2 z_l}{c_r(c_l + c_r)} + \frac{(c_l - c_r)(c_l^3 + 2c_l^2 c_r - 3c_l c_r^2 + 2c_r^3)z_l^2}{4c_r^3(c_l + c_r)}, \tag{9.51}$$

$$A_u^r \equiv T' = \frac{2\sqrt{c_l c_r}}{c_l + c_r} - i\frac{\sqrt{c_l}(c_l - c_r)^2 z_l}{c_r^{3/2}(c_l + c_r)} - \frac{\sqrt{c_l}(c_l - c_r)^2(c_l^2 - 4c_l c_r + c_r^2)z_l^2}{8c_r^{7/2}(c_l + c_r)}, \tag{9.52}$$

$$A_-^l = \frac{(c_l - c_r)\sqrt{z_l}}{D_-^l \sqrt{c_l}(c_l + c_r)} + \frac{(c_l - c_r)}{2D_-^l \sqrt{c_l}c_r(c_l + c_r)}[-c_r + i(2c_l - c_r)]z_l^2, \tag{9.53}$$

$$A_+^r = \frac{\sqrt{c_l}(-c_l + c_r)\sqrt{z_l}}{D_+^r c_r(c_l + c_r)} + \frac{\sqrt{c_l}(c_l - c_r)}{2D_+^r c_r^3(c_l + c_r)}[c_l^2 + i(c_l^2 + c_r^2 - c_l c_r)]z_l^2 \tag{9.54}$$

which implies $|R'|^2 + |T'|^2 = 1$, as required by unitarity.

The scattering modes $\phi_\omega^{v,in}$ and $\phi_\omega^{u,in}$, and the similarly constructed $\varphi_\omega^{v,in}$ and $\varphi_\omega^{u,in}$, constitute a complete "in" basis for our field operator $\hat{\phi}$, that can be then expanded as

$$\hat{\phi}(x,t) = \int_0^\infty d\omega [\hat{a}_\omega^{v,in} \phi_{v,r}^{in}(t,x) + \hat{a}_\omega^{u,in} \phi_{u,l}^{in}(t,x) + \hat{a}_\omega^{v,in\dagger} \varphi_{v,r}^{in*}(t,x)$$
$$+ \hat{a}_\omega^{u,in\dagger} \varphi_{u,l}^{in*}(t,x)]. \tag{9.55}$$

The "in" vacuum state $|0, in\rangle$ is defined as usual by $\hat{a}_\omega^{u,in}|0, in\rangle = 0$ and $\hat{a}_\omega^{v,in}|0, in\rangle = 0$. The N-phonons states that constitute the "in" basis of the Hilbert space are constructed by a repeated action of the creation operators $\hat{a}_\omega^{\dagger u, in}$ and $\hat{a}_\omega^{\dagger v, in}$ on the vacuum state.

While the "in" basis has been constructed using incoming scattering modes, an alternative "out" basis can be constructed starting from the outgoing scattering basis of the $\hat{\phi}$ field operator, composed of modes that emerge from the scattering region around $x = 0$ with unit amplitude on a wave propagating at $t = +\infty$ either rightwards towards $x = +\infty$ or leftwards towards $x = -\infty$.

We begin by defining the $\phi_\omega^{v, out}$ scattering mode: as it is sketched in Fig. 9.2, this is a linear combination of in-going right and left moving components with amplitudes A_u^l and A_v^r and decaying modes with amplitudes A_-^l and A_+^r. These coefficients are chosen in a way to give after scattering only a left moving v-mode of unit amplitude. This imposes the condition:

$$\begin{pmatrix} 1 \\ A_u^l \\ 0 \\ A_-^l \end{pmatrix} = M \begin{pmatrix} A_v^r \\ 0 \\ A_+^r \\ 0 \end{pmatrix} \tag{9.56}$$

that yields

$$A_u^l \equiv R'^* = \frac{c_r - c_l}{c_l + c_r} + \frac{i(c_l - c_r)^2 z_l}{c_r(c_l + c_r)} + \frac{(c_l - c_r)(c_l^3 + 2c_l^2 c_r - 3c_l c_r^2 + 2c_r^3)z_l^2}{4c_r^3(c_l + c_r)}, \tag{9.57}$$

$$A_v^r \equiv T'^* = \frac{2\sqrt{c_l c_r}}{c_l + c_r} + \frac{i\sqrt{c_l}(c_l - c_r)^2 z_l}{c_r^{3/2}(c_l + c_r)} - \frac{(c_l - c_r)^2(c_l^2 - 4c_l c_r + c_r^2)z_l^2}{8c_r^{7/2}(c_l + c_r)}, \tag{9.58}$$

$$A_-^l = \frac{(c_l - c_r)\sqrt{z_l}}{D_-^l \sqrt{c_l}(c_l + c_r)} - \frac{(c_l - c_r)z_l^2}{2D_-^l c_r(c_l + c_r)}[c_r + i(2c_l - c_r)], \tag{9.59}$$

$$A_+^r = \frac{\sqrt{c_l}(-c_l + c_r)\sqrt{z_l}}{D_+^r c_r(c_l + c_r)} + \frac{\sqrt{c_l}(c_l - c_r)z_l^2}{2D_+^r c_r^3(c_l + c_r)}[c_l^2 - i(c_l^2 + c_r^2 - c_r c_l)] \tag{9.60}$$

with $|R'^*|^2 + |T'^*|^2 = 1$.

The same procedure can be used to construct the mode $\phi_\omega^{u, out}$, by imposing that the out-going waves consist of a unit amplitude right moving u-mode only. In this case, the matching relations are

$$\begin{pmatrix} 0 \\ A_u^l \\ 0 \\ A_-^l \end{pmatrix} = M \begin{pmatrix} A_v^r \\ 1 \\ A_+^r \\ 0 \end{pmatrix} \tag{9.61}$$

with

$$A_u^l \equiv T^* = \frac{2\sqrt{c_l c_r}}{c_l + c_r} + \frac{i\sqrt{c_l}(c_l - c_r)^2 z_l}{c_r^{3/2}(c_l + c_r)} - \frac{\sqrt{c_l}(c_l - c_r)^2(c_l^2 - 4c_l c_r + c_r^2)z_l^2}{8c_r^{7/2}(c_l + c_r)},$$
(9.62)

$$A_v^r \equiv R^* = \frac{c_l - c_r}{c_l + c_r} + \frac{ic_l(c_l - c_r)^2 z_l}{c_r^2(c_l + c_r)} - \frac{c_l(c_l - c_r)(2c_l^3 - 3c_l^2 c_r + 2c_l c_r^2 + c_r^3)z_l^2}{4c_r^4(c_l + c_r)},$$
(9.63)

$$A_-^l = \frac{c_l(c_l - c_r)z_l}{D_-^l \sqrt{c_r}(c_l + c_r)} + \frac{(c_l - c_r)z_l^2}{2D_-^l c_r^{5/2}(c_l + c_r)} \left[-c_r^2 + i\left(c_l^2 + c_r^2 - c_l c_r\right)\right],$$
(9.64)

$$A_+^r = \frac{c_l(-c_l + c_r)z_l}{D_+^r c_r^{3/2}(c_l + c_r)} + \frac{c_l^2(c_l - c_r)z_l^2}{2D_+^r c_r^{7/2}(c_l + c_r)} \left[c_l + i(2c_r - c_l)\right].$$
(9.65)

In analogy to what was done for the "in" basis, this "out" basis can be used to obtain a decomposition of the $\hat{\phi}$ field operator as

$$\hat{\phi}(x,t) = \int_0^\infty d\omega \big[\hat{a}_\omega^{v,out} \phi_{v,l}^{out}(t,x) + \hat{a}_\omega^{u,out} \phi_{u,r}^{out}(t,x) + \hat{a}_\omega^{v,out\dagger} \varphi_{v,l}^{out*}(t,x) $$
$$+ \hat{a}_\omega^{u,out\dagger} \varphi_{u,r}^{out*}(t,x)\big]$$
(9.66)

in terms of the bosonic annihilation and creation operators for the out-going modes. This also leads to an alternative vacuum state defined by the conditions $\hat{a}_\omega^{u,out}|0,out\rangle = \hat{a}_\omega^{v,out}|0,out\rangle = 0$ and an alternative "out" basis of the Hilbert space.

9.4.3 Bogoliubov Transformation

As both the "in" and the "out" basis are complete, the "in" and "out" scattering modes can be related by the simple linear scattering relations

$$\phi_{v,r}^{in} = T\phi_{v,l}^{out} + R\phi_{u,r}^{out},$$
$$\phi_{u,l}^{in} = R'\phi_{v,l}^{out} + T'\phi_{u,r}^{out},$$
(9.67)

that can be summarized in terms of a unitary 2×2 scattering matrix S

$$S = \begin{pmatrix} T & R \\ R' & T' \end{pmatrix}.$$
(9.68)

Analogous relations hold for the φ_ω modes.

Expressed in terms of mode amplitudes, these scattering relations define a linear Bogoliubov transformation relating the annihilation and creation operators for the "out" modes to the ones of the "in" modes. In the specific case of the sub-sub interface considered in the present section, all u and v modes involved in the scattering process have positive norm, so there is no mixing of the annihilation and creation operators:

$$\hat{a}_\omega^{v,out} = T\hat{a}_\omega^{v,in} + R'\hat{a}_\omega^{u,in},$$
$$\hat{a}_\omega^{u,out} = R\hat{a}_\omega^{v,in} + T'\hat{a}_\omega^{u,in}. \qquad (9.69)$$

As a result, the Bogoliubov transformation trivially reduces to a unitary transformation of the "in" and "out" Hilbert space that conserves the number of excitations and, in particular, preserves the vacuum state: if the system is initially in the $|0, in\rangle$ state with no incoming particles, the number of outgoing particles will also be zero,

$$n_\omega^{v(u),out} = \langle 0, in|\hat{a}_\omega^{v(u),out\dagger}\hat{a}_\omega^{v(u),out}|0, in\rangle$$
$$= \langle 0, in|\left(T^*(R^*)\hat{a}_\omega^{v,in\dagger} + R'^*(T'^*)\hat{a}_\omega^{u,in\dagger}\right)$$
$$\times \left(T(R)\hat{a}_\omega^{v,in} + R'(T')\hat{a}_\omega^{u,in}\right)|0, in\rangle = 0. \qquad (9.70)$$

No phonon can be created, but only scattered at the horizon.

9.4.4 Density-Density Correlations

Correlation functions are a modern powerful tool to investigate the properties of strongly correlated atomic gases [49–55]. We shall concentrate our attention to the correlation pattern of the density fluctuations at equal time, defined as

$$G^{(2)}(t; x, x') \equiv \frac{1}{2n^2} \lim_{t \to t'} \langle in|\{\hat{n}_1(t, x), \hat{n}_1(t', x')\}|in\rangle, \qquad (9.71)$$

where $\{,\}$ denotes the anticommutator. In our configuration with a spatially uniform condensate density n, the density fluctuation operator $\hat{n}_1$ can be expanded in the in-going annihilation and creation operators as

$$\hat{n}^1 \equiv n\left(\hat{\phi}(x,t) + \hat{\phi}^\dagger(x,t)\right)$$
$$= n\int_0^\infty d\omega\left[\hat{a}_\omega^{v,in}\left(\phi_{v,r}^{in} + \varphi_{v,r}^{in}\right) + \hat{a}_\omega^{u,in}\left(\phi_{u,l}^{in} + \varphi_{u,l}^{in}\right) + \text{h.c.}\right], \qquad (9.72)$$

or, alternatively, one can use the "out" basis.

In either case, by evaluating $G^{(2)}$ on the vacuum state $|0, in\rangle = |0, out\rangle$, one finds for one point located to the left ($x < 0$) and the other to the right ($x > 0$) the expression

$$G^{(2)}(t; x, x') \simeq -\frac{\hbar}{2\pi mn(c_r + c_l)}\left[\frac{1}{(v_0 - c_l)(v_0 - c_r)(\frac{x}{c_l - v_0} + \frac{x'}{v_0 - c_r})^2}\right.$$

$$+ \frac{1}{(v_0+c_l)(v_0+c_r)(-\frac{x}{v_0+c_l}+\frac{x'}{v_0+c_r})^2}\bigg], \quad (9.73)$$

which has little significance, only showing correlations decreasing with the square of the distance weighted by the effective speed of sound in the different regions. The physical origin of these correlations is traced back to the repulsive interactions between particles in the gas.

9.5 Subsonic-Supersonic Configuration

9.5.1 The Modes and the Matching Matrix

The warm up exercise discussed in detail in the previous section has allowed to take confidence in the formalism. In this section, we shall consider the much more interesting case of the *acoustic black hole* configuration sketched in Fig. 9.3: taking again the flow to be in the negative x direction ($v_0 < 0$), we assume that the flow is sub-sonic $c_r > |v_0|$ in the upstream $x > 0$ sector, while it is super-sonic $c_l < |v_0|$ in the downstream $x < 0$ sector.

The analogy with a gravitational black hole is simply understood: long wavelength sound waves in the $x < 0$ supersonic region are dragged away by the flow and no longer able to propagate in the upstream direction. The outer boundary of the super-sonic region separating it from the sub-sonic one plays the role of the horizon: long wavelength sound waves can cross it only in the direction of the flow, and eventually get trapped inside the acoustic black hole. Even if this picture perfectly captures the dynamics of long wavelength Bogoliubov waves in the sonic window where the dispersion has the form Eq. (9.32), the supersonic correction that is visible in Eq. (9.31) introduces remarkable new effects as we shall see in this section.

The analysis of the dispersion relation and of the modes in the $x > 0$ subsonic region on the right of the horizon is the same as given in the previous section: two oscillating modes exist with real wave vectors $k^r_{u(v)}$ as well as two complex conjugate evanescent modes with $k^r_\pm$.

In the $x < 0$ supersonic region on the left of the horizon, the dispersion relation has a significantly different shape, as shown in the lower-left panel Fig. 9.3. In particular, it is immediate to see that there exists a threshold frequency ω_{max} above which the situation resembles the one of the subsonic regime: two oscillatory modes exist propagating in the downstream and upstream directions, respectively. Note that the upstream propagation occurs in spite of the super-sonic character of the underlying flow because of the super-luminal dispersion of Bogoliubov waves predicted by Eq. (9.31). Of course, this mode falls well outside the sonic region where the hydrodynamic approximation is valid. The threshold frequency ω_{max} is given by the maximum frequency of the negative norm Bogoliubov mode as indicated in the lower-left panel Fig. 9.3. In formulas, it corresponds to the Bogoliubov frequency

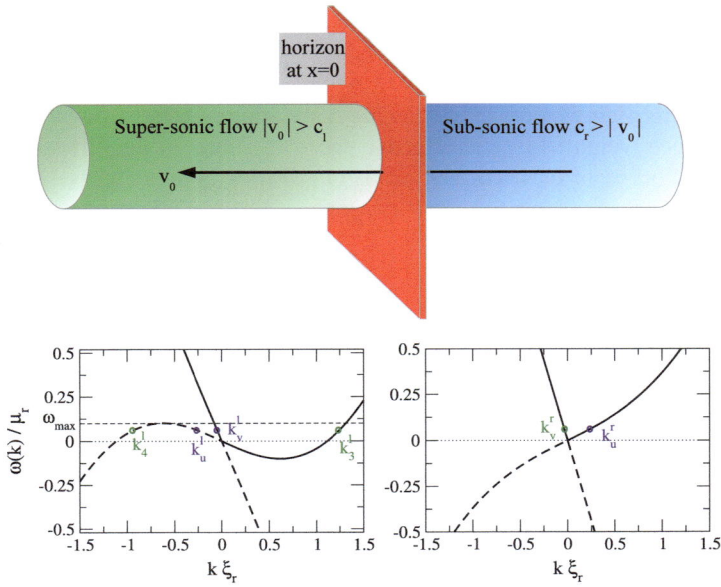

Fig. 9.3 *Upper panel*: sketch of the subsonic-supersonic flow configuration. *Low panels*: dispersion relation of Bogoliubov modes in the asymptotic regions away from the horizon

of the mode at a k_{max} value such that

$$k_{max} = -\frac{1}{\xi_l}\left[-2 + \frac{v_0^2}{2c_l^2} + \frac{|v_0|}{2c_l}\sqrt{8 + \frac{v_0^2}{c_l^2}}\right]^{1/2}. \qquad (9.74)$$

The $0 < \omega < \omega_{max}$ case is much more interesting: from Fig. 9.3, one sees that four real roots of the dispersion relation exist, corresponding to four oscillatory modes, two on the positive norm branch and two on the negative norm one. Two of these modes denoted as u, v lie in the small k region at

$$k_v = \frac{\omega}{v - c_l}\left[1 + \frac{c_l^3 z_l^2}{8(v_0 - c_l)^3} + O(z_l^2)\right], \qquad (9.75)$$

$$k_u = \frac{\omega}{v + c_l}\left[1 - \frac{c_l^3 z_l^2}{8(v_0 + c_l)^3} + O(z_l^2)\right] \qquad (9.76)$$

and have a hydrodynamic character. In contrast to the sub-sonic case, both of them propagate in the downstream direction with a negative group velocity $\frac{d\omega}{dk} < 0$: also the u mode, which propagates to the right in the frame comoving with the fluid, is dragged by the super-sonic flow and is forced to propagate in the left direction. Furthermore, while the k_v solution belongs as before to the positive norm branch, the k_u solution belongs now to the negative norm branch and the corresponding excitation quanta carry a negative energy $\omega < 0$.

The wavevector of the other two roots indicated as k_3 and k_4 in the figure is non-perturbative in ξ

$$k_{3,4} = \frac{\omega v_0}{c_l^2 - v_0^2}\left[1 - \frac{(c_l^2 + v_0^2)c_l^4 z_l^2}{4(c_l^2 - v_0^2)^3} + O(z_l^4)\right]$$

$$\pm \frac{2\sqrt{v_0^2 - c_l^2}}{c_l \xi_l}\left[1 + \frac{(c_l^2 + 2v_0^2)c_l^4 z_l^2}{8(c_l^2 - v_0^2)^3} + O(z_l^4)\right] \qquad (9.77)$$

and lies well outside the hydrodynamic region. Comparing these roots with Eq. (9.39), one realizes that $k_{3,4}$ are the analytic continuation for supersonic flow of the growing and decaying modes previously discussed for the subsonic regime. The k_3 mode belongs to the positive norm branch, while k_4 to the negative one; both of them have a positive group velocity and propagate in the upstream direction.

For $\omega < \omega_{max}$, the general solution of the modes equation in the super-sonic (left) region reads then

$$\phi_\omega^l = e^{-i\omega t}\left[A_v^l D_v^l e^{ik_v^l x} + A_u^l D_u^l e^{ik_u^l x} + A_3^l D_3^l e^{ik_3^l x} + A_4^l D_4^l e^{ik_4^l x}\right],$$

$$\varphi_\omega^l = e^{-i\omega t}\left[A_v^l E_v^l e^{ik_v^l x} + A_u^l E_u^l e^{ik_u^l x} + A_3^l E_3^l e^{ik_3^l x} + A_4^l E_4^l e^{ik_4^l x}\right],$$

while in the sub-sonic (right) region it reads

$$\phi_\omega^r = e^{-i\omega t}\left[A_v^r D_v^r e^{ik_v^r x} + A_u^r D_u^r e^{ik_u^r x} + A_+^r D_+^r e^{ik_+^r x} + A_-^r D_-^r e^{ik_-^r x}\right]. \qquad (9.78)$$

An analogous expression holds for φ_ω^r once we replace $D(\omega)$ with $E(\omega)$.

As we have discussed in the sub-sub case, the field amplitudes on the left and the right of the discontinuity point at $x = 0$ are related by

$$\begin{pmatrix} A_v^l \\ A_u^l \\ A_3^l \\ A_4^l \end{pmatrix} = M \begin{pmatrix} A_v^r \\ A_u^r \\ A_+^r \\ A_-^r \end{pmatrix}, \qquad (9.79)$$

where the matching matrix is $M = W_l^{-1} W_r$, written in terms of W_r given by Eq. (9.43) and

$$W_l = \begin{pmatrix} D_v^l & D_u^l & D_3^l & D_4^l \\ ik_v^l D_v^l & ik_u^l D_u^l & ik_3^l D_3^l & ik_4^l D_4^l \\ E_v^l & E_u^l & E_3^l & E_4^l \\ ik_v^l E_v^l & ik_u^l E_u^l & ik_3^l D_3^l & ik_4^l D_4^l \end{pmatrix}. \qquad (9.80)$$

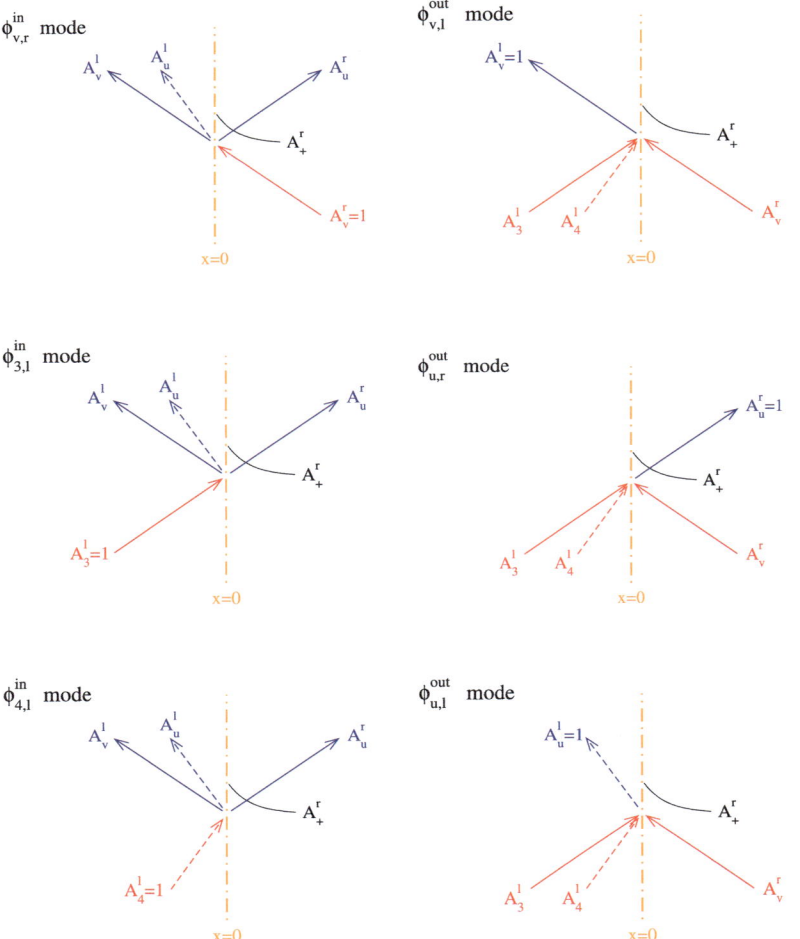

Fig. 9.4 Sketch of the Bogoliubov modes involved in the "in" (*left panels*) and "out" (*right panels*) basis. The mode labels refer to the dispersion shown in the lower panels of Fig. 9.3

9.5.2 The "In" and "Out" Basis

We can now proceed to construct the "in" scattering basis. Differently from the sub-sub case discussed in the previous section, there are now three "in" scattering modes which are associated with the processes sketched in Fig. 9.4.

The mode $\phi_\omega^{v,in}$ is defined as left-moving, unit amplitude, v wave originating in the $x > 0$ sub-sonic region propagating towards the horizon, which upon scattering generates in the subsonic region a reflected right-moving u wave of amplitude A_u^r and a spatially decaying wave of amplitude A_+^r. There are now two transmitted waves in the $x < 0$ supersonic region and both travel in the leftward direction along

9 Simple Models of Hawking Radiation in Atomic BECs

the flow. One is the standard transmitted v wave, with positive norm and amplitude A_v^l, the other is a negative norm u wave with amplitude A_u^l, the so-called *anomalous transmitted* wave.

To leading order in z_l, the corresponding amplitudes are

$$A_v^l = \sqrt{\frac{c_r}{c_l}} \frac{v_0 - c_l}{v_0 - c_r} = S_{vl,vr}, \tag{9.81}$$

$$A_u^r = \frac{v_0 + c_r}{v_0 - c_r} = S_{ur,vr}, \tag{9.82}$$

$$A_u^l = \sqrt{\frac{c_r}{c_l}} \frac{v_0 + c_l}{c_r - v_0} = S_{ul,vr}, \tag{9.83}$$

$$A_+^r = \frac{c_l \sqrt{z_l} \sqrt{c_r(v_0^2 - c_l^2)}}{\sqrt{2} D_+^r (v_0 - c_l)(c_r^2 - v_0^2)^{3/2}(c_r + c_l)} \left[\sqrt{c_r^2 - v_0^2} \left(v + \sqrt{v_0^2 - c_l^2}\right) \right.$$
$$\left. + i \left(v_0 \sqrt{v_0^2 - c_l^2} + v_0^2 - c_r^2\right) \right] = S_{+r,vr}. \tag{9.84}$$

Note the shorthand notation introduced in (9.81)–(9.84) to simply identify the incoming and outgoing channel: for example the matrix element $S_{ul,vr}$ indicates that the incoming channel (second index) is a v-mode entering from the right region, while the outgoing channel (first index) is a u-mode escaping in the left region. The conservation of the Bogoliubov norm translates into a unitary condition between the amplitudes of the propagating modes,

$$\left|A_v^l\right|^2 + \left|A_u^r\right|^2 - \left|A_u^l\right|^2 = 1, \tag{9.85}$$

where the minus sign comes from the negative norm u, l-mode.

As sketched in Fig. 9.4, the other two "in" scattering modes $\phi_\omega^{3,in}$ and $\phi_\omega^{4,in}$ are constructed in a similar way by imposing a unit amplitude in the k_3 or k_4 waves incident on the horizon from the left supersonic side. For the $\phi_\omega^{3,in}$ "in" scattering mode, the corresponding amplitudes are given by

$$A_v^l = \frac{(v_0^2 - c_l^2)^{3/4}(v_0 + c_r)}{c_l^{3/2} \sqrt{2z_l}(c_l + c_r)\sqrt{c_r^2 - v_0^2}} \left(\sqrt{c_r^2 - v_0^2} + i\sqrt{v_0^2 - c_l^2}\right) = S_{vl,3l}, \tag{9.86}$$

$$A_u^r = \frac{\sqrt{2c_r}(v_0^2 - c_l^2)^{3/4}(v_0 + c_r)}{c_l \sqrt{z_l}(c_r^2 - c_l^2)\sqrt{c_r^2 - v_0^2}} \left(\sqrt{c_r^2 - v_0^2} + i\sqrt{v_0^2 - c_l^2}\right) = S_{ur,3l}, \tag{9.87}$$

$$A_u^l = \frac{(v_0^2 - c_l^2)^{3/4}(v_0 + c_r)}{c_l^{3/2} \sqrt{2z_l}(c_l - c_r)\sqrt{c_r^2 - v_0^2}} \left(\sqrt{c_r^2 - v_0^2} + i\sqrt{v_0^2 - c_l^2}\right) = S_{ul,3l}, \tag{9.88}$$

$$A_+^r = \frac{(v_0^2 - c_l^2)^{1/4}}{2D_+^r(v_0^2 - c_r^2)} \left(v_0 - i\sqrt{c_r^2 - v_0^2}\right) = S_{+r,3l} \tag{9.89}$$

and satisfy the unitarity relation

$$|A_v^l|^2 + |A_u^r|^2 - |A_u^l|^2 = 1. \tag{9.90}$$

Remarkably the amplitudes for the propagating modes now diverge at small ω as $\frac{1}{\sqrt{\omega}}$.

For the $\phi_\omega^{4,in}$ "in" scattering mode,

$$A_v^l = \frac{(v_0^2 - c_l^2)^{3/4}(v_0 + c_r)}{c_l^{3/2}\sqrt{2z_l}(c_l + c_r)\sqrt{c_r^2 - v_0^2}}\left(\sqrt{c_r^2 - v_0^2} - i\sqrt{v_0^2 - c_l^2}\right) = S_{vl,4l}, \tag{9.91}$$

$$A_u^r = \frac{\sqrt{2c_r}(v_0^2 - c_l^2)^{3/4}(v_0 + c_r)}{c_l\sqrt{z_l}(c_r^2 - c_l^2)\sqrt{c_r^2 - v_0^2}}\left(\sqrt{c_r^2 - v_0^2} - i\sqrt{v_0^2 - c_l^2}\right) = S_{ur,4l}, \tag{9.92}$$

$$A_u^l = \frac{(v_0^2 - c_l^2)^{3/4}(v_0 + c_r)}{c_l^{3/2}\sqrt{2z_l}(c_l - c_r)\sqrt{c_r^2 - v_0^2}}\left(\sqrt{c_r^2 - v_0^2} - i\sqrt{v_0^2 - c_l^2}\right) = S_{ul,4l}, \tag{9.93}$$

$$A_+^r = \frac{(v_0^2 - c_l^2)^{1/4}(v_0^2 - c_l^2 + v_0\sqrt{v_0^2 - c_l^2})}{2D_+(c_r^2 - v_0^2)(c_l^2 - v_0^2 + v_0\sqrt{v_0^2 - c_l^2})}\left(v_0 - i\sqrt{c_r^2 - v_0^2}\right) = S_{+r,4l} \tag{9.94}$$

and the unitarity condition reads

$$|A_v^l|^2 + |A_u^r|^2 - |A_u^l|^2 = -1: \tag{9.95}$$

where the minus sign on the right-hand side comes from the fact that the incoming unit amplitude k_4 mode has negative norm. All together, the v, 3 and 4 "in" scattering modes form a basis in which the $\hat{\phi}$ field operator may be expanded.

The construction of the "out" basis proceeds along similar lines: one has the three $\phi_{v,l}^{out}$, $\phi_{u,r}^{out}$ and $\phi_{u,l}^{out}$ "out" scattering modes, where the $l(r)$ label near the superscript u indicates again the left (right) region of space. The corresponding scattering processes are depicted in Fig. 9.4.

As the corresponding amplitudes will not be needed in the following, we refer the reader to Ref. [21] for their explicit expression. As discussed in [6–8, 22], the field operator can then be equivalently expanded in the basis of the "in" scattering modes as

$$\hat{\phi} = \int_0^{\omega_{max}} d\omega [\hat{a}_\omega^{v,in}\phi_{v,r}^{in} + \hat{a}_\omega^{3,in}\phi_{3,l}^{in} + \hat{a}_\omega^{4,in\dagger}\phi_{4,l}^{in}$$
$$+ \hat{a}_\omega^{v,in\dagger}\varphi_{v,r}^{in*} + \hat{a}_\omega^{3,in\dagger}\varphi_{3,l}^{in*} + \hat{a}_\omega^{4,in}\varphi_{4,l}^{in*}], \tag{9.96}$$

or, equivalently on the basis of the "out" scattering ones. Note in particular the third term on the right-hand side, as the corresponding k_4 mode is a negative norm one,

9 Simple Models of Hawking Radiation in Atomic BECs

this term enters with a $\phi_{4,l}^{in}$ field multiplied by a creation $\hat{a}_\omega^{4,in\dagger}$ operator: as we shall see in the next sub-section, this simple fact is the key element leading to the emission of analog Hawking radiation by the horizon.

9.5.3 Bogoliubov Transformation

The "in" and "out" basis are now related by a 3×3 scattering matrix S relating the three incoming states to the three outgoing states. Explicitly

$$\phi_{v,r}^{in} = S_{vl,vr}\phi_{v,l}^{out} + S_{ur,vr}\phi_{u,r}^{out} + S_{ul,vr}\phi_{u,l}^{out}, \qquad (9.97)$$

$$\phi_{3,l}^{in} = S_{vl,3l}\phi_{v,l}^{out} + S_{ur,3l}\phi_{u,r}^{out} + S_{ul,3l}\phi_{u,l}^{out}, \qquad (9.98)$$

$$\phi_{4,l}^{in} = S_{vl,4l}\phi_{v,l}^{out} + S_{ur,4l}\phi_{u,r}^{out} + S_{ul,4l}\phi_{u,l}^{out}. \qquad (9.99)$$

Because of the negative norm of the ϕ_{ul}^{out} mode, conservation of the Bogoliubov norm imposes the modified unitarity condition $S^\dagger \eta S = S \eta S^\dagger$ with $\eta = \mathrm{diag}(1, 1, -1)$ and the scattering matrix S mixes positive and negative norm modes.

As a result, the Bogoliubov transformation relating the creation and destruction operators of the "in" and "out" scattering states is no longer trivial and mixes creation and destruction operators as follows

$$\begin{pmatrix} \hat{a}_\omega^{v,out} \\ \hat{a}_\omega^{ur,out} \\ \hat{a}_\omega^{ul,out\dagger} \end{pmatrix} = \begin{pmatrix} S_{vl,vr} & S_{vl,3l} & S_{vl,4l} \\ S_{ur,vr} & S_{ur,3l} & S_{ur,4l} \\ S_{ul,vr} & S_{ul,3l} & S_{ul,4l} \end{pmatrix} \begin{pmatrix} \hat{a}_\omega^{v,in} \\ \hat{a}_\omega^{3in} \\ \hat{a}_\omega^{4in\dagger} \end{pmatrix}. \qquad (9.100)$$

The non triviality of the Bogoliubov transformation has the crucial consequence that the "in" and "out" vacuum states no longer coincide $|0, in\rangle \neq |0, out\rangle$: while the $|0, in\rangle$ "in" vacuum state (defined as the state annihilated by the $\hat{a}_\omega^{(v,3,4),in}$ operators) contains no incoming phonons, it contains a finite amount of phonons in all three out-going modes due to a parametric conversion process taking place at the horizon,

$$n_\omega^{u,r} = \langle 0, in | \hat{a}_\omega^{ur,out\dagger} \hat{a}_\omega^{ur,out} | 0, in \rangle = |S_{ur,4l}|^2, \qquad (9.101)$$

$$n_\omega^{u,l} = \langle 0, in | \hat{a}_\omega^{ul,out\dagger} \hat{a}_\omega^{ul,out} | 0, in \rangle = |S_{ul,vr}|^2 + |S_{ul,3l}|^2, \qquad (9.102)$$

$$n_\omega^{v,l} = \langle 0, in | \hat{a}_\omega^{v,out\dagger} \hat{a}_\omega^{v,out} | 0, in \rangle = |S_{vl,4l}|^2. \qquad (9.103)$$

Note in particular the remarkable relation

$$n_\omega^{u,l} = |S_{ul,vr}|^2 + |S_{ul,3l}|^2 = |S_{ur,4l}|^2 + |S_{vl,4l}|^2 = n_\omega^{u,r} + n_\omega^{v,l}. \qquad (9.104)$$

The physical meaning of the above relations can be understood as follows. Suppose that at $t = -\infty$ we have prepared the system in the $|0, in\rangle$ vacuum state, so there are no incoming phonons. We are working in the Heisenberg picture of

Quantum Mechanics, so that $|0, in\rangle$ describes the state of our systems at all time. Now Eqs. (9.101)–(9.103) tell us that at late time in this state there will be outgoing quanta on both sides of the horizon: the vacuum has spontaneously emitted phonons. This occurs by converting vacuum fluctuations of the k_4 mode into real on shell Bogoliubov phonons (see lower left panel of Fig. 9.4) in the hydrodynamic region.

While processes involving particle creation in time-varying settings are well-known in quantum mechanics, e.g. the dynamical Casimir effect [23–39], the production of particles in a stationary background seems to contradict energy conservation. The solution of this puzzle lies in Eq. (9.104): besides the positive energy ur and vl phonons, there is also production of negative energy (ul) phonons, the so called "partners" which propagate down in the supersonic region. The number of these latter equals the number of the formers. This is how energy conservation and particle production coexist in our stationary systems. As particles are produced in pairs with opposite $\pm\omega$ frequencies, energy is conserved.

Now let us give a closer look at Eq. (9.101): according to this, an hypothetical observer sitting far away from the horizon in the subsonic region at $x \to +\infty$ will observe a flux of phonons coming from the horizon. This is the analogue of black hole Hawking radiation. The number of phonons of this kind emitted per unit time and per unit bandwidth is

$$\frac{dN_\omega^{u,r}}{dtd\omega} = |S_{ur,4l}|^2 \simeq \frac{(c_r + v_0)}{(c_r - v_0)} \frac{(v_0^2 - c_l^2)^{3/2}}{(c_r^2 - c_l^2)} \frac{2c_r}{c_l \xi_l \omega}. \tag{9.105}$$

The $\frac{1}{\omega}$ dependence of the above expression is reminiscent of the low frequency expansion of a thermal Bose distribution [22]

$$n_T(\omega) = \frac{1}{e^{\frac{k_B T}{\hbar \omega}} - 1} \simeq \frac{k_B T}{\hbar \omega} + \cdots \tag{9.106}$$

and one can try to identify the $\frac{1}{\omega}$ coefficient of (9.105) as an effective temperature

$$T = \frac{\hbar}{k_B} \frac{(c_r + v_0)}{(c_r - v_0)} \frac{(v_0^2 - c_l^2)^{3/2}}{(c_r^2 - c_l^2)} \frac{2c_r}{c_l \xi_l}. \tag{9.107}$$

As the surface gravity of our toy model with an abrupt discontinuity in the flow is formally infinite while the temperature remains finite, the connection of the analog model to the original gravitational framework seems to fail. However, to investigate the correspondence with the gravitational black holes, one has to consider more general and realistic velocity profiles where the transition from the subsonic region to the supersonic one is smooth enough to justify the hydrodynamical approximation. Accurate numerical calculations in this regime show that the emission is indeed thermal in this case and the temperature is to a good accuracy determined by the surface gravity κ of the associated black hole according to Eq. (9.21). As shown in [6–8, 42], the original Hawking's prediction for the emission temperature holds provided the spatial variation of the flow parameters occurs on a characteristic length scale longer than $\xi^{2/3}\kappa^{-1/3}$. Of course, the thermal spectrum is restricted to frequencies lower than the upper cut-off at ω_{max}: above this frequency, one in fact recovers the physics of the sub-sub interface where no emission takes place. These results, together with the full numerical simulation in [18] confirm that the emission of Hawking radiation is not an artifact of the hydrodynamical approximation and provide an independent validation of the model of phonon propagation based on the metric in Eq. (9.20).

Even in the most favorable configurations, realistic estimates of the Hawking temperature in atomic BECs give values of the order of 10 nK, that is one order of magnitude lower than the typical temperature of the condensates (100 nK). This makes the Hawking emission of Bogoliubov phonons in BECs a quite difficult effect to reveal in an actual experiment, as the interesting signal is masked by an overwhelming thermal noise.

A proposal to overcome this difficulty was put forward in [17]: as the pairs of Bogoliubov excitations produced by the Hawking process originate from the same vacuum fluctuation, their strong correlation is expected to be responsible for specific features in the correlation function of density fluctuations. This idea was soon confirmed by numerical simulations of the dynamics of atomic condensates in acoustic black hole configurations. The features analytically predicted in [17] are indeed visible in the density correlation pattern and, moreover, are robust with respect to a finite temperature. At present, this method represents the most promising strategy to experimentally detect the analog Hawking effect in atomic Bose-Einstein condensates. The first investigation into the potential power of density correlation techniques in the context of an analog dynamical Casimir effect in condensates has been recently reported in [41] along the lines of the theoretical proposal in [40].

9.5.4 Density-Density Correlations

The fact that the density correlation function in BECs exhibits characteristic peaks associated with phonon creation à la Hawking can be easily seen in our simple toy model. For simplicity, let us restrict our attention to the contribution to the density-density correlation function due to the "out" particles and consider the decomposition

$$\hat{n}^1(t,x) \simeq n \int_0^{\omega_{max}} \left[\hat{a}_\omega^{v,out} \left(\phi_{v,l}^{out} + \varphi_{v,l}^{out} \right) + \hat{a}_\omega^{ur,out} \left(\phi_{u,r}^{out} + \varphi_{u,r}^{out} \right) \right. \\ \left. + \hat{a}_\omega^{ul,out\dagger} \left(\phi_{u,l}^{out} + \varphi_{u,l}^{out} \right) + \text{h.c.} \right]. \tag{9.108}$$

Expanding the "out" creation and annihilation operators in terms of the "in" ones and using the approximate form of the S matrix elements given by Eq. (9.100), and finally evaluating expectation values on the $|0, in\rangle$ "in" vacuum state, one finds that the above expression describes correlations between the (ur) and (ul) particles and the (ur) and (vl) particles if the points x and x' are taken on opposite sides with

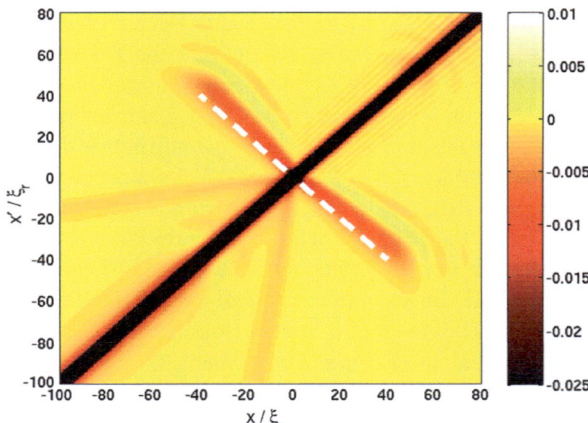

Fig. 9.5 Color plots of the rescaled density correlation function $(n_0 \xi_r) \times [G^{(2)}(x, x') - 1]$ a time $g_r n t/\hbar = 160$ after the switch-on of the black hole horizon. The calculation has been performed using the truncated-Wigner method of [18]. Black hole parameters: $|v_0|/c_l = 1.5$, $|v_0|/c_r = 0.75$. The *dashed white line* indicates the analytically expected position (9.110) of the negative peak in the density correlation signal

respect the horizon, while one finds (ul)–(vl) correlations if both points are inside the horizon. If both x, x' are located outside the horizon, correlations just show a monotonic decrease with distance as in the sub-sub case.

As in general one has $|S_{vl,4l}| \ll |S_{ul,4l}|$, the main contribution to the density correlation in the $x < 0$ and $x' < 0$ sector comes from the (ul)–(ur) term describing correlations between the Hawking phonon (ur) and its partner (ul). Integrating over all frequencies up to ω_{max}, one obtains term of the form [22]

$$G^{(2)}(x, x') \sim -\frac{1}{4\pi n} \frac{(v_0^2 - c_l^2)^{3/2}}{c_l(v_0 + c_l)(v_0 - c_r)(c_r - c_l)} \frac{\sin[\omega_{max}(\frac{x'}{v_0+c_r} - \frac{x}{v_0+c_l})]}{\frac{x'}{v_0+c_r} - \frac{x}{v_0+c_l}}.$$
(9.109)

From this expression, it is easy to see that the density-density correlation function has a negative value and is peaked along the half-line

$$\frac{x'}{v_0 + c_r} = \frac{x}{v_0 + c_l}.$$
(9.110)

The stationarity of the Hawking process is apparent in the fact that the peak value of Eq. (9.109) does not depend on the distance from the horizon.

The physical picture that emerges from this mathematical derivation is that pairs of (ul) and (ur) phonons are continuously created by the horizon at each time t and then propagate in opposite directions at speeds $v_{ul} = v_0 + c_l < 0$ for (ul) and $v_{ur} = v_0 + c_r > 0$ for (ur). At time Δt after their emission they are located at $x = v_{ul} \Delta t$ and $x' = v_{ur} \Delta t$, which explains the geometrical shape of the peak line Eq. (9.110) where correlations are strongest. A numerical evaluated example of the correlation function of density fluctuations is shown in Fig. 9.5: the dashed line indicates the expected position of the peak line Eq. (9.110). A detailed discussion of the other peaks (that are barely visible on the color scale of the figure) can be found in [18, 22].

9.5.5 Remarks

Let us try to summarize the results discussed in the present chapter. We have seen that for a stationary flowing BEC with a horizon-like boundary separating an upstream subsonic region from a downstream supersonic one, spontaneous emission of Bogoliubov phonons occurs at the horizon by converting zero-point quantum fluctuations into real and observable radiation quanta. The emitted radiation appears to an observer outside the horizon in the subsonic region to have an approximately thermal distribution: this is the analog Hawking effect in BECs.

Similarly to the gravitational context, where nothing can travel faster than light, the horizon has a well-defined meaning of surface of no return: no physical signal can travel from inside the black hole to the outside crossing the horizon in the outward direction. In the case of an atomic BECs, the "sonic" horizon is defined as the surface where the speed of sound c equals the velocity $|v_0|$ of the fluid: for the hydrodynamical u, v modes at low wavevector, the acoustic black hole exactly mimics what happens in the gravitational case: no long wavevector sound wave can cross the horizon in the upstream direction. On the other hand the dispersion of Bogoliubov modes in an atomic BEC shows significant super-luminal corrections: the higher the wavevector of the excitation, the larger its group velocity. As a result, the $k_{3,4}$ modes are able to travel in the upstream direction in the super-sonic region inside the horizon and therefore to escape from the black hole. In contrast to the hydrodynamic modes, they are not trapped inside and do not see any horizon: for them the gravitational analogy has no meaning.

Some authors have recently introduced wavevector dependent *rainbow metrics* to describe the propagation of different modes at different wave vectors and have defined several distinct concepts of horizon, such as the *phase horizon* and the *group horizon*. Our opinion is that these additional concepts may end up hiding the essence of Hawking radiation behind unessential details.

The key ingredient in order to have the emission of radiation in the "in" vacuum state is in fact the presence of negative energy states that allow the emission of a pair of quanta while conserving energy: this requires that the flow becomes supersonic within some spatial region. The main role of a horizon where the flow goes from sub- to super-sonic is in determining the thermal shape of the spectral distribution of the emitted radiation. To better appreciate this fundamental point, the next chapter will be devoted to a short discussion of configurations with a super-sonic flow on both sides of the interface: as it was first pointed out in [43], the spontaneous emission of radiation takes place in this case in spite of the total absence of a horizon: sound waves are always dragged by the super-sonic flow and can not propagate upstream. Because of the absence of an horizon, the resulting spectral distribution of the associated zero-point radiation is however very different from the thermal Hawking radiation, with a low-frequency tail dominated by a constant term instead of $1/\omega$.

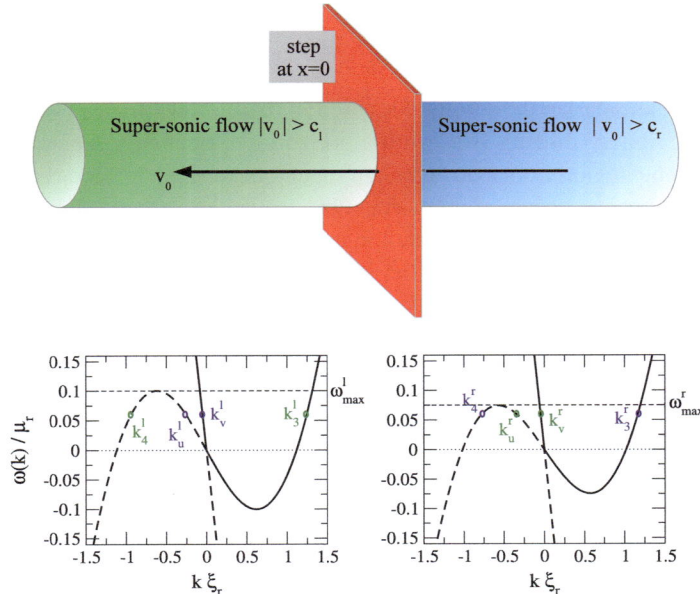

Fig. 9.6 *Upper panel*: sketch of the supersonic-supersonic flow configuration. *Low panels*: dispersion relation of Bogoliubov modes in the asymptotic regions away from the horizon

9.6 Supersonic-Supersonic Configuration

Consider a BEC undergoing supersonic motion at every point, with a sound velocity profile varying abruptly at $x = 0$: no sonic horizon is present in this setting and at all points long wavelength sound waves are dragged in the downstream direction by the underlying flowing fluid.

The dispersion relation pattern on either sides of the discontinuity is shown in the lower panels of Fig. 9.6. For $\omega < \omega_{max} = \min[\omega^l_{max}, \omega^r_{max}]$, one has four oscillatory solutions in both regions with real wavevectors. The k_u and k_v hydrodynamic solutions propagate in the downstream direction (i.e. to the left with negative v_0) while the large wavevector k_3 and k_4 solutions are able to propagate upstream. The $k_{v,3}$ and $k_{u,4}$ solutions correspond to positive and negative norm modes respectively.

The general solution of the mode equations in both regions is

$$\phi^{r(l)}_\omega = e^{-i\omega t}\big[A^{l(r)}_v D^{l(r)}_v e^{ik^{l(r)}_v x} + A^{l(r)}_u D^{l(r)}_u e^{ik^{l(r)}_u x} \\ + A^{l(r)}_3 D^{l(r)}_3 e^{ik^{l(r)}_3 x} + A^{l(r)}_4 D^{l(r)}_4 e^{ik^{l(r)}_4 x}\big]. \quad (9.111)$$

As usual, the left and right amplitudes are related by

$$\begin{pmatrix} A_v^l \\ A_u^l \\ A_3^l \\ A_4^l \end{pmatrix} = M \begin{pmatrix} A_v^r \\ A_u^r \\ A_3^r \\ A_4^r \end{pmatrix}, \tag{9.112}$$

the matching matrix is given by $M = W_l^{-1} W_r$, where

$$W_{l(r)} = \begin{pmatrix} D_v^{l(r)} & D_u^{l(r)} & D_3^{l(r)} & D_4^{l(r)} \\ ik_v^{l(r)} D_v^{l(r)} & ik_u^{l(r)} D_u^{l(r)} & ik_3^{l(r)} D_3^{l(r)} & ik_4^{l(r)} D_4^{l(r)} \\ E_v^{l(r)} & E_u^{l(r)} & E_3^{l(r)} & E_4^{l(r)} \\ ik_v^{l(r)} E_v^{l(r)} & ik_u^{l(r)} E_u^{l(r)} & ik_3^{l(r)} D_3^{l(r)} & ik_4^{l(r)} D_4^{l(r)} \end{pmatrix}. \tag{9.113}$$

As illustrated in Fig. 9.7, the "in" basis is here defined by four incoming waves: two of them (u, v) are incident on the discontinuity from the right (left panels on the first and second rows); the two others (3, 4), from the left (left panels on the third and fourth rows). The "out" basis is defined along the same lines as sketched in the four panels of the right column.

The field operator can be expanded either in the "in" or in the "out" basis as

$$\hat{\phi} = \int_0^{\omega_{max}} d\omega \big[\hat{a}_\omega^{v,in} \phi_{v,r}^{in} + \hat{a}_\omega^{u,in\dagger} \phi_{u,r}^{in} + \hat{a}_\omega^{3,in} \phi_{3,l}^{in} + \hat{a}_\omega^{4,in\dagger} \phi_{4,l}^{in}$$
$$+ \hat{a}_\omega^{v,in\dagger} \varphi_{v,r}^{in*} + \hat{a}_\omega^{u,in} \varphi_{u,r}^{in*} + \hat{a}_\omega^{3,in\dagger} \varphi_{3,l}^{in*} + \hat{a}_\omega^{4,in} \varphi_{4,l}^{in*} \big] \tag{9.114}$$

or

$$\hat{\phi} = \int_0^{\omega_{max}} d\omega \big[\hat{a}_\omega^{v,out} \phi_{v,l}^{out} + \hat{a}_\omega^{u,out\dagger} \phi_{u,l}^{out} + \hat{a}_\omega^{3,out} \phi_{3,r}^{out} + \hat{a}_\omega^{4,out\dagger} \phi_{4,r}^{out}$$
$$+ \hat{a}_\omega^{v,out\dagger} \varphi_{v,l}^{out*} + \hat{a}_\omega^{u,out} \varphi_{u,l}^{out*} + \hat{a}_\omega^{3,out\dagger} \varphi_{3,r}^{out*} + \hat{a}_\omega^{4,out} \varphi_{4,r}^{out*} \big]. \tag{9.115}$$

The "in" and "out" basis are related by

$$\phi_{v,r}^{in} = S_{vl,vr} \phi_{v,l}^{out} + S_{ul,vr} \phi_{u,l}^{out} + S_{3r,vr} \phi_{3,r}^{out} + S_{4r,vr} \phi_{4,r}^{out}, \tag{9.116}$$

$$\phi_{u,r}^{in} = S_{vl,ur} \phi_{v,l}^{out} + S_{ul,ur} \phi_{u,l}^{out} + S_{3r,ur} \phi_{3,r}^{out} + S_{4r,ur} \phi_{4,r}^{out}, \tag{9.117}$$

$$\phi_{3,l}^{in} = S_{vl,3l} \phi_{v,l}^{out} + S_{ul,3l} \phi_{u,l}^{out} + S_{3r,3l} \phi_{3,r}^{out} + S_{4r,3l} \phi_{4,r}^{out}, \tag{9.118}$$

$$\phi_{4,l}^{in} = S_{vl,4l} \phi_{v,l}^{out} + S_{ul,4l} \phi_{u,l}^{out} + S_{3r,4l} \phi_{3,r}^{out} + S_{4r,4l} \phi_{4,r}^{out} \tag{9.119}$$

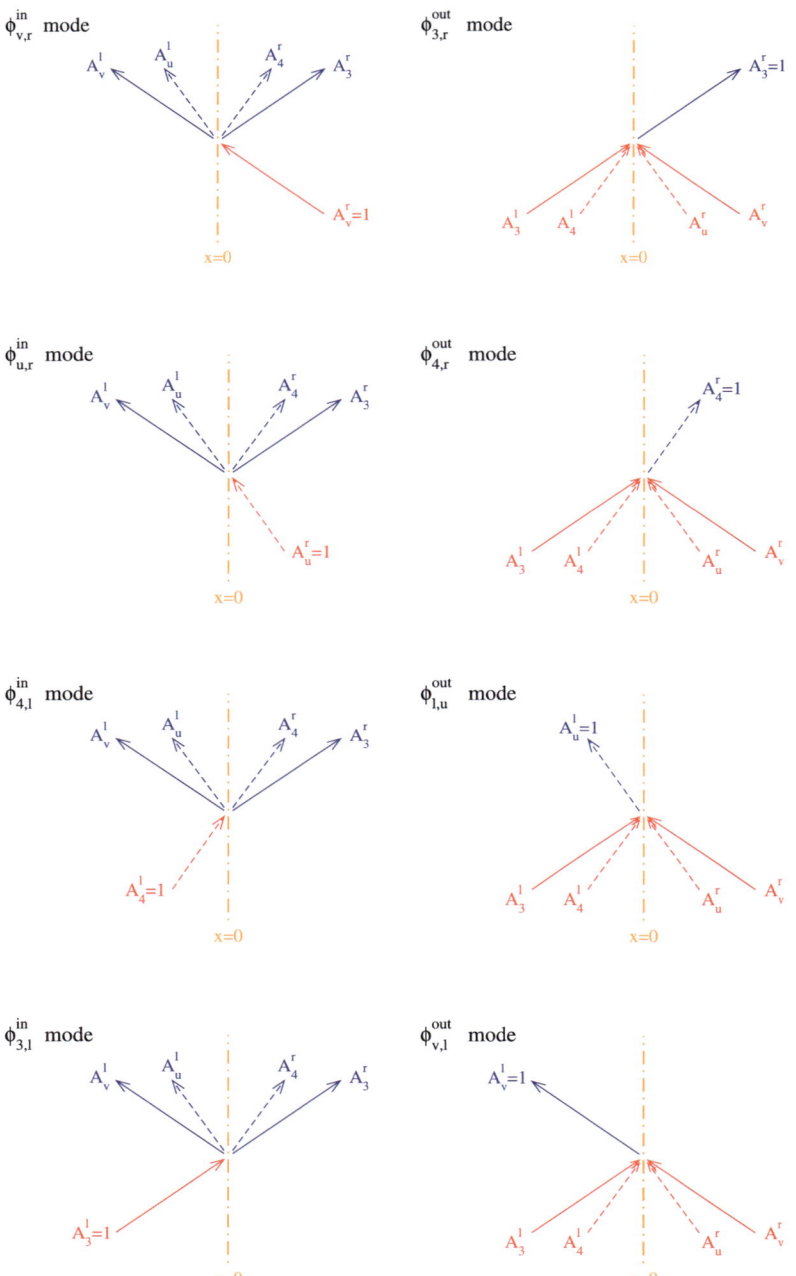

Fig. 9.7 Sketch of the Bogoliubov modes involved in the "in" (*left panels*) and "out" (*right panels*) basis. The mode labels refer to the dispersion shown in the lower panels of Fig. 9.6

and the corresponding relation between the "in" and "out" annihilation and creation operators is

$$\begin{pmatrix} \hat{a}_\omega^{v,out} \\ \hat{a}_\omega^{u,out\dagger} \\ \hat{a}_\omega^{3,out} \\ \hat{a}_\omega^{4,out\dagger} \end{pmatrix} = \begin{pmatrix} S_{vl,vr} & S_{vl,ur} & S_{vl,3l} & S_{vl,4l} \\ S_{ul,vr} & S_{ul,ur} & S_{ul,3l} & S_{ul,4l} \\ S_{3r,vr} & S_{3r,ur} & S_{3r,3l} & S_{3r,4l} \\ S_{4r,vr} & S_{4r,ur} & S_{4r,3l} & S_{4r,4l} \end{pmatrix} \begin{pmatrix} \hat{a}_\omega^{v,in} \\ \hat{a}_\omega^{u,in\dagger} \\ \hat{a}_\omega^{3,in} \\ a_\omega^{4,in\dagger} \end{pmatrix}. \quad (9.120)$$

Explicit expressions for the corresponding amplitudes are listed in the Appendix.

Here we see again that the S matrix mixes creation and annihilation operators. As a consequence, the $|0, in\rangle$ and $|0, out\rangle$ vacua do not coincide: in particular the "in" vacuum $|0, in\rangle$ state with no incident quanta leads to a finite amount of out-going particles that can be detected: in the left region, corresponding to the u, v modes; in the right region, the 3, 4 modes.

More precisely

$$n_\omega^{v,l} = \langle 0, in | \hat{a}_\omega^{v,out\dagger} \hat{a}_\omega^{v,out} | 0, in \rangle = |S_{vl,ur}|^2 + |S_{vl,4l}|^2 \quad (9.121)$$

$$n_\omega^{u,l} = \langle 0, in | \hat{a}_\omega^{u,out\dagger} \hat{a}_\omega^{u,out} | 0, in \rangle = |S_{ul,vr}|^2 + |S_{ul,3l}|^2 \quad (9.122)$$

$$n_\omega^{3,r} = \langle 0, in | \hat{a}_\omega^{3,out\dagger} \hat{a}_\omega^{3,out} | 0, in \rangle = |S_{3r,ur}|^2 + |S_{3r,4l}|^2 \quad (9.123)$$

$$n_\omega^{4,r} = \langle 0, in | \hat{a}_\omega^{4,out\dagger} \hat{a}_\omega^{4,out} | 0, in \rangle = |S_{4r,vr}|^2 + |S_{4r,3l}|^2 \quad (9.124)$$

and unitarity of the S matrix imposes that

$$n_\omega^{v,l} + n_\omega^{3,r} = n_\omega^{u,l} + n_\omega^{4,r}: \quad (9.125)$$

the number of positive energy particles equals the number of negative energy ones. From the explicit expressions for the S matrix elements listed in the Appendix, it is immediate to see that all spectral distributions $n_\omega^{v,l}$, $n_\omega^{u,l}$, $n_\omega^{3,r}$ and $n_\omega^{4,r}$ at low frequencies are dominated by constant terms, in stark contrast with the $1/\omega$ shape of the thermal Hawking radiation.

9.7 Conclusions

In this chapter, we have given an introductory review of Hawking radiation effects in atomic Bose-Einstein condensates. By focussing our attention on a simple toy model based on a piecewise uniform flow interrupted by sharp interfaces, we have made use of the standard Bogoliubov theory of dilute condensates to obtain analytical predictions for the quantum vacuum emission of phonons by the interface: a necessary and sufficient condition for this emission to occur is super-sonic flow. While the low-frequency part of the emission follows an approximately thermal form for a black-hole interface separating a sub-sonic upstream region from a super-sonic downstream one, a completely different spectrum is found for flows that do not exhibit any horizon and are super-sonic everywhere.

The interest of our development is manifold: on one hand, our analytical treatment provides an intuitive understanding of Hawking radiation based on a Bogoliubov generalization of the scattering of waves by square potentials in a one dimensional Schrödinger equation. On the other hand, our derivation is however completely "ab initio", based on the fundamental microscopic quantum description of the BEC without any recourse to the gravitational analogy. As a result, it does not depend on the hydrodynamic approximation that underlies the introduction of the effective metric and shows that the Hawking effect in atomic BECs is not at all an artifact of the low wavelength (hydrodynamical) approximation: no transplanckian problem is present which may cast doubts on the derivation, rather our derivation shows that the transplanckian problem is itself an artifact of the hydrodynamical approximation.

The intense theoretical and experimental activity that is currently in progress makes us confident that the existence of analog Hawking radiation will be soon experimentally confirmed. The robustness of Hawking radiation with respect to the microscopic details of the condensed-matter system would be a strong indication that, in spite of that lack of quantum, microscopic, description of gravity, Hawking's prediction of black hole radiation with its important thermodynamical implications is a real milestone in our understanding of Nature.

Acknowledgements Continuous stimulating discussions with S. Finazzi, R. Parentani and N. Pavloff are warmly acknowledged. I.C. acknowledges partial financial support from ERC via the QGBE grant.

Appendix

In this appendix, we give the explicit expressions for S Matrix coefficients, to first order in the limit of small ω, for the supersonic-supersonic condition treated in Sect. 9.6. Note in particular how those involved in the vacuum emission ($S_{vl,ur}$, $S_{ul,vr}$, $S_{3r,4l}$, $S_{4r,3l}$) grow as $\sqrt{\omega}$ at low ω, while ($S_{4r,vr}$, $S_{3r,ur}$, $S_{ul,3l}$, $S_{vl,4l}$) tend to constants.

$$S_{vl,vr} = \sqrt{\frac{v_0 + c_l}{v_0 - c_l}} \frac{(c_r^2 - c_l^2)\sqrt{\omega \xi_l}}{2\sqrt{2}(v_0^2 - c_l^2)^{1/4}(v_0^2 - c_r^2)},$$

$$S_{ul,vr} = \sqrt{\frac{v_0 - c_l}{v_0 + c_l}} \frac{(c_r^2 - c_l^2)\sqrt{\omega \xi_l}}{2\sqrt{2}(v_0^2 - c_l^2)^{1/4}(v_0^2 - c_r^2)},$$

$$S_{3r,vr} = \frac{\sqrt{v_0^2 - c_l^2} + \sqrt{v_0^2 - c_r^2}}{2(v_0^2 - c_l^2)^{1/4}(v_0^2 - c_r^2)^{1/4}},$$

$$S_{4r,vr} = \frac{\sqrt{v_0^2 - c_l^2} - \sqrt{v_0^2 - c_r^2}}{2(v_0^2 - c_l^2)^{1/4}(v_0^2 - c_r^2)^{1/4}},$$

$$S_{vl,ur} = \sqrt{\frac{v_0 + c_l}{v_0 - c_l}} \frac{(c_l^2 - c_r^2)\sqrt{\omega \xi_l}}{2\sqrt{2}(v_0^2 - c_l^2)^{1/4}(v_0^2 - c_r^2)},$$

$$S_{ul,ur} = \sqrt{\frac{v_0 - c_l}{v_0 + c_l}} \frac{(c_r^2 - c_l^2)\sqrt{\omega \xi_l}}{2\sqrt{2}(v_0^2 - c_l^2)^{1/4}(v_0^2 - c_r^2)},$$

$$S_{3r,ur} = \frac{\sqrt{v_0^2 - c_r^2} - \sqrt{v_0^2 - c_l^2}}{2(v_0^2 - c_l^2)^{1/4}(v_0^2 - c_r^2)^{1/4}},$$

$$S_{4r,ur} = \frac{\sqrt{v_0^2 - c_l^2} + \sqrt{v_0^2 - c_r^2}}{2(v_0^2 - c_l^2)^{1/4}(v_0^2 - c_r^2)^{1/4}}, \qquad (9.126)$$

$$S_{vl,3l} = \frac{c_l + c_r}{2\sqrt{c_l c_r}},$$

$$S_{ul,3l} = i\frac{c_l - c_r}{2\sqrt{c_l c_r}},$$

$$S_{3r,3l} = \frac{(v_0^2 - c_r^2)^{1/4}(c_l^2 - c_r^2)\sqrt{c_l \xi_l \omega}}{2\sqrt{2c_r}(c_r - v_0)(v_0^2 - c_l^2)},$$

$$S_{4r,3l} = \frac{(v_0^2 - c_r^2)^{1/4}(c_l^2 - c_r^2)\sqrt{c_l \xi_l \omega}}{2\sqrt{2c_r}(c_r - v_0)(v_0^2 - c_l^2)},$$

$$S_{vl,4l} = \frac{c_l - c_r}{2\sqrt{c_l c_r}},$$

$$S_{ul,4l} = i\frac{c_l + c_r}{2\sqrt{c_l c_r}},$$

$$S_{3r,4l} = \frac{(v_0^2 - c_r^2)^{1/4}(c_l^2 - c_r^2)\sqrt{c_l \xi_l \omega}}{2\sqrt{2c_r}(c_r + v_0)(c_l^2 - v_0^2)},$$

$$S_{4r,4l} = i\frac{(v_0^2 - c_r^2)^{1/4}(c_r^2 - c_l^2)\sqrt{c_l \xi_l \omega}}{2\sqrt{2c_r}(c_r + v_0)(c_l^2 - v_0^2)}.$$

References

1. Hawking, S.W.: Commun. Math. Phys. **43**, 199 (1975)
2. Unruh, W.G.: Phys. Rev. Lett. **46**, 1351 (1981)
3. Barceló, C., Liberati, S., Visser, M.: Living Rev. Relativ. **8**, 12 (2005). Available at URL:http://www.livingreviews.org/lrr-2005-12

4. Birrell, N.D., Davies, P.C.W.: Quantum Fields in Curved Space. Cambridge University Press, Cambridge (1982)
5. Jacobson, T.: Phys. Rev. D **44**, 1731 (1991)
6. Macher, J., Parentani, R.: Phys. Rev. A **80**, 043601 (2009)
7. Macher, J., Parentani, R.: Phys. Rev. A **80**, 043601 (2009)
8. Finazzi, S., Parentani, R.: Phys. Rev. D **83**, 084010 (2011)
9. Mayoral, C., Recati, A., Fabbri, A., Parentani, R., Balbinot, R., Carusotto, I.: New J. Phys. **13**, 025007 (2011)
10. Weinfurtner, S., Tedford, E.W., Penrice, M.C.J., Unruh, W.G., Lawrence, G.A.: Phys. Rev. Lett. **106**, 021302 (2011)
11. Belgiorno, F., Cacciatori, S.L., Clerici, M., Gorini, V., Ortenzi, G., Rizzi, L., Rubino, E., Sala, V.G., Faccio, D.: Phys. Rev. Lett. **105**, 203901 (2010)
12. Schützhold, R., Unruh, W.G.: Phys. Rev. Lett. **107**, 149401 (2011)
13. Belgiorno, F., Cacciatori, S.L., Clerici, M., Gorini, V., Ortenzi, G., Rizzi, L., Rubino, E., Sala, V.G., Faccio, D.: Phys. Rev. Lett. **107**, 149402 (2011)
14. Liberati, S., Prain, A., Visser, M.: arXiv:1111.0214
15. Finazzi, S., Carusotto, I.: arXiv:1204.3603
16. Unruh, W.G., Schutzhold, R.: arXiv:1202.6492
17. Balbinot, R., Fabbri, A., Fagnocchi, S., Recati, A., Carusotto, I.: Phys. Rev. A **78**, 021603 (2008)
18. Carusotto, I., Fagnocchi, S., Recati, A., Balbinot, R., Fabbri, A.: New J. Phys. **10**, 103001 (2008)
19. Garay, L.J., Anglin, J.R., Cirac, J.I., Zoller, P.: Phys. Rev. Lett. **85**, 4643 (2000)
20. Garay, L.J., Anglin, J.R., Cirac, J.I., Zoller, P.: Phys. Rev. A **63**, 023611 (2001)
21. Mayoral, C., Fabbri, A., Rinaldi, M.: Phys. Rev. D **83**, 124047 (2011)
22. Recati, A., Pavloff, N., Carusotto, I.: Phys. Rev. A **80**, 043603 (2009)
23. Fulling, S.A., et al.: Proc. R. Soc. Lond. Ser. A **348**, 393 (1976)
24. Lambrecht, A., et al.: Phys. Rev. Lett. **77**, 615 (1996)
25. Lambrecht, A., et al.: Eur. Phys. J. D **3**, 95 (1998)
26. Crocce, M., et al.: Phys. Rev. A **66**, 033811 (2002)
27. Uhlmann, M., et al.: Phys. Rev. Lett. **93**, 193601 (2004)
28. Yablonovitch, E.: Phys. Rev. Lett. **62**, 1742 (1989)
29. Braggio, C., et al.: Europhys. Lett. **70**, 754 (2005)
30. De Liberato, S., et al.: Phys. Rev. Lett. **98**, 103602 (2007)
31. Carusotto, I., et al.: Phys. Rev. A **77**, 063621 (2008)
32. De Liberato, S., et al.: Phys. Rev. A **80**, 053810 (2009)
33. Dodonov, V.V., et al.: Phys. Rev. A **47**, 4422 (1993)
34. Law, C.K.: Phys. Rev. A **49**, 433 (1994)
35. Artoni, M., et al.: Phys. Rev. A **53**, 1031 (1996)
36. Johansson, J.R., et al.: Phys. Rev. Lett. **103**, 147003 (2009)
37. Johansson, J.R., et al.: Phys. Rev. Lett. **103**, 147003 (2009)
38. Wilson, C.M., et al.: Phys. Rev. Lett. **105**, 233907 (2010)
39. Wilson, C.M., Johansson, G., Pourkabirian, A., Simoen, M., Johansson, J.R., Duty, T., Nori, F., Delsing, P.: Nature **479**, 376 (2011)
40. Carusotto, I., Balbinot, R., Fabbri, A., Recati, A.: Eur. Phys. J. D **56**, 391 (2010)
41. Jaskula, J.C., et al.: Private communication
42. Finazzi, S., Parentani, R.: arXiv:1202.6015 [gr-qc]
43. Finazzi, S., Parentani, R.: J. Phys. Conf. Ser. **314**, 012030 (2011)
44. Finazzi, S., Parentani, R.: New J. Phys. **12**, 095015 (2010)
45. Pitaevskii, L., Stringari, S.: Bose-Einstein Condensation. Oxford University Press, Oxford (2003)
46. Castin, Y.: In: Kaiser, R., Westbrook, C., David, F. (eds.) Coherent Atomic Matter Waves. Lecture Notes of les Houches Summer School. Springer, Berlin (2001)
47. Larré, P.-E., Recati, A., Carusotto, I., Pavloff, N.: Phys. Rev. A **85**, 013621 (2012)

48. Kamchatnov, A.M., Pavloff, N.: Phys. Rev. A **85**, 033603 (2012)
49. Greiner, M., Regal, C.A., Stewart, J.T., Jin, D.S.: Phys. Rev. Lett. **94**, 110401 (2005)
50. Fölling, S., Gerbier, F., Widera, A., Mandel, O., Gericke, T., Bloch, I.: Nature **434**, 481 (2005)
51. Rom, T., et al.: Nature **444**, 733 (2006)
52. Perrin, A., Chang, H., Krachmalnicoff, V., Schellekens, M., Boiron, D., Aspect, A., Westbrook, C.I.: Phys. Rev. Lett. **99**, 150405 (2007)
53. Jeltes, T., McNamara, J.M., Hogervorst, W., Vassen, W., Krachmalnicoff, V., Schellekens, M., Perrin, A., Chang, H., Boiron, D., Aspect, A., Westbrook, C.I.: Nature **445**, 402 (2007)
54. Hofferberth, S., Lesanovsky, I., Schumm, T., Imambekov, A., Gritsev, V., Demler, E., Schmiedmayer, J.: Nat. Phys. **4**, 489 (2008)
55. Öttl, A., Ritter, S., Köhl, M., Esslinger, T.: Phys. Rev. Lett. **95**, 090404 (2005)

Chapter 10
Transformation Optics

Ulf Leonhardt

Abstract Transformation optics applies ideas from Einstein's general theory of relativity in optical and electrical engineering for designing devices that can do the (almost) impossible: invisibility cloaking, perfect imaging, levitation, and the creation of analogues of the event horizon. This chapter gives an introduction to this field requiring minimal prerequisites.

10.1 Introduction

According to Einstein's general theory of relativity, the geometry of spacetime is curved by the momentum and energy of macroscopic objects. This curvature is what we perceive as gravity, because it influences the motion of particles such as Newton's apple falling from a tree in the spacetime geometry curved by Earth or the planets circling around in the spacetime geometry curved by the Sun. Gravity also influences the propagation of waves; the most striking demonstration of which is gravitational lensing where light from distant stars or galaxies is deflected and focused in the spacetime geometry created by other stars or galaxies. Gravity is universal, because the geometry of space and time sets the scene for everything, particle and wave alike.

Analogues of gravity occur when the geometry of spacetime appears to be altered by other means than momentum and energy. The most natural example of analogue gravity is the propagation of light in media. The medium—a piece of glass, the water in a vase or any other transparent substance—distorts images much the same way stars and galaxies distort light. We may say, with some justification to be given in this chapter, that media apparently alter the geometry of spacetime for light. This geometry differs from the natural spacetime geometry of gravity: the medium establishes a virtual geometry different from the natural geometry of physical space. The virtual geometry is created by a completely different physical process than the

---

U. Leonhardt (✉)
School of Physics and Astronomy, University of St. Andrews, North Haugh,
St. Andrews KY16 9SS, UK
e-mail: ulf@st-andrews.ac.uk

D. Faccio et al. (eds.), *Analogue Gravity Phenomenology*,
Lecture Notes in Physics 870, DOI 10.1007/978-3-319-00266-8_10,
© Springer International Publishing Switzerland 2013

Fig. 10.1 Einstein, Einwell, Maxwell. Transformation optics combines ideas from Einstein's general relativity and Maxwell's electromagnetism. It particular, it uses transformations of space—like the transformation between Einstein and Maxwell shown in the picture

geometry of physical space and it is also not universal, but restricted to certain physical phenomena. In the case of light in media, the virtual geometry differs from the real geometry only for light (and, typically, only for light within a certain frequency range).

In this chapter, we show how and when virtual geometries arise for light, or electromagnetic waves in general. For this we combine ideas from two of the most beautiful theories of physics, Maxwell's electromagnetism and Einstein's general relativity such that they become transformable into each other (Fig. 10.1). This connection between general relativity and electromagnetism in media is not new, it dates back to ideas by Gordon [1] published in 1923 and Tamm [2, 3] that appeared around that time, and further back to Newton who allegedly toyed with the idea that gravity is mediated by a medium before settling for Newtonian gravity [4] and also to the enigmatic genius of Fermat [5]. New applications of these ideas are in electrical and optical engineering, as design concepts for novel devices that do the (almost) impossible, for example invisibility cloaking and perfect imaging. This new research area with old and deep roots, called transformation optics [6–11], has been regarded as

10 Transformation Optics

one of the most fascinating research insights of the last decade [12]. In this chapter, we derive the foundations of this area and explain a few of the key applications. We also show how transformation optics is related to one of the recurring themes of this book, the physics of the event horizon.

10.2 Maxwell's Electromagnetism

10.2.1 Maxwell's Equations

Let us begin at the beginning, with Maxwell's equations in empty, flat space in Cartesian coordinates:

$$\nabla \cdot \boldsymbol{E} = 0 \qquad \text{GAUSS'S LAW,}$$

$$\nabla \times \boldsymbol{B} = \frac{1}{c^2} \frac{\partial \boldsymbol{E}}{\partial t} \qquad \text{AMPÈRE'S LAW}$$

$$\text{WITH MAXWELL'S DISPLACEMENT CURRENT,} \qquad (10.1)$$

$$\nabla \times \boldsymbol{E} = -\frac{\partial \boldsymbol{B}}{\partial t} \qquad \text{FARADAY'S LAW OF INDUCTION,}$$

$$\nabla \cdot \boldsymbol{B} = 0 \qquad \text{ABSENCE OF MAGNETIC MONOPOLES.}$$

As usual, c denotes the speed of light in vacuum. Throughout this chapter we use SI units for the electromagnetic fields. Now suppose we change the spatial coordinates, for example we use spherical coordinates r, θ, ϕ instead of the Cartesian x, y, z. The differentials of the new coordinates appear in a different way in the line element ds than the Cartesian differentials, for example as

$$ds^2 = dx^2 + dy^2 + dz^2 = dr^2 + r^2 d\theta^2 + r^2 \sin^2\theta \, d\phi^2 \qquad (10.2)$$

in spherical coordinates. In general, curved coordinates x^i contribute to the line element as

$$ds^2 = g_{ij} dx^i dx^j \qquad (10.3)$$

where we sum over repeated indices (running from 1 to 3). The g_{ij} usually depend on the x^i, as the line element (10.2) of the spherical coordinates shows. They constitute the metric tensor with determinant g and matrix inverse g^{ij},

$$g \equiv \det(g_{ij}), \qquad g^{ij} \equiv (g_{ij})^{-1}. \qquad (10.4)$$

Differential geometry [11] tells us how to express the divergences and curls in Maxwell's equations (10.1) in terms of curved coordinates:

$$\frac{1}{\sqrt{g}} \partial_i \sqrt{g} g^{ij} E_j = 0 \qquad \text{GAUSS'S LAW,}$$

$$\varepsilon^{ijk}\partial_j B_k = \frac{1}{c^2}\frac{\partial g^{ij}E_j}{\partial t} \qquad \text{AMPÈRE'S LAW}$$

WITH MAXWELL'S DISPLACEMENT CURRENT, (10.5)

$$\varepsilon^{ijk}\partial_j E_k = -\frac{\partial g^{ij}B_j}{\partial t} \qquad \text{FARADAY'S LAW OF INDUCTION,}$$

$$\frac{1}{\sqrt{g}}\partial_i \sqrt{g}g^{ij}B_j = 0 \qquad \text{ABSENCE OF MAGNETIC MONOPOLES,}$$

where the Levi-Civita tensor ε^{ijk}, appearing in the curls, is given in terms of the completely antisymmetric symbols $[ijk]$ as [11]:

$$\varepsilon^{ijk} = \pm\frac{1}{\sqrt{g}}[ijk]. \qquad (10.6)$$

The $\pm$ sign depends on the handedness of the coordinate system: "+" in right-handed systems and "−" in left-handed systems.

Maxwell's equations (10.5) are not only valid in a flat space expressed in curved coordinates, but also in genuine curved spaces.[1] The reason is the following: Maxwell's equations (10.5) are first-order partial differential equations containing maximally first derivatives of the metric tensor g_{ij}. Now, a theorem from differential geometry [11, 13] says that, for any given point, we can always construct a local Cartesian coordinate system where $g_{ij} = \delta_{ij}$ and $\partial_k g_{ij} = 0$ at that point, regardless how curved the geometry is. If the space is curved, these local Cartesian systems do not form a single, global Cartesian frame, but rather represent a patchwork of frames that are not consistent. The inconsistency is caused by the spatial curvature. However, there is always a local coordinate transformation from each local frame to the global frame of the curved manifold. As Maxwell's equations depend maximally on first derivatives of g_{ij} we can assume them in the form (10.1) in each local Cartesian frame and then transform to the general form (10.5) in the global frame. Therefore the form (10.5) describes electromagnetism in curved space as well.

10.2.2 The Medium of a Geometry

Consider now a case similar to gravitational lensing[2] where we assume a given spatial geometry. In the following we show how this geometry appears as a medium. For this we express Maxwell's equations (10.5) for the quantities E (electric field

---

[1] We consider purely spatial geometries first and then, in Sect. 10.5, we generalise our theory to spacetime geometries.

[2] In gravitational lensing, gravity alters primarily the measure of time, but due to the conformal invariance of electromagnetism (see Sect. 10.5.2) this is equivalent to altering the measure of space.

strength), **D** (dielectric displacement), **H** (magnetic field) and **B** (magnetic induction) familiar from the macroscopic electromagnetism in media,

$$\nabla \cdot \boldsymbol{D} = 0 \qquad \text{GAUSS'S LAW,}$$

$$\nabla \times \boldsymbol{H} = \frac{\partial \boldsymbol{D}}{\partial t} \qquad \text{AMPÈRE'S LAW}$$

$$\text{WITH MAXWELL'S DISPLACEMENT CURRENT,} \qquad (10.7)$$

$$\nabla \times \boldsymbol{E} = -\frac{\partial \boldsymbol{B}}{\partial t} \qquad \text{FARADAY'S LAW OF INDUCTION,}$$

$$\nabla \cdot \boldsymbol{B} = 0 \qquad \text{ABSENCE OF MAGNETIC MONOPOLES.}$$

Considering Gauss's law, we can write it as $\nabla \cdot \boldsymbol{D} = \partial_i D^i = 0$ with

$$D^i = \varepsilon_0 \varepsilon^{ij} E_j \tag{10.8}$$

and $\varepsilon^{ij} \propto \sqrt{g} g^{ij}$. Here ε_0 denotes the electric permittivity of the vacuum. Let us see whether this definition of the dielectric displacement **D** is consistent with the other place **D** occurs in Maxwell's equations (10.7), Ampère's law with Maxwell's displacement current:

$$\varepsilon^{ijk} \partial_j B_k = \pm \frac{1}{\sqrt{g}} [ijk] \partial_j B_k = \frac{1}{c^2} \frac{\partial g^{ij} E_j}{\partial t}. \tag{10.9}$$

If we write

$$H_k = \varepsilon_0 c^2 B_k \tag{10.10}$$

and

$$\varepsilon^{ij} = \pm \sqrt{g} g^{ij} \tag{10.11}$$

we obtain Ampère's law for **D** given by definition (10.8). Using the same arguments for the remaining two Maxwell equation we get

$$B^i = \mu_0 \mu^{ij} H_j \tag{10.12}$$

with the magnetic permeability of the vacuum

$$\mu_0 = \frac{1}{\varepsilon_0 c^2} \tag{10.13}$$

and the relative magnetic permeability

$$\mu^{ij} = \pm \sqrt{g} g^{ij}. \tag{10.14}$$

In the formalism developed here [11], **E** and **H** are the fundamental fields, whereas **D** and **B** are derived from E_i and H_i by raising the index with g^{ij} and multiplication by $\sqrt{g}$. Mathematically, the **E** and **H** are one-forms, whereas the **D** and **B** are vector densities with respect to the spatial geometry [11]. The fact that the ε^{ij} and μ^{ij} are matrices that depend on two indices indicates that the medium representing the geometry g_{ij} is anisotropic in general. We also see that a spatial geometry appears as a medium with

$$\varepsilon^{ij} = \mu^{ij}. \tag{10.15}$$

In electrical engineering, such media are called impedance-matched (to the vacuum).

10.2.3 The Geometry of a Medium

The converse is also true: impedance-matched media, *i.e.* media satisfying the condition (10.15), can be understood as spatial geometries, for the following argument. We calculate the determinant $\det \varepsilon$ of the matrix ε^{ij} from relation (10.11) and obtain

$$\det \varepsilon = \pm \sqrt{g}. \tag{10.16}$$

Consequently, we can write g^{ij} in terms of ε^{ij} and, by virtue of impedance matching (10.15), of μ^{ij} as well, as

$$g^{ij} = \frac{\varepsilon^{ij}}{\det \varepsilon} = \frac{\mu^{ij}}{\det \mu}. \tag{10.17}$$

Spatial geometries appear as impedance-matched media and impedance-matched media make virtual geometries. Impedance matching establishes an exact virtual geometry where electromagnetic fields are identical in all aspects to such fields in a real geometry. Without impedance matching, the geometric picture is not exact. In particular, the propagation of light in non-impedance-matched anisotropic media depends on the polarisation, a phenomenon called birefringence. However, non-impedance-matched media can still be used for establishing geometries in planar media for specific polarisations (see e.g. the experiments [14–18]), because in such cases not all components of the ε^{ij} and μ^{ij} tensors are needed. Furthermore, in optically isotropic media (where $\varepsilon^{ij} = \varepsilon \delta_{ij}$ and $\mu^{ij} = \mu \delta_{ij}$) the geometric picture gives an excellent approximation for electromagnetic waves within the validity range of geometrical optics [11].

10.3 Spatial Transformations

In general, the virtual geometry of light in media is curved. To give a simple example, a lens focuses parallel light rays in the focal point; parallels are thus no longer parallel, but meet, which violates Euclid's parallel axiom of flat space. Curved space is common place in optics. It is much harder to create a virtual geometry that is flat. What would it do? As any flat geometry can be reduced to Cartesian coordinates

10 Transformation Optics

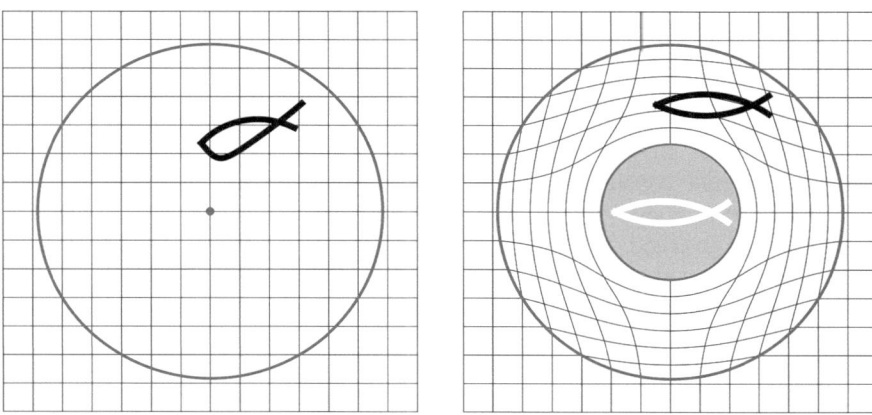

Fig. 10.2 Transformation of space. Optical materials appear to change the perception of space; objects (fish) in physical space (*right picture*) appear at positions (*left picture*) different from where they actually are. Suppose the medium performs a coordinate transformation from physical space (*right*) to virtual space (*left*) and vice versa. Virtual space is empty and so light propagates along straight lines. In virtual space, we may draw a coordinate system as a rectangular grid of light rays (*left grid*). In physical space, the light rays are curved; the coordinate grid of virtual space is transformed into a curved coordinate system in physical space (*right grid*). As the coordinate transformation only changes space within a circle, this *circle* marks the boundary of the optical material used to transform space. We see that the images of the fish are distorted in virtual space, because the coordinate transformation illustrated here is not uniform. Moreover, the white fish has completely disappeared, because it was swimming within a region of physical space (*grey*) that, in virtual space, is contracted to a single, invisible point. Such an optical material makes an invisibility device

by a coordinate transformation, the material would just perform a transformation of space where each point of physical space appears to be at a position in virtual space that may deviate from real one. If the new coordinates agree with the old ones outside of the device, that is made of the medium, we would not see the difference between propagation in the medium and empty, flat space. In short, the device would be completely invisible.

10.3.1 Invisibility Cloaking

Such invisible devices could be used to make other things invisible, too: they can be turned into invisibility devices as follows. Suppose the device performs the following transformation (Fig. 10.2): an extended region in physical space (Fig. 10.2 right) in condensed into a single point in virtual space (Fig. 10.2 left). Anything inside this region has thus become as small as a single point, invisibly small. Everything inside is hidden and, as the device itself is invisible, the very act of hiding is hidden as well. The transformation (Fig. 10.2) makes a perfect cloaking device [8].

Such a cloaking device has been demonstrated for microwaves [14] or, to be more precise, an approximation without impedance matching was made [14]. However, this device did only operate for one polarisation, because it used a medium that is not impedance-matched, and for one frequency only, for a fundamental reason [9] that applies to all purely transformation-based cloaking devices [8]. This reason is easy to understand: consider a light ray that just straddles the invisible point in virtual space. As the device performs a spatial transformation from virtual to physical space, the light propagation in both spaces is synchronised. Therefore the light must go around the invisible region in precisely the time it takes to pass a single point in virtual space: zero time. Consequently, the speed of light must tend to infinity at the inner surface of the cloaking device [9, 19]. By speed of light we mean the phase velocity here, whereas the group velocity turns out to approach zero [19]. Relativistic causality does not prohibit an infinity phase velocity in a medium, but it allows it only for a single frequency—a single colour—and thus in a purely stationary regime where nothing changes and no new information is transferred. Any change would cause distortions, which defeats the point of a cloaking device;[3] one might as well use a hologram of the background.

Perfect invisibility is impossible, but this does not prevent invisibility that is good enough. One might be content with deforming a surface by conformal [7] or quasiconformal transformations [20], which does not make an object disappear altogether, but makes it optically flat; the fugu in Fig. 10.3 is turned into a flatfish. One could then use conventional camouflage to disguise the flat object (as flatfish are masters of). Or one might use non-Euclidean cloaking devices where virtual space is not flat, but curved in an appropriate way [7, 21, 22]. In this case the speed of light is finite in the device (and can be made slower than c [22]) but the price to be paid is a time delay of the light making the detour in the cloaking device. Furthermore, relativistic causality is of little concern to the cloaking of sound waves, because the normal speed of sound is several orders of magnitude slower than c. Near-perfect acoustical cloaking over a broad band of frequencies is possible and has been demonstrated [23].

10.3.2 Transformation Media

What does it take to build a cloaking device and similar transformation devices? Let us work out the electromagnetic properties of devices required for transforming space. We use x^i to denote the coordinates of physical space and $x^{i'}$ for virtual space (the prime at the index does not mean that we simply use a different index variable but shall indicate the different coordinates of virtual space). As the device performs a transformation from physical to virtual space and vice versa, the $x^{i'}$ of virtual space are thus functions $x^{i'}(x^i)$ of the physical-space coordinates x^i—the

---

[3] It might be reassuring to know that perfect deception is impossible; the truth will always appear in the end.

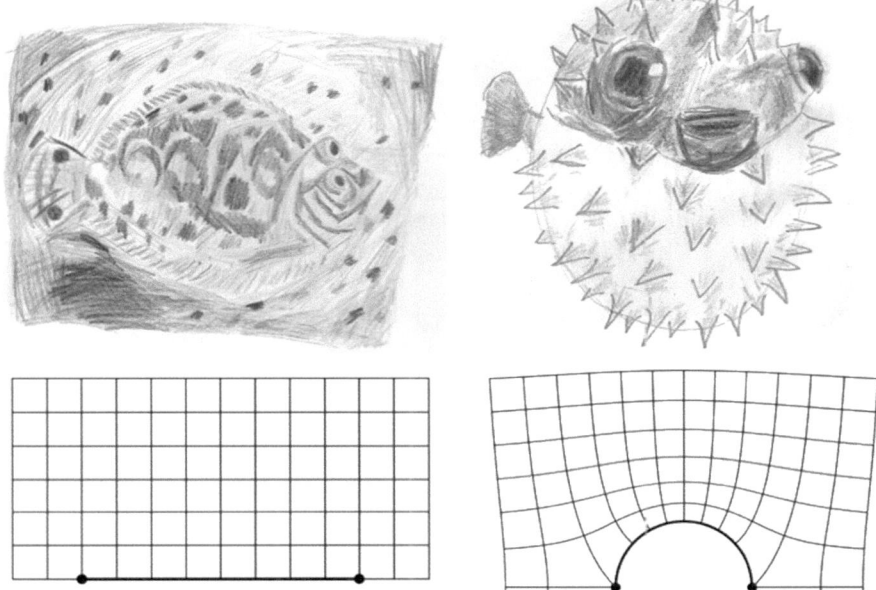

Fig. 10.3 From fugu to flatfish. A coordinate transformation may turn a voluminous object in physical space (fugu) into a flat object (flatfish). Credit: Maria Leonhardt

device performs a coordinate transformation. Suppose that virtual space is flat and empty, and that we describe it in Cartesian coordinates. In this case we obtain for the line element:

$$ds^2 = \delta_{i'j'} dx^{i'} dx^{j'} = \delta_{i'j'} \frac{\partial x^{i'}}{\partial x^i} \frac{\partial x^{j'}}{\partial x^j} dx^i dx^j \tag{10.18}$$

and thus, according to definition (10.3),

$$g_{ij} = \Lambda_i^{i'} \delta_{i'j'} \Lambda_j^{j'} \quad \text{with} \quad \Lambda_i^{i'} \equiv \frac{\partial x^{i'}}{\partial x^i}. \tag{10.19}$$

From this expression we get for the matrix inverse of g_{ij}:

$$g^{ij} = \Lambda_{i'}^i \delta^{i'j'} \Lambda_{j'}^j \quad \text{with} \quad \Lambda_{i'}^i \equiv \frac{\partial x^i}{\partial x^{i'}}, \tag{10.20}$$

or, in matrix notation,

$$\mathbf{g}^{-1} = \mathbf{\Lambda}\mathbf{\Lambda}^T \tag{10.21}$$

where g denotes the matrix g_{ij} and

$$\Lambda \equiv \left(\frac{\partial x^i}{\partial x^{i'}} \right). \tag{10.22}$$

Now we can calculate the matrices $\boldsymbol{\varepsilon}$ and $\boldsymbol{\mu}$ of the ε^{ij} and μ^{ij} according to the recipe (10.11) and (10.14). There we need the determinant g of g_{ij}, which is $g = (\det \Lambda)^{-2}$ according to formula (10.21). We thus obtain

$$\boldsymbol{\varepsilon} = \boldsymbol{\mu} = \frac{\Lambda \Lambda^T}{\det \Lambda}. \tag{10.23}$$

Equation (10.23) formulates a simple recipe for calculating the required electromagnetic properties of a spatial transformation device. This recipe is valid in Cartesian coordinates, where both virtual and physical space are described by Cartesian grids, but it can be easily extended to curved coordinates [9, 11].

10.3.3 Perfect Imaging with Negative Refraction

Apart from invisibility cloaking, another prominent application of transformation optics is perfect imaging [9]. Imagine a device that performs the following transformation in Cartesian coordinates

$$x = x(x'), \qquad y = y', \qquad z = z', \tag{10.24}$$

where $x(x')$ is folded as shown in Fig. 10.4. We see in Fig. 10.4 that in the fold of the function $x(x')$ each point x' in virtual space has three faithful images in physical space, so the electromagnetic field at these three points is the same as that at the point x'. Electromagnetic fields at each of the three points in physical space are therefore perfectly imaged at the other two: the device is a perfect lens. The transformation (10.24) has the transformation matrix $\Lambda = \text{diag}(dx/dx', 1, 1)$ and we find from the recipe (10.23):

$$\boldsymbol{\varepsilon} = \boldsymbol{\mu} = \text{diag}\left(\frac{dx}{dx'}, \frac{dx'}{dx}, \frac{dx'}{dx} \right). \tag{10.25}$$

Inside the device, *i.e.* inside the fold in the transformation of Fig. 10.4, the derivative dx'/dx becomes negative and the coordinate system changes handedness. The electromagnetic left-handedness of such a material appears through a transformation to a left-handed coordinate system; the material is called a left-handed material or also a material with negative refraction. When the negative slope in the transformation is $dx'/dx = -1$, Eq. (10.25) gives a perfect lens made of an isotropic material with $\varepsilon = \mu = -1$ (otherwise the material (10.25) is anisotropic). As Fig. 10.4 shows, the imaging range is equal to the thickness of the lens in this case.

Fig. 10.4 Perfect lens. Negatively refracting perfect lenses employ transformation media. The *top figure* shows a suitable coordinate transformation from the physical x axis to the electromagnetic x', the *lower figure* illustrates the corresponding device. The inverse transformation from x' to x is either triple or single-valued. The triple-valued segment on the physical x axis corresponds to the focal region of the lens: any source point has two images, one inside the lens and one on the other side. Since the device facilitates an exact coordinate transformation, the images are perfect

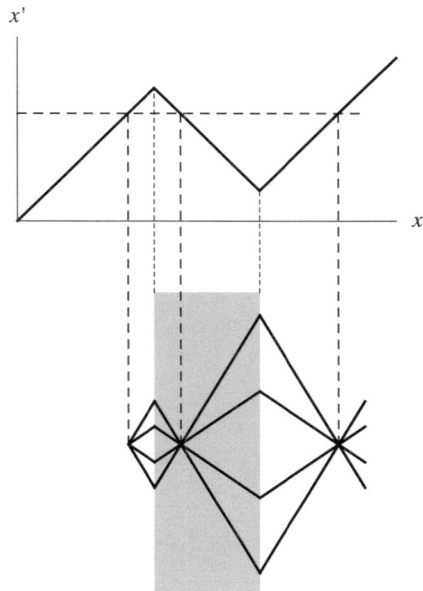

Perfect lensing was first analysed through the imaging of evanescent waves in a slab of negatively-refracting material [24]. These are waves that may carry images finer than the optical resolution limit. Various aspects of this idea have been subject to a considerable theoretical debate (see Ref. [25]) but experiments have confirmed negative refraction (see e.g. Ref. [26]). Sub-resolution imaging was observed for a "poor-man's perfect lens" [27] where the lens is effectively implemented by a sub-wavelength sheet of silver. Our pictorial argument leads to a simple intuitive explanation of why such lenses are indeed perfect. It also reveals some of the practical limitations of perfect imaging by negative refraction.

We have seen that if the imaging device performs the spatial transformation (10.24) illustrated in Fig. 10.4, the electromagnetic field is identical in three separate regions of space. How is this possible? The electromagnetic field cannot instantly hop from one region to another—this is forbidden by relativistic causality, but it can settle to identical field structures over time in a stationary regime. But this implies that the stationary response of the electromagnetic material must be very different from the instantaneous response: the material must be dispersive. Dispersion is always accompanied by dissipation and the dissipation turns out [28] to severely reduce the resolution of imaging by negative refraction in left-handed materials. Therefore only "poor man's perfect lenses" have worked [27] where the imaging distance is a mere fraction of the wavelength. Nevertheless, the tantalising ideas of negative refraction have inspired the entire research area of metamaterials and transformation optics [12].

Fig. 10.5 Casimir effect of left-handed metamaterials. (**A**) illustrates a material with $\varepsilon = \mu = -1$ sandwiched between two mirrors. (**B**) shows how the medium transforms the Casimir cavity of size a in physical x space into a cavity in x' space of size a' according to Eq. (10.26). The attractive Casimir force in x' space moves the mirrors further apart in x space: the Casimir effect in physical space is repulsive

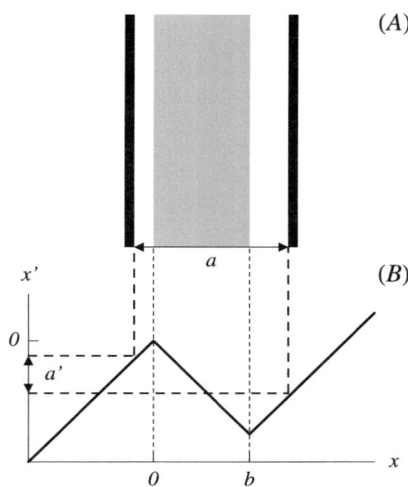

10.3.4 Quantum Levitation

Left-handed materials might also exhibit some interesting quantum effects. In particular, they may alter the Casimir force. The Casimir force is a force between dielectric materials triggered by the fluctuations of the quantum vacuum. The Casimir force is closely related to the van der Waals force—both forces are manifestations of the same physical mechanism. Like the van der Waals force, the Casimir force is usually attractive. These forces are responsible for the stickiness of dielectric materials at close distances. The archetype of the Casimir force is the attraction between two perfect mirrors [29]. Suppose now that between two perfect planar mirrors of distance a we insert a slab of left-handed material with $\varepsilon = \mu = -1$ and thickness b (Fig. 10.5). In virtual space, the transformation medium has done its deed and only the mirrors are left with the Casimir force acting upon them. The mirrors are at the virtual distance

$$a' = a - 2b. \tag{10.26}$$

For $a < 2b$ the distance is negative, which simply means that the two virtual mirrors have swapped sides. More importantly, when the two mirrors are attracted by the Casimir force in virtual space their distance $|a'|$ decreases, but then their distance in physical space $a = 2b + a'$ increases, because a' is negative with falling magnitude. This means that the Casimir force has become repulsive.

We obtain from Casimir's original result [29] the vacuum stress in x direction of [30]

$$\sigma_{xx} = -\frac{\hbar c \pi^2}{240 a'^4}. \tag{10.27}$$

When $a \sim 2b$ the repulsive Casimir force may become very strong such that it could levitate macroscopic objects on vacuum fluctuations, on, literally, nothing [30]. Note that, although our picture of quantum levitation is intuitive and simple, it is also a bit too simple, because the Casimir force is a broad-band phenomenon and so the theory implicitly assumes that the left-handed material remains left-handed over a broad frequency range, which, as we already know, is impossible. More realistic calculations [30] show that some gain is needed in the left-handed material, but, in one way or another, the idea of quantum levitation might still fly.[4]

10.4 Curved Space

Spatial transformations are simple and intuitive in theory, but often difficult to implement in practice. In particular, for the most interesting transformations such as invisibility cloaking and perfect imaging, fundamental problems prevent their practical realisation in a meaningful way, as we have seen. The alternative to spatial transformations is the implementation of a curved virtual space, which is the standard case in isotropic media anyway (even non-impedance-matched isotropic media appear to electromagnetic waves as geometries, as long as the approximation of geometrical optics is valid [11]). The theory of light in curved space is harder, but the experiments are much easer than in implementations of flat space. In some cases, curved spaces have extraordinary optical properties that makes their practical use in optical devices highly desirable. Let us discuss such a case that also represents the simplest curved space in 3D: the surface of the 4D hypersphere. There we can use a lot of the intuition we have about the surface of the ordinary 3D sphere to predict interesting physical phenomena without calculation, just by drawing pictures.

10.4.1 Einstein's Universe and Maxwell's Fish Eye

Our case is closely related to a famous cosmological model due to Einstein [33] that happens to be wrong for the Universe, but, when turned into an electromagnetic device, may be very useful in down-to-Earth applications. Einstein assumed that the Universe is static and for this case derived an exact solution of the equations of general relativity that describe how matter curves spacetime. Astronomical observations have shown however that the Universe is not static but expanding, yet this should not deter us from turning Einstein's solution into a practical device. In Einstein's static Universe light propagates as if it were confined to the 3D surface of a 4D hypersphere in $\{X, Y, Z, W\}$ space with

$$X^2 + Y^2 + Z^2 + W^2 = a^2 \qquad (10.28)$$

---

[4] A repulsive Casimir force was observed [31] between three materials with $\varepsilon_1 < \varepsilon_2 < \varepsilon_3$ over a sufficiently broad frequency range [32], *i.e.* in materials but not in empty space yet.

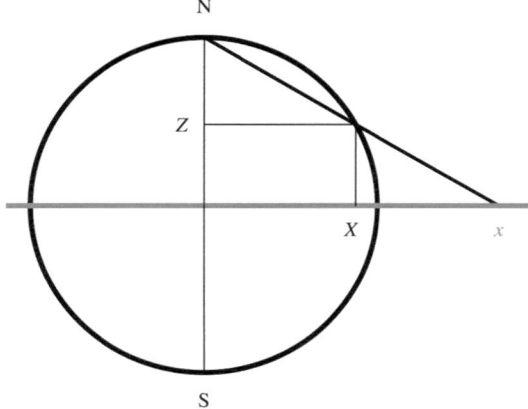

Fig. 10.6 Stereographic projection. Points on the sphere (or hypersphere) are projected to the plane (or hyperplane) as follows. A line is drawn through the North Pole N and the point on the sphere. Where this line intersects the plane cut through the Equator lies the projected point. The picture shows a cut through the sphere and plane where a point on the sphere is characterised by the coordinates $\{X, Z\}$ and the point on the plane by x

where the constant a describes the radius of the hypersphere. Suppose that the hypersurface (10.28) is the virtual space of an optical device. Furthermore, the Cartesian coordinates $\{x, y, z\}$ of physical space shall be connected to the virtual hypersphere by stereographic projection (Fig. 10.6):

$$x = \frac{X}{1 - W/a}, \qquad y = \frac{Y}{1 - W/a}, \qquad z = \frac{Z}{1 - W/a}. \qquad (10.29)$$

One easily verifies that the virtual-space coordinates are given by the following inverse stereographic projection:

$$X = \frac{2x}{1 + r^2/a^2}, \qquad Y = \frac{2y}{1 + r^2/a^2}, \qquad Z = \frac{2z}{1 + r^2/a^2}, \qquad W = a\frac{r^2 - a^2}{r^2 + a^2} \qquad (10.30)$$

where r denotes the radius in physical space with

$$r^2 = x^2 + y^2 + z^2. \qquad (10.31)$$

In order to deduce the effective geometry in physical space and hence the medium required to implement it, we express the line element in virtual space in terms of the differentials in physical space with the help of the inverse stereographic projection (10.30). We obtain

$$ds^2 = dX^2 + dY^2 + dZ^2 + dW^2 = n^2(dx^2 + dy^2 + dz^2) \qquad (10.32)$$

with the radius-dependent prefactor

$$n = \frac{2}{1+r^2/a^2}. \qquad (10.33)$$

From the line element (10.32) we read off the metric tensor g_{ij}, its determinant g and its matrix inverse g^{ij} as

$$g_{ij} = n^2 \delta_{ij}, \qquad g = n^6, \qquad g^{ij} = n^{-2}\delta_{ij}. \qquad (10.34)$$

According to relations (10.11) and (10.14) this spatial geometry corresponds to a medium with

$$\boldsymbol{\varepsilon} = \boldsymbol{\mu} = n\mathbb{1}. \qquad (10.35)$$

As $\boldsymbol{\varepsilon}$ and $\boldsymbol{\mu}$ are proportional to the unity matrix $\mathbb{1}$ the medium is optically isotropic, and it has the refractive-index profile (10.33). Isotropic media are usually the easiest to implement, which is the reason why we have chosen the stereographic projection (10.29) and not any other mapping from virtual to physical space.

The index profile (10.33) of Einstein's static Universe [33] was written down by Maxwell as a student at Trinity College Cambridge [34]. Maxwell was not aware of its relation to the stereographic projection of a sphere—that was discovered in optics by Luneburg [35] much later—Maxwell was simply fascinated by the extraordinary optical properties of a device with the profile (10.33). It reminded him of the eye of fish and therefore such a device is called Maxwell's fish eye.

10.4.2 Perfect Imaging with Positive Refraction

Two properties of the fish eye Maxwell found particularly fascinating: (1) light goes in circles and (2) all light rays from any point meet at a corresponding image point. These properties turn out to be simple mathematical consequences of the light propagation on the virtual hypersphere and the stereographic projection to physical space. Let us, instead of the 4D hypersphere, imagine an ordinary 3D sphere—the hypersphere is not much different (Fig. 10.7). The light rays on the sphere are the geodesics, the great circles. Now, the stereographic projection always maps circles on the sphere to circles in physical space [11] (some degenerate into lines, *i.e.* circles with infinite radius). From this follows property (1) of Maxwell's fish eye—light goes in circles. Property (2) is the easiest to understand. Consider the great circles of light rays emitted from a point P on the virtual sphere. All great circles from a given point P must intersect at the antipodal point. To see this, just rotate P to the North Pole of the sphere. In this case the great circles are the lines of longitude, and they all meet at the two poles. Therefore, the geodesics from the North Pole intersect at the South Pole. If we rotate the point P back to its original position the rotated South Pole turns into the antipodal point. Now, as the stereographic projections maps the surface of the hypersphere to physical space, the trajectories of light rays are simply the projections of the great circles. Therefore they

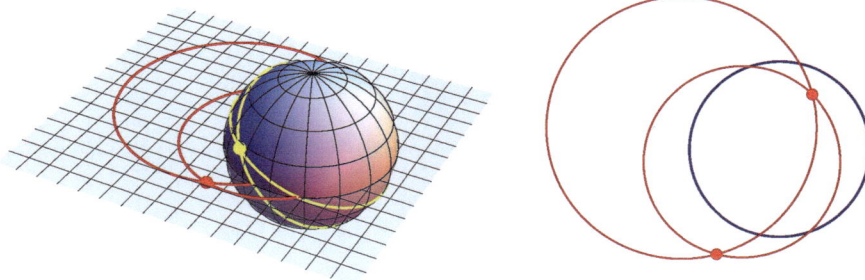

Fig. 10.7 Light propagation in Einstein's universe and Maxwell's fish eye. The virtual space (*left*) is a sphere or hypersphere—Einstein's universe. Light follows the geodesics, the great circles here. All light rays emitted from one point must come together again at the antipodal point. The physical space (*right*) is Maxwell's fish eye with the refractive-index profile (10.33). As the stereographic projection (Fig. 10.6) maps circles into circles, light goes in circles in physical space. Moreover, all light rays emitted at an arbitrary given point must focus at a corresponding image point. Maxwell's fish eye makes a perfect imaging device

all must meet in physical space as well, at the stereographic projection of the antipodal point. All light rays emitted from any point in Maxwell's fish eye meet at the corresponding image point. Devices where all light rays from any point within an object region intersect at the points of the image region are called absolute optical instruments [36].

Maxwell's fish eye is an absolute optical instrument and it has other curious optical properties, too—*e.g.* light goes in circles—but it has never been built in its original form (10.33) for two good reasons that are connected to each other. First, the profile of Maxwell's fish eye fills the entire physical space. Second, for $r \to \infty$ the refractive index (10.33) tends to zero, *i.e.* the speed of light becomes infinite at infinity. The two reasons are connected, because Maxwell's fish eye represents the geometry of a finite space, the virtual hypersphere, in an infinitely extended space. There the speed of light must go to infinity for keeping propagation times finite. However, there is a remedy [37, 38] that solves both problems in one stroke. Imagine we place a mirror around the Equator of the virtual sphere. The mirror would create the illusion that the light propagates in the entire virtual sphere, whereas in reality it is confined to one of the Hemispheres, say the Southern Hemisphere. In physical space, the Southern Hemisphere corresponds to the region with $r \leq a$ where the refractive index ranges only from 1 at $r = a$ to 2 in the centre and the device is finite now. Therefore, Maxwell's fish eye with a mirror can be built, and recently it has been built [39, 40] in 2D. In three-dimensional space the mirror should be a spherical shell at $r = a$ that encloses the index profile (10.33) of the fish eye.

As long as ray optics is concerned, absolute optical instruments like Maxwell's fish eye create perfect images, because all light rays from all object points faithfully arrive at the corresponding image points. However, the resolution of optical instruments is normally restricted by the wave nature of light [36] and cannot be made much finer than the wavelength. Is perfect imaging possible with Maxwell's fish

10 Transformation Optics

eye? It is wise to consider this problem in virtual space, on the sphere (representing the virtual hypersphere for the 3D fish eye). Any source can be regarded as a collection of point sources, so it suffices to investigate the wave produced by a single point source of arbitrary position on the sphere. A wave propagates from the point of emission round the sphere and focuses at the antipodal point; this corresponds with emission from a point in the plane of the actual device and focusing at the image point in physical space. The wave propagating round the virtual sphere would come to the antipodal point and focus there. Because of the symmetry of the sphere, the initially outgoing wave from the source points turns into an ingoing wave at the image point, *i.e.* a time-reversed outgoing wave. However, the time reversal is only complete if one essential element is present at the image point: a reversed source, a drain. The drain at the image point is something natural in imaging where one wishes to detect an image, for example by photochemical reactions or in a CCD array. The drain represents a detector. Without the detector the image is not infinitely sharp, but limited by the wavelength. The perfect image may appear, but only if one looks.

The crucial point of perfect imaging is that, given a choice of detectors in the image area, the light localises at the correct ones. The fact that the correct light localisation naturally happens in Maxwell's fish eye is also understandable if we imagine the absorption in an array of detectors as the time reverse of the emission by a collection of point sources. Given perfect time symmetry, the light must settle down at the image points that correspond to the actual source points and avoid the ones corresponding to potential source points that did not emit. In this way, a sharp image is formed with a resolution given by the cross section of the detectors and not by the wave nature of light. Perfect imaging is possible with positive refraction [37, 38]. This idea has stirred up controversy [41] but experiments [40, 42] indicate that it works.

10.4.3 Casimir Force

Maxwell's fish eye may also give new hope for a fundamental problem in physics [43]. Casimir suggested an intriguing model that could explain the stability of charged particles and the value of the fine structure constant [44]. The argument goes as follows: Imagine the particle as an electrically charged hollow sphere. Two forces are acting upon it: the electrostatic repulsion and the force of the quantum vacuum, the Casimir force—presumed to be attractive [29]. The stress σ of the quantum vacuum on a spherical shell of radius a must be given by a dimensionless constant times $\hbar c/a^4$ on purely dimensional grounds—the quantum stress is an energy density proportional to $\hbar$, and $\hbar c/a^4$ carries indeed the units of an energy density. Now, the electrostatic energy of the sphere is proportional to the square e^2 of its charge and is also inversely proportional to a^4 [45]. Therefore, an attractive Casimir force balances the electrostatic repulsion regardless of how small a is, provided $e^2/(\hbar c)$ assumes a certain value given by the strength of the Casimir force. This

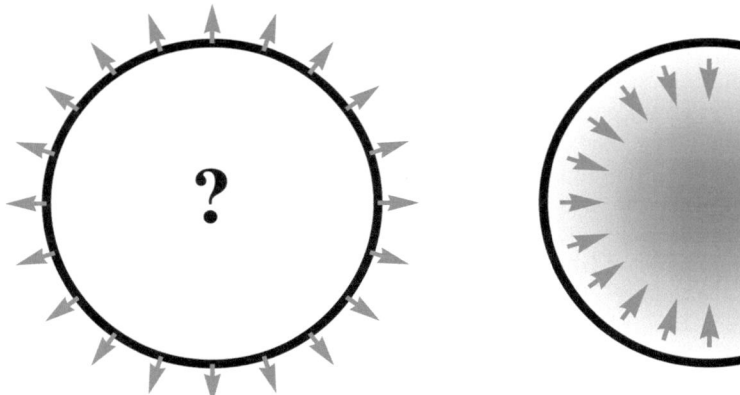

Fig. 10.8 The Casimir force on a spherical shell (*left*) is repulsive [46], or is it? We assumed the shell (*right*) to be filled with the medium of Maxwell's fish eye and found an attractive Casimir stress in the material. The *shades of grey* indicate the profile of the medium (10.33)

strength depends on the internal structure of the particle—the fact that it is a spherical shell—but not on its size, which could be imperceptibly small. Casimir's model, however crude, could simultaneously explain the fine-structure constant $e^2/(\hbar c)$ and the stability of charged elementary particles!

All one needs to do is calculate the Casimir force on a spherical shell, but such calculations are notoriously difficult. After a marathon struggle with special functions, Boyer succeeded in numerically computing the force for an infinitely conducting, infinitely thin shell and found a surprising result [46] that shattered Casimir's idea: the vacuum force is repulsive and so cannot possibly balance the electrostatic repulsion. Boyer's heroic calculation was confirmed in a sophisticated and elegant paper by Milton, DeRaad and Schwinger [47]. The spherical shell has become the archetype for a shape that causes Casimir repulsion. However, doubts have been lingering about whether the repulsive force of the shell may be an artefact of the simple model used, for the following reason: the bare stress of the quantum vacuum is always infinite and this infinity is removed by regularization procedures. The most plausible regularization involves considering the relative stress between or inside macroscopic bodies. But an infinitely thin sphere does not represent an extended macroscopic body, nor multiple bodies. Suppose the physically relevant vacuum stress of an extended spherical shell tends to infinity in the limit when the shell becomes infinitely thin and infinitely conducting. In this case the regularization would remove this physically significant infinity, producing a finite result that may very well have the wrong sign.

Consider a minor modification of Casimir's model (Fig. 10.8). Imagine that the spherical shell (though still infinitely conducting and infinitely thin) is no longer hollow, but filled with the medium (10.33) of Maxwell's fish eye. In this case one can derive an exact solution for the Casimir stress, because the fish eye corresponds to a very simple virtual space with a high degree of symmetry, the hypersphere. One finds for the stress tensor [43]

$$\sigma = -\frac{\hbar c \mathbb{1}}{\pi^2 a^4 n (1 - r^2/a^2)^4}.\tag{10.36}$$

The stress is isotropic, negative and falls monotonically, so the Casimir-force density $\nabla \cdot \sigma$ is always attractive in our model. Close to the mirror the stress and the force in the material tend to infinity. Since the material (10.33) of Maxwell's fish eye represents only a modest modification of empty space with refractive index n_1, it is likely that the Casimir stress close to a perfect spherical mirror around a ball of index n_1 is infinite as well, even when $n_1 = 1$, *i.e.* for a hollow perfect mirror. An imperfect mirror, on the other hand, may lead to a finite and possibly attractive vacuum force. In any case, our present calculation shows that the perfect spherical mirror produces an artefact, a diverging Casimir force density, in a relatively normal material, which casts new doubts on the claim of Casimir repulsion for the hollow sphere [46, 47] and gives new hope for Casimir's fascinating explanation [44] of the fine-structure constant and the stability of elementary charged particles.[5]

10.5 Spacetime Media

So far we considered only spatial geometries for light—we showed that they appear as impedance-matched media and that impedance-matched media appear as spatial geometries. We discussed applications of spatial transformations and curved virtual space such as invisibility cloaking and perfect imaging. Let us now include time and expend our theory to spacetime geometries.

10.5.1 Spacetime Geometries

We distinguish spacetime coordinates x^α by Greek indices running for 0 to 3 where the 0-th coordinate refers to time, whereas purely spatial coordinates are indicated by Latin indices. For example. $x^\alpha = \{ct, x, y, z\}$ are the Galileian coordinates of Minkowski spacetime. The lines in spacetime are world lines, they describe the trajectories of particles in space and time. The line element ds/c with

$$ds^2 = -c^2 dt^2 + dx^2 + dy^2 + dz^2 \tag{10.37}$$

in Minkowski spacetime, measures the proper time experienced by a particle on its way. The line element ds characterises the spacetime geometry. The square of ds is

---

[5]Of course, in a more realistic theory the charged particle should not be regarded as being a classical object interacting with the quantum vacuum, as in Casimir's case [44], but rather as a self-consistent quantum structure.

a quadratic form of the increments of the spacetime coordinates,

$$ds^2 = g_{\alpha\beta} dx^\alpha dx^\beta. \tag{10.38}$$

As in the case of spatial geometries, we denote the determinant of $g_{\alpha\beta}$ by g and the matrix inverse by $g^{\alpha\beta}$; note that g is usually negative. For example, in Minkowski spacetime, $g_{\alpha\beta} = \eta_{\alpha\beta}$ with

$$\eta_{\alpha\beta} = \text{diag}(-1, 1, 1, 1) = \eta^{\alpha\beta} \tag{10.39}$$

and $g^{\alpha\beta} = \eta^{\alpha\beta}$ and $g = -1$. We know that purely spatial geometries act as impedance-matched media on electromagnetic fields; how do genuine spacetime geometries appear?

10.5.2 Magneto-Electric Media

Plebanski [48] deduced a description of electromagnetism in spacetime geometries that closely resembles the familiar form of constitutive equations. Here we will simply state Plebanski's result; readers interested in its derivation are referred to the Appendices of Refs. [9, 11]. Plebanski's constitutive equations are

$$\boldsymbol{D} = \varepsilon_0 \boldsymbol{\varepsilon} \boldsymbol{E} + \frac{\boldsymbol{w}}{c} \times \boldsymbol{H}, \qquad \boldsymbol{B} = \mu_0 \boldsymbol{\mu} \boldsymbol{H} - \frac{\boldsymbol{w}}{c} \times \boldsymbol{E} \tag{10.40}$$

with the electromagnetic properties

$$\varepsilon^{ij} = \mu^{ij} = \mp \frac{\sqrt{-g}}{g_{00}} g^{ij}, \qquad w_i = \frac{g_{0i}}{g_{00}}. \tag{10.41}$$

The constitutive equations (10.40) and (10.41) show that spacetime geometries with $g_{0i} \neq 0$ mix electric and magnetic fields; they appear as magneto-electric media (also known as bi-anisotropic media [49]). The magneto-electric coupling vector $\boldsymbol{w}$ has the physical dimensions of a velocity. We will show in the next subsection that $\boldsymbol{w}$ is related to the local velocity of a moving medium. In a moving medium, the material responds to the electric and magnetic fields in locally co-moving frames where the medium appears to be at rest. There it is described by the constitutive equations (10.8) and (10.12). The Lorentz transformations to such locally co-moving frames mix electric and magnetic fields, which gives Plebanski's constitutive equations.

The dielectric tensors $\boldsymbol{\varepsilon}$ and $\boldsymbol{\mu}$ of Eq. (10.41) closely resemble the tensors (10.11) and (10.14) of purely spatial geometries, except that g and g_{00} are negative. Furthermore, only the ratio of $\sqrt{-g} g^{ij}$ and g_{00} matters to electromagnetic fields, and so does the ratio of g_{0i} and g_{00} in the magneto-electric coupling, which reflects an important property of light (and electromagnetic radiation in general) known as conformal invariance. It originates from the fact that the proper time of a light ray is zero—light does not experience time. Therefore light does not recognise the magnitude of the line element (10.38), only the relative contributions of the increments dx^α to ds^2. This implies that we can multiply the line element (10.38) by an arbitrary non-vanishing function of the spacetime coordinates without any effect on the propagation of light. Thus suppose we make the following change

$$g_{\alpha\beta} \to \Omega^2 g_{\alpha\beta}. \tag{10.42}$$

As $g^{\alpha\beta} \to \Omega^{-2} g^{\alpha\beta}$ and $g \to \Omega^8 g$, Plebanski's constitutive equations (10.41) are invariant; hence electromagnetism is conformally invariant.

10.5.3 Moving Media

Let us discuss an instructive example of a medium that corresponds to a spacetime geometry, the moving isotropic impedance-matched medium. We denote the local velocities of the medium by $\boldsymbol{u}$ where $\boldsymbol{u}$ may vary. For a given spacetime point x^α we can always erect a locally co-moving frame where $\boldsymbol{u}$ vanishes at that point (where the medium is locally at rest). As the medium is impedance-matched we can describe it in the locally co-moving frame by a spacetime geometry with

$$g_{\alpha\beta} = \mathrm{diag}\left(-1, n^2, n^2, n^2\right). \tag{10.43}$$

In view of the conformal invariance of electromagnetism we can replace this $g_{\alpha\beta}$ by

$$g_{\alpha\beta} = \mathrm{diag}\left(-n^{-2}, 1, 1, 1\right). \tag{10.44}$$

Using the definition (10.39) of the metric tensor of Minkowski spacetime and introducing

$$u_\alpha = (-1, 0, 0, 0) \tag{10.45}$$

we can write $g_{\alpha\beta}$ as

$$g_{\alpha\beta} = \eta_{\alpha\beta} + \left(1 - n^{-2}\right) u_\alpha u_\beta. \tag{10.46}$$

Now, the Lorentz transformation from the locally co-moving frame back to the laboratory frame maintains the spacetime geometry (10.37) of Minkowski space and hence leaves $\eta_{\alpha\beta}$ invariant. Additionally, we write u_α in terms of quantities with a simple geometrical meaning in spacetime:

$$u_\alpha = \eta_{\alpha\beta} u^\beta, \qquad u^\alpha = \frac{dx^\alpha}{ds} \tag{10.47}$$

where ds refers to the Minkowski line element (10.37). The u^α form the local four-velocity of the medium. In the co-moving frame $x^\alpha = (ct, 0, 0, 0)$ and $ds = cdt$ and so the four-velocity agrees with expression (10.45). As ds is a Lorentz invariant, the four-velocity behaves like x^α under a Lorentz transformations to the laboratory frame. Applying formula (10.38) and $\boldsymbol{u} = \mathrm{d}\boldsymbol{x}/\mathrm{d}t$ we obtain the explicit expressions

$$u^\alpha = \frac{(1, \boldsymbol{u}/c)}{\sqrt{1 - u^2/c^2}}, \qquad u_\alpha = \frac{(-1, \boldsymbol{u}/c)}{\sqrt{1 - u^2/c^2}}. \qquad (10.48)$$

In this way we can easily express the tensor (10.46) in the laboratory frame. This $g_{\alpha\beta}$ describes the spacetime geometry established for light by the moving medium. It was discovered by Gordon [1] in 1923 and independently rediscovered several times [50–52]. The matrix inverse $g^{\alpha\beta}$ of $g_{\alpha\beta}$ is

$$g^{\alpha\beta} = \eta^{\alpha\beta} + (1 - n^2) u^\alpha u^\beta, \qquad (10.49)$$

as one easily verifies by calculating the matrix product $g_{\alpha\gamma} g^{\gamma\beta}$ that gives δ^β_α, the unity matrix. With these expressions we can calculate the dielectric properties of moving media. They are particularly instructive in the limit of low velocities. In this case we obtain from Plebanski's constitutive equations (10.41):

$$\boldsymbol{\varepsilon} = \boldsymbol{\mu} \approx n\mathbb{1}, \qquad \boldsymbol{w} \approx (n^2 - 1) \boldsymbol{u}. \qquad (10.50)$$

We see that the magneto-electric coupling vector $\boldsymbol{w}$ is proportional to the velocity. The proportionally factor is the susceptibility $n^2 - 1$ that vanishes in empty space when $n = 1$. In the case of large velocities, $\boldsymbol{w}$ and also $\boldsymbol{\varepsilon} = \boldsymbol{\mu}$ depend in a more complicated way on the velocity of the moving medium [11].

10.5.4 Spacetime Transformations

Suppose the medium moves in one direction only, say the z direction (but possibly with varying velocity) and that n may also only vary in z. We will show that the light propagation in z direction is equivalent to a transformation in space and time; the one-dimensionally moving medium appears as a spacetime transformation medium [9]. Our starting point is Gordon's spacetime geometry (10.49). If $\boldsymbol{u}$ has only a z-component u and all other components vanish we have

$$g^{\alpha\beta} = \begin{pmatrix} \frac{u^2 - c^2 n^2}{c^2 - u^2} & 0 & 0 & \frac{(1-n^2)cu}{c^2 - u^2} \\ 0 & 1 & 0 & 0 \\ 0 & 0 & 1 & 0 \\ \frac{(1-n^2)cu}{c^2 - u^2} & 0 & 0 & \frac{c^2 - n^2 u^2}{c^2 - u^2} \end{pmatrix}. \qquad (10.51)$$

We introduce the new coordinates t' and z' defined by

$$t' \mp \frac{z'}{c} = t - \int \frac{\mathrm{d}z}{v_\pm}. \qquad (10.52)$$

Here $v_\pm$ denotes the relativistic addition of the velocity of light in the medium in either positive and negative direction, $\pm c/n$, and the velocity of the medium, u:

$$v_\pm = \frac{u \pm c/n}{1 \pm u/(cn)}. \tag{10.53}$$

Gordon's $g^{\alpha\beta}$ tensor appears in the new coordinates as

$$g^{\alpha'\beta'} = \Lambda^{\alpha'}_\alpha g^{\alpha\beta} \Lambda^{\beta'}_\beta \tag{10.54}$$

with the transformation matrix

$$\Lambda^{\alpha'}{}_\alpha = \begin{pmatrix} 1 & 0 & 0 & \frac{(n^2-1)cu}{c^2-n^2u^2} \\ 0 & 1 & 0 & 0 \\ 0 & 0 & 1 & 0 \\ 0 & 0 & 0 & \frac{n(c^2-u^2)}{c^2-n^2u^2} \end{pmatrix}. \tag{10.55}$$

The result is the diagonal matrix

$$g^{\alpha'\beta'} = \text{diag}\left(-\frac{n^2(c^2-u^2)}{c^2-n^2u^2}, 1, 1, \frac{n^2(c^2-u^2)}{c^2-n^2u^2}\right) \tag{10.56}$$

with the inverse

$$g_{\alpha'\beta'} = \text{diag}\left(-\frac{c^2-n^2u^2}{n^2(c^2-u^2)}, 1, 1, \frac{c^2-n^2u^2}{n^2(c^2-u^2)}\right). \tag{10.57}$$

The determinant of the metric is

$$g' = -\frac{(c^2-n^2u^2)^2}{n^4(c^2-u^2)^2}. \tag{10.58}$$

The metric $g_{\alpha'\beta'}$ describes the geometry in virtual spacetime. To find out how this geometry appears as a medium we use Plebanski's constitutive equations (10.41) in virtual spacetime, with primed instead of unprimed tensors. Since $g_{\alpha'\beta'}$ is diagonal, the magneto-electric coupling vector $\boldsymbol{w}'$ vanishes: in virtual spacetime the medium is at rest. For the dielectric tensors we obtain

$$\boldsymbol{\varepsilon}' = \boldsymbol{\mu}' = \text{diag}(1, 1, \varepsilon'_{zz}) \tag{10.59}$$

with some ε'_{zz} we do not need to specify here. Since electromagnetic waves propagating in the z-direction are polarised in the x, y plane their electromagnetic fields only experience the x and y components of the dielectric tensors. Consequently, for one-dimensional wave propagation virtual spacetime is empty, waves are free here. In virtual spacetime, left and right-moving wave packets are functions of either $t' + z'/c$ or $t' - z'/c$; in physical spacetime they are modulated wavepackets according to the transformation (10.52). Instead of the velocity of light in vacuum they experience the relativistic velocity addition (10.53) of the speed of light in the medium and the velocity of the medium.

Suppose that at some place, say $z = 0$, the velocity of the moving medium reaches the speed of light in the medium,

$$|u(0)| = \frac{c}{n(0)}. \qquad (10.60)$$

Without loss of generality, we assume that $u < 0$ around $z = 0$ (the medium moves from the right to the left) and consider electromagnetic waves propagating against the flow as wave packets with velocity v_+ that, according to the addition (10.53) of velocities, vanishes at $z = 0$. Linearising v_+ at $z = 0$,

$$v_+ = \alpha z, \qquad (10.61)$$

we see that the integral in the virtual coordinates (10.52) develops a logarithmic singularity. Virtual spacetime appears as two branches, one corresponding to $z < 0$ and the other to $z > 0$, both being two independent Minkowski spaces. From this follows that right-propagating light confined to either the region where $z < 0$ or $z > 0$ will remain there; right-propagating light cannot cross $z = 0$, the horizon, where the velocity of the medium reaches the speed of light.

How to create a horizon for light in practice? It seems rather hopeless for a realistic medium to reach the speed of light, because although normal media reduce the velocity of light, they do not reduce c by much; light is still rather fast. However [54], one can mimic a moving medium with an intense pulse of light in a suitable material with nonlinear optical response [53]. In such a case, the refractive index of the material n may get an additional contribution that is proportional to the intensity of the pulse (which is called Kerr effect). The pulse enhances the refractive index as if an additional piece of material were added to the medium. As this effective piece of material is made by light, it naturally moves at the speed of light in the material. Consider two light fields now: the pulse and the probe light that experiences the pulse as a contribution to the refractive index. One can control the speed of both the pulse and the probe using the frequency and polarisation dependence of the refractive index [54] and by the lateral profiles [55] of the light fields; different frequencies, polarisations and lateral profiles propagate at different speeds. The place where the velocity of the probe is sufficiently reduced by the pulse to reach the speed of the pulse makes a horizon [54, 55].

Acknowledgements I am most grateful for the discussions I had with many distinguished scientists about geometry, light and a wee bit of magic. In particular, I thank my group at St Andrews and wish them well, Simon Horsley, Susanne Kehr, Thomas Philbin, Sahar Sahebdivan, and William Simpson. My work has been supported by the University of St Andrews, the Engineering and Physical Sciences Research Council and the Royal Society.

References

1. Gordon, W.: Ann. Phys. (Leipz.) **72**, 421 (1923) (in German)

2. Tamm, I.Y.: J. Russ. Phys.-Chem. Soc. **56**, 2–3 (1924) (in Russian)
3. Tamm, I.Y.: J. Russ. Phys.-Chem. Soc. **56**, 3–4 (1925) (in Russian)
4. Penrose, R.: Private communication
5. Mahoney, M.S.: The Mathematical Career of Pierre de Fermat, pp. 1601–1665. Princeton University Press, Princeton (1994)
6. Greenleaf, A., Lassas, M., Uhlmann, G.: Math. Res. Lett. **10**, 685 (2003)
7. Leonhardt, U.: Science **312**, 1777–1780 (2006)
8. Pendry, J.B., Schurig, D., Smith, D.R.: Science **312**, 1780 (2006)
9. Leonhardt, U., Philbin, T.G.: New J. Phys. **8**, 247 (2006)
10. Chen, H., Chan, C.T., Sheng, P.: Nat. Mater. **9**, 387 (2010)
11. Leonhardt, U., Philbin, T.G.: Geometry and Light: The Science of Invisibility. Dover, Mineola (2010)
12. Service, R., Cho, A.: Science **330**, 1622 (2010)
13. Schutz, B.F.: A First Course in General Relativity. Cambridge University Press, Cambridge (2009)
14. Schurig, D., Mock, J.J., Justice, B.J., Cummer, S.A., Pendry, J.B., Starr, A.F., Smith, D.R.: Science **314**, 977 (2006)
15. Ma, Y.G., Ong, C.K., Tyc, T., Leonhardt, U.: Nat. Mater. **8**, 639 (2009)
16. Zhang, B., Luo, Y., Liu, X., Barbastathis, G.: Phys. Rev. Lett. **106**, 033901 (2011)
17. Chen, X., Luo, Y., Zhang, J., Jiang, K., Pendry, J.B., Zhang, S.: Nat. Commun. **2**, 176 (2011)
18. Ma, Y.G., Sahebdivan, S., Ong, C.K., Tyc, T., Leonhardt, U.: New J. Phys. **13**, 033016 (2011)
19. Chen, H., Chan, C.T.: J. Appl. Phys. **104**, 033113 (2008)
20. Li, J., Pendry, J.B.: Phys. Rev. Lett. **101**, 203901 (2008)
21. Leonhardt, U., Tyc, T.: Science **323**, 110 (2009)
22. Perczel, J., Tyc, T., Leonhardt, U.: New J. Phys. **13**, 083007 (2011)
23. Zhang, S., Xia, C., Fang, N.: Phys. Rev. Lett. **106**, 024301 (2011)
24. Pendry, J.B.: Phys. Rev. Lett. **85**, 3966 (2000)
25. Minkel, J.R.: Phys. Rev. Focus **9**, 23 (2002)
26. Valentine, J., Zhang, S., Zentgraf, T., Ulin-Avila, E., Genov, D.A., Bartal, G., Zhang, X.: Nature **455**, 376 (2008)
27. Fang, N., Lee, H., Sun, C., Zhang, X.: Science **308**, 534 (2005)
28. Stockman, M.I.: Phys. Rev. Lett. **98**, 177404 (2007)
29. Casimir, H.B.G.: Proc. K. Ned. Akad. Wet. **51**, 793 (1948)
30. Leonhardt, U., Philbin, T.G.: New J. Phys. **9**, 254 (2007)
31. Munday, J.N., Capasso, F., Parsegian, V.A.: Nature **457**, 170 (2009)
32. Dzyaloshinskii, I.E., Lifshitz, E.M., Pitaevskii, L.P.: Adv. Phys. **10**. 165 (1961)
33. Einstein, A.: Sitzungsber. K. Preuss. Akad. Wiss. (Berlin) **142** (1917)
34. Maxwell, J.C.: Camb. Dublin Math. J. **8**, 188 (1854)
35. Luneburg, R.K.: Mathematical Theory of Optics. University of California Press, Berkeley (1964)
36. Born, M., Wolf, E.: Principles of Optics. Cambridge University Press, Cambridge (1999)
37. Leonhardt, U.: New J. Phys. **11**, 093040 (2009)
38. Leonhardt, U., Philbin, T.G.: Phys. Rev. A **81**, 011804 (2010); **84**, 049902 (2011)
39. Gabrielli, L.H., Lipson, M.: J. Opt. **13**, 024010 (2011)
40. Ma, Y.G., Sahebdivan, S., Ong, C.K., Tyc, T., Leonhardt, U.: New J. Phys. **13**, 033016 (2011)
41. Tyc, T., Zhang, X.: Nature **480**, 42 (2011)
42. Ma, Y.G., Sahebdivan, S., Ong, C.K., Tyc, T., Leonhardt, U.: New J. Phys. (in press)
43. Leonhardt, U., Simpson, W.M.R.: Phys. Rev. D **84**, 081701 (2011)
44. Casimir, H.B.G.: Physica **19**, 846 (1956)
45. Jackson, J.D.: Classical Electrodynamics. Wiley, New York (1998)
46. Boyer, T.H.: Phys. Rev. **174**, 1764 (1968)
47. Milton, K.A., DeRaad, L.L., Schwinger, J.: Ann. Phys. (N.Y.) **115**, 388 (1978)
48. Plebanski, J.: Phys. Rev. **118**, 1396 (1960)

49. Serdyukov, A., Semchenko, I., Tretyakov, S., Sihvola, A.: Electromagnetics of Bi-anisotropic Materials. Gordon and Breach, Amsterdam (2001)
50. Quan, P.M.: C. R. Acad. Sci. (Paris) **242**, 465 (1957)
51. Quan, P.M.: Arch. Ration. Mech. Anal. **1**, 54 (1957/58)
52. Leonhardt, U., Piwnicki, P.: Phys. Rev. A **60**, 4301 (1999)
53. Boyd, R.W.: Nonlinear Optics. Academic Press, San Diego (1992)
54. Philbin, T.G., Kuklewicz, C., Robertson, S., Hill, S., Konig, F., Leonhardt, U.: Science **319**, 1367 (2008)
55. Belgiorno, F., Cacciatori, S.L., Clerici, M., Gorini, V., Ortenzi, G., Rizzi, L., Rubino, E., Sala, V.G., Faccio, D.: Phys. Rev. Lett. **105**, 203901 (2010)

Chapter 11
Laser Pulse Analogues for Gravity

Eleonora Rubino, Francesco Belgiorno, Sergio Luigi Cacciatori, and Daniele Faccio

Abstract Intense pulses of light may be used to create an effective flowing medium which mimics certain properties of black hole physics. It is possible to create the analogues of black and white hole horizons and a photon emission is predicted that is analogous to Hawking radiation. We give an overview of the current state of the art in the field of analogue gravity with laser pulses and of its implications and applications for optics.

11.1 Introduction

"Analogue gravity" is the study of phenomena traditionally associated to gravitation and general relativity, by means of analogue models that can be realised in very different physical systems and that do not directly rely on gravity at all [1]. Following recent developments in the field, in this Chapter we attempt to give a brief overview of how laser-pulses may be used to create the analogue of an astrophysical event horizon. Analogues, as a tool to study the physics of one system in an often apparently and completely different system, are certainly not a novelty. Just to name one recent example, certain features of electron behaviour in solid state physics have been reproduced using light propagation in specifically engineered media, e.g. photonic crystals or optical waveguide arrays [2]. Using these and related systems it has been possible to access, using classical physics, a variety of quantum phenomena ranging from Bloch oscillations [3, 4] to Anderson localisation [5]. Some of these

---

E. Rubino · S.L. Cacciatori
Dipartimento di Scienza e Alta Tecnologia, Università dell'Insubria, Via Valleggio 11, 22100 Como, Italy

F. Belgiorno
Dipartimento di Matematica, Politecnico di Milano, Piazza Leonardo 32, 20133 Milano, Italy

D. Faccio (✉)
School of Engineering and Physical Sciences, SUPA, Heriot-Watt University, Edinburgh EH14 4AS, UK
e-mail: d.faccio@hw.ac.uk

results have also led to significant technological developments in laser physics and are a clear example of why we should study, or attempt to study analogue models.

In order to better focus on the analogue gravity framework, where a moving background is able to induce a cinematic horizon in the sense specified above, it can be useful to recall that even the gravitational case can be interpreted in terms of a flow, according to the so-called "river model", which allows a more straightforward link with the physical situation occurring in analogue models. We sketch in the following the main steps.

The basic metric that describes a black hole, first derived by Schwarzschild, was later cast in different form by Painlevé and Gullstrand [6, 7]. The Painlevé-Gullstrand metric for a 2D black hole is

$$ds^2 = c^2 dt^2 - (dr - V dt)^2, \tag{11.1}$$

where

$$V = -\sqrt{\frac{2GM}{r}}. \tag{11.2}$$

Note that (11.1) corresponds to the t–r part of a physical (4D) black hole. Based on these equations we may interpret space as if it were a fluid that is flowing with velocity V. This is the basis of the "river model" [8] that allows an intuitively appealing yet mathematically correct understanding of how analogue models for gravity work.

Figures 11.1(a) and (b) schematically show how space flows and falls into an astrophysical black hole or, under time reversal, emerges out of what is called a white hole. The flowing river of space moves in a Galilean fashion through a flat Galilean background space [8] and we may define a special point, the Schwarzschild radius, for which $r_S = 2GM/c^2$ and the flow velocity equals c: beyond this point the flow exceeds the speed of light. Therefore, because objects moving through the river must obey the laws of special relativity and their speed cannot exceed c, it is not possible to escape out of the black hole once inside r_S, nor is it possible to penetrate inside the white hole beyond r_S. On the basis of this reasoning, one may therefore attempt to construct a laboratory analogue using light in a flowing medium that recreates a flowing river of space. The original proposal by Unruh based on acoustic waves in a flowing fluid has a metric that can be reduced to a form similar to Eq. (11.1) in which c represents the speed of sound [9].

11.1.1 White Holes

In analogue models and, in particular, in the experimental setting which aims to verify theoretical predictions emerging from analogue gravity models, an important role is reserved to white holes. We therefore recall some of their properties in what follows.

Black holes are relatively well-known objects whilst white holes, also solutions to the Einstein equations, are less studied due mostly to the fact that there is no obvious mechanism by which a gravitational white hole may form. A black hole will

11 Laser Pulse Analogues for Gravity

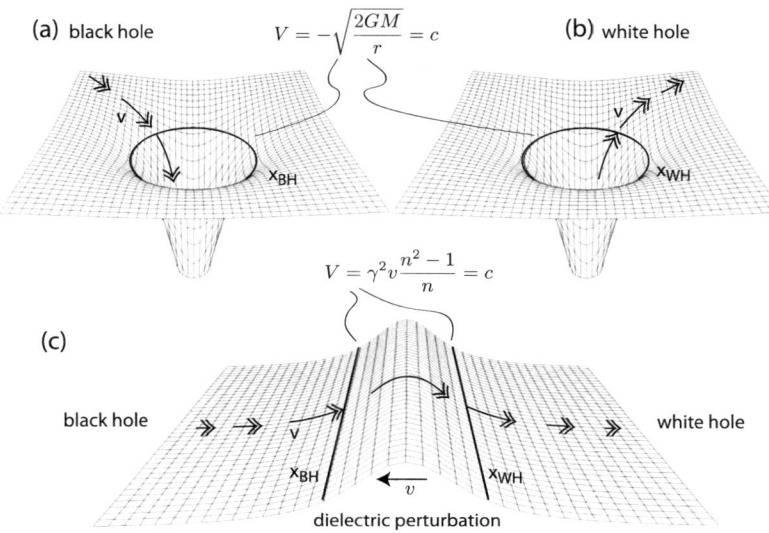

Fig. 11.1 The flowing river of space in an astrophysical black (**a**) and white (**b**) hole. The black and white hole horizons are the points at which space flow becomes equal to c. (**c**) Depicts the effective geometry induced by a 1-dimensional dielectric perturbation moving from right to left. In the frame comoving with the perturbation, space flows as shown by the *arrows*. *Longer arrows* indicate larger velocities

trap any incoming matter, or light and forms as a natural consequence of the gravitational collapse of super-massive stars. A white hole may eject particles or light until it burns out and, most importantly it does not appear as the result of a gravitational collapse. Hawking specifically addressed this problem [10] and pointed out that the very nature of Hawking radiation implies that the black hole is at thermal equilibrium. Then, by the ergodic assumption, the system is equally likely to pass through all possible states if observed for a long enough time, including the time-reversal of any of those states. In other words, to an external observer a white hole and a black hole are completely indistinguishable and the emission of a white hole is the same as that of a black hole with the same mass [10, 11]. So we are justified in studying one or the other kind of hole (black or white), based on what is most convenient in a given context. Indeed, if we are unlikely to observe white holes in the cosmos, these objects actually turn out to be more the rule than the exception in the context of analogue gravity.

11.2 Analogue Gravity with Optics in Moving Media

The study of light in moving media is certainly not a novelty and has a relatively long history. As far back as 1818, well before Einstein introduced the theory of special relativity, Fresnel discovered theoretically that the speed of light v_ϕ in a uniform and flowing medium depends on the flow velocity v:

$$v_\phi = \frac{c}{n} + \left(1 - \frac{1}{n^2}\right)v. \tag{11.3}$$

Not many years passed and in 1851 Fizeau experimentally verified this prediction. Only after Einstein developed his theory of general relativity, was it possible to make the first connection between electromagnetic waves propagating in dielectrics and in gravitational fields. The first work in this sense was developed by Gordon who derived the so-called Gordon metric that describes how light propagates in a moving dielectric medium and to which we will refer below [12]. A complete review is found in for example in [1].

Indeed, propagation of light in moving dielectric media may be described in terms of geodesics in an effective metric that depends on the refractive index, without any further reference to the original picture involving Maxwell equations. All relevant dielectric properties are reabsorbed in the effective metric which is actually seen by photons. This can be done at the level of both physical and geometrical optics. We also note that, in order to find out an horizon in the sense specified in the previous section (Sect. 11.1), it is required to have a spatial inhomogeneity in the refractive index, as in the acoustic models a spatial inhomogeneity in the velocity field is required. Beyond Hawking-like phenomena in stationary dielectrics, a more recent review of optics in nonstationary media has been given by Shvartsburg [13] with particular emphasis on the interaction of light with a rapidly varying ionisation or plasma. Within a similar context, Rosanov has studied the "parametric" Doppler effect in which light interacts with a medium that is at rest but has time-varying or moving parameters [14, 15]. A tightly correlated phenomenon is the so-called time refraction, first introduced by Mendonça by drawing a parallelism between the behaviour of light at a spatial boundary and a time-varying boundary [16, 17]: as is well-known and summarised by Snell's relations, when light traverses a spatial (and time-stationary) boundary separating two media with different refractive indices, its wave-vector is modified. In a similar fashion, if the refractive index changes in time, i.e. there is a "temporal boundary", then the frequency of light is modified. These ideas were then extended to account for quantum effects and, of particular interest in this context, including also the quantum vacuum and excitation of entangled photon pairs [18–20].

Over the last 10 years a number of papers have returned to the problem of light in moving media within the specific context of analogue gravity and with the explicit goal of evaluating the analogy between these systems and gravitational black holes [21–32]. However, it was only recently that Leonhardt proposed an idea by which an experimentally accessible layout may actually create the conditions required to observe the analogue of an horizon for light propagating in a dielectric medium [33].

There is a potential drawback affecting a priori any optical analogue gravity model whose aim is to induce a kinematical horizon for light. It is extremely difficult to imagine a method by which we may actually induce a flow of matter at speeds that are close to the speed of light, as required in order to recreate an analogue optical horizon. One early proposal by Leonhardt attempted to bypass this obstacle by using so-called slow light [22, 34]: with electromagnetically induced transparency or metamaterials it is possible to slow light down to small fractions of c. However, the slowing down of light in these systems primarily affects only the *group* velocity whilst it was later realised that the important quantity for observing particle creation at a horizon is the *phase* velocity [34]. Leonhardt recently solved this issue by suggesting that the medium itself need not flow at all. All that is necessary is a localised perturbation of the refractive index that travels at speeds close to c. So the medium itself remains at rest in the laboratory frame and by using nonlinear optics (as described in Sect. 11.6) we may create an ultrafast perturbation. Just to fix ideas, we assume the perturbation to be Gaussian-shaped with positive δn, so that the surrounding refractive index has some background value $n = n_0$ and then this gradually increases up to a maximum value $n = n_0 + \delta n$. This perturbation of the dielectric medium is then made to move at a velocity v that is very close to the speed of light in the medium c/n_0.

We may modify Fresnel's relation, Eq. (11.3) to account for the fact that the perturbation is now localised: $v_\phi = c/n_0 + [1 - 1/(n_0 + \delta n)^2]v$: we can see from this that an increase δn in the local refractive index is indeed equally perceived by light as a local increase in the flow velocity, i.e. both lead to a slowing down of the light pulse. If $v \gtrsim c/n_0$, the perturbation will catch up with the light pulse or, in the comoving frame, the light pulse will be gradually sucked inwards. As the pulse is sucked in, the refractive index (or space flow velocity) increases and the speed at which the pulse falls in towards the perturbation increases. The light pulse will eventually pass the point x_{BH} of no return at which $v = c/(n_0 + \delta n_{BH})$. This point is the analogue of a black hole horizon. In a similar fashion, one may consider the trailing edge of a perturbation with $v \lesssim c/n_0$: light approaching from behind will catch up with the perturbation. As it penetrates within the higher refractive index region it will be slowed down by the higher refractive index or, equivalently, by the faster space flow. The pulse will then be blocked at the point x_{WH} for which $v = c/(n_0 + \delta n_{WH})$. This is to all effects a time-reversed version of the black hole horizon, i.e. it is the analogue of a white hole horizon.

11.3 Dielectric Metrics and Hawking Radiation

The relevant metric in the dielectric analogue context is the Gordon metric [12, 21, 30, 35, 36],

$$ds^2 = \frac{c^2}{[n(x-vt)]^2}dt^2 - dx^2, \qquad (11.4)$$

where the travelling dielectric perturbation is described by $n(x - vt) = n_0 + \delta n(x - vt)$. We choose for simplicity of description a smooth dielectric perturbation δn with a Gaussian profile, but, for our purposes, a more general framework can be introduced [35]. We may rewrite this in the perturbation comoving frame by means of a boost: $t' = \gamma(t - \frac{v}{c^2}x)$, $x' = \gamma(x - vt)$, so that

$$ds^2 = \gamma^2 \frac{c^2}{n^2}\left[1 - \left(\frac{nv}{c}\right)^2\right]dt'^2 + 2\gamma^2 \frac{v}{n^2}(1-n^2)dt'dx' - \gamma^2\left[1 - \left(\frac{v}{nc}\right)^2\right]dx'^2. \tag{11.5}$$

The primed coordinates, here and in the rest of this chapter, indicate comoving coordinates.

It is easy to show that the above metric (11.5) is static, i.e. it has a time translation symmetry; moreover it is possible to find global coordinates such that no "cross terms" between spatial and time coordinates appear in the metric (see e.g. [37]), and which may be associated with a further coordinate system which carries (11.5) in a form which is explicitly static [35]. If we define

$$dt' = d\tau - \alpha(x')dx', \tag{11.6}$$

with

$$\alpha(x') = \frac{g_{01}(x')}{g_{00}(x')}, \tag{11.7}$$

then the metric becomes

$$ds^2 = \frac{c^2}{n^2(x')} g_{\tau\tau}(x')d\tau^2 - \frac{1}{g_{\tau\tau}(x')}dx'^2, \tag{11.8}$$

where

$$g_{\tau\tau}(x') := \gamma^2\left(1 + n(x')\frac{v}{c}\right)\left(1 - n(x')\frac{v}{c}\right). \tag{11.9}$$

There is a remarkable resemblance of the τ, x'-part of the metric with a standard static spherically symmetric metric in general relativity in the so-called Schwarzschild gauge, aside from the important difference represented by the factor $\frac{c^2}{n^2}$ replacing c^2. It is indeed possible to follow an approach similar to that followed by Painlevé and Gullstrand and recast the metric in a form that is identical to Eq. (11.1), with a medium flow velocity that naturally does not depend on a mass or gravitational constant but rather on the refractive index profile and speed, $V = \gamma v(n^2 - 1)/n$. Similarly to the gravitational case, a horizon is formed when $V = c$ and depending on the direction of the flow of space, the analogue of a black or white hole is formed (see Fig. 11.1).

Equivalently, the horizons are determined by the condition $g_{\tau\tau} = 0$, and exists when

$$\frac{c}{n_0 + \delta n} < v < \frac{c}{n_0}. \tag{11.10}$$

The external region corresponds to $x < x_{WH}$ and to $x > x_{BH}$. The leading horizon $x = x_{BH}$ is a black hole horizon, whereas the trailing one $x = x_{WH}$ is a white hole horizon. These points are indicated also in Fig. 11.1(c) which illustrates the geometry of a one-dimensional δn perturbation. The arrows indicate the equivalent flow of space: interestingly, the flow of space is such that a single perturbation may recreate the analogue of both a black hole, on the leading edge, and a white hole on the trailing edge.

Equation (11.10) may be read in two different ways: for a given background index and perturbation amplitude, only perturbations with a certain velocity will give rise to an analogue horizon. Alternatively, for a given perturbation velocity and amplitude, only those frequencies that propagate with a refractive index that satisfies Eq. (11.10) will experience the effect of the horizon. Indeed, in general n_0 varies with frequency ω due to material dispersion. We note that although relation (11.10) was not originally derived from a dispersive theory, recent models that account also for dispersion arrive at exactly the same equation (see e.g. [20]) where $n_0 = n_0(\omega)$ is the medium phase index. This equation therefore represents the fundamental relation against which one may compare measurements in order to verify if any observed radiation may be related to the presence of an horizon. For example, one may vary the velocity and/or the perturbation amplitude and search for consistency with Eq. (11.10).

From the comoving-frame metric (11.5) we may deduce the equivalent of a surface gravity at the horizon which is found to be [35]

$$\kappa = \gamma^2 v^2 \frac{dn}{dx}\bigg|_H \quad (11.11)$$

where the refractive index perturbation gradient is evaluated at the horizon H. This surface gravity may be associated to a temperature for the radiation emitted from the analogue horizon. Indeed, the particularity of Hawking radiation is that it is predicted to exhibit a blackbody spectrum with a temperature given by,

$$T_H = \frac{\hbar \kappa}{2\pi c k_B} \quad (11.12)$$

where $\hbar$ is the reduced Planck constant, c is the speed of light in vacuum and k_B is Boltzmann's constant.

Laser pulse induced perturbations may be extremely steep, with a rise from n_0 to $n_0 + \delta n$ over a distance of the order or 1 μm or even less. This leads to temperatures, measured in the comoving frame, that are easily of the order of 1–10 K, i.e. many orders of magnitude higher than in any other system proposed to date.

These formulas only show that *if* Hawking radiation is emitted by the analogue horizon, then it is expected to have a certain temperature. A full quantum electrodynamical model of the perturbation accounting for the interaction with quantum vacuum, such as that developed in Ref. [35] is required in order to show that Hawking radiation is actually emitted from the horizon. The model starts by considering the electromagnetic vacuum states in the absence of any perturbation and then compares these with those in the presence of the travelling perturbation. The so-called Bogoliubov coefficients, that express the new states in function of the old states, can be used to directly evaluate the number of photons produced in such a scenario. The result clearly shows a logarithmic divergence of the phase of the electromagnetic field under conditions identical to Eq. (11.10) [35].

We can sketch the main idea as follows. Let us consider a situation where there is no dielectric perturbation. This condition represents our initial vacuum state (or "IN" vacuum state), which is characterized by a suitable set of quantum operators corresponding to creation and annihilation of particles. As we switch the laser on, very rapid nonlinear effects give rise to the dielectric perturbation within the dielectric sample, and a new vacuum state arises, which we call "OUT" vacuum state. We can associate with it a new set of creation and annihilation operators, which are apt to the OUT vacuum state. Particle creation occurs since the IN vacuum state is not a vacuum state for the OUT number operator (constructed with the OUT creation and annihilation operators), which represents an observable at large times. In other terms, if $N_k^{out} = a^{out+}_k a^{out}_k$ is the number operator for the OUT particles in the quantum state labeled by k, we obtain

$$\langle 0\, in|N_k^{out}|0\, in\rangle > 0, \tag{11.13}$$

which indicates that particle creation occurs when passing from the IN to the OUT state.

From a physical point of view, compared to the common S-matrix approach of quantum field theory, the scheme does not rely on perturbation theory and is associated with linear maps, called Bogoliubov transformations, which relate the IN creation and annihilation operators to the OUT ones. The latter approach is typical for the case of quantum field theory in external fields and in curved spacetime [38–40].

As pointed out by Hawking in his seminal paper, in a collapse situation what is really relevant in determining the late time behavior of Hawking radiation is the set of field modes which, as traced back from future null infinity, pile up near the future horizon, with a logarithmic divergence in the phase of the field which is at the root of the thermality of black hole radiation. In particular, $\langle 0\, in|N_k^{out}|0\, in\rangle$ displays a Planckian behaviour which is corrected by a grey-body factor associated with the potential barrier scattering for field modes. For the case of our analogue black hole, created by the laser pulse traveling through the medium, the field modes presents an analogous logarithmic phase divergence, and thermality of the spectrum can be still deduced, even without referring to the geometrical analogy (see [35]). Indeed, this emission is found to follow the expected blackbody dependence with temperature

$$T = \frac{1}{\gamma}\frac{1}{1-(v/c)n_0\cos\theta}T_H \tag{11.14}$$

where T_H is evaluated from Eqs. (11.12) and (11.11), and θ is the angle of the direction of observation with respect to the propagation axis of the perturbation. The multiplicative factor in Eq. (11.14) is simply the Doppler shift that transforms the temperature from the comoving frame to the laboratory frame. When viewed from the forward direction, $\theta = 0$ deg, the temperature measured in the laboratory frame is therefore predicted to be of the order 1000 K or more [33, 35, 41].

All of previous analysis relies on a model which does not include the effects of optical dispersion. At the same time, the introduction of dispersion effects is quite nontrivial, because dispersion relations are modified in a very significant way and

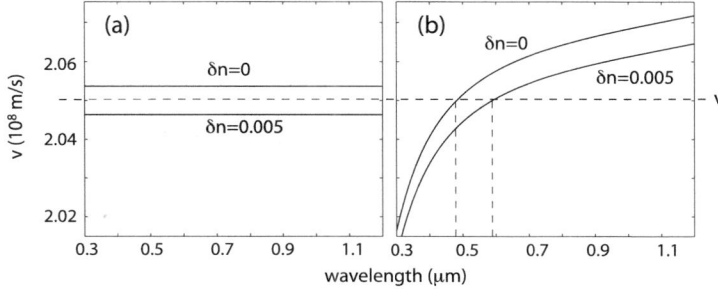

Fig. 11.2 (**a**) Light phase velocity with no dispersion far away from the perturbation ($\delta n = 0$) and at the peak of the perturbation with $\delta n = 0.005$. The *dashed line* indicates a value of the perturbation v for which Hawking radiation will cover the full predicted blackbody spectrum. (**b**) Light phase velocity with dispersion: fused silica glass dispersion has been used in this example. For the same conditions as in (**a**), only a very restricted bandwidth of wavelengths will be emitted in the form of Hawking radiation

the geometrical transcription, which works very well in the non-dispersive case, becomes much more problematic in presence of dispersion. Indeed, the so-called rainbow metrics have to be introduced, which are not properly metrics in the sense of common (pseudo)Riemannian geometry [1]. On the one hand, nontrivial dispersion relations tend to jeopardize a geometrical approach; on the other, they represent a substantial tool for a correct interpretation of the physics beyond analogue Hawking radiation. In the following subsection, the problem is discussed in detail for the case of a dielectric perturbation moving in a dielectric medium.

11.3.1 The Role of Dispersion

In order to create an effective flowing medium we must perform experiments in a dielectric material of some kind within which we generate the travelling perturbation. In general in any medium in which we decide to perform these experiments, the refractive index will vary as function of frequency or, equivalently wavelength. This leads to a qualitative deviation from the ideal (dispersion-less) case analysed by Hawking.

In Fig. 11.2(a) we show the phase velocity of light with no dispersion (it is thus constant at all wavelengths) far away from the perturbation ($\delta n = 0$) and at the peak of the perturbation with $\delta n = 0.005$. If the perturbation velocity is tuned anywhere in between these two velocity values, for example to the value indicated by the dashed line, then the whole spectrum (shaded area) experiences an horizon and Hawking radiation will cover the full predicted blackbody spectrum. In Fig. 11.2(b) we show the same situation but now including dispersion: fused silica glass dispersion has been used in this example. For the same conditions as in (a), only a restricted bandwidth of wavelengths will be emitted in the form of Hawking radiation (shaded area) and

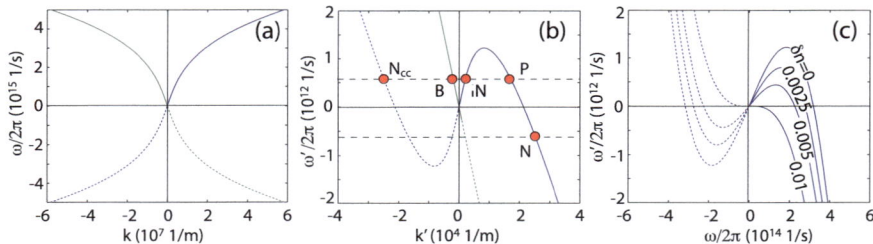

Fig. 11.3 Dispersion curves for the numerical simulations in Figs. 11.4(a) and 11.5(b). (**a**) Dispersion curves in the lab. reference frame. (**b**) Dispersion curves in the comoving frame. The *full-circles* indicate the position of the IN, P and N mode frequencies and the *horizontal dashed lines* correspond to $\pm\omega'_{IN}$. In (**c**) we plot the comoving frequency, $\omega'/2\pi$, as a function of the laboratory frame frequency, $\omega/2\pi$ for various values of δn

the full blackbody spectrum will not be visible. On the one hand dispersion appears therefore to "ruin" the spectrum. But on the other it provides us with a remarkably effective method to test the presence of an horizon: the emitted spectrum depends critically on the velocity of the perturbation. If this can be controlled, then we have a very simple method by which we may compare precise spectral measurements against the straightforward prediction of Eq. (11.10).

Dispersion has a further important consequence on Hawking radiation. In the absence of dispersion, radiation would accumulate at the white hole horizon for an infinite time. This would also lead to an infinite phase divergence and a consequently infinite blue-shift of the incoming frequency. Likewise, any emission observed far from a black hole horizon must originate from an infinitely blue-shifted vacuum fluctuations close to the horizon. Hawking radiation would therefore seem to originate from wavelengths that close to the horizon were smaller than the Planck scale, $\sim 10^{-35}$ m. This in turn raises some doubts regarding the validity of the actual prediction of Hawking radiation as the laws of physics are expected to change radically at these length scales (the so-called transplanckian problem). Dispersion completely avoids this issue: as light accumulates on the white hole horizon it is blue shifted. However, in most dielectrics blue-shifted wavelengths will travel slower than red-shifted wavelengths implying that at a certain point the blue-shift will become such that the light will slow down and finally detach itself from the perturbation well before the Planck scale is reached. The study of analogue Hawking radiation therefore occurs in a regime in which transplanckian issues are of no concern or relevance.

In Fig. 11.3(a) we show a typical dispersion relation that we will be dealing with. This is a simple quadratic dispersion curve, $n_0 = A + B\omega^2$ where $A = 1.44$ and $B = 10^{-31}$ s^2. Such a simplified dispersion relation can actually match the real dispersion of, for example, fused silica glass in the visible and near-infrared spectral region with sufficient precision to capture all of the necessary physics and is a very good model for the dispersion of diamond over practically the whole spectrum, even down to very low frequencies. The solid/dashed curves in Fig. 11.3 indicate the dispersion branches that have positive/negative frequency in the laboratory reference frame. A common practice is to adopt the dispersion curves in the comoving frame rather than in the laboratory reference frame. We pass from one reference frame to the other using the Doppler relations

$$\omega' = \gamma(\omega - vk),$$
$$k' = \gamma\left(k - \frac{v}{c^2}\omega\right) \quad (11.15)$$

where $k = \omega[n(\omega) + \delta n]/c$. The laboratory frame dispersion curves in Fig. 11.3(a) transform in the comoving frame (for $v = 0.99c/n$ and $\delta n = 0$) as shown in Fig. 11.3(b). An alternative and useful representation with ω' as a function of ω is shown in Fig. 11.3(c). We note that for increasing δn or v, the comoving frame dispersion curves will bend further downwards, as shown in Fig. 11.3(c). The comoving frame dispersion curves are particularly convenient due to the fact that frequency ω' is a conserved quantity. This can be demonstrated on very simple grounds by deriving the Hamiltonian for a spacetime varying medium and by showing that ω' is a constant of motion [16]. An alternative route is based on a scattering model. The perturbation acts as a scattering defect and the scattered light will have a wave-vector $k_{out}(\omega)$ that is related to the input wave-vector $k_{IN}(\omega_{IN})$ by momentum conservation [42]:

$$k_{out}(\omega) = k_{IN}(\omega_{IN}) + \frac{\omega - \omega_{IN}}{v}. \quad (11.16)$$

This momentum conservation relation may be transformed into the comoving frame using Eq. (11.15) and leads to

$$\omega' = \omega'_{IN}. \quad (11.17)$$

In other words, momentum conservation in the laboratory reference frame is equivalent to frequency conservation in the comoving frame. We may therefore use this to predict how a input probe wave is transformed during the interaction with the perturbation by simply looking for the intersections of the comoving dispersion relation with a horizontal line that passes through ω'_{IN}. We call these intersection points "modes", in the sense that they identify specific modes of the electromagnetic field and are described by well-defined ω and k values (or ω' and k' in the comoving reference frame). In particular, in the following we will continuously refer to the input mode as the "IN" mode and the positive or negative frequency Hawking modes generated at the horizon as the "P" and "N" modes. In Fig. 11.3(b) the horizontal dashed line intersects the dispersion curve in two points for positive frequencies. The IN mode has positive group velocity $v'_g = d\omega'/dk'$ and thus approaches the perturbation, i.e. it is moving forward on the comoving frame. The scattered mode, indicated with "P", occurs with a negative gradient and is thus reflected backwards from the perturbation (note that in the laboratory frame, both IN and P modes will be moving in the forward direction). A third mode is possible and is indicated with "B": this is just the IN mode that is propagating backwards and is usually not considered as only the forward propagating IN mode is excited. Finally, a fourth mode is available, namely the intersection with $-\omega'_{IN}$: this gives what we will call the negative mode, indicated with "N". This mode is allowed as it is the complex conjugate of the "N_{cc}" mode that lies at $+\omega'_{IN}$. However, we prefer to consider the N mode rather than the N_{cc} mode as the former has positive frequency in the laboratory reference frame and will correspond to the mode actually measured in experiments.

We note that mode conversion from the IN mode to the P mode has been observed in a wide variety of settings (although this may not be immediately apparent due to the different terminology with respect to that used here):

1. In optical fibres it is possible to excite a soliton pulse as the result of a balance between nonlinear (intensity induced) frequency broadening and dispersion. However, at high input intensities high order solitons are generated that then breakup as the result of an instability and shed blue-shifted light that is usually called a dispersive wave or Cherenkov radiation (not to be confused with the Cherenkov radiation generated by superluminal charged particles) [43]. The dispersive wave emission obeys a momentum conservation law which, neglecting a nonlinear phase correction term, is identical to Eq. (11.16) [43–46] and thus falls under the same general explanation presented here;
2. In higher dimensions, e.g. in 2D waveguides or in bulk media, self-focusing and self-induced spatiotemporal reshaping of the input pulse lead to the formation of so-called X-waves. X-waves are characterised by two hyperbolic branches in the spectrum when viewed in angle-frequency coordinates: one branch passes through ω_{IN} and the other passes through the P mode frequency at zero angle. A more general description may be given in which the whole X-wave is actually expressed using only Eq. (11.16) [42, 47–50].

The P mode therefore emerges as an ubiquitous feature in nonlinear optics and it owes this ubiquity to the fact that it is simply a restatement of momentum conservation. The same reasoning may of course be applied to the negative N mode: this mode too is a result of momentum conservation and should therefore be expected.

However, before the development of analogue gravity models, no (optical) measurements had ever been performed reporting the existence of this mode. Moreover, the dispersion relations simply tell us *which* are the allowed modes, but do not tell us *if* the modes will actually be created. The P and N modes together form the classical analogues of the Hawking pairs emitted from an horizon, yet they also naturally emerge as from a classical analysis of the Maxwell equations.

The existence of these modes is a very general feature that is related only to the form of the dispersion relation and to the existence of a natural comoving reference frame in which frequency conservation leads to the excitation of negative frequencies. Recently this same reasoning has been applied to surface waves travelling in flowing water. By creating a gradient in the water flow, negative frequencies were observed, generated at the horizon [51]. These first measurements were then further developed and led to the first demonstration of *stimulated* Hawking emission [52]. The horizon is stimulated by a probe wave and the amplitudes of the emitted P and N waves are measured. In 2012, the first measurements have also confirmed the existence of these "classical" counterparts, i.e. stimulated P an N emission also in the optical analogue described here [57].

According to the theory, Hawking radiation in both the gravitational and analogue context will be characterised by precise relations that link the norms of the emitted waves (normalised with respect to the norm of the input wave $|IN|^2$) [52–55]:

$$|P|^2 - |N|^2 = 1, \quad (11.18)$$

$$\frac{|N|^2}{|P|^2} = e^{-\alpha\omega}. \quad (11.19)$$

Equation (11.18) implies that $|P|^2 + |N|^2 > 1$ and the Hawking effect will lead to amplification. Equation (11.19) imposes a strict relation between the two modes: it can be easily verified that indeed (11.18) and (11.19) imply a thermal emission for the N mode, $|N|^2 = 1/[\exp(\alpha\omega) - 1]$, where α may be linked to a black body temperature:

$$T_H = \frac{\hbar}{\alpha k_B}. \quad (11.20)$$

Gerlach gave a description of the black hole horizon in terms of a *parametric amplifier* [56]. In the absence of a probe pulse the horizon will amplify vacuum fluctuations but it will of course likewise convert and amplify any classical probe pulse that is sent onto it [11]. Parametric amplification is a well studied phenomenon in wave physics, in particular in the context of nonlinear optics. Very efficient excitation and amplification of vacuum states is achieved for example using crystals with a second order, or so-called $\chi^{(2)}$ nonlinearity [58], i.e. crystals that exhibit a nonlinear response that scales with the square of the input electric field. This same mechanism is actually the most widely used and robust method for generating quantum correlated photon pairs that have then in turn been used to test quantum theories and develop quantum information transmission and manipulation. The photon distribution of the radiation excited by these optical methods is also thermal. However there is a fundamental difference with respect to Hawking radiation: the thermal emission obtained by standard nonlinear optical parametric processes is thermal with a *different* temperature characterising each mode (i.e. frequency in the monochromatic limit) [59, 60]. Conversely, Hawking emission is composed of radiation that has exactly the same temperature T_H over the whole spectrum [56, 59]. Parametric amplification at a horizon in the form of Hawking radiation is therefore a very specific and peculiar effect that is quite unlike usual nonlinear optical parametric amplification.

11.4 Numerical Simulations of One-Dimensional Dielectric White Holes

An interesting question raised by these findings and predictions is "what does Hawking radiation correspond to in the framework of the Maxwell equations?". As should be expected, analogue Hawking radiation does emerge from Maxwell's equations yet it is a new and unexpected effect that has not been predicted before.

We performed numerical simulations using the Finite-Difference-Time-Domain technique applied to the discretised Maxwell equations [61]. The equations solved are

$$\frac{\partial E_y}{\partial x} = -\mu \frac{\partial H_z}{\partial t},$$
$$\frac{\partial H_z}{\partial x} = -\frac{\partial (\varepsilon E_y)}{\partial t} \quad (11.21)$$

where $\varepsilon = \varepsilon(x - vt) = \sqrt{n_0 + \delta n(x - vt)}$ is the medium permittivity. We underline that there are no nonlinearities involved in these equations: only linear propagation is simulated and the travelling refractive index perturbation is included in ε. In our simulations we took a super-Gaussian form for the perturbation: $\delta n = \delta n_{max} \exp[-((x - vt)/\sigma)^m]$ where m is an even integer. In these studies we do not directly verify the emission of radiation from the vacuum state, we rather stimulate the horizon by an incoming classical, coherent, probe pulse and we study how this pulse evolves. The study of the stimulated Hawking emission (SHE) is extremely important since allows us to gather information on the underlying Hawking radiation mechanism. As mentioned above, we will only focus attention on the white hole, i.e. on how the probe pulse interacts with the steep, trailing edge of the perturbation.

Figure 11.4 shows an example of such a simulation. Dispersive effects are introduced through numerical dispersion that depends on the grid resolution and may thus be controlled [62]. Figure 11.6(a) shows the dispersion in the comoving frame relative to the simulations in Fig. 11.4. The perturbation has $v = 0.99c/n_0$, maximum amplitude $\delta n_{max} = 0.01$ and super-Gaussian order $m = 26$ so that δn raises from 0 to $\sim \delta n_{max}$ over a distance ~ 1 μm. The input probe pulse is taken with initial wavelength 4 μm and is placed behind the perturbation. This is equivalent to studying the evolution of a mode that attempts to enter a white hole. Figures 11.4(a)–(d) show the electric field profile at various propagation distances within a window centred on the perturbation: the input probe pulse catches up with the perturbation where it is blocked at the horizon and frequency shifted until it finally starts to lag behind (due to dispersion that decreases the pulse velocity with decreasing wavelength). Figures 11.4(e)–(h) show the spectra relative to each electric field profile: the input spectral peak (IN) is transformed into two distinct peaks (P and N). We underline once more that this is a purely *linear* simulation, i.e. the observed frequency conversion is not the result of an optical interaction involving e.g. $\chi^{(2)}$ or $\chi^{(3)}$ nonlinearities. Rather, this frequency conversion finds a simple explanation in terms of the generation of positive and negative Hawking modes: in Fig. 11.6(a) we show the modes on the dispersion curve at the horizon. As can be seen, both the P and N modes lie on the curve and both conserve ω'_{IN}. Moreover, in the following section (Sect. 11.5), we repeat such a simulation for many different input frequencies and then, by taking the ratio of the photon numbers in the two output modes, $R = |N|^2/|P|^2$, we verify that these satisfy relation (11.19), i.e. the emission is thermal with the same temperature over the whole spectrum of input frequencies [63].

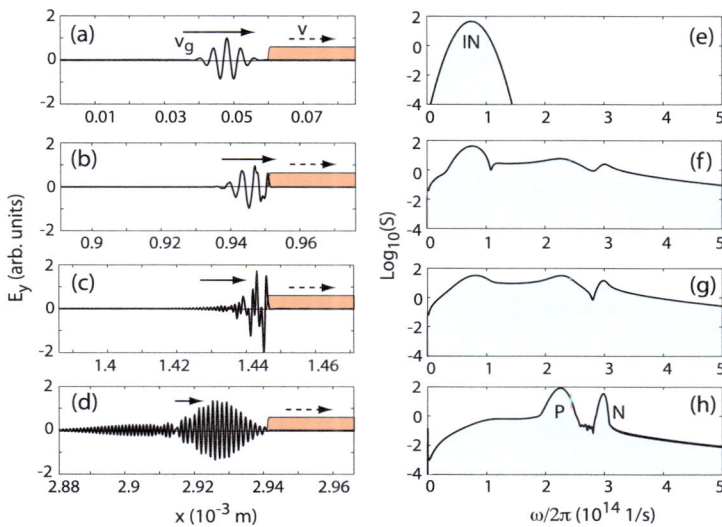

Fig. 11.4 Numerical simulations of stimulated Hawking radiation. (**a**)–(**d**) Evolution of the electric field of an input few-cycle laser pulse with an initial wavelength of 4 µm. The *shaded area* indicates the refractive index perturbation. The *arrows* indicate the qualitative amplitude of the velocities (solid arrow for the probe pulse, dashed arrow for the perturbation). The spectra (S) relative to each of these graphs are shown in (**e**)–(**h**) in logarithmic scale: the input spectral peak is indicated with "IN". The output spectrum clearly exhibits two distinct blue shifted peaks, the positive and negative Hawking modes indicated with "P" and "N" in (**h**)

Figure 11.5 shows the result of a simulation with the same input parameters as in Fig. 11.4 with the only difference that the input wavelength is now 2 µm and the probe pulse is placed inside the perturbation which is now moving faster than the pulse. This is equivalent to study the evolution of a mode that exits a white hole. As the pulse exits the perturbation, it is frequency converted, as before, to a P and N mode. The dispersion curves corresponding to this simulation are shown in Fig. 11.6(b): note that because the input probe pulse has a phase velocity that is lower than the perturbation velocity, in the comoving frame it now has initial negative (comoving) frequency.

11.5 Stimulated Hawking Emission and Amplification

In this section we study in more detail the stimulated process SHE that has been introduced in the previous numerical section (Sect. 11.4). Although the stimulated case clearly distinguishes from the spontaneous Hawking radiation (which originates from the vacuum fluctuations and has been studied in the experimental analogue contest, see Sect. 11.7) it bears in common important features with it, and may thus be useful to understand the underlying physics. In particular we want to shed

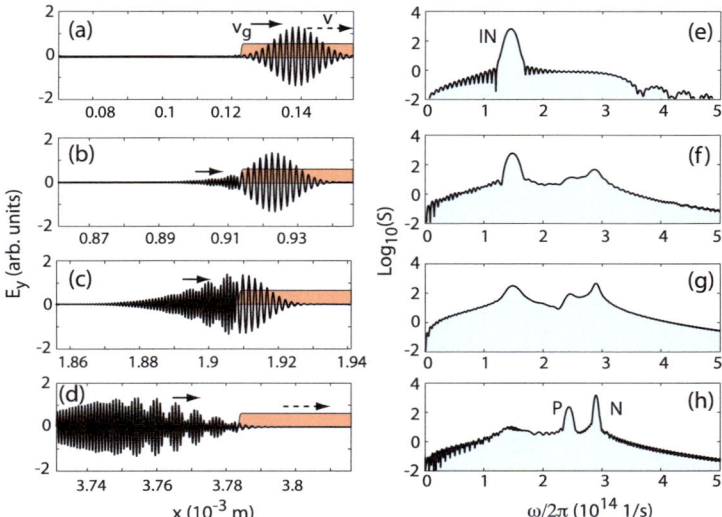

Fig. 11.5 Numerical simulations of stimulated Hawking radiation. (**a**)–(**d**) Evolution of the electric field of a input few-cycle laser pulse with an initial wavelength of 2 μm. The input pulse now starts from inside the perturbation with *lower* initial group velocity $v_g < v$ and thus exits the perturbation, passing through the white hole horizon. The spectra (S) relative to each of these graphs are shown in (**e**)–(**h**) in logarithmic scale

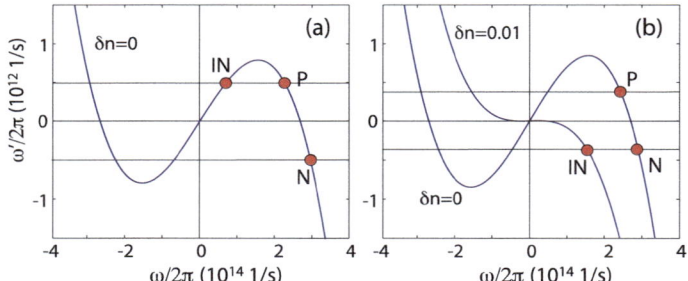

Fig. 11.6 Dispersion curves for the numerical simulations in Figs. 11.4(a) and 11.5(b). We plot the comoving frequency, $\omega'/2\pi$, as a function of the laboratory frame frequency, $\omega/2\pi$. The full–circles indicate the position of the IN, P and N mode frequencies

some light on the photon amplification process and on the thermality of the emitted radiation.

The conversion and amplification of the two output modes, P and N, may be better understood in the contest of a *generalized* Manley-Rowe relation. Indeed, in the geometrical optics approximation, when j new frequencies are generated starting from an input seed pulse at frequency ω_0, the photon number of the output waves,

Fig. 11.7 Numerical results for the stimulated Hawking emission. Normalized photon number evolutions for the input mode, $|IN|^2$ (*black dotted curve*), total outgoing photons, $|P|^2+|N|^2$ (*blue solid*), and difference photons, $|P|^2-|N|^2$ (*red dashed*). Input seed wavelength $\lambda_{IN} = 5$ μm, DP velocity $v = 0.97c$ and DP amplitude $\delta n_{max} = 0.09$

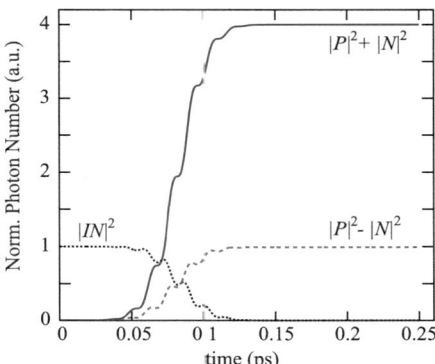

I_j, is linked to the photon number of the input mode, I_0, by the following generalized Manley-Rowe relation [64–66]:

$$\text{sign}(\gamma_0) I_0 = \sum_{j>0} \text{sign}(\gamma_j) I_j, \quad (11.22)$$

where, for $\alpha = 0, 1, \ldots, j$, γ_α is defined as $\gamma_\alpha = (v_\phi^2 - v v_\phi)_\alpha$, v is the velocity of the moving perturbation DP and v_ϕ is the mode phase velocity. We note that photon amplification occurs as far as $\gamma_j < 0$ for some $j > 0$ [64]. Indeed, the total number of photons in this case, obtained as $\sum_{j>0} I_j$, exceeds I_0.

In particular we are interested in the case $j = 2$, i.e. when there is creation of a pair of new converted modes that, in the reference frame comoving with the DP, have positive $\omega' > 0$ (P-mode), and negative $\omega' < 0$ (N-mode), frequencies. In this specific case, and noting that

$$\text{sign}(\gamma_j) = \text{sign}(\omega_j - v k_j) = \text{sign}(\omega'_j), \quad (11.23)$$

Eq. (11.22) assumes the interesting form of Eq. (11.18), which indeed states that photon amplification is tied hand in hand with the generation of negative frequencies.

We can see this effect also in the numerical simulations described in Sect. 11.4. Considering for example the interaction scheme shown in Fig. 11.4, we may plot the photon number evolution with time. Figure 11.7 shows the numerical results for an input seed wavelength $\lambda_{IN} = 5$ μm, DP velocity $v = 0.97c$ and DP amplitude $\delta n_{max} = 0.09$. The normalized photon number in the IN-mode (black dotted curve) decreases along propagation and after ∼0.12 ps is fully converted in the P and N modes. At the end of the interaction the photon *difference*, $|P|^2 - |N|^2$ (red dashed curve), is shown to indeed conserve the photon number, i.e. $|P|^2 - |N|^2 = 1$, while the total photon number, $|P|^2 + |N|^2$ (blue solid curve), clearly shows an increase of 4 times with respect to the IN-mode, thus indicating the presence of amplification.

Moreover, for the stimulated case, it is possible to study the thermal behavior for the emitted radiation, predicted by relation (11.19). Figure 11.8(a) shows the spectral dependence of the ratio R between the number of photon in the negative

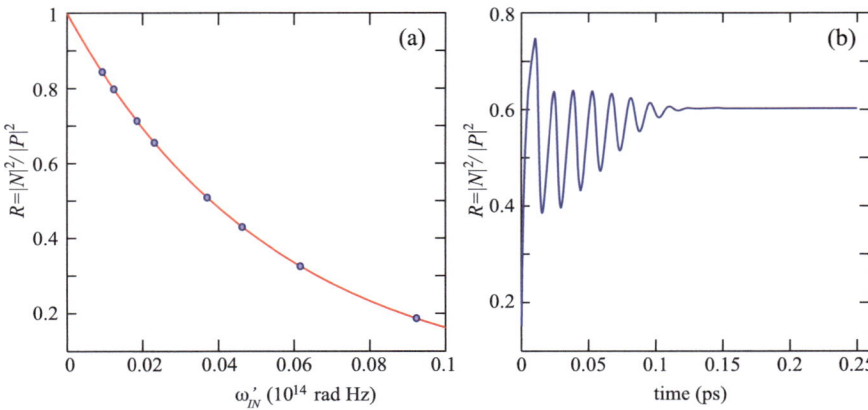

Fig. 11.8 (a) ratio $R = |N|^2/|P|^2$ versus input comoving frequency ω'_{IN}, by varying λ_{IN} between 2 and 20 μm (*blue circles*) and (**b**) R versus time, for the same simulation as in Fig. 11.7. The data in (**a**) show a clear exponential dependence with best fit $R = 0.99\exp(-18.2 \times 10^{-14}\omega')$ (*red solid curve*), corresponding to a blackbody emission with temperature $T_H = 263.4$ kelvin. DP velocity $v = 0.97c$ and DP amplitude $\delta n_{max} = 0.09$

and in the positive mode, $R = |N|^2/|P|^2$, as a function of the input comoving frequency ω'_{IN}. This set of simulations was performed with a fixed DP velocity $v = 0.97c$ and a varying input wavelength λ_{IN}, thus ensuring that the whole output spectrum experiences the same horizon and, from relation (11.11) and (11.12), would be amplified at the same temperature T_H [63]. Indeed, the blue circles in Fig. 11.8(a) show that R follows an exponential law with best fit (red solid curve) given by $R = 0.99\exp(-18.2 \times 10^{-14}\omega')$: for $\omega'_{IN} = 0$, we have $R \sim 1$ to a very good approximation, as should be expected, and from the decay constant we may estimate the Hawking temperature T_H, by substituting in relation (11.20). This corresponds to a blackbody emission with temperature $T_H = 263.4$ kelvin. Impressively, this numerical evaluation resulted in perfect agreement with the one predicted with the quantum field theory model, Eq. (11.12), where the gradient of the perturbation is evaluated for a position of the horizon corresponding to $\delta n_H \sim 0.032$.

We also note that R is characterized by an oscillating behavior during the interaction and it reaches an asymptotic value only after the conversion is complete. For example, in Fig. 11.8(b) we plot R as a function of the interaction time for the same simulation as in Fig. 11.7. In this case, it is evident that an extrapolation of a value of R before waiting for the full conversion would underestimate the Hawking temperature. Indeed, for the plot of Fig. 11.8(a) we took the asymptotic values of R, which are the only that can determine a correct evaluation for T_H.

Remains to study how T_H varies as a function of the dielectric perturbation amplitude, δn_{max}, so to verify the accuracy of the perturbative model for smaller values of δn_{max}.

11.6 Creating an Effective Moving Medium with a Laser Pulse

So far the discussion has referred to a generic refractive index perturbation without actually mentioning how this may be generated. The idea originally proposed by Leonhardt [33] is based on the Kerr effect [58]: a sufficiently intense laser pulse propagating in any isotropic medium such as a gas, liquid or amorphous solid will excite a nonlinear polarization response:

$$P = \varepsilon_0 \left(\chi^{(1)} + \chi^{(3)} E^2 \right) E, \quad (11.24)$$

where $\chi^{(1)}$ is the linear susceptibility that is related to the linear refractive index $n_0 = \sqrt{1 + \chi^{(1)}}$, $\chi^{(3)}$ is the third order nonlinear susceptibility (also know as the Kerr nonlinearity) and $E = |E|\cos(\omega t)$ is the electric field of the intense laser pulse that excites the medium. The third order polarization term may thus be re-written as $P_{NL} = 1/4 \chi^{(3)} |E|^3 \cos(3\omega t) + 3/4 \chi^{(3)} |E|^3 \cos(\omega t)$. The first term oscillates at 3ω, i.e. it acts as a source for third harmonic frequency generation and may be neglected (unless the beams or the medium are specifically engineered so as to enhance this process). The second term oscillates at the input frequency ω and from Eq. (11.24) we define an effective refractive index $n = \sqrt{1 + \chi^{(1)} + 3/4 \chi^{(3)} |E|^2} \simeq n_0 + n_2 I$, where the nonlinear index is $n_2 = (3/8) \chi^{(3)}/n_0$ and $I = |E|^2$. The intensity profile of a laser pulse usually has a Gaussian-like form, i.e. $I = I(x - vt) = \exp[-((x - vt)/\sigma)^2]$, where the pulse speed will be given by the group velocity of light in the medium, $v = v_g = d\omega/dk$. In other words, an intense laser pulse propagating in a nonlinear Kerr medium will create a refractive index perturbation $\delta n = n_2 I$ that travels close to the speed of light.

We note that the reasoning above may be generalised without loss of generality to the case in which the Kerr medium is excited by an intense laser pulse and the resulting perturbation acts upon a second, weak probe pulse [33, 58, 63]. There are various methods by which the Kerr medium may be excited to induce a refractive index perturbation. However, successful measurements of Hawking radiation do impose some additional constraints:

1. The intensity profile should be stationary during propagation in order to recreate stationary excitation conditions. This is not a trivial requirement due to the fact that the same Kerr effects that generate the perturbation also lead to back-reaction on the pump laser pulse and detrimental effects such as pulse splitting, self-focusing and white light generation;
2. The existence and frequency range of an horizon is completely determined by the perturbation velocity as seen in Eq. (11.10). So ideally we would like to control the speed v.

Bearing this in mind, a few experimental setups have been proposed and are summarised in Fig. 11.9. Leonhardt's original proposal was based on the use of optical solitons propagating in highly nonlinear photonic crystal fibres. These are fibres with micro-structured cores that on the one hand allow one to tightly confine light within the core region, so as to increase the pulse intensity and amplitude of the perturbation, and on the other allow to engineer the dispersion relation and thus

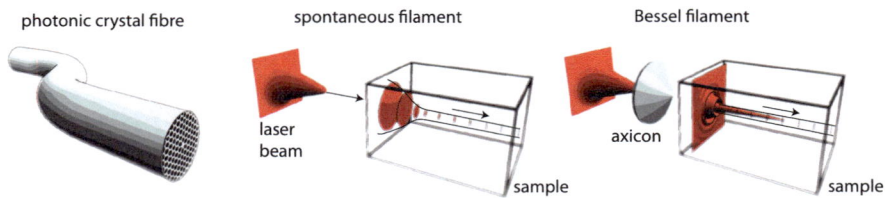

Fig. 11.9 Sketch of three different methods employed to generate intense laser pulses with quasi-stationary propagation over long distances. Photonic crystal fibres have micro-structured cores that allow to both tightly confine light and control the material dispersion. Spontaneous filaments are obtained by loosely focusing an intense laser pulse in a bulk Kerr medium, e.g. fused silica glass

the group velocity of the stationary soliton. The δn_{max} generated by these solitons is usually of the order of 10^{-4}. Experiments using fibre solitons therefore satisfy our list of requirements and indeed the first evidence of horizon-related frequency conversion was experimentally observed using such fibres in 2008 [33].

Another option is to attempt to harness the nonlinear propagation of laser pulses in bulk media. One possibility that has been proposed [41, 63, 67] is to use so called filaments. The term filament, or light filament, denotes the formation of a dynamical structure with an intense core that is able to propagate over extended distances much larger than the typical diffraction length while keeping a narrow beam size without the help of any external guiding mechanism [68]. These sub-diffractive "light-bullets" may be either generated spontaneously of by pre-shaping the laser beam. Spontaneous filaments are born as a consequence of a spontaneous spatio-temporal reshaping of an input Gaussian-shaped laser pulse when it is loosely focused into a nonlinear Kerr medium. If the input power is larger than a certain critical threshold power, the pulse will self-focus as a result of the spatially varying refractive index perturbation that the pulse generates and that acts in a similar fashion to a focusing lens. At the same time the spectrum is broadened and both the transverse and longitudinal profiles of the pulse are simultaneously reshaped until a quasi-stationary (or "dynamical") state is formed. Spontaneous filaments exhibit a number of features that are attractive for the generation of optical horizons: (i) they are extremely simple to obtain and are characterized by a very intense peak, e.g. $I \sim 10^{12}$–10^{13} W/cm^2 that propagates over distances of the order of 2–10 diffraction lengths. In fused silica $n_2 \sim 3 \times 10^{16}$ cm^2/W so we may have $\delta n_{max} \sim 10^{-3}$. (ii) Along the longitudinal coordinate the pulse may split in two and the trailing pulse will be drastically shortened and exhibit an extremely sharp shock front on the trailing edge. Moreover, the trailing peak propagates with a velocity that is significantly slower than the input pulse group velocity. By controlling the input conditions, e.g. wavelength, focusing, power, it is possible to control to a certain extent both the steepening effects and the peak velocity.

Figure 11.10(a) shows an example of the numerically simulated evolution of the longitudinal profile of a 1.055 µm, 100 fs long laser pulse that undergoes filamentation in fused silica. The pulse splits in two daughter pulses: the rear pulse has

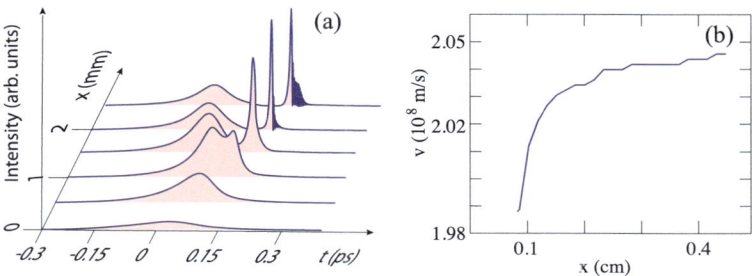

Fig. 11.10 (a) Example of the numerically simulated longitudinal profile of a spontaneous filament at various propagation distances. (b) Evolution of the of the intense trailing peak velocity

significantly higher intensity and exhibits a self-steepened shock front on its trailing edge with a nearly single optical duration. Figure 11.10(b) shows the velocity of this peak over a longer propagation distance. As can be seen v varies and gradually accelerates. This feature has been used in experiments: by selecting the emission at different x it is possible to study the behavior at different v [63, 67].

A further possibility that has been proposed is the use of Bessel filaments. In this case the transverse profile of the input laser beam is re-shaped into a Bessel-like pattern using a conical lens (also called axicon). The axicon transforms the laser beam by redirecting light along a cone at an azimuthally symmetric angle θ towards the optical axis. Light propagating towards the axis will therefore interfere and the resulting interference pattern will be a non-diffracting Bessel pattern. Moreover, simple geometric considerations show that central Bessel peak propagates along the x-axis with velocity $v = v_g/\cos\theta$. Therefore, by simply changing the angle of the axicon, it is possible to control the propagation velocity of the Bessel pulse. We note that the spontaneous filament leads to a trailing intense peak that is slowed down with respect to the input pulse group velocity, v_g whilst the Bessel pulse travels faster.

11.7 Experiments: *Spontaneous Emission from a Moving Perturbation*

Laser-pulse induced analogues are the first analogue systems in which experiments were carried out to investigate and probe the quantum properties of analogue Hawking emission. These experiments were initiated in 2008 by the work of Philbin et al. [33]: a soliton was created inside a photonic crystal fibre and blue-shifting of part of the soliton radiation was observed as a consequence of the interaction with the self-generated white hole horizon. This was a purely classical effect but clearly demonstrated the possibility to generate horizons in dielectric media using intense laser pulses.

This idea was later extended to filament induced perturbations. The spectral transformations of a filament pulse are significantly richer than in a fibre due to

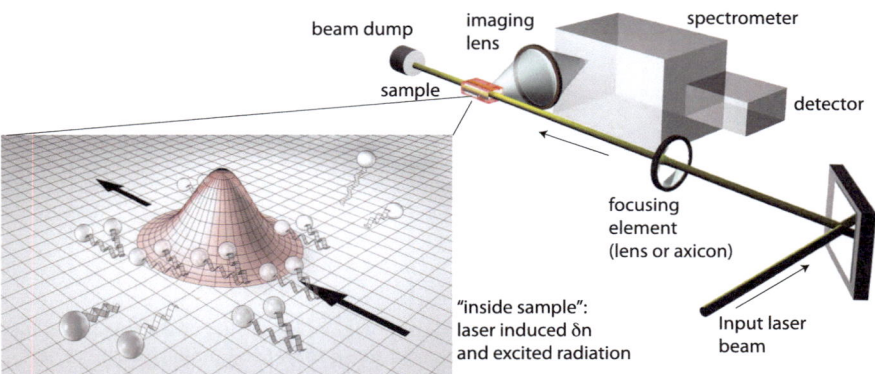

Fig. 11.11 Sketch of the experimental layout. The input laser beam is focused into the Kerr sample (fused silica glass) using either a 20 cm focal length lens (to induce spontaneous filamentation) or an axicon with a 20 degree cone base angle (to create a Bessel filament). The *inset* shows a schematic representation of the Kerr sample: the perturbation propagates following the arrows and photon pairs, corresponding to the negative and positive Hawking pairs, are excited

the additional transverse degree of freedom. Very specific hyperbolic-shaped spectra are observed in angle-wavelength coordinates. These features were shown to be reproduced and predicted very precisely within the framework of a model based on the metric (11.5), thus confirming that by using the basic mathematical tools of general relativity it is possible to capture relevant details of optical pulse propagation in the presence of a travelling perturbation [41]. These first experiments were then adapted so as to search for signatures of spontaneous Hawking radiation.

The experimental layout is depicted in Fig. 11.11: the input laser pulse is provided by a regeneratively amplified, 10 Hz repetition rate Nd:glass laser. The pulse duration is 1 ps and energy is varied between $\sim$50 µJ and 1.2 mJ. The input pulse is sent on to a 2 cm long sample of pure fused silica after a reshaping of the beam by means of the focusing element (lens or axicon). Indeed filaments were formed by either loosely focusing the input pulse with a 20 cm focal length lens (spontaneous filaments) or by replacing the lens with an axicon so as to generate a $\theta \sim 7$ deg Bessel filament. Light emitted in the forward direction is extremely intense with an average photon number of the order 10^{15} photons/pulse. Considering that Hawking radiation is unlikely to provide more that 1 photon/pulse, a huge and extremely challenging suppression of the laser pump pulse radiation is required. The photons emitted from the sample were therefore collected at 90 degrees rather than in the forward direction: the spectrum is then recorded with an imaging spectrometer coupled to a 16 bit, cooled CCD camera. This arrangement was chosen in order to strongly suppress or eliminate any additional nonlinear effects (e.g. four wave mixing and self phase modulation). By employing both spontaneous and Bessel filaments in separate measurements it was possible to measure radiation emitted from perturbation with various velocities.

The results for the spontaneous filament, adapted from Refs. [63, 67], are shown in Fig. 11.12. In particular, Fig. 11.12(a) shows the full spectrum (black curve), inte-

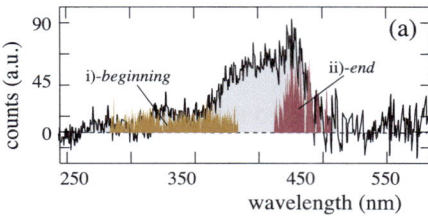
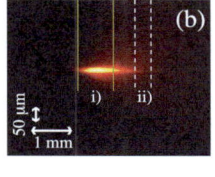

Fig. 11.12 Spectra generated by the spontaneous filament (**a**) and image of the filament from the side, at 90 deg (**b**). The coloured areas in (**a**) indicate spectra measured for two different positions of the imaging spectrometer input slit, i.e., (*i*) *beginning* and (*ii*) *end* section of the filament, highlighted also in (**b**) by the *vertical lines*. The *black curve* in (**a**) shows the spectrum measured with the input split fully open

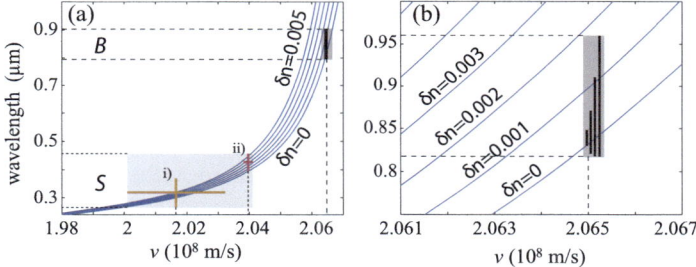

Fig. 11.13 (**a**) Measured radiation wavelength for various perturbation velocities. The *blue curves* show the predicted velocity dependence of the radiation wavelengths. The Bessel measurement is expanded in (**b**): the four different vertical bars indicated four different spectrum bandwidths obtained by increasing the input laser pulse intensity. The *bars* are slightly displaced horizontally for clarity but they actually all have the same $v = 2.065 \times 10^8$ m/s

grated over 3600 laser shots, obtained by keeping the input slit of the spectrometer fully open, in order to collect photons emitted from the whole filament. The filament, imaged at 90 deg, is shown in Fig. 11.12(b). The coloured areas in 11.12(a) indicate spectra measured for two different positions of the imaging spectrometer input slit, i.e. when collecting light from either the (i) beginning, or the (ii) end region of the filament, highlighted also in (b) by the vertical lines. Indeed, as shown in Fig. 11.10(b), the pulse velocity varies in propagation, so that a certain range of velocity values is covered during propagation, with low velocity at the beginning of the filament and higher velocity at the end. This has the effect of broadening the emission window, which is predicted from Eq. (11.10) to be between 270 and 450 nm. Remarkably, different sections of the filament emit only at selected portions of the overall spectrum, and this peculiar behavior is in quantitative agreement with the predictions of Eq. (11.10).

These results are summarised also in Fig. 11.13(a): the solid curves represent the predicted Hawking emission wavelengths based on Eq. (11.10), for increasing δn. The two shaded areas summarise the results from the spontaneous and Bessel filament measurements (indicated with S and B, respectively): the vertical extension indicates the measured spectral bandwidth (at half maximum) and the horizontal bars indicate the velocity variation of the perturbation in the case of spontaneous filaments. The measurements follow the expected dependence and indicate a tight correlation between the wavelength, or colour of the emitted radiation, and the perturbation velocity thus supporting the claim that this radiation is indeed emitted from the analogue horizon. The measurement indicated with B refers to the Bessel filament and is shown in more detail in Fig. 11.13(b). In this case there is no spread of the perturbation velocity and it was possible to probe the emission for increasing peak intensities, i.e. increasing δn. The bandwidth of the emitted radiation increases with intensity and increases predominantly to longer wavelengths. This is precisely in agreement, also at a quantitative level [63, 67], with Eq. (11.10).

On the basis of these results, these measurements have been proposed as the first experimental evidence of Hawking-like emission from an analogue horizon [63, 67]. However, this is not generally considered to be conclusive evidence and further measurements are called for in order to verify some open issues: the measured photon numbers, of the order of 0.1–0.01 photons/pulse appear to be too high to be accounted for on the basis of a blackbody emission. Yet other models [20] appear to give predictions that confirm the measurements. Moreover, the theory predicts that photons will be generated in correlated pairs. The photons collected at 90 deg are likely to have suffered strong scattering after emission in the forward direction and it will therefore be very unlikely to observe any kind of pair-correlation with such a setup. Other experimental layouts, e.g. based on fibres which will confine the photon pairs along the same direction, are therefore required.

There are however more critical aspects that need further consideration before the observed emission can be truly compared to analogue Hawking radiation. Some of these measurements were performed in regimes in which the laser pulse induced perturbation did not present a group-velocity horizon for any possible co-propagating frequencies. Group-velocity horizons are considered to be a fundamental ingredient for the black hole analogy to be robust—indeed, it is only in the presence of such a horizon that one can guarantee a true and complete blocking of light at the horizon and hence the presence of two causally disconnected regions. The optical analogue therefore seems to hint that emission occurs irrespectively of the presence of such a blocking horizon. This in itself is not a contradiction and models are being developed in which such an emission is predicted [69, 70]. However, these models treat the observed emission as an analogue of spontaneous emission from a cosmological expansion (see also the Chapter by R. Schützhold and W. Unruh) rather than as true Hawking emission. There are therefore many subtleties within these measurements that need to be carefully considered before any definite claims can be made. Future models will need to consider the full role of dispersion "ab initio" in the quantum field theory description and additional measurements are required in order to further investigate the presence or absence of the fundamental ingredients required for an full analogy with Hawking radiation.

11.8 Conclusions and Perspectives

Laser pulses have been demonstrated in a variety of settings to generate analogue white holes and horizons that transform light according to the predictions of models that are derived within the context of general relativity. There is an intrinsic beauty in spacetime geometries, that same beauty that pushed Einstein to develop, and others to accept, the theory of general relativity [71]. The extension to the study of the "geometry of light" [72], is no exception in this sense. It is possibly not clear to date how far-reaching this extension will be and if it will allow to gain further insight into quantum gravity theories or black hole physics. However, analogies between black hole kinematics and flowing media are certainly extending the limits of our understanding across various disciplines, e.g. waves in fluids, acoustics and optics. And thanks to the common underlying tools derived from general relativity, discoveries developed in one field apply also to the others and a deeper insight is achieved by directly comparing similar or different behaviours across the various disciplines.

The optical analogue is still in its infancy and we expect a strong development in the next years. A number of recent proposals are based either directly on the presence of an horizon or on the same technology required to build an horizon. For example Demircan et al. have studied an optical transistor that acts through an optical event horizon [73], McCall et al. have proposed a temporal cloaking device that uses pulses that split and then recombine thus cancelling out portions of history [74] and Ginis et al. have proposed a frequency converter based on a metamaterial analogue of cosmological expansion [75]. Stimulated Hawking mode conversion as seen in the numerical simulations presented here and in literature [63] represents a novel kind of optical amplifier and is awaiting experimental demonstration. Such an amplification mechanism could then in turn lead to the first "black hole laser" whereby a wave is trapped between two horizons that form a cavity: at each reflection from the white hole horizon light is amplified through a stimulated Hawking process with a resulting laser-like behaviour [76, 77].

These are just a few ideas and possibilities. There are certainly many more that await to be investigated and brought to light through the blending of general relativity with flowing media.

References

1. Barceló, C., Liberati, S., Visser, M.: Living Rev. Relativ. **14**(3), 1 (2011)
2. Joannopoulos, J., Johnson, S., Winn, J., Meade, R.: Photonic Crystals: Molding the Flow of Light, 2nd edn. Princeton University Press, New Jersey (2008)
3. Martin de Sterke, C., Bright, J., Krug, P., Hammon, T.: Phys. Rev. E **57**(2), 2365 (1998)
4. Peschel, U., Pertsch, T., Lederer, F.: Opt. Lett. **23**(21), 1701 (1998)
5. Wiersma, D.S., Bartolini, P., Lagendijk, A., Righini, R.: Nature **390**, 671 (1997)
6. Painlevé, P.: C. R. Acad. Sci. **173**, 677 (1921)
7. Gullstrand, A.: Ark. Mat. Astron. Fys. **16**, 1 (1922)
8. Hamilton, A., Lisle, J.: Am. J. Phys. **76**, 519 (2008)
9. Unruh, W.: Phys. Rev. Lett. **46**(21), 1351 (1981)

10. Hawking, S.: Phys. Rev. D **13**, 191 (1976)
11. Bekenstein, J., Meisels, A.: Phys. Rev. D **15**(10), 2775 (1977)
12. Gordon, W.: Ann. Phys. **377**, 421 (1923)
13. Shvartsburg, A.: Phys. Usp. **48**, 797 (2005)
14. Rosanov, N.: Opt. Spectrosc. **106**, 430 (2009)
15. Rosanov, N.: Opt. Spectrosc. **106**, 742 (2009)
16. Mendonça, J.: Theory of Photon Acceleration. Series in Plasma Physics. Institute of Physics, Bristol (2001)
17. Mendonça, J., Shukla, P.: Phys. Scr. **65**, 160 (2002)
18. Mendonça, J., Guerreiro, A., Martins, A.: Phys. Rev. A **62**, 033805 (2000)
19. Mendonça, J., Guerreiro, A.: Phys. Rev. A **72**, 063805 (2005)
20. Guerreiro, A., Ferreira, A., Mendonça, J.: Phys. Rev. A **83**, 0523025 (2011)
21. Leonhardt, U., Piwnicki, P.: Phys. Rev. A **60**, 4301 (1999)
22. Leonhardt, U., Piwnicki, P.: Phys. Rev. Lett. **84**, 822 (2000)
23. Schützhold, R., Plunien, G., Soff, G.: Phys. Rev. Lett. **88**, 061101 (2002)
24. Brevik, I., Halnes, G.: Phys. Rev. D **65**, 024005 (2001)
25. De Lorenci, V., Klippert, R.: Phys. Rev. D **65**, 064027 (2002)
26. De Lorenci, V., Souza, M.: Phys. Lett. B **512**, 417 (2001)
27. Novello, M., Salim, J.: Phys. Rev. D **63**, 083511 (2001)
28. Marklund, M., Anderson, D., Cattani, F., Lisak, M., Lundgren, L.: Am. J. Phys. **70**, 680 (2002)
29. De Lorenci, V., Klippert, R., Obukhov, Y.N.: Phys. Rev. D **68**, 061502 (2003)
30. Novello, M., Bergliaffa, B.P., Santiago, E., Salim, J., De Lorenci, V., Klippert, R.: Class. Quantum Gravity **20**, 859 (2003)
31. Unruh, W., Schützhold, R.: Phys. Rev. D **68**, 024008 (2003)
32. Schützhold, R., Unruh, W.: Phys. Rev. Lett. **95**, 031301 (2005)
33. Philbin, T.G., Kuklewicz, C., Robertson, S., Hill, S., König, F., Leonhardt, U.: Science **319**(5868), 1367 (2008)
34. Novello, M., Visser, M., Volovik, G.E.: Artificial Black Holes. World Scientific, Singapore (2002)
35. Belgiorno, F., Cacciatori, S., Ortenzi, G., Rizzi, L., Gorini, V., Faccio, D.: Phys. Rev. D **83**, 024015 (2011)
36. Cacciatori, S., Belgiorno, F., Gorini, V., Ortenzi, G., Rizzi, L., Sala, V., Faccio, D.: New J. Phys. **12**, 095021 (2010)
37. Wald, R.: General Relativity. University of Chicago Press, Chicago (1984)
38. Birrell, N., Davies, P.: Quantum Fields in Curved Space. Cambridge University Press, Cambridge (1982)
39. Fulling, S.: Aspects of Quantum Field Theory in Curved Space-Time. Cambridge University Press, Cambridge (1989)
40. Parker, L., Toms, D.: Quantum Field Theory in Curved Space-Time: Quantized Fields and Gravity. Cambridge University Press, Cambridge (2009)
41. Faccio, D., Cacciatori, S., Gorini, V., Sala, V., Averchi, A., Lotti, A., Kolesik, M., Moloney, J.: Europhys. Lett. **89**, 34004 (2010)
42. Faccio, D., Averchi, A., Couairon, A., Kolesik, M., Moloney, J., Dubietis, A., Tamosauskas, G., Polesana, P., Piskarskas, A., Di Trapani, P.: Opt. Express **15**(20), 13077 (2007)
43. Akhmediev, N., Karlsson, M.: Phys. Rev. A **51**(3), 2602 (1995)
44. Dudley, J., Genty, G., Coen, S.: Rev. Mod. Phys. **78**(4), 1135 (2006)
45. Dudley, J., Taylor, J.: Supercontinuum Generation in Optical Fibres, 1st edn. Cambridge University Press, Cambridge (2010)
46. Kolesik, M., Tartara, L., Moloney, J.: Phys. Rev. A **82**, 045802 (2010)
47. Kolesik, M., Faccio, D., Wright, E., Di Trapani, P., Moloney, J.: Opt. Lett. **34**, 286 (2009)
48. Kolesik, M., Wright, E., Moloney, J.: Phys. Rev. Lett. **92**, 253901 (2004)
49. Faccio, D., Porras, M., Dubietis, A., Bragheri, F., Couairon, A., Di Trapani, P.: Phys. Rev. Lett. **96**, 93901 (2006)

50. Faccio, D., Averchi, A., Lotti, A., Kolesik, M., Moloney, J., Couairon, A., Di Trapani, P.: Phys. Rev. A **78**, 033825 (2008)
51. Rousseaux, G., Mathis, C., Maïssa, P., Philbin, T., Leonhardt, U.: New J. Phys. **10**, 053015 (2008)
52. Weinfurtner, S., Tedford, E., Penrice, M., Unruh, W., Lawrence, G.: Phys. Rev. Lett. **106**, 021302 (2011)
53. Hawking, S.: Nature **248**(5443), 30 (1974)
54. Hawking, S.: Commun. Math. Phys. **43**, 199 (1975)
55. Unruh, W.: Phys. Rev. D **14**(4), 870 (1976)
56. Gerlach, U.: Phys. Rev. D **14**, 1479 (1976)
57. Rubino, E., McLenaghan, J., Kehr, S.C., Belgiorno, F., Townsend, D., Rohr, S., Kuklewicz, C.E., Leonhardt, U., Konig, F., Faccio, D.: Phys. Rev. Lett. **108**, 253901 (2012)
58. Boyd, R.W.: Nonlinear Optics, 3rd edn. Academic Press, New York (2008)
59. Leonhardt, U.: Quantum Optics, 1st edn. Cambridge University Press, Cambridge (2010)
60. Walls, D., Milburn, G.: Quantum Optics, 1st edn. Springer, Berlin (1994)
61. Yee, K.: IEEE Trans. Antennas Propag. **14**, 302 (1966)
62. Taflove, A., Hagness, S.: Computational Electrodynamics: The Finite-Difference Time-Domain Method, 3rd edn. Artech House, Norwood (2005)
63. Rubino, E., Belgiorno, F., Cacciatori, S.L., Clerici, M., Gorini, V., Ortenzi, G., Rizzi, L., Sala, V.G., Kolesik, M., Faccio, D.: New J. Phys. **13**, 085005 (2011)
64. Ostrovskii, L.A.: Sov. Phys. JETP **34**, 293 (1972)
65. Sorokin, Y.M.: Radiophys. Quantum Electron. **15**, 36 (1972)
66. Kravtsov, Y.A., Ostrovskii, L.A., Stepanov, N.S.: Proc. IEEE **62**, 1492 (1974)
67. Belgiorno, F., Cacciatori, S.L., Clerici, M., Gorini, V., Ortenzi, G., Rizzi, L., Rubino, E., Sala, V.G., Faccio, D.: Phys. Rev. Lett. **105**, 203901 (2010)
68. Couairon, A., Mysyrowicz, A.: Phys. Rep. **441**(2–4), 47 (2007)
69. Unruh, W.G., Schützhold, R.: arXiv:1202.6492v3 (2012)
70. Liberati, S., Prain, A.S., Visser, M.: Phys. Rev. D **85**, 084014 (2012)
71. Dirac, P.: Sci. Am. **5**, 47 (1963)
72. Leonhardt, U., Philbin, T.: Geometry and Light. Dover, New York (2010)
73. Demircan, A., Amiranashvili, S., Steinmeyer, G.: Phys. Rev. Lett. **106**, 163901 (2011)
74. McCall, M., Favaro, A., Kinsler, P., Boardman, A.: J. Opt. **13**, 024003 (2011)
75. Ginis, V., Tassin, P., Craps, B., Veretennicoff, I.: Opt. Express **18**(5), 5350 (2010)
76. Corley, S., Jacobson, T.: Phys. Rev. D **59**, 124011 (1999)
77. Faccio, D., Arane, T., Leonhardt, U.: Class. Quantum Gravity **29**, 224009 (2012)

Chapter 12
An All-Optical Event Horizon in an Optical Analogue of a Laval Nozzle

Moshe Elazar, Shimshon Bar-Ad, Victor Fleurov, and Rolf Schilling

Abstract The formal analogy between the propagation of coherent light in a medium with Kerr nonlinearity and the flow of a dissipationless liquid is exploited in a demonstration of an all-optical event horizon in an optical analogue of the aeronautical Laval nozzle. This establishes a unique experimental platform, in which one can observe and study very unusual dynamics of classical and quantum fluctuations, and in particular an analogue of the Hawking radiation emitted by astrophysical black holes. We present a detailed theoretical analysis of these dynamics, and demonstrate experimentally the formation of such an event horizon in a suitably-shaped waveguide structure.

12.1 Introduction

Event horizons are well known in the context of astrophysics and cosmology. Less familiar is the analogy between an astrophysical event horizon and the sonic horizon in transonic fluid flow. The analogy is not restricted to the mere existence of the horizon, but also pertains to the prediction that a thermal spectrum of sound waves should be emitted from a sonic horizon, similarly to Hawking radiation. This is especially significant since Hawking radiation and the associated Black-hole evaporation are one of the most impressive phenomena at the intersection of general relativity and quantum field theory.

The quantum nature of the physical vacuum implies that a black hole, defined classically as an object from which even light cannot escape, in fact has characteristic temperature and entropy [1], and, moreover, emits thermal radiation [2]. The suggestion that a thermal spectrum of sound waves would also be emitted from a sonic horizon, or "Mach horizon" (the fluid counterpart of the horizon, where the fluid velocity equals the sound velocity) [3], was based on the observation that the

---

M. Elazar · S. Bar-Ad · V. Fleurov
Raymond and Beverly Sackler School of Physics and Astronomy, Tel-Aviv University, Tel-Aviv 69978, Israel

R. Schilling
Johannes Gutenberg University, Mainz, Germany

derivation of Hawking uses only the linear wave equation in curved spacetime, and not the Einstein equations. The same conditions for wave propagation arise when considering sound propagation in a fluid when the background flow is non-trivial, and in particular when the background flow is a stationary accelerating transonic flow.

An accelerating fluid may thus lend itself to laboratory experiments on the physics of black holes and black hole evaporation. Since direct experimentation with black holes is hardly possible, experiments using analogous phenomena in physical systems where the "high-energy" (short-wavelength) physics is known are a very exciting proposition. This sets the stage for an extensive activity aiming to create laboratory black hole analogues. A sonic horizon was investigated in several water tank experiments [4–6], and recently also in an experiment with a Bose-Einstein condensate (BEC) [7]. An optical analogue of Hawking radiation was investigated in filamentation experiments in glass [8], while a "white hole" horizon, involving total back-reflection of probe light from a moving soliton, was studied in optical fibers [9]. A rotating black hole experiment involving nonlinear optics was also proposed [10]. Similar ideas were developed for BEC [11–14], superfluid ^{3}He-A [15], degenerate Fermi liquids [16], SQUID array transmission lines [17], and media with singularities in the electric permittivity and magnetic permeability [18].

In what follows we describe in detail another approach to "analogue gravity"— an all-optical experiment based on laser light propagation in a distinctive nonlinear waveguide, which is analogous to a Laval nozzle [19]. The latter is an important apparatus for transonic acceleration of fluids, and is a well-known device in the context of aerodynamics: Typically, a high-pressure hot gas undergoes isentropic expansion through a convergent-divergent nozzle, accelerating above the local sound velocity (*i.e.* Mach $= 1$) at the nozzle throat (the point where the nozzle is the narrowest). Such a device was briefly mentioned by Unruh [3], and was later employed in water tank experiments, and discussed by others in the context of BEC [12, 20] and general isentropic fluids [21]. Using the same idea, extended to the realm of nonlinear optics, we explore an all-optical analogue of the Laval nozzle. As we show below, when a medium with a defocusing (*i.e.* repulsive) Kerr nonlinearity is confined within a properly shaped channel (*i.e.* transverse refractive index profile), incident light moving at a small angle relative to the longitudinal (z) axis of the channel may propagate transversally in a way that resembles the accelerating flow of an equivalent fluid, which may be called "luminous fluid". Thus the coordinate z plays the role of time, the transverse component of the wave-vector plays the role of velocity, the propagation of light is mapped to a flow with a finite transverse velocity, and a change of the propagation angle corresponds to acceleration of the fluid. This approach has proved to be an extremely powerful one when applied to the problem of coherent tunneling [22–27], and has also been used to model fluid flow around an obstacle [28] and dispersive shock waves in defocussing media [29–35]. In the optical analogue of the Laval nozzle an incoming "subsonic" (*i.e.* low-incidence) laser beam "accelerates" (*i.e.* changes its direction) while traversing the nozzle, reaching a critical velocity, which is equivalent to the sound velocity in a real fluid, at the nozzle throat, and exiting the nozzle at a "supersonic" velocity. Significantly, classical "sound" waves which may be excited at the throat, and manage to escape the nozzle on its subsonic side, but are washed away by the supersonic flow that exits the nozzle on the other side, show many similarities to the quantum Hawking radiation which is emitted from the event horizon of a black hole. This analogue of Hawking radiation has a unique optical signature which can be readily detected, and has an important advantage over other analogue gravity experiments in that supersonic "velocities" are very easily obtained. It thus presents a new and promising platform for analogue gravity experiments.

This chapter is organized as follows: We first present a theoretical analysis that shows the mapping from the nonlinear optical representation to its hydrodynamic counterpart, and the conditions under which an incoming plane wave is tilted while traversing a properly shaped channel, in analogy to transonic acceleration of a real fluid. We then describe an experimental realization of a horizon, which forms when a beam confined in a nonlinear channel escapes through a small slot cut along the side of the channel. Different regimes of operation are studied, corresponding to transonic and supersonic acceleration. Next we analyze classical and quantum fluctuations on the background transonic flow, in the vicinity of the horizon. We show that these fluctuations have a thermal spectrum with a characteristic temperature that is analogous to the Hawking temperature. Furthermore, similar to Hawking radiation, fluctuations that propagate against the background flow break into a part which is carried away with the supersonic flow (*i.e.* "falls" into the black hole), and another part, which penetrates into the subsonic region (*i.e.* "escapes" from the black hole). Finally we discuss constraints on the possibility of observing the classical fluctuations experimentally.

12.2 Transonic Flow of a Luminous Fluid

The propagation of weakly nonlinear coherent optical waves is described by the nonlinear Schrödinger (NLS) equation

$$i\frac{\partial A}{\partial z} = -\frac{1}{2\beta_0}\nabla^2 A + U(x, y)A + \lambda |A|^2 A. \tag{12.1}$$

It assumes the paraxial approximation, according to which the electric field of the light wave is written as $E = A(x, y, t; z)e^{-i\beta_0 z}$, where $A(x, y, t; z)$ is the weakly z dependent complex amplitude of the light propagating in the z direction. The time coordinate t is converted into the τ coordinate [36] which describes the shape of the wave packet in the moving coordinate system. Hence the Laplace operator in Eq. (12.1) is

$$\nabla^2 = \partial_x^2 + \partial_y^2 \pm \partial_\tau^2$$

with the $+$ sign for anomalous and $-$ sign for normal dispersion. $U(x, y)$ is the equivalent external potential created by spatial variation of the refraction index in the medium and does not depend on τ.

Now we briefly sketch the fundamental result [3] as applied to coherent light propagation in a medium with a Kerr nonlinearity. The Madelung transformation [37] $A = f e^{i\phi}$ (see also [10, 38, 39]) maps the NLS Eq. (12.1) onto two equations,

$$\partial_z \rho + \nabla \cdot [\rho \boldsymbol{v}] = 0, \tag{12.2}$$

$$\partial_z \boldsymbol{v} + \frac{1}{2}\nabla \boldsymbol{v}^2 = -\frac{1}{\beta_0}\nabla(V_{qu} + U + \lambda\rho) \tag{12.3}$$

for an equivalent fluid with density $\rho = f^2$ and velocity $\beta_0 \boldsymbol{v} = -\nabla \phi$. For an incident continuous wave the coordinate τ is redundant. The term $V_{qu} = -\frac{1}{2\beta_0}\frac{\nabla^2 f}{f}$ is reflects the dispersion, especially at short wave length, and is analogous to the so called "quantum potential" (QP) which appears in the corresponding quantum mechanical problem. Here the problem is classical, but we may conditionally call this term QP with $\hbar = 1$. The light wave number β_0 plays the role of a "particle mass", and the velocity $\boldsymbol{v}$ is dimensionless.

The simplest way to analyze Eqs. (12.2) and (12.3) is first to neglect the QP and then to linearize these equations with respect to small fluctuations $\rho - \rho_0 = \rho_0 \psi$ and $\phi - \phi_0 = \varphi$ around a steady solution $\rho_0(x, y)$ and $\varphi_0(x, y)$. This results in a Klein-Gordon equation

$$(-g)^{-1/2}\partial_\mu(-g)^{1/2}g^{\mu\nu}\partial_\nu \varphi = 0, \tag{12.4}$$

in a space whose curvature is determined by the metric $g^{\mu\nu}$ with the infinitesimal interval

$$d\sigma^2 = g_{\mu\nu}dx^\mu dx^\nu$$

$$= \sqrt{\frac{\beta_0}{\lambda\rho_0}}\left[d\boldsymbol{r}^2 - 2dz\boldsymbol{v}_0 \cdot d\boldsymbol{r} - \left(\frac{\lambda\rho_0}{\beta_0} - v_0^2\right)dz^2\right] \tag{12.5}$$

where $g^{\mu\mu'}g_{\mu'\nu} = \delta^\mu_\nu$ and g is the determinant of the metric. Ordinary "sound waves" in the effective luminous fluid arise as solutions of Eq. (12.4) around the equilibrium solution $\rho_0 = const$, $\phi_0 = const$ and $\boldsymbol{v}_0 = 0$, that exists at $U = 0$, so that $\lambda\rho_0/\beta_0 = s^2$ is the squared sound velocity. The nonlinearity coefficient must be positive, $\lambda > 0$, otherwise the "sound velocity" becomes imaginary and various instabilities, such as collapsing solitons [40] arise. Equation (12.4) may include corrections due to the QP, which may be of importance for short waves. One can readily estimate the scale at which these corrections can become important by comparing the first and third terms on the RHS of Eq. (12.1). This gives us the characteristic length $l_{nl}^{-2} = \lambda\rho_0\beta_0$, which we call here the nonlinearity length. It corresponds to the well-know healing length in BEC.

Equation (12.5) is analogous to the equation proposed by Unruh [3] for a description of fluctuations on the background of a transonic flow in an isentropic fluid. Therefore the conclusions made in Ref. [3] can be applied to our optical system as well. In particular, an effect that is analogous to Hawking radiation from the vicinity of the horizon of a black hole should, in principle, also be observed here. However,

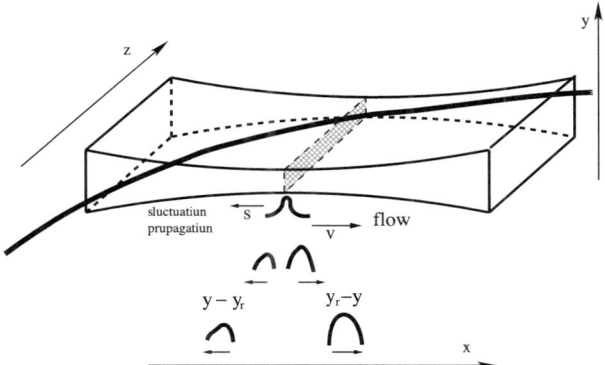

Fig. 12.1 A cartoon of a laser beam propagating inside the optical Laval nozzle. The beam, with a small initial angle relative to the z axis, bends more towards the x direction as it propagates along the z direction ("time"). The bend in the cartoon is strongly exaggerated. A fluctuation is schematically shown near the Mach horizon as it is cut in two parts which propagate in opposite directions

the approximation we have made, neglecting the QP, may introduce significant limitations on the observability of the effect. Such limitations due to the contribution of the QP will be discussed later on. First we will discuss, within the framework of this approximation, a possible scenario under which transonic acceleration of such a "luminous liquid" may be observed.

A conceivable way to create such a transonic flow is to use a Laval nozzle [41, 42], *i.e.* a vessel with a variable cross-section $S(x)$, whose application to other condensed-matter analogues was discussed in Refs. [12, 21]. The flow velocity initially increases with decreasing $S(x)$, until it reaches the sound velocity at the narrowest part of the vessel, called the throat. Further acceleration of the supersonic fluid is obtained by increasing $S(x)$. Here we analyze the transonic flow for the optical analogue of the Laval nozzle shown in Fig. 12.1. The incident laser beam, nearly parallel to the z axis, is tilted in the x direction at an angle $\arctan(k_x/\beta_0)$, *i.e.* the initial phase of the complex electric field amplitude $A(x, y)$ is $\varphi = k_x x$. k_x/β_0 plays the role of "flow velocity" in the x direction. Acceleration of the flow within the channel of variable width (along the y-axis) corresponds to bending of the beam (*i.e.* an increase of k_x), as shown in Fig. 12.1. It also leads to small corrections to β_0 that are quadratic in $k_x/\beta_0 \ll 1$ (We assume that the wavelength $1/\beta_0$ is smaller than all the other relevant scales.)

Reference [19] considered such a flow, bounded by walls of hyperbolic shape

$$\frac{y^2}{a^2} - \frac{x^2}{b^2} = 1 \tag{12.6}$$

with sliding boundary conditions. This type of description will certainly work in the vicinity of the throat, even if the vessel profile has a general form, not necessarily exactly hyperbolic.

It follows from the symmetry of the system that there is a streamline along the straight line that separates the two branches of the hyperbola at $y = 0$. The x dependence of the flow velocity along this line has the form

$$v_0(x) = \bar{s}(1 + \alpha x), \qquad \rho_0 = \bar{\rho}(1 - \alpha x) \qquad (12.7)$$

where overlined quantities refer to the throat at $x = 0$. The coefficient α characterizes the spatial acceleration of the flow in the vicinity of the throat, and the particular value $\alpha = 1/(\sqrt{3}c)$, with $c = \sqrt{a^2 + b^2}$, is obtained for the hyperbolic shape of the throat discussed above. Since this shape is generic, especially close to the throat, we do not expect a strong variation of this coefficient. The above two equations satisfy the 1D continuity equation $\partial_x(\rho_0(x)v_0(x)) = 0$ up to terms of order $O(\alpha x)$. The same equations can be used for the curved streamlines lying close to the symmetry line ($y = 0$) if we neglect small corrections of order $O((\alpha y)^2)$. The sound velocity of the luminous liquid is

$$\lambda \rho(x) = \beta_0 s^2(x),$$

and the corresponding values of these quantities at the throat are related as $\lambda \bar{\rho}_0 = \beta_0 \bar{s}^2$.

12.3 An Experimental Horizon

We study the flow through an optical Laval nozzle experimentally by launching a continuous wave laser beam into an appropriately shaped waveguide with reflective walls, filled with a Kerr-type defocusing nonlinear material. The experimental challenge here is to create conditions of steady flow with a subsonic input velocity. Since the velocity is k_x/β_0 and the sound velocity is $s^2 = \lambda \rho/\beta_0$, the above input conditions imply a small input angle of the beam and a high nonlinearity and/or input intensity. (Note that, for a given angle, low and high intensities correspond to supersonic and subsonic flow conditions, respectively, which is somewhat unintuitive.) However, an unavoidable consequence of these conditions is a strong self-defocusing of the beam, and as a result the wave packet which traverses the nozzle is an expanding "droplet" of liquid, with a tendency of the power density in the cavity to decrease with increasing input power. Furthermore, while the peak intensity of the droplet may correspond to subsonic flow, it is always surrounded by supersonic flow (in contrast to the usual case in hydrodynamics), and when confined to a Laval nozzle such as the one described in Fig. 12.1, the fluid flows from the throat towards both sides of the nozzle. It is thus impossible to generate the steady sonic background flow conditions stipulated by the theory in a simple waveguide with a convergent-divergent cross-section formed by two convex walls, as in Fig. 12.1. To circumvent this problem we use an alternative waveguide design, based on a light-pipe of circular cross-section drilled in an aluminum block, with a groove of triangular cross-section, cut along the side of the channel, acting as the divergent section of the nozzle (see Fig. 12.2). The total length of each light pipe is

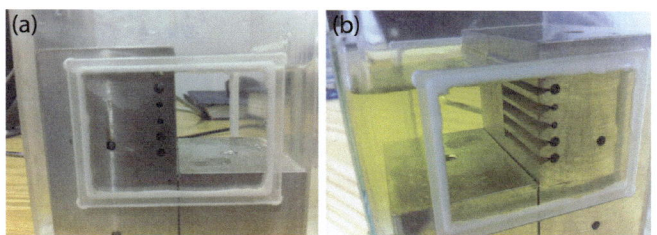

Fig. 12.2 Images of the waveguide structures used in the experiment. Panel (**a**) shows the input plane, with six circular holes that are the input ports of six lightpipes of different dimensions. Panel (**b**) shows the exit plane of the same waveguide structure, and the grooves of triangular cross-section, cut half-way along the sides of the channels, and forming the divergent sections of the nozzles

$L = 67$ mm, and the groove extends over the second half of this length. This design is intended to "trap" the expanding beam and confine it in a homogeneous, high-density and low-velocity mode, thus preparing it for ejection through the groove, and is akin to the configuration of a rocket engine: a high-pressure gas is first loaded into a combustion chamber, and is then expelled through the nozzle. The aluminum block, with several nozzles of different diameters, aperture sizes and groove opening angles, is enclosed in a plexiglass cell with glass windows, which is sealed and filled with iodine-doped ethanol. The nonlinear index variations result from optical absorption by the iodine, which in turn leads to thermally-induced changes of the index of refraction—a non-local nonlinearity which slightly washes out the thermal gradients [32]. The nonlinearity $\lambda \rho$ can be expressed, in terms of the nonlinearly-induced refractive index change δn, as $\delta n \beta_0 / n_0$, where n_0 is the linear refractive index of the material [32]. The corresponding dimensionless sound velocity is then $s^2 = \delta n / n_0$, meaning that the input beam is subsonic for $k_x / \beta_0 < \sqrt{\delta n / n_0}$. We use a continuous-wave frequency-doubled YAG laser (532 nm), and focus the beam to a ∼0.5 mm waist at the input of a waveguide. The input power is varied by means of the laser controller, in order to avoid thermal effects in variable-density filters. Images of the exit plane of the waveguide are recorded by means of a CCD camera. In all cases images were acquired after stabilization of the thermal gradients. The images we present are unprocessed except for background subtraction.

Figure 12.3 presents images of the exit plane of two of the waveguides, and the corresponding power density cross-sections along the nozzle axis, for an input power (2 Watts) that is sufficiently high to completely fill the waveguides (at an iodine concentration of ∼40 ppm). Figure 12.3(a) shows a 2 mm diameter waveguide, and Fig. 12.3(c) shows a 3 mm diameter waveguide, both having a ∼0.5 mm opening, (*i.e.* nozzle throat). Figures 12.3(b) and 12.3(d) are the power density cross-sections corresponding to Figs. 12.3(a) and 12.3(c), respectively, obtained by summation over 12 CCD lines at the center of each nozzle. Figure 12.3(e) shows the free expansion of the beam when it propagates outside the waveguide structure. Figure 12.3(f) shows a reference image of the 3 mm diameter waveguide, obtained with incoherent light and with the laser beam blocked. Figures 12.3(a)–(d)

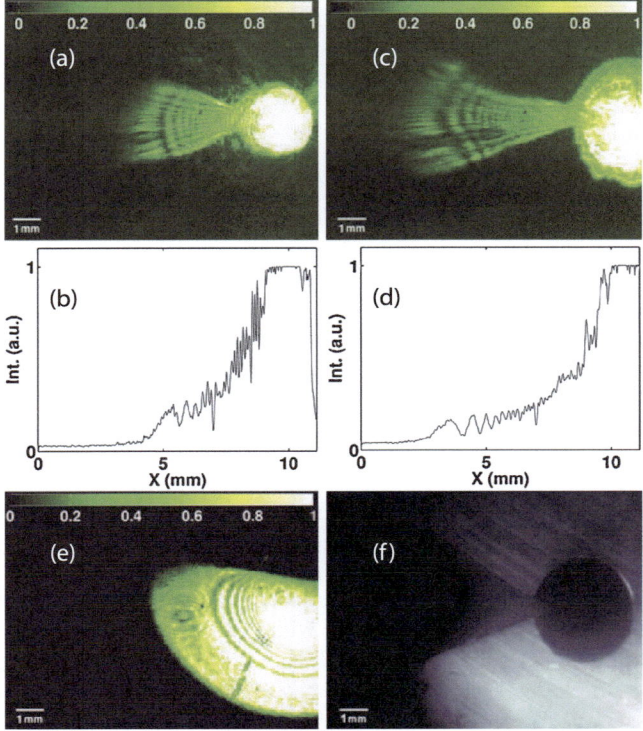

Fig. 12.3 Images of the exit plane of the waveguides and corresponding power density cross-sections along the nozzle axis, for a 2 Watt input power and an iodine concentration of ∼40 ppm. Panels (**a**) and (**b**) show data for a 2 mm diameter waveguide, panels (**c**) and (**d**) show data for a 3 mm diameter waveguide, and panel (**e**) shows the free expansion of the beam outside the waveguide structure. Panel (**f**) is a reference image of the 3 mm diameter waveguide

clearly show the jets of luminous liquid ejected from the nozzles as the beam propagates through the waveguides. Note that the jets extend farther than the edge of the beam undergoing free expansion (Fig. 12.3(e); A detailed analysis is presented in Fig. 12.4). Furthermore, there is a sharp drop of the density as the jet exits the nozzles, which is clearly seen in the images and in the power density cross-sections. This demonstrates that the luminous liquid is accelerating at the nozzle throat rather than gradually expanding through the opening. Finally, while the confined beam propagates along the waveguide walls at a very slow (*i.e.* subsonic) velocity, the following analysis shows that the jet of luminous liquid is indeed supersonic: The dimensionless velocity of the jet outside the waveguide is first calculated from its extension in the transverse direction, deduced from the images. The relation is simply $v = \Delta x/\Delta z = 2\Delta x/L \sim 0.1$, where Δx is the transverse distance from the nozzle throat to the edge of the jet, and $\Delta z = L/2$ is the distance along the z axis that the same part of the jet has propagated by the time it reached the exit plane. This velocity should be compared to the local sound velocity, which can be estimated

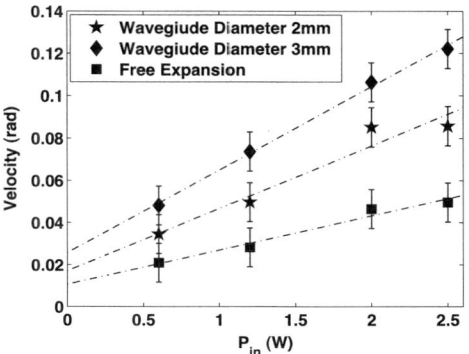

Fig. 12.4 The measured jet velocities as a function of input power for an iodine concentration of ~40 ppm. Data are shown for two waveguides and for free expansion of the beam outside the waveguide structure. The calculation of velocities is explained in the text. The *lines* are guides to the eye

by analyzing the light intensity distribution in the exit plane and the rate of expansion of the freely-expanding (*i.e.* self-defocusing) beam. The latter, deduced from Fig. 12.3(e), allows us to calculate $\lambda\rho_0$ and the corresponding sound velocity at the input. The former in turn allows us to deduce the sound velocity which corresponds to the lower density of the jet, taking into account the expansion of the beam in the light-pipe, the relative intensities of the jet and inside the light-pipe, and measured losses. This calculation gives a local sound velocity in the jet on the order of 1×10^{-3} or less, meaning that the local Mach number is >100. This clearly establishes that the luminous liquid undergoes transonic acceleration and forms a "sonic" horizon as it expands through the nozzle.

Figure 12.4 shows the dimensionless velocities ($v = 2\Delta x/L$) of the jets emanating from the 3 mm and 2 mm nozzles, as a function of the input intensity, for a fixed iodine concentration (~40 ppm). Also shown is the velocity at the envelope of the freely-expanding beam, which we estimate as $dx/dz \approx \Delta x'/L$. (In this case we measure $\Delta x'$ from the center of the beam, which we determine from low-intensity measurements; This velocity corresponds to the asymptotic expansion angle, obtained for $L \gg 1/\lambda\rho_0$, *i.e.* when the propagation distance is much longer than the characteristic defocusing distance, and is a reasonable estimate for intensities >1 Watt.) Figure 12.4 clearly shows that the velocity of the jet is higher than that of the freely-expanding beam throughout the experimental intensity range, in spite of the fact that the initial conditions for the free expansion involve higher pressures (the equivalent of pressure in a luminous liquid is $\frac{1}{2}\lambda\rho^2$). On the other hand, Fig. 12.4 shows that the jet emanating from the 2 mm waveguide is slower than that ejected from the 3 mm nozzle, although, for a given input intensity, the pressure in the latter is supposed to be lower. This discrepancy can result from the fact that the opening in the 2 mm waveguide subjects a larger angle, resulting in a less directional jet (compare Figs. 12.3(a) and (c)). We also measure higher losses (due to scattering and absorption) in the smaller waveguide, so in fact the power densities in the two waveguides are comparable. Finally, the non-locality of the nonlinearity may have a stronger effect on the 2 mm waveguide.

Measurements at lower nonlinearities illustrate another regime of operation of the nozzles. Figure 12.5 shows a set of images of the exit plane of the test cell for

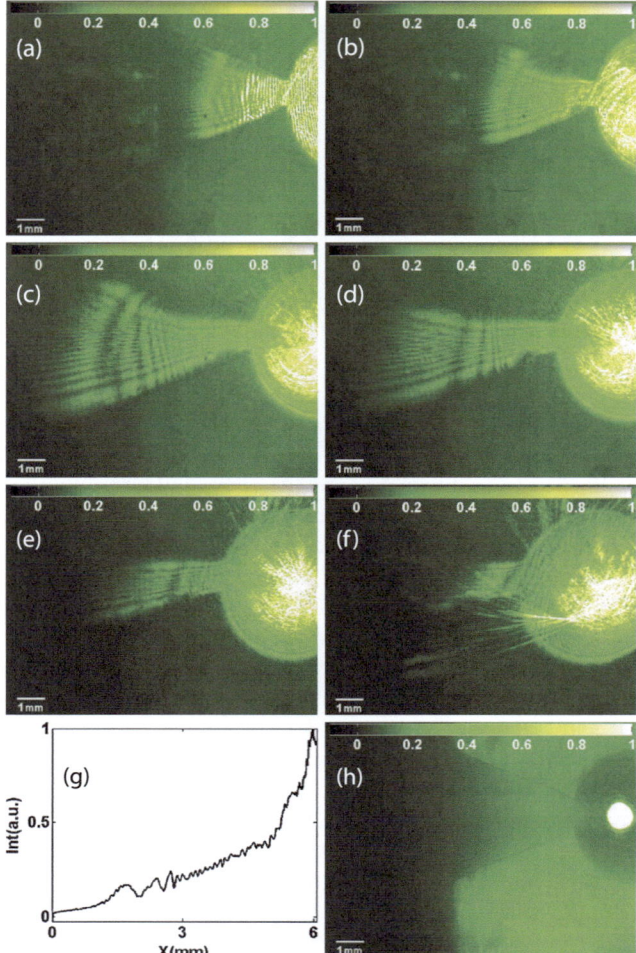

Fig. 12.5 Acceleration of supersonic input beams through a 3 mm nozzle. The input power is 1.5 Watts, and the iodine concentration is ∼20 ppm. Under these conditions the defocusing is not sufficiently strong to completely fill the waveguide. Panels (**a**)–(**f**) show images of the exit plane of the waveguide for six different positions of the nozzle throat, as the nozzle is moved, along its axis, relative to the fixed beam position. The distances between the beam axis and the throat are (**a**) 0.33 mm, (**b**) 0.54 mm, (**c**) 1.02 mm, (**d**) 1.25 mm, (**e**) 1.59 mm, and (**f**) 1.94 mm. Panel (**g**) shows the power density cross-section, along the nozzle axis, in image (**c**). Panel (**h**) is an image of a low intensity beam, used for calibration of the throat—beam axis distance

self-defocusing (*i.e.* beam expansion) that is not sufficiently strong to completely fill the 3 mm waveguide. (The images were obtained with an input power of 1.5 Watts and an iodine concentration of only ∼20 ppm.) Six images, Figs. 12.5(a)–(f) were acquired at six different positions of the nozzle throat, as the nozzle was moved,

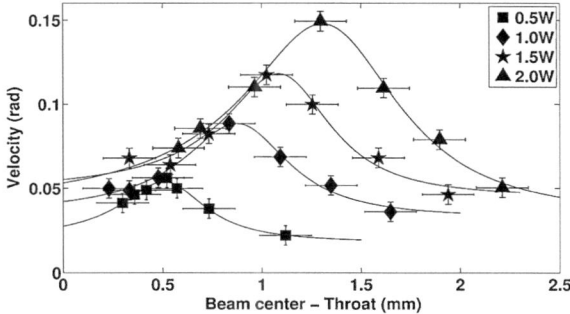

Fig. 12.6 A plot of the jet velocity as a function of the distance between the beam axis and the 3 mm nozzle throat, at four input powers, with an iodine concentration of ∼20 ppm. Under these conditions the defocusing is not sufficiently strong to completely fill the waveguide. The calculation of velocities is explained in the text, and the *curves* are guides to the eye

along its axis, relative to the fixed beam position. Figure 12.5(g) is a power density cross-section, along the nozzle axis, of Fig. 12.5(c), and Fig. 12.5(h) is an image of a low intensity beam, used for calibration of the distance between the beam axis and the nozzle throat. It is evident from the images that the extension of the jet, which is proportional to the jet velocity, strongly depends on the displacement of the nozzle throat. Figure 12.6 summarizes the dependence of the jet velocity on the displacement and the input power. Note that as the input power is increased the optimum acceleration is obtained when the beam axis is moved farther away from the throat (at an input power of 2 Watts the beam axis is then close to the center of the waveguide). A comparison of the data in Fig. 12.6 with the divergences of the freely-expanding beams, measured separately for the same input powers, shows that the optimum acceleration at the nozzle is obtained when the envelope of the freely-expanding beam coincides with the nozzle throat halfway through the waveguide (*i.e.* at $z = L/2$). Under these conditions the luminous liquid entering the nozzle is already moving at a supersonic velocity, and it is accelerated further in the divergent section of the nozzle. The smooth power density cross-section shown in Fig. 12.5(g) supports this interpretation (compare this to the sharp density gradients at the throat in Fig. 12.3). In this case the nozzle operates in a regime that is not typical of a Laval nozzle, though.

12.4 Fluctuations

12.4.1 Classical Straddled Fluctuations

Now we will discuss the dynamics of fluctuations on the background of the transonically accelerating flow of luminous liquid. This flow is characterized by the density and velocity spatial distributions $\rho_0(x) = |A_0(x)|^2$ and $v_0(x) = -\frac{1}{\beta_0}\partial_x\varphi_0(x)$. We also limit ourselves to the $1+1$ case described by the propagation length z ("time") and the perpendicular coordinate x. The equations of motion for the fluctuations are straightforward deduced by linearizing Eqs. (12.2) and (12.3) with respect to the stationary transonic solution $A_0(x) = f_0 e^{-i\varphi_0(x)}$. One obtains

$$(\partial_z + v_0 \partial_x)\chi - \frac{1}{\beta_0} \frac{1}{f_0^2} \partial_x (f_0^2 \partial_x \xi) = 0, \tag{12.8}$$

$$(\partial_z + v_0 \partial_x)\xi + \frac{1}{4\beta_0} \frac{1}{f_0^2} \partial_x (f_0^2 \partial_x \chi) - \lambda f_0^2 \chi = 0 \tag{12.9}$$

where the quantities

$$\chi = \frac{1}{f_0}\left[e^{-i\varphi_0}\psi^\dagger + e^{i\varphi_0}\psi\right],$$
$$\xi = \frac{1}{2if_0}\left[e^{-i\varphi_0}\psi^\dagger - e^{i\varphi_0}\psi\right] \tag{12.10}$$

are related to the density and velocity fluctuations by

$$\delta\mathbf{v}(x,z) = -\frac{1}{\beta_0}\partial_x \xi(x,z),$$
$$\delta\rho(x,z) = \rho_0(x)\chi(x,z) \tag{12.11}$$

with $\rho_0 = f_0^2$.

A straightforward way to analyze Eqs. (12.8) and (12.9) is to neglect the second term in Eq. (12.9) (*i.e.* the QP), solve the resulting equation with respect to χ and substitute it into Eq. (12.8). That is how we arrive at the Klein-Gordon equation (12.4). Assuming in addition that the flow acceleration is linear near the horizon, i.e. $\rho_0 = \bar{\rho}(1 - \alpha x)$ and $v_0(x) = \bar{s}(1 + \alpha x)$, one obtains, up to linear terms in x,

$$\{\partial_z^2 + 2\bar{s}(1+\alpha x)\partial_z \partial_x + \alpha \bar{s}\partial_z + 3\alpha \bar{s}^2 \partial_x (x\partial_x)\}\xi = 0. \tag{12.12}$$

As shown above the spatial acceleration is $\alpha = 1/(c\sqrt{3})$ in the particular case of a hyperbolic profile of the nozzle throat.

This equation has two sets of eigenfunctions, which at $|x| \to 0$ behave like

$$\xi_1 \propto e^{\gamma_0 \ln(-x) - ivz}, \qquad \xi_2 \propto e^{ik_2(v)x - ivz} \tag{12.13}$$

where

$$\gamma_0 = i\frac{2v}{3\bar{s}\alpha}, \qquad k_2(v) = \frac{v}{\bar{s}}\frac{v+i\bar{s}\alpha}{2v - 3i\bar{s}\alpha}.$$

The correspondingly eigenfunctions for the density fluctuations are

$$\chi_1 \propto e^{(\gamma_0-1)\ln(-x)-ivz}, \qquad \chi_2 \propto e^{ik_2(v)x-ivz} \tag{12.14}$$

These two solutions [19, 20, 43, 44] are the propagating plane waves in the Bogoliubov spectrum, strongly distorted by the accelerating background flow. The solution ξ_1 corresponds to the excitation "attempting" to propagate against the flow in the vicinity of the Mach horizon, which cuts it into two parts moving in opposite directions away from the horizon. The solution ξ_2 is a fluctuation propagating with the flow, and therefore has a shape quite similar to that of a plane wave. In the high frequency limit this mode propagates at twice the sound velocity.

In contrast to ξ_2 and χ_2, the solutions ξ_1 and χ_1 are singular since they possess a branching point at $x = 0$. This singular behavior is typical of the eigenfunctions of the Klein-Gordon equation in the vicinity of the horizon of a real black hole (see e.g. Ref. [45] and references therein) and leads to the celebrated Hawking radiation. Regularization in this case requires knowledge of the physics at the Planck length scale (see e.g. Ref. [46]). In the luminous liquid the part of Planck length is played by the nonlinearity length l_{nl} (which is equivalent to the healing length in BEC). In order to regularize χ_1 (and also ξ_1) near the horizon we have to take into account the QP in Eq. (12.9) (see the next subsection).

As an illustration of this special behavior of the fluctuations near the Mach horizon, formed in the throat of the hyperbolic Laval nozzle, we consider a "straddled" fluctuation

$$f(x,z) = \int dv g(v) e^{-ivz} \xi_1(x) = \Phi(x,z) G(x,z) H(x) \qquad (12.15)$$

with the Gaussian spectral density

$$g(v) = \frac{1}{\Gamma\sqrt{2\pi}} \exp\left\{-\frac{(v-v_0)^2}{2\Gamma^2}\right\}$$

of width Γ around a positive "frequency" v_0. Such a fluctuation is composed exclusively of the "left moving" normal modes $\xi_1(x)$, singular at $x = 0$. Any "right moving" normal mode $\xi_2(x)$ would escape the vicinity of the horizon at twice the sound velocity and can therefore be neglected. The "time" evolution (i.e. z dependence) of such a wave packet is determined by the envelope function

$$G(x,z) = \exp\left\{-\frac{\Gamma^2}{2}\left[\frac{2c}{\sqrt{3}\bar{s}}\ln|x| - z\right]^2\right\}.$$

The factor

$$H(x) = \exp\left\{\left[\frac{2\pi c v_0}{\sqrt{3}\bar{s}} + \frac{2\pi^2 \Gamma^2 c^2}{3\bar{s}^2}\right]\theta(x)\right\}$$

($\theta(x)$ is the Heaviside step function) is equal to one at $x < 0$ (subsonic region) and is larger than one at $x > 0$ (supersonic region). It appears due to the branching point of the function $\ln(-x)$ at $x = 0$.

The phase factor

$$\Phi(x,z) = \exp\left\{i\left[\frac{2c}{\sqrt{3}\bar{s}}\ln|x| - z\right]\left[v_0 + \pi\frac{2\Gamma^2 c}{\sqrt{3}\bar{s}}\theta(x)\right]\right\}$$

in (12.15) indicates that the "wave vector" is now proportional to $\ln|x|$, i.e. diverges at $x \to 0$. It can be interpreted as the wave length tending to zero in this limit, which corresponds to a "blue shift" when approaching the horizon. The "frequencies" of these oscillations on both sides of the horizon strongly vary with $|x|$. This is reminiscent of "time slowing" near the horizon of a real black hole.

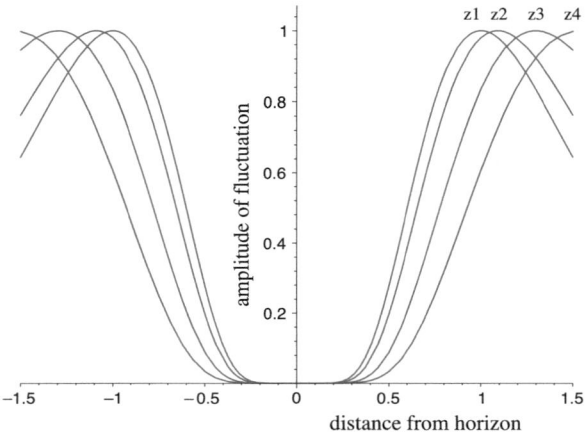

Fig. 12.7 Propagation of a straddled wave packet to the left and right of the Mach horizon. The normalized propagation distances are $z_1 = 0$, $z_2 = 0.1$, $z_3 = 0.3$, $z_4 = 0.5$, and $\Gamma = 2$

Figure 12.7 shows the dynamics of the envelope function $G(x, z)$. Here the coordinate x is measured in units c whereas the propagation distance z is measured in units $c/\bar{s}$. Correspondingly the units $\bar{s}/c$ are used for ν and Γ. The figure shows how the maxima of the fluctuation amplitude propagate on both sides of the Mach horizon, with zero amplitude on the horizon. The part residing to the right side of the horizon (the supersonic region, $x > 0$) should be multiplied by the factor $H(x) > 1$ at $x > 0$ (not shown in Fig. 12.7). Although the discussion here is classical the parameter $2\pi c v_0/(\sqrt{3}\bar{s}) = \hbar v_0/2T_H$ appearing in the function $H(x)$ is directly related to the Hawking temperature T_H characterizing the spectrum of the spontaneous radiation from the horizon in the quantum problem. It will be discussed below. Therefore the ratio of amplitudes of a classical fluctuation on the left and the right of the horizon directly measures the Hawking temperature.

12.4.2 Regularization of Fluctuations Near the Mach Horizon

Here we will discuss the behavior of the fluctuations in the nearest vicinity of the sonic horizon. The eigenfunctions (12.13) and (12.14) are obtained neglecting the QP contribution (the second term in Eq. (12.9)). One can readily see, by substituting the singular eigenfunction (12.14) that the neglected term in fact diverges at $|x| \to 0$, meaning that there is always a narrow region close to the horizon where the above approximation does not hold. The QP introduces dispersion effects at high enough frequencies corresponding to small length scales. The extent to which short range dispersion influences Hawking radiation is a question that has been extensively dealt with over the past two decades. In order to deduce quantitative results, *e.g.* for the flux and the spectrum, the mode equations based on specific dispersion laws were solved numerically [47–51] and analytically [14, 20, 52–58]. Some of these works use the framework of general relativity [47, 48, 52–55] and the others consider analogue gravity [20, 49–51, 56–58]. They provide clear evidence that Hawking radiation is basically a low frequency phenomenon and is robust, although counter examples were presented [57] (see Ref. [59] for more details).

The equations for fluctuations with full account of the QP contribution are analyzed as follows. We apply the same assumptions concerning the coordinate dependence of the velocity and density near the Laval nozzle throat as in the previous subsection. The Laplace representation

$$\chi(x,z) = e^{-i\nu z} \int_C dk \chi_k e^{ikx} \tag{12.16}$$

is used where the contour C is chosen in such a way that the integral converges [60]. Here and below we omit the subscript x when denoting the wave vector of the fluctuations in the x direction. ξ_k is defined similarly to (12.16). Then Eqs. (12.8) and (12.9) become

$$(-i\nu + i\bar{s}k)\chi_k - \alpha\bar{s}\partial_k(k\chi_k) + i\alpha k \frac{1}{\beta_0}\xi_k + \frac{1}{\beta_0}k^2\xi_k = 0, \tag{12.17}$$

$$-\frac{1}{4\beta_0}\left[i\alpha k + k^2\right]\chi_k - \beta_0\bar{s}^2(\chi_k - i\alpha\partial_k\chi_k) + (-i\nu + i\bar{s}k)\xi_k - \alpha\bar{s}\partial_k(k\xi_k) = 0 \tag{12.18}$$

where the terms $O(\alpha^2/k^2)$ have been omitted. This approximation is equivalent to considering a region $\alpha|x| \ll 1$ near the horizon. We can now explicitly solve Eq. (12.17) with respect to ξ_k and substitute the result into Eq. (12.18). We then get the first order differential equation

$$\partial_k \ln \chi_k = \frac{1}{i\alpha k(2\bar{\nu} - 3k - i\alpha)}\left[-\frac{l_h^2}{2}(i\alpha k + k^2)^2 - k^2 + (\bar{\nu} - k)^2\right] \tag{12.19}$$

for the function χ_k, where $\bar{\nu} = \nu/\bar{s}$. It can be readily solved to within a term independent of k,

$$\ln \chi_k = \gamma_1 \ln(k) + \gamma_2 \ln\left(k - \frac{2}{3}\bar{\nu} - \frac{i}{3}\alpha\right) + \Lambda(k, \bar{\nu}) \tag{12.20}$$

where

$$\gamma_1 = \frac{1}{4} - \frac{i\bar{\nu}}{2\alpha}, \quad \gamma_2 = -\frac{1}{4} - i\frac{1}{6\alpha}\bar{\nu} - \frac{4i}{81\alpha}l_{nl}^2\bar{\nu}^3 + \frac{14}{81}l_{nl}^2\bar{\nu}^2$$

and

$$\Lambda(k,\bar{\nu}) = \frac{l_{nl}^2}{\alpha}\left\{-\frac{i}{18}k^3 + \frac{5}{36}\alpha k^2 - \frac{i}{18}\bar{\nu}k^2 - \frac{2i}{27}\bar{\nu}^2 k + \frac{4}{27}\bar{\nu}\alpha k\right\}. \tag{12.21}$$

This means that now we know the Laplace transforms of the eigenfunctions and what is left is to carry out the back Laplace transform (12.16) in order to have the eigenfunctions in real space. This procedure is described in detail in Refs. [43, 44], where a discussion of the relevant literature is also presented. Using the saddle point integration we can see that eigenfunctions (12.13) emerge if we neglect the $\Lambda(k, \nu)$ term, *i.e.* at $l_{nl} \to 0$, which is equivalent to neglecting the QP. However, the singular eigenfunctions now have a somewhat different form, namely

$$\chi_s \propto x^{\gamma-1}, \tag{12.22}$$

where the parameter

$$\gamma = -\gamma_1 - \gamma_2 = \frac{2i\nu}{3\alpha\bar{s}} + \frac{4i}{81}\frac{l_{nl}^2\nu^3}{\alpha\bar{s}^3} - \frac{2}{27}\frac{l_{nl}^2\nu^2}{\bar{s}^2} \tag{12.23}$$

differs from the parameter γ_0 in Eqs. (12.13) and (12.14) for ξ_1 and χ_1, respectively, since it contains small corrections due to the QP. That is why the notation χ_s is used instead of χ_1. This type of singular behavior of the eigenfunction typically occurs near the horizon and is a crucial ingredient in the formation of Hawking radiation (see *e.g.* Refs. [45, 61]).

Accounting for the $\Lambda(k, \nu)$ term is necessary when considering large wave vectors, $l_{nl}^2 k^3/\alpha \gg 1$, *i.e.* for $|x| \ll l_r = l_{nl}/(\alpha l_{nl})^{1/3}$ [43, 44]. At such small distances from the horizon the QP contribution leads to a regularization of the singular eigenfunction (12.22). On the other hand the whole derivation is carried out under the assumption that $|x| \ll \min\{1/\alpha, 1/\nu\}$. Since the singular eigenfunction (12.22) is important for the formation of Hawking radiation, we always need a spatial window $l_r \ll |x| \ll \min\{1/\alpha, 1/\nu\}$ in order to have a chance to observe the radiation effect. It is also interesting that the authors of Ref. [51] found *empirically* a length scale d_ξ with the same dependence on α and l_h as that of l_r (cf. their Eq. (24), where κ_K and $1/\Lambda$ correspond to our α and l_h, respectively). However, it is not clear whether l_r and d_ξ are identical.

12.4.3 Quantization and the Hawking Temperature

The analysis carried out so far follows from the NLS equation (12.1) which is deduced from the classical Maxwell equations for an electromagnetic wave propagating in a nonlinear Kerr medium in the paraxial approximation. It means that the above analysis is essentially classical. Although the results are indicative of how the quantum Hawking radiation will emerge, the issue of quantization of the fluctuations still needs to be addressed.

We start with a discussion of Eqs. (12.8) and (12.9), which can be represented as Euler-Lagrange equations corresponding to the Lagrangian density

$$\mathscr{L} = \frac{1}{2}f_0^2(\chi\partial_z\xi - \xi\partial_z\chi) + \frac{1}{2}f_0^2 v_0(\chi\partial_x\xi - \xi\partial_x\chi) - \frac{1}{2}\lambda f_0^4 \chi^2$$
$$- \frac{1}{2\beta_0}f_0^2(\partial_x\xi)^2 - \frac{1}{8\beta_0\partial_x}f_0^2(\partial_x\chi)^2 \tag{12.24}$$

(see also the discussion in the Appendix of Ref. [50]). This Lagrangian density allows us to define the canonical momenta

$$p_\chi = \frac{\delta}{\delta \partial_z \chi}\left[\mathscr{L} - \frac{1}{2}\partial_z(f_0\chi\xi)\right] = -f_0^2\xi, \qquad (12.25)$$

$$p_\xi = \frac{\delta}{\delta \partial_z \xi}\left[\mathscr{L} + \frac{1}{2}\partial_z(f_0\chi\xi)\right] = f_0^2\chi \qquad (12.26)$$

conjugate to each of the fields χ and ξ. Together they form two canonical pairs, each of which can be used to derive the Hamiltonian density

$$\begin{aligned}\mathscr{H} &= -f_0\partial_z\chi\xi - \left[\mathscr{L} - \frac{1}{2}\partial_z(f_0\chi\xi)\right] = f_0\chi\partial_z\xi - \left[\mathscr{L} + \frac{1}{2}\partial_z(f_0\chi\xi)\right]\\ &= -\frac{1}{2}f_0^2 v_0(\chi\partial_x\xi - \xi\partial_x\chi) + \frac{1}{2}\lambda f_0^4\chi^2 + \frac{1}{2\beta_0}f_0^2(\partial_x\xi)^2 + \frac{1}{8\beta_0}f_0^2(\partial_x\chi)^2.\end{aligned}$$
$$(12.27)$$

While the derivation of Eq. (12.27) depends on the choice of the canonical pair, the final Hamiltonian does not depend on it. Equations (12.8) and (12.9) are now just the pair of Hamilton equations of motion for the Hamiltonian density (12.27). The manipulations of adding or subtracting a full derivative indicate that the variations of the Lagrangian with respect to $\partial_z\chi$ and $\partial_z\xi$ are not uniquely defined. However, this feature of the Lagrangian (12.24) does not affect the Euler-Lagrange equations (12.8) and (12.9) but is important when defining the canonical momentum and Hamiltonian. It means that we have to choose which of the functions χ or ξ plays the part of the canonical coordinate. Then the other one will enter the definition of the conjugate canonical momentum. It is emphasized that we can use only one of these pairs at a time, but not both of them simultaneously. The physical result is independent of the choice.

If ξ is chosen as the canonical coordinate then the quantization condition reads

$$\left[p_\xi(x,z), \xi(x',z)\right] = \left[f_0^2\chi(x,z), \xi(x',z)\right] = \hbar\delta(x-x') \qquad (12.28)$$

where $[\cdots,\cdots]$ denotes the commutator of two operators. The other choice of the canonical pair would obviously yield the same quantization condition.

It is worthwhile to note that here the commutation condition is defined for two operators at equal values of the propagation length z (our "time"), contrary to the standard quantization procedure where the commutation is defined at equal times. It is not an unusual situation when considering light propagation in the paraxial approximation (see e.g. the discussion in Ref. [62]).

Now we have to define the norm of the fluctuation eigenfunctions. It is convenient to do this by first rewriting the equations (12.8) and (12.9) as

$$\frac{i}{2}f_0^2\sigma_y\partial_z\vartheta + \frac{i}{2}f_0^2 v_0\partial_z\sigma_y\partial_x\vartheta - \frac{1}{2}\lambda f_0^4(1+\sigma_z)\vartheta$$
$$+ \frac{1}{4\beta_0}\partial_x(f_0^2\partial_x\vartheta) = 0 \qquad (12.29)$$

where the two component field

$$\vartheta = \begin{pmatrix}\frac{1}{\sqrt{2}}\chi\\ \sqrt{2}\xi\end{pmatrix} \qquad (12.30)$$

has been introduced. We also extend the definition of these two fields and consider them complex. Then the corresponding Lagrangian density becomes

$$\mathscr{L} = \frac{i}{4} f_0^2 [\vartheta^\dagger \sigma_y (\hat{D}\vartheta) - (\hat{D}\vartheta^\dagger)\sigma_y \vartheta] - \frac{1}{2}\lambda f_0^4 \vartheta^\dagger (1+\sigma_z)\vartheta - \frac{1}{4\beta_0} f_0^2 (\nabla\vartheta^\dagger)\nabla\vartheta \tag{12.31}$$

where σ_i are Pauli matrices and $\hat{D} = \partial_z + v_0 \partial_x$. Now we can readily deduce the continuity equation

$$\partial_z \varrho + \partial_x j = 0$$

where the quantity $\varrho = f_0^2 \vartheta^\dagger \sigma_y \vartheta$ plays the role of the density of this two component field, whereas

$$j = f_0^2 v_0 \vartheta^\dagger \sigma_y \vartheta - i \frac{1}{2\beta_0} f_0^2 [\vartheta^\dagger \partial_x \vartheta - (\partial_x \vartheta^\dagger)\vartheta] \tag{12.32}$$

is the corresponding current density. Then the integral

$$\int dx f_0^2 \vartheta^\dagger \sigma_y \vartheta = -i \int dx f_0^2 (\chi^* \xi - \xi^* \chi) \tag{12.33}$$

is a conserved quantity, which can be used in order to normalize solutions (eigenfunctions) of Eqs. (12.8) and (12.9) or to define a scalar product of two such solutions. Similarly to the well-known Klein-Gordon norm, Eq. (12.33) is not positively defined. Positive and negative frequencies are considered separately in order to resolve this problem in the case of the Klein-Gordon equation (see *e.g.* Ref. [46]). A similar procedure can be applied in this case as well.

Now we are in a position to calculate the frequency spectrum of the Hawking radiation emanating from the Mach horizon. We use an approach similar to that proposed by Damour *et al.* [45, 61]. The central point in this approach is the calculation of the norm of a straddled fluctuation, which clearly demonstrates how a negative frequency state is cut into negative and positive frequency states propagating in opposite directions away from the horizon. The norm (12.33) reads

$$\langle \vartheta_s, \vartheta_s \rangle = \int dx \varrho_s(x) = -i \int dx f_0^2 (\chi_s^* \xi_s - \xi_s^* \chi_s) \tag{12.34}$$

for the pair of eigenfunctions (ξ_s, χ_s). We have to deal here with the two component function (12.30), of which χ_s calculated above is only one component. The second component ξ_s can be found now, say, by solving Eq. (12.9). Neglecting the contribution of the QP in Eq. (12.9) we get

$$\chi_s = \frac{1}{\lambda f_0^2} \hat{D}\xi_s \tag{12.35}$$

so that

$$\xi_s \propto x^\gamma$$

for $\min\{1/\nu, 1/\alpha\} \gg |x| \gg l_r$.

12 An All-Optical Event Horizon in an Optical Analogue of a Laval Nozzle

The two component field density reads

$$\varrho_s = -i\left\{\left[(\partial_z \xi_s^*)\xi_s - \xi_s^* \partial_z \xi_s\right] + v_0(x)\left[(\partial_x \xi_s^*)\xi_s - \xi_s^* \partial_x \xi_s\right]\right\} \quad (12.36)$$

where the coefficient λ is absorbed in the normalization factor of ξ_s. One can readily see that the definition of the norm (12.34) coincides in this approximation with the Klein-Gordon scalar product in the corresponding curved space. The local coordinate transformation

$$d\tilde{z} = dz + \frac{v_0(x)dx}{s^2(x) - v_0^2(x)}, \qquad d\tilde{x} = dx \quad (12.37)$$

proposed in Ref. [3] allows us to rewrite the density in the form

$$\tilde{\varrho}_s = -i\frac{s^2(\tilde{x})}{s^2(\tilde{x}) - v_0^2(\tilde{x})}\left[(\partial_{\tilde{z}} \xi_s^*)\xi_s - \xi_s^* \partial_{\tilde{z}} \xi_s\right], \quad (12.38)$$

which has different signs in the subsonic $v(\tilde{x}) < s(\tilde{x})$ and supersonic $v(\tilde{x}) > s(\tilde{x})$ regions.

Next we calculate the norm using the density (12.38). The corresponding integral is separated into two regions—outside the "black hole" $\tilde{x} < -l_r$ (left), and inside it $\tilde{x} > l_r$ (right). We may neglect the contribution of the narrow region $|x| < l_r$ and reduce the norm to

$$\langle \vartheta_s, \vartheta_s \rangle \approx \int_{-\infty}^{-l_r} d\tilde{x} \tilde{\varrho}_s + \int_{+l_r}^{\infty} d\tilde{x} \tilde{\varrho}_s$$
$$= \langle \vartheta_s, \vartheta_s \rangle_{left} + \langle \vartheta_s, \vartheta_s \rangle_{right}. \quad (12.39)$$

The function $\tilde{\varrho}_s$ in Eq. (12.38) diverges at $|\tilde{x}| \to 0$, which could have resulted in a divergent contribution of the neglected region of integration. However, it is important to emphasize that Eq. (12.38) holds only outside the narrow region near the Mach horizon, i.e. for $|\tilde{x}| > l_r$. Within this region, $|\tilde{x}| < l_r$, we have to go back to Eq. (12.34), which gives a non-divergent result. Although infinities are indicated as the integration limits in Eq. (12.39), the principal contributions come from the regions $\min\{1/\alpha, 1/\nu\} > |x| > l_r$. Since both χ_s and ξ_s are regular at $|x| < l_r$ we will get only a small correction to the norm (12.39) from this narrow region, which can be neglected as long as $l_r \ll \min\{1/\alpha, 1/\nu\}$.

The integrals on the two sides of the horizon (left and right) approximately obey the relation

$$\langle \vartheta_s, \vartheta_s \rangle_{left} = -e^{2\pi \operatorname{Im}\gamma} \langle \vartheta_s, \vartheta_s \rangle_{right}$$

due to the analytical properties of the function ξ_s. The total negative norm (12.39) is cut into the left moving positive frequency state and the right moving negative frequency state. The left state propagates against the flow "outside the black hole". Its relative weight is

$$N(\nu) = \left(e^{2\pi \operatorname{Im}\gamma} - 1\right)^{-1}, \quad (12.40)$$

where

$$\operatorname{Im}\gamma = \frac{2\nu}{3\alpha}\left[1 + \frac{2}{27}\frac{l_{nl}^2 \nu^2}{\bar{s}^2}\right]$$

contains a correction due to the QP, proportional to the third power of the frequency (wave number in the z direction). Thus $N(\nu)$ deviates from the Planck distribution for black body radiation. This result can also be understood as a black body radiation spectrum with the frequency-dependent temperature

$$T_H(\nu) = \frac{T_H(0)}{1 + \frac{2}{27}\frac{\nu^2 l_{nl}^2}{\bar{s}^2}}.$$

For frequencies that are not very high, $\nu \ll \bar{s}/l_r$, this dependence may be neglected, and $N(\nu)$ becomes the standard Planck distribution with the effective Hawking temperature

$$T_H(0) = \frac{3\hbar \bar{s} \alpha}{4\pi k_B},$$

where $\hbar$ has been re-introduced.

12.5 Discussion

While the nonlinearity length $l_{nl} = 1/\sqrt{\lambda \rho_0 \beta_0} \sim 30$ μm in the experiment is sufficiently short to give way to fluctuations with a linear dispersion relation, the challenge is to create an equivalent Hawking temperature that would be high enough to measure experimentally. Note that this is not a real temperature, but it nevertheless sets a constraint on the minimum light intensity (and sound velocity) required in the waveguide: As explained above, the ratio of amplitudes of the two parts of a classical fluctuation—the part which is carried away with the supersonic flow (*i.e.* "falls" into the black hole) and the part which penetrates into the subsonic region (*i.e.* "escapes" from the black hole), is $\exp\{\frac{2\pi c \nu_0}{\sqrt{3\bar{s}}} + \frac{2\pi^2 \Gamma^2 c^2}{3\bar{s}^2}\} > 1$. For small enough Γ this ratio is on the order of unity when $2\pi c \nu_0 \approx \sqrt{3\bar{s}}$, where c is a characteristic length scale of the nozzle ($c = \sqrt{a^2 + b^2}$ for the hyperbolically-shaped throat). The same condition can also be written $l_H \approx 2l_0$, where $l_H = \hbar/T_H = 4\pi c/\sqrt{3\bar{s}}$ and $l_0 = 1/\nu_0$ are characteristic length scales of the horizon and the fluctuations, respectively. Fulfillment of this condition would allow one to observe both parts of a classical fluctuation in the experiment, and to directly measure the Hawking temperature. In the experiment described above l_H is on the order of a few meters ($c \approx 1 \times 10^{-3}$ m, $\bar{s} \approx 1 \times 10^{-3}$), while l_0 must be on the order of a few centimeters (the length of the cavity). This may still allow observation of the part of a fluctuation which is carried away with the supersonic flow, but the part which penetrates into the subsonic region will most likely be submerged in noise. Therefore l_H must be decreased by two orders of magnitude. Note, however, that for given input intensity and nonlinear coefficient the factor $c/\bar{s}$ in the expression for l_H grows like c^2. We therefore estimate that an order of magnitude smaller cavity will be sufficient for observing both parts of a straddled classical fluctuation. The requirement for a slow rate of acceleration [51] can be met by a refined, smoother waveguide cross-section, compared to the rudimentary prototype that we have used for demonstration purposes.

12.6 Summary

In summary, we demonstrate the possibility of creating an optical analogue of the Laval nozzle, as a new platform for analogue gravity experiments. The challenge for future investigations will be to study the dynamics of straddled fluctuations, which may be either quantum or classical, and even artificially created. The equivalent of the Hawking temperature enters as an important parameter characterizing all types of fluctuations. This temperature is measured in units of momentum, rather than energy, which corresponds to a wavelength that exceeds the width of nozzle throat by about three orders of magnitude, and is in principle accessible to experimental measurements.

Acknowledgements Support of Israeli Science Foundation Grant No. 944/05 and of United States—Israel Binational Science Foundation Grant No. 2006242 is acknowledged.

References

1. Bekenstein, J.D.: Phys. Rev. D **7**, 2333 (1973)
2. Hawking, S.W.: Commun. Math. Phys. **43**, 199 (1975)
3. Unruh, W.G.: Phys. Rev. Lett. **46**, 1351 (1981)
4. Rousseaux, G., Mathis, C., Maissa, P., Philbin, T.G., Leonhardt, U.: New J. Phys. **10**, 053015 (2008)
5. Rousseaux, G., Maissa, P., Mathis, C., Coulet, P., Philbin, T.G., Leonhardt, U.: New J. Phys. **12**, 095018 (2010)
6. Weinfurtner, S., Tedford, E.W., Penrice, M.C.J., Unruh, W.G., Lawrence, G.A.: Phys. Rev. Lett. **106**, 021302 (2011)
7. Lahav, O., Itah, A., Blumkin, A., Gordon, C., Rinott, S., Zayats, A., Steinhauer, J.: Phys. Rev. Lett. **105**, 240401 (2010)
8. Belgiorno, F., Cacciatori, S.L., Clerici, M., Gorini, V., Ortenzi, G., Rizzi, L., Rubino, E., Sala, V.G., Faccio, D.: Phys. Rev. Lett. **105**, 203901 (2010)
9. Philbin, T.G., Kuklewicz, C., Robertson, S., Hill, S., König, F., Leonhardt, U.: Science **319**, 1367 (2008)
10. Marino, F.: Phys. Rev. A **78**, 063804 (2008)
11. Barceló, C., Liberati, S., Visser, M.: Phys. Rev. A **68**, 053613 (2003)
12. Barceló, C., Liberati, S., Visser, M.: Int. J. Mod. Phys. A **18**, 3735 (2003)
13. Carusotto, I., Fagnocchi, S., Recati, A., Balbinot, R., Fabbri, A.: New J. Phys. **10**, 103001 (2008)
14. Recati, A., Pavloff, N., Carusotto, I.: Phys. Rev. A **80**, 043603 (2009)
15. Jacobson, T.A., Volovik, G.E.: Phys. Rev. D **58**, 064021 (1998)
16. Giovanazzi, S.: Phys. Rev. Lett. **94**, 061302 (2005)
17. Nation, P.D., Blencowe, M.P., Rimberg, A.J., Buks, E.: Phys. Rev. Lett. **103**, 087004 (2009)
18. Reznik, B.: Phys. Rev. D **62**, 044044 (2000)
19. Fouxon, I., Farberovich, O.V., Bar-Ad, S., Fleurov, V.: Europhys. Lett. **92**, 14002 (2010)
20. Leonhardt, U., Kiss, T., Öhberg, P.: J. Opt. B, Quantum Semiclass. Opt. **5**, S42 (2003)
21. Sakagami, M., Ohashi, A.: Prog. Theor. Phys. **107**(6), 1267 (2002)
22. Fleurov, V., Soffer, A.: Europhys. Lett. **72**, 287 (2005)
23. Dekel, G., Fleurov, V., Soffer, A., Stucchio, C.: Phys. Rev. A **75**, 043617 (2007)
24. Dekel, G., Farberovich, V., Fleurov, V., Soffer, A.: Phys. Rev. A **81**, 063638 (2010)
25. Barak, A., Peleg, O., Stucchio, C., Soffer, A., Segev, M.: Phys. Rev. Lett. **100**, 153901 (2008)

26. Wan, W.J., Muenzel, S., Fleischer, J.W.: Phys. Rev. Lett. **104**, 073903 (2010)
27. Cohen, E., Muenzel, S., Fleischer, J., Soffer, A., Fleurov, V.: arXiv:1107.0627
28. Frisch, T., Pomeau, Y., Rica, S.: Phys. Rev. Lett. **69**, 1644 (1992)
29. El, G.A., Geogjaev, V.V., Gurevich, A.V., Krylov, A.L.: Physica D **87**, 186 (1995)
30. Wan, W., Jia, S., Fleischer, J.W.: Nat. Phys. **3**, 46 (2007)
31. Jia, S., Wan, W., Fleischer, J.W.: Phys. Rev. Lett. **99**, 223901 (2007)
32. Barsi, C., Wan, W., Sun, C., Fleischer, J.W.: Opt. Lett. **32**, 2930 (2007)
33. Hakim, V.: Phys. Rev. E **55**, 2835 (1997)
34. Leszczyszyn, A.M., El, G.A., Gladush Yu, G., Kamchatnov, A.M.: Phys. Rev. A **79**, 063608 (2009)
35. Gladush, Yu.G., El, G.A., Gammal, A., Kamchatnov, A.M.: Phys. Rev. A **75**, 033619 (2007)
36. Agrawal, G.P.: Nonlinear Fiber Optics. Academic Press, New York (1995)
37. Madelung, E.: Z. Phys. **40**, 322 (1927)
38. Marburger, J.H.: Prog. Quantum Electron. **4**, 35 (1975)
39. Dekel, G., Fleurov, V., Soffer, A., Stucchio, C.: Phys. Rev. A **75**, 043617 (2007)
40. Silberberg, Y.: Opt. Lett. **15**, 1283 (1990)
41. Sedov, L.I.: Two Dimensional Problems in Hydrodynamics and Aerodynamics. Interscience, New York (1965)
42. Landau, L.D., Lifshits, E.M.: Fluid Mechanics. Pergamon, Oxford (1987)
43. Fleurov, V., Schilling, R.: Phys. Rev. A **85**, 045602 (2012)
44. Fleurov, V., Schilling, R.: arXiv:1105.0799
45. Damour, T., Lilley, M.: arXiv:0802.4169v1 [hep-th]
46. Birrel, N.D., Davies, P.C.W.: Quantum Fields in Curved Spaces. Cambridge University Press, Cambridge (1984)
47. Unruh, W.G.: Phys. Rev. D **51**, 2827 (1995)
48. Corley, S.: Phys. Rev. D **55**, 6155 (1997)
49. Macher, J., Parentani, R.: Phys. Rev. D **79**, 124008 (2009)
50. Macher, J., Parentani, R.: Phys. Rev. A **80**, 043601 (2009)
51. Finazzi, S., Parentani, R.: Phys. Rev. D **83**, 084010 (2011)
52. Brout, R., Massar, S., Parentani, R., Spindel, Ph.: Phys. Rev. D **52**, 4559 (1995)
53. Corley, S., Jacobson, T.: Phys. Rev. D **54**, 1568 (1996)
54. Corley, S.: Phys. Rev. D **57**, 6280 (1998)
55. Corley, S., Jacobson, T.: Phys. Rev. D **59**, 124011 (1999)
56. Leonhardt, U., Kiss, T., Öhberg, P.: Phys. Rev. A **67**, 033602 (2003)
57. Unruh, W.G., Schützhold, R.: Phys. Rev. D **71**, 024028 (2005)
58. Schützhold, R., Unruh, W.G.: Phys. Rev. D **78**, 041504R (2008)
59. Barceló, C., Liberati, S., Visser, M.: Living Rev. Relativ. **8**, 12 (2005)
60. Bleistein, N., Handelsman, R.A.: Asymptotic Expansions of Integrals. Dover, New York (1986)
61. Damour, T., Ruffini, R.: Phys. Rev. D **14**, 332 (1976)
62. Aiello, A., Marquardt, C., Leuchs, G.: Phys. Rev. A **81**, 053838 (2010)

Chapter 13
Lorentz Breaking Effective Field Theory and Observational Tests

Stefano Liberati

Abstract Analogue models of gravity have provided an experimentally realizable test field for our ideas on quantum field theory in curved spacetimes but they have also inspired the investigation of possible departures from exact Lorentz invariance at microscopic scales. In this role they have joined, and sometime anticipated, several quantum gravity models characterized by Lorentz breaking phenomenology. A crucial difference between these speculations and other ones associated to quantum gravity scenarios, is the possibility to carry out observational and experimental tests which have nowadays led to a broad range of constraints on departures from Lorentz invariance. We shall review here the effective field theory approach to Lorentz breaking in the matter sector, present the constraints provided by the available observations and finally discuss the implications of the persisting uncertainty on the composition of the ultra high energy cosmic rays for the constraints on the higher order, analogue gravity inspired, Lorentz violations.

13.1 Introduction

Our understanding of the fundamental laws of Nature is based at present on two different theories: the Standard Model of Fundamental Interactions (SM), and classical General Relativity (GR). However, in spite of their phenomenological success, SM and GR leave many theoretical questions still unanswered. First of all, since we feel that our understanding of the fundamental laws of Nature is deeper (and more accomplished) if we are able to reduce the number of degrees of freedom and coupling constants we need in order to describe it, many physicists have been trying to construct unified theories in which not only sub-nuclear forces are seen as different aspects of a unique interaction, but also gravity is included in a consistent manner.

Another important reason why we seek for a new theory of gravity comes directly from the gravity side. We know that GR fails to be a predictive theory in some regimes. Indeed, some solutions of Einstein's equations are known to be singular at some points, meaning that in these points GR is not able to make any prediction.

S. Liberati (✉)
SISSA, Via Bonomea 265, 34136 Trieste, Italy
e-mail: liberati@sissa.it

D. Faccio et al. (eds.), *Analogue Gravity Phenomenology*,
Lecture Notes in Physics 870, DOI 10.1007/978-3-319-00266-8_13,
© Springer International Publishing Switzerland 2013

Moreover, there are apparently honest solutions of GR equations predicting the existence of time-like closed curves, which would imply the possibility of traveling back and forth in time with the related causality paradoxes. Finally, the problem of black-hole evaporation seems to clash with Quantum Mechanical unitary evolution.

This long list of puzzles spurred an intense research toward a quantum theory of gravity that started almost immediately after Einstein's proposal of GR and which is still going on nowadays. The quantum gravity problem is not only conceptually challenging, it has also been an almost metaphysical pursue for several decades. Indeed, we expect QG effects at experimentally/observationally accessible energies to be extremely small, due to suppression by the Planck scale $M_{Pl} \equiv \sqrt{\hbar c/G_N} \simeq 1.22 \times 10^{19}$ GeV/c^2. In this sense it has been considered (and it is still considered by many) that only ultra-high-precision (or Planck scale energy) experiments would be able to test quantum gravity models.

It was however realized (mainly over the course of the past decade) that the situation is not as bleak as it appears. In fact, models of gravitation beyond GR and models of QG have shown that there can be several of what we term low energy "relic signatures" of these models, which would lead to deviation from the standard theory predictions (SM plus GR) in specific regimes. Some of these new phenomena, which comprise what is often termed "QG phenomenology", include:

- Quantum decoherence and state collapse [1]
- QG imprint on initial cosmological perturbations [2]
- Cosmological variation of couplings [3, 4]
- TeV Black Holes, related to extra-dimensions [5]
- Violation of discrete symmetries [6]
- Violation of spacetime symmetries [7]

In this lecture I will focus upon the phenomenology of violations of spacetime symmetries, and in particular of Local Lorentz invariance (LLI), a pillar both of quantum field theory as well as GR (LLI is a crucial part of the Einstein Equivalence Principle on which metric theories of gravity are based).

13.2 A Brief History of an Heresy

Contrary to the common trust, ideas about the possible breakdown of LLI have a long standing history. It is however undeniable that the last twenty years have witnessed a striking acceleration in the development both of theoretical ideas as well as of phenomenological tests before unconceivable. We shall here present an incomplete review of these developments.

13.2.1 The Dark Ages

The possibility that Lorentz invariance violation (LV) could play a role again in physics dates back by at least sixty years [8–13] and in the seventies and eighties there was already a well established literature investigating the possible phenomenological consequences of LV (see e.g. [14–19]).

The relative scarcity of these studies in the field was due to the general expectation that new effects were only to appear in particle interactions at energies of order the Planck mass M_{Pl}. However, it was only in the nineties that it was clearly realized that there are special situations in which new effects could manifest also at lower energy. These situations were termed "Windows on Quantum Gravity".

13.2.2 Windows on Quantum Gravity

At first glance, it appears hopeless to search for effects suppressed by the Planck scale. Even the most energetic particles ever detected (Ultra High Energy Cosmic Rays, see, e.g., [20, 21]) have $E \lesssim 10^{11}$ GeV $\sim 10^{-8} M_{Pl}$. However, even tiny corrections can be magnified into a significant effect when dealing with high energies (but still well below the Planck scale), long distances of signal propagation, or peculiar reactions (see, e.g., [7] for an extensive review).

A partial list of these *windows on QG* includes:

- sidereal variation of LV couplings as the lab moves with respect to a preferred frame or direction
- cumulative effects: long baseline dispersion and vacuum birefringence (e.g. of signals from gamma ray bursts, active galactic nuclei, pulsars)
- anomalous (normally forbidden) threshold reactions allowed by LV terms (e.g. photon decay, vacuum Cherenkov effect)
- shifting of existing threshold reactions (e.g. photon annihilation from Blazars, ultra high energy protons pion production)
- LV induced decays not characterized by a threshold (e.g. decay of a particle from one helicity to the other or photon splitting)
- maximum velocity (e.g. synchrotron peak from supernova remnants)
- dynamical effects of LV background fields (e.g. gravitational coupling and additional wave modes)

It is difficult to assign a definitive "paternity" to a field, and the so called Quantum Gravity Phenomenology is no exception in this sense. However, among the papers commonly accepted as seminal we can cite the one by Kostelecký and Samuel [22] that already in 1989 envisaged, within a string field theory framework, the possibility of non-zero vacuum expectation values (VEV) for some Lorentz breaking operators. This work led later on to the development of a systematic extension of the SM (what was later on called "minimal standard model extension" (mSME)) incorporating all possible Lorentz breaking, power counting renormalizable, operators (i.e. of mass dimension ≤ 4), as proposed by Colladay and Kostelecký [23]. This provided a framework for computing in effective field theory the observable consequences for many experiments and led to much experimental work setting limits on the LV parameters in the Lagrangian (see e.g. [24] for a review).

Another seminal paper was that of Amelino-Camelia and collaborators [25] which highlighted the possibility to cast observational constraints on high energy violations of Lorentz invariance in the photon dispersion relation by using the aforementioned propagation over cosmological distance of light from remote astrophysical sources like gamma ray bursters (GRBs) and active galactic nuclei (AGN). The field of phenomenological constraints on quantum gravity induced LV was born.

Finally, we should also mention the influential papers by Coleman and Glashow [26–28] which brought the subject of systematic tests of Lorentz violation to the attention of the broader community of particle physicists.

Let me stress that this is necessarily an incomplete account of the literature which somehow pointed a spotlight on the investigation of departures from Special Relativity. Several papers appeared in the same period and some of them anticipated many important results, see e.g. [29, 30], which however at the time of their appearance were hardly noticed (and seen by many as too "exotic").

In the years 2000 the field reached a concrete maturity and many papers pursued a systematization both of the framework as well as of the available constraints (see e.g. [31–33]). In this sense another crucial contribution was the development of an effective field theory approach also for higher order (mass dimension greater than four), naively non-power counting renormalizable, operators.[1] This was firstly done for dimension 5 operators in QED [37] by Myers and Pospelov and later on extended to dimension 6 operators by Mattingly [38].

Why all this attention to Lorentz breaking tests developed in the late nineties and in the first decade of the new century? I think that the answer is twofold as it is related to important developments coming from experiments and observation as well as from theoretical investigations. It is a fact that the zoo of quantum gravity models/scenarios with a low energy phenomenology had a rapid growth during those years. This happened mainly under the powerful push of novel puzzling observations that seemed to call for new physics possibly of gravitational origin. For example, in cosmology these are the years of the striking realization that our universe is undergoing an accelerated expansion phase [39, 40] which apparently requires a new exotic cosmological fluid, called dark energy, which violates the strong energy condition (to be added to the already well known, and still mysterious, dark matter component).

Also in the same period high energy astrophysics provided some new puzzles, first with the apparent absence of the Greisen-Zatsepin-Kuzmin (GZK) cut off [41, 42] (a suppression of the high-energy tail of the UHECR spectrum due to UHECR interaction with CMB photons) as claimed by the Japanese experiment AGASA [43], later on via the so called TeV-gamma rays crisis, i.e. the apparent detection of a reduced absorption of TeV gamma rays emitted by AGN [44]. Both these "crises"

---

[1] Anisotropic scaling [34–36] techniques were recently recognized to be the most appropriate way of handling higher order operators in Lorentz breaking theories and in this case the highest order operators are indeed crucial in making the theory power counting renormalizable. This is why we shall adopt sometime the expression "naively non renormalizable".

later on subsided or at least alternative, more orthodox, explanations for them were advanced. However, their undoubtedly boosted the research in the field at that time.

It is perhaps this past "training" that made several exponents of the quantum gravity phenomenology community the among most ready to stress the apparent incompatibility of the recent CERN–LNGS based experiment OPERA [45] measure of superluminal propagation of muonic neutrinos and Lorentz EFT (see e.g. [46–49]. There is now evidence that the Opera measurement might be flawed due to unaccounted experimental errors and furthermore it seems to be refuted by a similar measurement of the ICARUS collaboration [50]. Nonetheless, this claim propelled a new burst of activity in Lorentz breaking phenomenology which might still provide useful insights for future searches.

Parallel to these exciting developments on the experimental/observational side, also theoretical investigations provided new motivations for Lorentz breaking searches and constraints. Indeed, specific hints of LV arose from various approaches to Quantum Gravity. Among the many examples are the above mentioned string theory tensor VEVs [22] and spacetime foam models [25, 51–54], then semiclassical spin-network calculations in Loop QG [55], non-commutative geometry [56–58], some brane-world backgrounds [59].

Indeed, during the last decades there were several attempts to formulate alternative theories of gravitation incorporating some form of Lorentz breaking, from early studies [60–64] to full-fledged theories such as the Einstein–Aether theory [65–67] and Hořava–Lifshitz gravity [35, 68, 69] (which in some limit can be seen as a UV completion of the Einstein–Aether framework [70]).

Finally, a relevant part of this story is related to the vigorous development in the same years of the so called condensed matter analogues of "emergent gravity" [71] which is the main topic of this school. Let us then consider these models in some detail and discuss some lesson that can be drawn from them.

13.3 Bose–Einstein Condensates as an Example of Emergent Local Lorentz Invariance

Analogue models for gravity have provided a powerful tool for testing (at least in principle) kinematical features of classical and quantum field theories in curved spacetimes [71]. The typical setting is the one of sound waves propagating in a perfect fluid [72, 73]. Under certain conditions, their equation can be put in the form of a Klein-Gordon equation for a massless particle in curved spacetime, whose geometry is specified by the acoustic metric. Among the various condensed matter systems so far considered, Bose–Einstein condensate (BEC) [74, 75] had in recent years a prominent role for their simplicity as well as for the high degree of sophistication achieved by current experiments. In a BEC system one can consider explicitly the quantum field theory of the quasi-particles (or phonons), the massless excitations over the condensate state, propagating over the condensate as the analogue of a quantum field theory of a scalar field propagating over a curved effective spacetime described by the acoustic metric. It provides therefore a natural framework to explore different aspects of quantum field theory in various interesting curved backgrounds (for example quantum aspects of black hole physics [76, 77] or the analogue of the creation of cosmological perturbations [78–81]) or even, and more relevantly for our discussion here, emerging spacetime scenarios.

In BEC, the effective emerging metric depends on the properties of the condensate wave-function. One can expect therefore the gravitational degrees of freedom to be encoded in the variables describing the condensate wave-function [75], which is solution of the well known Bogoliubov–de Gennes (BdG) equation. The dynamics of gravitational degrees of freedom should then be inferred from this equation, which is essentially non-relativistic. The "emerging matter", the quasi-particles, in the standard BEC, are phonons, i.e. massless excitations described at low energies by a relativistic (we shall see in which sense) wave equation, however, at high energies, the emergent nature of the underlying spacetime becomes evident and the relativistic structure of the equation broken. Let's see this in more detail as a conceptual exercise and for highlighting the inspirational role played in this sense by analogue models of gravity.

13.3.1 The Acoustic Geometry in BEC

Let us start by very briefly reviewing the derivation of the acoustic metric for a BEC system, and show that the equations for the phonons of the condensate closely mimic the dynamics of a scalar field in a curved spacetime. In the dilute gas approximation, one can describe a Bose gas through a quantum field $\hat{\Psi}$ satisfying

$$i\hbar \frac{\partial}{\partial t}\hat{\Psi} = \left(-\frac{\hbar^2}{2m}\nabla^2 + V_{\text{ext}}(\mathbf{x}) + \kappa(a)\hat{\Psi}^\dagger \hat{\Psi}\right)\hat{\Psi}. \tag{13.1}$$

m is the mass of the atoms, a is the scattering length for the atoms and κ parametrises the strength of the interactions between the different bosons in the gas. It can be re-expressed in terms of the scattering length a as

$$\kappa(a) = \frac{4\pi a \hbar^2}{m}. \tag{13.2}$$

As usual, the quantum field can be separated into a macroscopic (classical) condensate and a fluctuation: $\hat{\Psi} = \psi + \hat{\varphi}$, with $\langle \hat{\Psi} \rangle = \psi$. Then, by adopting the self-consistent mean field approximation

$$\hat{\varphi}^\dagger \hat{\varphi} \hat{\varphi} \simeq 2\langle \hat{\varphi}^\dagger \hat{\varphi}\rangle \hat{\varphi} + \langle \hat{\varphi}\hat{\varphi}\rangle \hat{\varphi}^\dagger, \tag{13.3}$$

one can arrive at the set of coupled equations:

$$i\hbar \frac{\partial}{\partial t}\psi(t,\mathbf{x}) = \left(-\frac{\hbar^2}{2m}\nabla^2 + V_{\text{ext}}(\mathbf{x}) + \kappa n_c\right)\psi(t,\mathbf{x})$$
$$+ \kappa\{2\tilde{n}\psi(t,\mathbf{x}) + \tilde{m}\psi^*(t,\mathbf{x})\}; \tag{13.4}$$

13 Lorentz Breaking Effective Field Theory and Observational Tests

$$i\hbar \frac{\partial}{\partial t}\hat{\varphi}(t, \mathbf{x}) = \left(-\frac{\hbar^2}{2m}\nabla^2 + V_{\text{ext}}(\mathbf{x}) + \kappa 2 n_T\right)\hat{\varphi}(t, \mathbf{x})$$
$$+ \kappa m_T \hat{\varphi}^\dagger(t, \mathbf{x}). \tag{13.5}$$

Here

$$n_c \equiv |\psi(t, \mathbf{x})|^2; \qquad m_c \equiv \psi^2(t, \mathbf{x}); \tag{13.6}$$

$$\tilde{n} \equiv \langle \hat{\varphi}^\dagger \hat{\varphi}\rangle; \qquad \tilde{m} \equiv \langle \hat{\varphi}\hat{\varphi}\rangle; \tag{13.7}$$

$$n_T = n_c + \tilde{n}; \qquad m_T = m_c + \tilde{m}. \tag{13.8}$$

In general one will have to solve both equations for ψ and $\hat{\varphi}$ simultaneously. The equation for the condensate wave function ψ is closed only when the back-reaction effects due to the fluctuations are neglected. (The back-reaction being hidden in the quantities $\tilde{m}$ and $\tilde{n}$.) This approximation leads then to the so-called Gross–Pitaevskii equation and can be checked *a posteriori* to be a good description of dilute Bose–Einstein condensates near equilibrium configurations.

Adopting the Madelung representation for the wave function ψ of the condensate

$$\psi(t, \mathbf{x}) = \sqrt{n_c(t, \mathbf{x})} \exp[-i\theta(t, \mathbf{x})/\hbar], \tag{13.9}$$

and defining an irrotational "velocity field" by $\mathbf{v} \equiv \nabla\theta/m$, the Gross–Pitaevskii equation can be rewritten as a continuity equation plus an Euler equation:

$$\frac{\partial}{\partial t}n_c + \nabla \cdot (n_c \mathbf{v}) = 0, \tag{13.10}$$

$$m\frac{\partial}{\partial t}\mathbf{v} + \nabla\left(\frac{mv^2}{2} + V_{\text{ext}}(t, \mathbf{x}) + \kappa n_c - \frac{\hbar^2}{2m}\frac{\nabla^2(\sqrt{n_c})}{\sqrt{n_c}}\right) = 0. \tag{13.11}$$

These equations are completely equivalent to those of an irrotational and inviscid fluid apart from the existence of the so-called quantum potential

$$V_{\text{quantum}} = -\hbar^2 \nabla^2 \sqrt{n_c}/(2m\sqrt{n_c}), \tag{13.12}$$

which has the dimensions of energy.

If we write the mass density of the Madelung fluid as $\rho = mn_c$, and use the fact that the flow is irrotational we can write the Euler equation in the more convenient Hamilton–Jacobi form:

$$m\frac{\partial}{\partial t}\theta + \left(\frac{[\nabla\theta]^2}{2m} + V_{\text{ext}}(t, \mathbf{x}) + \kappa n_c - \frac{\hbar^2}{2m}\frac{\nabla^2\sqrt{n_c}}{\sqrt{n_c}}\right) = 0. \tag{13.13}$$

When the gradients in the density of the condensate are small one can neglect the quantum stress term leading to the standard hydrodynamic approximation.

Let us now consider the quantum perturbations above the condensate. These can be described in several different ways, here we are interested in the "quantum acoustic representation"

$$\hat{\varphi}(t, \mathbf{x}) = e^{-i\theta/\hbar} \left(\frac{1}{2\sqrt{n_c}} \hat{n}_1 - i \frac{\sqrt{n_c}}{\hbar} \hat{\theta}_1 \right), \tag{13.14}$$

where $\hat{n}_1, \hat{\theta}_1$ are real quantum fields. By using this representation Eq. (13.5) can be rewritten as

$$\partial_t \hat{n}_1 + \frac{1}{m} \nabla \cdot (n_1 \nabla \theta + n_c \nabla \hat{\theta}_1) = 0, \tag{13.15}$$

$$\partial_t \hat{\theta}_1 + \frac{1}{m} \nabla \theta \cdot \nabla \hat{\theta}_1 + \kappa(a) n_1 - \frac{\hbar^2}{2m} D_2 \hat{n}_1 = 0. \tag{13.16}$$

Here D_2 represents a second-order differential operator obtained from linearizing the quantum potential. Explicitly:

$$D_2 \hat{n}_1 \equiv -\frac{1}{2} n_c^{-3/2} [\nabla^2 (n_c^{+1/2})] \hat{n}_1 + \frac{1}{2} n_c^{-1/2} \nabla^2 (n_c^{-1/2} \hat{n}_1). \tag{13.17}$$

The equations we have just written can be obtained easily by linearizing the Gross–Pitaevskii equation around a classical solution: $n_c \to n_c + \hat{n}_1$, $\phi \to \phi + \hat{\phi}_1$. It is important to realize that in those equations the back-reaction of the quantum fluctuations on the background solution has been assumed negligible. We also see in Eqs. (13.15), (13.16), that time variations of V_{ext} and time variations of the scattering length a appear to act in very different ways. Whereas the external potential only influences the background Eq. (13.13) (and hence the acoustic metric in the analogue description), the scattering length directly influences both the perturbation and background equations. From the previous equations for the linearised perturbations it is possible to derive a wave equation for $\hat{\theta}_1$ (or alternatively, for $\hat{n}_1$). All we need is to substitute in Eq. (13.15) the $\hat{n}_1$ obtained from Eq. (13.16). This leads to a PDE that is second-order in time derivatives but infinite order in space derivatives—to simplify things we can construct the symmetric 4×4 matrix

$$f^{\mu\nu}(t, \mathbf{x}) \equiv \begin{pmatrix} f^{00} & \vdots & f^{0j} \\ \cdots\cdots & \cdot & \cdots\cdots\cdots \\ f^{i0} & \vdots & f^{ij} \end{pmatrix}. \tag{13.18}$$

(Greek indices run from 0–3, while Roman indices run from 1–3.) Then, introducing (3 + 1)-dimensional spacetime coordinates

$$x^\mu \equiv (t; x^i) \tag{13.19}$$

the wave equation for θ_1 is easily rewritten as

$$\partial_\mu (f^{\mu\nu} \partial_\nu \hat{\theta}_1) = 0. \tag{13.20}$$

Where the $f^{\mu\nu}$ are *differential operators* acting on space only. Now, if we make a spectral decomposition of the field $\hat{\theta}_1$ we can see that for wavelengths larger than $\xi = \hbar/mc_{\text{sound}}$ (ξ corresponds to the "healing length" and $c_{\text{sound}}(a, n_c)^2 = \frac{\kappa(a)n_c}{m}$), the terms coming from the linearization of the quantum potential (the D_2) can be neglected in the previous expressions, in which case the $f^{\mu\nu}$ can be approximated by scalars, instead of differential operators. Then, by identifying

$$\sqrt{-g}\,g^{\mu\nu} = f^{\mu\nu}, \tag{13.21}$$

the equation for the field $\hat{\theta}_1$ becomes that of a (massless minimally coupled) quantum scalar field over a curved background

$$\Delta\theta_1 \equiv \frac{1}{\sqrt{-g}}\partial_\mu\left(\sqrt{-g}\,g^{\mu\nu}\partial_\nu\right)\hat{\theta}_1 = 0, \tag{13.22}$$

with an effective metric of the form

$$g_{\mu\nu}(t, \mathbf{x}) \equiv \frac{n_c}{mc_{\text{sound}}(a, n_c)}\begin{pmatrix} -\{c_{\text{sound}}(a, n_c)^2 - v^2\} & \vdots & -v_j \\ \cdots\cdots\cdots\cdots\cdots & \cdot & \cdots\cdots \\ -v_i & \vdots & \delta_{ij} \end{pmatrix}. \tag{13.23}$$

Here the magnitude $c_{\text{sound}}(n_c, a)$ represents the speed of the phonons in the medium:

$$c_{\text{sound}}(a, n_c)^2 = \frac{\kappa(a)n_c}{m}, \tag{13.24}$$

and v_i is the velocity field of the fluid flow,

$$v_i = \frac{1}{m}\nabla_i\theta. \tag{13.25}$$

13.3.2 Lorentz Violation in BEC

It is interesting to consider the case in which the above "hydrodynamical" approximation for BECs does not hold. In order to explore a regime where the contribution of the quantum potential cannot be neglected we can use the so called *eikonal* approximation, a high-momentum approximation where the phase fluctuation $\hat{\theta}_1$ is itself treated as a slowly-varying amplitude times a rapidly varying phase. This phase will be taken to be the same for both $\hat{n}_1$ and $\hat{\theta}_1$ fluctuations. In fact, if one discards the unphysical possibility that the respective phases differ by a time varying quantity, any time-independent difference can be safely reabsorbed in the definition of the (complex) amplitudes $\mathscr{A}_\theta, \mathscr{A}_\rho$. Specifically, we shall write

$$\hat{\theta}_1(t, \mathbf{x}) = \text{Re}\{\mathscr{A}_\theta \exp(-i\phi)\}, \tag{13.26}$$

$$\hat{n}_1(t, \mathbf{x}) = \text{Re}\{\mathscr{A}_\rho \exp(-i\phi)\}. \tag{13.27}$$

As a consequence of our starting assumptions, gradients of the amplitude, and gradients of the background fields, are systematically ignored relative to gradients of ϕ. Note however, that what we are doing here is not quite a "standard" eikonal approximation, in the sense that it is not applied directly on the fluctuations of the field $\psi(t, \mathbf{x})$ but separately on their amplitudes and phases ρ_1 and ϕ_1. We can then adopt the notation

$$\omega = \frac{\partial \phi}{\partial t}; \qquad k_i = \nabla_i \phi. \tag{13.28}$$

Then the operator D_2 can be approximated as

$$D_2 \hat{n}_1 \approx -\frac{1}{2} n_c^{-1} k^2 \hat{n}_1. \tag{13.29}$$

A similar result holds for D_2 acting on $\hat{\theta}_1$. That is, under the eikonal approximation we effectively replace the *operator* D_2 by the *function*

$$D_2 \to -\frac{1}{2} n_c^{-1} k^2. \tag{13.30}$$

For the matrix $f^{\mu\nu}$ this effectively results in the replacement

$$f^{00} \to -\left[\kappa(a) + \frac{\hbar^2 k^2}{4mn_c}\right]^{-1}, \tag{13.31}$$

$$f^{0j} \to -\left[\kappa(a) + \frac{\hbar^2 k^2}{4mn_c}\right]^{-1} \frac{\nabla^j \theta_0}{m}, \tag{13.32}$$

$$f^{i0} \to -\frac{\nabla^i \theta_0}{m}\left[\kappa(a) + \frac{\hbar^2 k^2}{4mn_c}\right]^{-1}, \tag{13.33}$$

$$f^{ij} \to \frac{n_c \delta^{ij}}{m} - \frac{\nabla^i \theta_0}{m}\left[\kappa(a) + \frac{\hbar^2 k^2}{4mn_c}\right]^{-1} \frac{\nabla^j \theta_0}{m}. \tag{13.34}$$

(As desired, this has the net effect of making $f^{\mu\nu}$ a matrix of numbers, not operators.) The physical wave equation (13.20) now becomes a nonlinear dispersion relation

$$f^{00}\omega^2 + (f^{0i} + f^{i0})\omega k_i + f^{ij} k_i k_j = 0. \tag{13.35}$$

After substituting the approximate D_2 into this dispersion relation and rearranging, we see (remember: $k^2 = \|k\|^2 = \delta^{ij} k_i k_j$)

$$-\omega^2 + 2v_0^i \omega k_i + \frac{n_c k^2}{m}\left[\kappa(a) + \frac{\hbar^2}{4mn_c}k^2\right] - (v_0^i k_i)^2 = 0. \tag{13.36}$$

That is (with $v_0^i = \frac{1}{m}\nabla_i \theta_0$)

$$(\omega - v_0^i k_i)^2 = \frac{n_c k^2}{m}\left[\kappa(a) + \frac{\hbar^2}{4mn_c}k^2\right]. \tag{13.37}$$

Introducing the speed of sound c_{sound} this takes the form:

$$\omega = v_0^i k_i \pm \sqrt{c_{\text{sound}}^2 k^2 + \left(\frac{\hbar}{2m}k^2\right)^2}. \tag{13.38}$$

We then see that BEC is a paradigmatic framework where a spacetime geometry emerges at low energies and Lorentz invariance is as an accidental (never exact) symmetry. This symmetry is naturally broken at high energies and appears eminently in modified dispersion relations for the quasi-particles living above the condensate background.

13.4 Modified Dispersion Relations and Their Naturalness

As mentioned before, not only analogue models but also several QG scenarios played an important role in motivating search for departures from Lorentz invariance and in most of these models, LV enters through modified dispersion relations of the sort (13.38). These relations can be cast in the general form

$$E^2 = p^2 + m^2 + f(E, p; \mu; M), \tag{13.39}$$

where the low energy speed of light $c = 1$; E and p are the particle energy and momentum, respectively; μ is a particle-physics mass-scale (possibly associated with a symmetry breaking/emergence scale) and M denotes the relevant QG scale. Generally, it is assumed that M is of order the Planck mass: $M \sim M_{\text{Pl}} \approx 1.22 \times 10^{19}$ GeV, corresponding to a quantum (or emergent) gravity effect. Note that we assumed a preservation of rotational invariance by QG physics and that only boost invariance is affected by Planck-scale corrections. This does not need to be the case (see however [33] for a discussion about this assumption) and constraints on possible breakdown of rotational invariance have been considered in the literature (especially in the context of the minimal standard model extension). We assume it here only for simplicity and clarity in assessing later the available constraints on the EFT framework.

Of course, once given (13.39) the natural thing to do is to expand the function $f(E, p; \mu; M)$ in powers of the momentum (energy and momentum are basically indistinguishable at high energies, although they are both taken to be smaller than the Planck scale),

$$E^2 = p^2 + m^2 + \sum_{i=1}^{\infty} \tilde{\eta}_i p^i, \tag{13.40}$$

where the lowest order LV terms (p, p^2, p^3, p^4) have primarily been considered [7].[2]

About this last point some comments are in order. In fact, from a EFT point of view the only relevant operators should be the lowest order ones, i.e. those of mass dimension 3, 4 corresponding to terms of order p and p^2 in the dispersion relation. Situations in which higher order operators "weight" as much as the lowest order ones are only possible at the cost of a severe, indeed arbitrary, fine tuning of the coefficients $\tilde{\eta}_i$.

However, we do know by now (see further discussion below) that current observational constraints are tremendous on dimension 3 operators and very severe on dimension 4 ones. This is kind of obvious, given that these operators would end up modifying the dispersion relation of elementary particles at low energies. Dimension 3 operator would dominate at $p \to 0$ while the dimension 4 ones would generically induce a, species dependent, constant shift in the limit speed for elementary particles.

Of course one might be content to limit oneself to the study of just these terms but we stress that emergent gravity scenarios, e.g. inspired by analogue gravity models, or QG gravity models, strongly suggest that if the origin of the breakdown of Lorentz invariance is rooted in the UV behaviour of gravitational physics then it should be naturally expected to become evident only at high energies. So one would then predict a hierarchy of LV coefficients of the sort

$$\tilde{\eta}_1 = \eta_1 \frac{\mu^2}{M}, \qquad \tilde{\eta}_2 = \eta_2 \frac{\mu}{M}, \qquad \tilde{\eta}_3 = \eta_3 \frac{1}{M}, \qquad \tilde{\eta}_4 = \eta_4 \frac{1}{M^2}. \qquad (13.41)$$

In characterizing the strength of a constraint one can then refer to the η_n without the tilde, so to compare to what might be expected from Planck-suppressed LV. In general one can allow the LV parameters η_i to depend on the particle type, and indeed it turns out that they *must* sometimes be different but related in certain ways for photon polarization states, and for particle and antiparticle states, if the framework of effective field theory is adopted.

13.4.1 The Naturalness Problem

While the above hierarchy (13.41) might seem now a well motivated framework within which asses our investigations, it was soon realized [83] that it is still quite unnatural from an EFT point of view. The reason is pretty simple: in EFT radiative corrections will generically allow the percolation of higher dimension Lorentz

---

[2] I disregard here the possible appearance of dissipative terms [82] in the dispersion relation, as this would correspond to a theory with unitarity loss and to a more radical departure from standard physics than that envisaged in the framework discussed herein (albeit a priori such dissipative scenarios are logically consistent and even plausible within some quantum/emergent gravity frameworks).

violation to the lowest dimension terms due to the coupling and self-couplings of particles [83]. In EFT loop integrals will be naturally cut-off at the EFT breaking scale, if such scale is as well the Lorentz breaking scale the two will basically cancel leading to unsuppressed, couplings dependent, contributions to the propagators. Hence radiative corrections will not allow a dispersion relation with only p^3 or p^4 Lorentz breaking terms but will automatically induce extra unsuppressed LV terms in p and p^2 which will be naturally dominant.

Several ideas have been advanced in order to justify such a "naturalness problem" (see e.g. [33]), it would be cumbersome to review here all the proposals, but one can clearly see that the most straightforward solution for this problem would consist in breaking the degeneracy between the EFT scale and the Lorentz breaking one. This can be achieved in two alternative ways.

13.4.1.1 A New Symmetry

Most of the aforementioned proposals implicitly assume that the Lorentz breaking scale is the Planck scale. One then needs the EFT scale (which can be naively identified with what we called previously μ) to be different from the Planck scale and actually sufficiently small so that the lowest order "induced" coefficients can be suppressed by suitable small ratios of the kind μ^p/M^q where p, q are some positive powers.

A possible solution in this direction can be provided by introducing what is commonly called a "custodial symmetry" something that forbids lower order operators and, once broken, suppress them by introducing a new scale. The most plausible candidate for this role was soon recognized to be Super Symmetry (SUSY) [84, 85]. SUSY is by definition a symmetry relating fermions to bosons i.e. matter with interaction carriers. As a matter of fact, SUSY is intimately related to Lorentz invariance. Indeed, it can be shown that the composition of at least two SUSY transformations induces spacetime translations. However, SUSY can still be an exact symmetry even in presence of LV and can actually serve as a custodial symmetry preventing certain operators to appear in LV field theories.

The effect of SUSY on LV is to prevent dimension ≤ 4, renormalizable LV operators to be present in the Lagrangian. Moreover, it has been demonstrated [84, 85] that the renormalization group equations for Supersymmetric QED plus the addition of dimension 5 LV operators à la Myers & Pospelov [37] do not generate lower dimensional operators, if SUSY is unbroken. However, this is not the case for our low energy world, of which SUSY is definitely not a symmetry.

The effect of soft SUSY breaking was also investigated in [84, 85]. It was found there that, as expected, when SUSY is broken the renormalizable operators are generated. In particular, dimension κ ones arise from the percolation of dimension $\kappa + 2$ LV operators.[3] The effect of SUSY soft-breaking is, however, to introduce a suppression of order $m_s^2/M_{\rm Pl}$ ($\kappa = 3$) or $(m_s/M_{\rm Pl})^2$ ($\kappa = 4$), where $m_s \simeq 1$ TeV is the

---

[3] We consider here only $\kappa = 3, 4$, for which these relationships have been demonstrated.

scale of SUSY soft breaking. Although, given present constraints, the theory with $\kappa = 3$ needs a lot of fine tuning to be viable, since the SUSY-breaking-induced suppression is not enough powerful to kill linear modifications in the dispersion relation of electrons, if $\kappa = 4$ then the induced dimension 4 terms are suppressed enough, provided $m_s < 100$ TeV. Current lower bounds from the Large Hadron Collider are at most around 950 GeV for the most simple models of SUSY [86] (the so called "constrained minimal supersymmetric standard model", CMSSM).

Finally, it is also interesting to note that the analogue model of gravity can be used as a particular implementation of the above mentioned mechanism for avoiding the so called naturalness problem via a custodial symmetry. This was indeed the case of multi-BEC [87, 88].

13.4.1.2 Gravitational Confinement of Lorentz Violation

The alternative to the aforementioned scenario is to turn the problem upside down. One can in fact assume that the Lorentz breaking scale (the M appearing in the above dispersion relations) is not set by the Planck scale while the latter is the EFT breaking scale. If in addition one starts with a theory which has higher order Lorentz violating operators only in the gravitational sector, then one can hope that the gravitational coupling $G_N \sim M_{\rm Pl}^{-2}$ will let them "percolate" to the matter sector however it will do so introducing factors of the order $(M/M_{\rm Pl})^2$ which can become strong suppression factors if $M \ll M_{\rm Pl}$. This is basically the idea at the base of the work presented in [89] which applies it to the special case of Horařa–Lifshitz gravity. There it was shown that indeed a workable low energy limit of the theory can be derived through this mechanism which apparently is fully compatible with extant constraints on Lorentz breaking operators in the matter sector. We think that this new route deserves further attention and should be more deeply explored in the future.

13.5 Dynamical Frameworks

Missing a definitive conclusion about the naturalness problem, the study of LV theories has basically proceeded by considering separately extensions of the Standard Model based on naively power counting renormalizable operators or non-renormalizable operators (at some given mass dimension). In what follows we shall succinctly describe these frameworks before to discuss theoretical alternatives.

13.5.1 SME with Renormalizable Operators

Most of the research in EFT with only renormalizable (*i.e.* mass dimension 3 and 4) LV operators has been carried out within the so called (minimal) SME [22]. It consists of the standard model of particle physics plus all Lorentz violating renormalizable operators (*i.e.* of mass dimension ≤4) that can be written without changing the field content or violating gauge symmetry. The operators appearing in the SME can be conveniently classified according to their behaviour under CPT. Since the most common particles used to cast constraints on LV are photons and electrons, a prominent role is played by LV QED.

If we label by ± the two photon helicities, we can write the photon dispersion relation as [90]

$$E = (1 + \rho \pm \sigma)|\mathbf{p}| \tag{13.42}$$

where ρ and σ depend on LV parameters appearing in the LV QED Lagrangian, as defined in [7]. Note that the dependence of the dispersion relation on the photon helicity is due to the fact that the SME generically also contemplates the possibility of a breakdown of rotational invariance.

We already gave (see Sect. 13.4) motivations for assuming rotation invariance to be preserved, at least in first approximation, in LV contexts. If we make this assumption, we obtain a major simplification of our framework, because in this case all LV tensors must reduce to suitable products of a time-like vector field, which is usually called u^α and, in the preferred frame, is assumed to have components $(1, 0, 0, 0)$. Then, the rotational invariant LV operators are

$$-bu_\mu \bar\psi \gamma_5 \gamma^\mu \psi + \frac{1}{2} i c u_\mu u_\nu \bar\psi \gamma^\mu \overleftrightarrow{D}^\nu \psi + \frac{1}{2} i d u_\mu u_\nu \bar\psi \gamma_5 \gamma^\mu \overleftrightarrow{D}^\nu \psi \tag{13.43}$$

for electrons and

$$-\frac{1}{4}(k_F)_\kappa \eta_{\lambda\mu} u_\nu F^{\kappa\lambda} F^{\mu\nu} \tag{13.44}$$

for photons.

The high energy ($M_{\text{Pl}} \gg E \gg m$) dispersion relations for QED can be expressed as (see [7] and references therein for more details)

$$E_{\text{el}}^2 = m_e^2 + p^2 + f_e^{(1)} p + f_e^{(2)} p^2 \quad \text{electrons}, \tag{13.45}$$

$$E_\gamma^2 = \left(1 + f_\gamma^{(2)}\right) p^2 \quad \text{photons} \tag{13.46}$$

where $f_e^{(1)} = -2bs$, $f_e^{(2)} = -(c - ds)$, and $f_\gamma^{(2)} = k_F/2$ with $s = \pm 1$ the helicity state of the electron [7]. The positron dispersion relation is the same as (13.45) replacing $p \to -p$, this will change only the $f_e^{(1)}$ term.

We notice here that the typical energy at which a new phenomenology should start to appear is quite low. In fact, taking for example $f_e^{(2)} \sim O(1)$, one finds that the corresponding extra-term is comparable to the electron mass m precisely at $p \simeq m \simeq 511$ keV. Even worse, for the linear modification to the dispersion relation, we would have, in the case in which $f_e^{(1)} \simeq O(1)$, that $p_{\text{th}} \sim m^2/M_{\text{Pl}} \sim 10^{-17}$ eV. (Notice that this energy corresponds by chance to the present upper limit on the photon mass, $m_\gamma \lesssim 10^{-18}$ eV [91].) As said, this implied strong constraints on the parameters and was a further motivation for exploring the QG preferred possibility of higher order Lorentz violating operators and consequently try to address the naturalness problem.

13.5.2 Dimension Five Operators SME

An alternative approach within EFT is to study non-renormalizable operators. Nowadays it is widely accepted that the SM could just be an effective field theory and in this sense its renormalizability is seen as a consequence of neglecting some higher order operators which are suppressed by some appropriate mass scale. It is a short deviation from orthodoxy to imagine that such non-renormalizable operators can be generated by quantum gravity effects (and hence be naturally suppressed by the Planck mass) and possibly associated to the violation of some fundamental spacetime symmetry like local Lorentz invariance.

Myers & Pospelov [37] found that there are essentially only three operators of dimension five, quadratic in the fields, that can be added to the QED Lagrangian preserving rotation and gauge invariance, but breaking local LI.[4]

These extra-terms, which result in a contribution of $O(E/M_{Pl})$ to the dispersion relation of the particles, are the following:

$$-\frac{\xi}{2M_{Pl}} u^m F_{ma}(u \cdot \partial)(u_n \tilde{F}^{na}) + \frac{1}{2M_{Pl}} u^m \bar{\psi} \gamma_m (\zeta_1 + \zeta_2 \gamma_5)(u \cdot \partial)^2 \psi, \quad (13.47)$$

where $\tilde{F}$ is the dual of F and ξ, $\zeta_{1,2}$ are dimensionless parameters. All these terms also violate the CPT symmetry. More recently, this construction has been extended to the whole SM [92].

From (13.47) the dispersion relations of the fields are modified as follows. For the photon one has

$$\omega_\pm^2 = k^2 \pm \frac{\xi}{M_{Pl}} k^3, \quad (13.48)$$

(the $+$ and $-$ signs denote right and left circular polarisation), while for the fermion (with the $+$ and $-$ signs now denoting positive and negative helicity states)

$$E_\pm^2 = p^2 + m^2 + \eta_\pm \frac{p^3}{M_{Pl}}, \quad (13.49)$$

with $\eta_\pm = 2(\zeta_1 \pm \zeta_2)$. For the antifermion, it can be shown by simple "hole interpretation" arguments that the same dispersion relation holds, with $\eta_\pm^{af} = -\eta_\mp^f$

---

[4]Actually these criteria allow the addition of other (CPT even) terms, but these would not lead to modified dispersion relations (they can be thought of as extra, Planck suppressed, interaction terms) [92].

where af and f superscripts denote respectively anti-fermion and fermion coefficients [33, 93].

As we shall see, observations involving very high energies can thus potentially cast $O(1)$ and stronger constraint on the coefficients defined above. A natural question arises then: what is the theoretically expected value of the LV coefficients in the modified dispersion relations shown above?

This question is clearly intimately related to the meaning of any constraint procedure. Indeed, let us suppose that, for some reason we do not know, because we do not know the ultimate high energy theory, the dimensionless coefficients $\eta^{(n)}$, that in principle, according to the Dirac criterion, should be of order $O(1)$, are defined up to a dimensionless factor of $m_e/M_{\rm Pl} \sim 10^{-22}$. (This could well be as a result of the integration of high energy degrees of freedom.) Then, any constraint of order larger than 10^{-22} would be ineffective, if our aim is learning something about the underlying QG theory.

This problem could be further exacerbated by renormalization group effects, which could, in principle, strongly suppress the low-energy values of the LV coefficients even if they are $O(1)$ at high energies. Let us, therefore, consider the evolution of the LV parameters first. Bolokhov & Pospelov [92] addressed the problem of calculating the renormalization group equations for QED and the Standard Model extended with dimension-five operators that violate Lorentz Symmetry.

In the framework defined above, assuming that no extra physics enters between the low energies at which we have modified dispersion relations and the Planck scale at which the full theory is defined, the evolution equations for the LV terms in Eq. (13.47) that produce modifications in the dispersion relations, can be inferred as

$$\frac{d\zeta_1}{dt} = \frac{25}{12}\frac{\alpha}{\pi}\zeta_1, \qquad \frac{d\zeta_2}{dt} = \frac{25}{12}\frac{\alpha}{\pi}\zeta_2 - \frac{5}{12}\frac{\alpha}{\pi}\xi, \qquad \frac{d\xi}{dt} = \frac{1}{12}\frac{\alpha}{\pi}\zeta_2 - \frac{2}{3}\frac{\alpha}{\pi}\xi, \tag{13.50}$$

where $\alpha = e^2/4\pi \simeq 1/137$ ($\hbar = 1$) is the fine structure constant and $t = \ln(\mu^2/\mu_0^2)$ with μ and μ_0 two given energy scales. (Note that the above formulae are given to lowest order in powers of the electric charge, which allows one to neglect the running of the fine structure constant.)

These equations show that the running is only logarithmic and therefore low energy constraints are robust: $O(1)$ parameters at the Planck scale are still $O(1)$ at lower energy. Moreover, they also show that η_+ and η_- cannot, in general, be equal at all scales.

13.5.3 Dimension Six Operators SME

If CPT ends up being a fundamental symmetry of nature it would forbid all of the above mentioned operators (hence pushing at further high energies the emergence of Lorentz breaking physics). It makes then sense to consider dimension six, CPT even, operators which furthermore do give rise to dispersion relations of the kind appearing in the above mentioned BEC analogue gravity example.

The CPT even dimension 6 LV terms have only recently been computed [38] through the same procedure used by Myers & Pospelov for dimension 5 LV. The known fermion operators are

$$-\frac{1}{M_{\rm Pl}}\bar{\psi}(u\cdot D)^2\bigl(\alpha_L^{(5)}P_L+\alpha_R^{(5)}P_R\bigr)\psi,$$

$$-\frac{i}{M_{\rm Pl}^2}\bar{\psi}(u\cdot D)^3(u\cdot\gamma)\bigl(\alpha_L^{(6)}P_L+\alpha_R^{(6)}P_R\bigr)\psi, \quad (13.51)$$

$$-\frac{i}{M_{\rm Pl}^2}\bar{\psi}(u\cdot D)\Box(u\cdot\gamma)\bigl(\tilde{\alpha}_L^{(6)}P_L+\tilde{\alpha}_R^{(6)}P_R\bigr)\psi,$$

where $P_{R,L}$ are the usual left and right spin projectors $P_{R,L}=(1\pm\gamma^5)/2$ and where D is the usual QED covariant derivative. All coefficients α are dimensionless because we factorize suitable powers of the Planck mass.

The known photon operator is

$$-\frac{1}{2M_{\rm Pl}^2}\beta_\gamma^{(6)} F^{\mu\nu}u_\mu u^\sigma(u\cdot\partial)F_{\sigma\nu}. \quad (13.52)$$

From these operators, the dispersion relations of electrons and photons can be computed, yielding

$$E^2-p^2-m^2=\frac{m}{M_{\rm Pl}}\bigl(\alpha_R^{(5)}+\alpha_L^{(5)}\bigr)E^2+\alpha_R^{(5)}\alpha_L^{(5)}\frac{E^4}{M_{\rm Pl}^2}$$

$$+\frac{\alpha_R^{(6)}E^3}{M_{\rm Pl}^2}(E+sp)+\frac{\alpha_L^{(6)}E^3}{M_{\rm Pl}^2}(E-sp), \quad (13.53)$$

$$\omega^2-k^2=\beta^{(6)}\frac{k^4}{M_{\rm Pl}^2}, \quad (13.54)$$

where m is the electron mass and where $s=\sigma\cdot\mathbf{p}/|\mathbf{p}|$ is the helicity of the electrons. Also, notice that a term proportional to E^2 is generated.

Because the high-energy fermion states are almost exactly chiral, we can further simplify the fermion dispersion relation Eq. (13.54) (we pose $R=+, L=-$)

$$E^2=p^2+m^2+f_\pm^{(4)}p^2+f_\pm^{(6)}\frac{p^4}{M_{\rm Pl}^2}. \quad (13.55)$$

Being suppressed by a factor of order $m/M_{\rm Pl}$, we will drop in the following the quadratic contribution $f_\pm^{(4)}p^2$, indeed this can be safely neglected, provided that $E>\sqrt{mM_{\rm Pl}}$. Let me stress however, that this is exactly an example of a dimension 4 LV term with a natural suppression, which for electron is of order $m_e/M_{\rm Pl}\sim 10^{-22}$. Therefore, any limit larger than 10^{-22} placed on this term would not have to be considered as an effective constraint. To date, the best constraint for a rotational invariant electron LV term of dimension 4 is $O(10^{-16})$ [94].

Coming back to Eq. (13.55), it may seem puzzling that in a CPT invariant theory we distinguish between different fermion helicities. However, although they are CPT invariant, some of the LV terms displayed in Eq. (13.52) are odd under P and T.

CPT invariance allows us to determine a relationship between the LV coefficients of the electrons and those of the positrons. Indeed, to obtain these we must consider that, by CPT, the dispersion relation of the positron is given by (13.54), with the replacements $s \to -s$ and $p \to -p$. This implies that the relevant positron coefficients $f_{\text{positron}}^{(6)}$ are such that $f_{e_\pm^+}^{(6)} = f_{e_\mp^-}^{(6)}$, where $e_\pm^+$ indicates a positron of positive/negative helicity (and similarly for the $e_\pm^-$).

13.5.4 Other Frameworks

Picking up a well defined dynamical framework is sometimes crucial in discussing the phenomenology of Lorentz violations. In fact, not all the above mentioned "windows on quantum gravity" can be exploited without adding additional information about the dynamical framework one works with. Although cumulative effects exclusively use the form of the modified dispersion relations, all the other "windows" depend on the underlying dynamics of interacting particles and on whether or not the standard energy-momentum conservation holds. Thus, to cast most of the constraints on dispersion relations of the form (13.40), one needs to adopt a specific theoretical framework justifying the use of such deformed dispersion relations.

The previous discussion mainly focuses on considerations based on Lorentz breaking EFTs. This is indeed a conservative framework within which much can be said (e.g. reaction rates can still be calculated) and from an analogue gravity point of view it is just the natural frame to work within. Nonetheless, this is of course not the only dynamical framework within which a Lorentz breaking kinematics can be cast. Because the EFT approach is nothing more than a highly reasonable, but rather arbitrary "assumption", it is worth studying and constraining additional models, given that they may evade the majority of the constraints discussed in this review.

13.5.4.1 D-Brane Models

We consider here the model presented in [52, 54], in which modified dispersion relations are found based on the Liouville string approach to quantum spacetime [95]. Liouville-string models of spacetime foam [95] motivate corrections to the usual relativistic dispersion relations that are first order in the particle energies and that correspond to a vacuum refractive index $\eta = 1 - (E/M_{\text{Pl}})^\alpha$, where $\alpha = 1$. Models with quadratic dependences of the vacuum refractive index on energy: $\alpha = 2$ have also been considered [59].

In particular, the D-particle realization of the Liouville string approach predicts that only gauge bosons such as photons, not charged matter particles such as electrons, might have QG-modified dispersion relations. This difference may be traced to the facts that [96] excitations which are charged under the gauge group are represented by open strings with their ends attached to the D-brane [97], and that only neutral excitations are allowed to propagate in the bulk space transverse to the brane. Thus, if we consider photons and electrons, in this model the parameter η is forced to be null, whereas ξ is free to vary. Even more importantly, the theory is CPT even, implying that vacuum is not birefringent for photons ($\xi_+ = \xi_-$).

13.5.4.2 Doubly Special Relativity

Lorentz invariance of physical laws relies on only few assumptions: the principle of relativity, stating the equivalence of physical laws for non-accelerated observers, isotropy (no preferred direction) and homogeneity (no preferred location) of spacetime, and a notion of precausality, requiring that the time ordering of co-local events in one reference frame be preserved [98–105].

All the realizations of LV we have discussed so far explicitly violate the principle of relativity by introducing a preferred reference frame. This may seem a high price to pay to include QG effects in low energy physics. For this reason, it is worth exploring an alternative possibility that keeps the relativity principle but that relaxes one or more of the above postulates.

For example, relaxing the space isotropy postulate leads to the so-called Very Special Relativity framework [106], which was later on understood to be described by a Finslerian-type geometry [107–109]. In this example, however, the generators of the new relativity group number fewer than the usual ten associated with Poincaré invariance. Specifically, there is an explicit breaking of the $O(3)$ group associated with rotational invariance.

One may wonder whether there exist alternative relativity groups with the same number of generators as special relativity. Currently, we know of no such generalization in coordinate space. However, it has been suggested that, at least in momentum space, such a generalization is possible, and it was termed "doubly" or "deformed" (to stress the fact that it still has 10 generators) special relativity, DSR. Even though DSR aims at consistently including dynamics, a complete formulation capable of doing so is still missing, and present attempts face major problems. Thus, at present DSR is only a kinematic theory. Nevertheless, it is attractive because it does not postulate the existence of a preferred frame, but rather deforms the usual concept of Lorentz invariance in the following sense.

Consider the Lorentz algebra of the generators of rotations, L_i, and boosts, B_i:

$$[L_i, L_j] = \iota\epsilon_{ijk}L_k; \quad [L_i, B_j] = \iota\epsilon_{ijk}B_k; \quad [B_i, B_j] = -\iota\epsilon_{ijk}L_k \quad (13.56)$$

(Latin indices $i, j, \ldots$ run from 1 to 3) and supplement it with the following commutators between the Lorentz generators and those of translations in spacetime (the momentum operators P_0 and P_i):

$$[L_i, P_0] = 0; \qquad [L_i, P_j] = \imath \epsilon_{ijk} P_k; \tag{13.57}$$

$$[B_i, P_0] = \imath f_1\left(\frac{P}{\kappa}\right) P_i; \tag{13.58}$$

$$[B_i, P_j] = \imath \left[\delta_{ij} f_2\left(\frac{P}{\kappa}\right) P_0 + f_3\left(\frac{P}{\kappa}\right) \frac{P_i P_j}{\kappa}\right], \tag{13.59}$$

where κ is some unknown energy scale. Finally, assume $[P_i, P_j] = 0$. The commutation relations (13.58)–(13.59) are given in terms of three unspecified, dimensionless structure functions f_1, f_2, and f_3, and they are sufficiently general to include all known DSR proposals—the DSR1 [110], DSR2 [111, 112], and DSR3 [113]. Furthermore, in all the DSRs considered to date, the dimensionless arguments of these functions are specialized to

$$f_i\left(\frac{P}{\kappa}\right) \to f_i\left(\frac{P_0}{\kappa}, \frac{\sum_{i=1}^{3} P_i^2}{\kappa^2}\right), \tag{13.60}$$

so rotational symmetry is completely unaffected. For the $\kappa \to +\infty$ limit to reduce to ordinary special relativity, f_1 and f_2 must tend to 1, and f_3 must tend to some finite value.

DSR theory postulates that the Lorentz group still generates spacetime symmetries but that it acts in a non-linear way on the fields, such that not only is the speed of light c an invariant quantity, but also that there is a new invariant momentum scale κ which is usually taken to be of the order of M_{Pl}. Note that DSR-like features are found in models of non-commutative geometry, in particular in the κ-Minkowski framework [114, 115], as well as in non-canonically non commutative field theories [116].

Concerning phenomenology, an important point about DSR in momentum space is that in all three of its formulations (DSR1 [110], DSR2 [111, 112], and DSR3 [113]) the component of the four momentum having deformed commutation with the boost generator can always be rewritten as a non-linear combination of some energy-momentum vector that transforms linearly under the Lorentz group [117]. For example in the case of DSR2 [111, 112] one can write s

$$E = \frac{-\pi_0}{1 - \pi_0/\kappa}; \tag{13.61}$$

$$p_i = \frac{\pi_i}{1 - \pi_0/\kappa}. \tag{13.62}$$

It is easy to ensure that while π satisfies the usual dispersion relation $\pi_0^2 - \pi^2 = m^2$ (for a particle with mass m), E and p_i satisfy the modified relation

$$\left(1 - m^2/\kappa^2\right) E^2 + 2\kappa^{-1} m^2 E - p^2 = m^2. \tag{13.63}$$

Furthermore, a different composition for energy-momentum now holds, given that the composition for the physical DSR momentum p must be derived from the standard energy-momentum conservation of the pseudo-variable π and in general implies non-linear terms. A crucial point is that due to the above structure if a threshold reaction is forbidden in relativistic physics then it is going to be still forbidden by DSR. Hence many constraints that apply to EFT do not apply to DSR.

Despite its conceptual appeal, DSR is riddled with many open problems. First, if DSR is formulated as described above—that is, only in momentum space—then it is an incomplete theory. Moreover, because it is always possible to introduce the new variables π_μ, on which the Lorentz group acts in a linear manner, the only way that DSR can avoid triviality is if there is some physical way of distinguishing the pseudo-energy $\epsilon \equiv -\pi_0$ from the true-energy E, and the pseudo-momentum π from the true-momentum p. If not, DSR is no more than a nonlinear choice of coordinates in momentum space.

In view of the standard relations $E \leftrightarrow \imath\hbar\partial_t$ and $p \leftrightarrow -\imath\hbar\nabla$ (which are presumably modified in DSR), it is clear that to physically distinguish the pseudo-energy ϵ from the true-energy E, and the pseudo-momentum π from the true-momentum p, one must know how to relate momenta to position. At a minimum, one needs to develop a notion of DSR spacetime.

In this endeavor, there have been two distinct lines of approach, one presuming commutative spacetime coordinates, the other attempting to relate the DSR feature in momentum space to a non commutative position space. In both cases, several authors have pointed out major problems. In the case of commutative spacetime coordinates, some analyses have led authors to question the triviality [118] or internal consistency [119–121] of DSR. On the other hand, non-commutative proposals [57] are not yet well understood, although intense research in this direction is under way [122]. Finally we cannot omit the recent development of what one could perhaps consider a spin-off of DSR that is Relative Locality, which is based on the idea that the invariant arena for classical physics is a curved momentum space rather than spacetime (the latter being a derived concept) [123].

DSR and Relative Locality are still a subject of active research and debate (see e.g. [124–127]); nonetheless, they have not yet attained the level of maturity needed to cast robust constraints.[5] For these reasons, in the next sections we focus upon LV EFT and discuss the constraints within this framework.

---

[5]Note however, that some knowledge of DSR phenomenology can be obtained by considering that, as in Special Relativity, any phenomenon that implies the existence of a preferred reference frame is forbidden. Thus, the detection of such a phenomenon would imply the falsification of both special and doubly-special relativity. An example of such a process is the decay of a massless particle.

13.6 Experimental Probes of Low Energy LV: Earth Based Experiments

The world as we see it seems ruled by Lorentz invariance to a very high degree. Hence, when seeking tests of Lorentz violations one is confronted with the challenge to find or very high precision experiments able to test Special Relativity or observe effects which might be sensitive to tiny deviations from standard LI. Within the ansatz we lied down in the previous sections it is clear that the first route is practical only when dealing with low energy violations of Lorentz invariance as systematically described by the minimal Standard Model extension (mSME) while astrophysical tests, albeit much less precise, are the choice too for testing LV induced by higher order operators. Let us then briefly review the main experimental tools used so far in order to perform precision tests of Lorentz invariance in laboratory (for more details see e.g. [7, 24, 128]).

13.6.1 Penning Traps

In a Penning trap a charged particle can be localized for long times using a combination of static magnetic and electric fields. Lorentz violating tests are based on monitoring the particle cyclotron motion in the magnetic field and Larmor precession due to the spin. In fact the relevant frequencies for both these motions are modified in the mSME and Penning traps can be made very sensitive to differences in these frequencies.

13.6.2 Clock Comparison Experiments

Clock comparison experiments are generally performed by considering two atomic transition frequencies (which can be considered as two clocks) in the same point in space. The basic idea is that as the clocks move in space, they pick out different components of the Lorentz violating tensors in the mSME. This would yield a sidereal drift between the two clocks. Measuring the difference between the frequencies over long periods, allows to cast very high precision limits on the parameters in the mSME (generally for protons and neutrons).

13.6.3 Cavity Experiments

In cavity experiments one casts constraints on the variation of the cavity resonance frequency as its orientation changes in space. While this is intrinsically similar to clock comparison experiments, these kind of experiments allows to cast constraints also on the electromagnetic sector of the mSME.

13.6.4 Spin Polarized Torsion Balance

The electron sector of the mSME can be effective constrained via spin-torsion balances. An example is an octagonal pattern of magnets which is constructed so to have an overall spin polarization in the octagon's plane. Four of these octagons are suspended from a torsion fiber in a vacuum chamber. This arrangement of the magnets give an estimated net spin polarization equivalent to $\approx 10^{23}$ aligned electron spins. The whole apparatus is mounted on a turntable. As the turntable moves Lorentz violation in the mSME produces an interaction potential for non-relativistic electrons which induces a torque on the torsion balance. The torsion fiber is then twisted by an amount related to the relevant LV coefficients.

13.6.5 Neutral Mesons

In the mSME one expects an orientation dependent change in the mass difference e.g. of neutral kaons. By looking for sidereal variations or other orientation effects one can derive bounds on each component of the relevant LV coefficients.

13.7 Observational Probes of High Energy LV: Astrophysical QED Reactions

Let us begin with a brief review of the most common types of reaction exploited in order to give constraints on the QED sector.

For definiteness, we refer to the following modified dispersion relations:

$$E_\gamma^2 = k^2 + \xi_\pm^{(n)} \frac{k^n}{M_{\text{Pl}}^{n-2}} \quad \text{Photon,} \tag{13.64}$$

$$E_{el}^2 = m_e^2 + p^2 + \eta_\pm^{(n)} \frac{p^n}{M_{\text{Pl}}^{n-2}} \quad \text{Electron-Positron,} \tag{13.65}$$

where, in the EFT case, we have $\xi^{(n)} \equiv \xi_+^{(n)} = (-)^n \xi_-^{(n)}$ and $\eta^{(n)} \equiv \eta_+^{(n)} = (-)^n \eta_-^{(n)}$.

13.7.1 Photon Time of Flight

Although photon time-of-flight constraints currently provide limits several orders of magnitude weaker than the best ones, they have been widely adopted in the astrophysical community. Furthermore they were the first to be proposed in the seminal paper [25]. More importantly, given their purely kinematical nature, they may be applied to a broad class of frameworks beyond EFT with LV. For this reason, we provide a general description of time-of-flight effects, elaborating on their application to the case of EFT below.

In general, a photon dispersion relation in the form of (13.64) implies that photons of different colors (wave vectors k_1 and k_2) travel at slightly different speeds. Let us first assume that there are no birefringent effects, so that $\xi_+^{(n)} = \xi_-^{(n)}$. Then, upon propagation on a cosmological distance d, the effect of energy dependence of the photon group velocity produces a time delay

$$\Delta t^{(n)} = \frac{n-1}{2} \frac{k_2^{n-2} - k_1^{n-2}}{M_{\text{Pl}}^{n-2}} \xi^{(n)} d, \qquad (13.66)$$

which clearly increases with d and with the energy difference as long as $n > 2$. The largest systematic error affecting this method is the uncertainty about whether photons of different energy are produced simultaneously in the source.

So far, the most robust constraints on $\xi^{(3)}$, derived from time of flight differences, have been obtained within the D-brane model (discussed in Sect. 13.5.4.1) from a statistical analysis applied to the arrival times of sharp features in the intensity at different energies from a large sample of GRBs with known redshifts [129], leading to limits $\xi^{(3)} \leq O(10^3)$. A recent example illustrating the importance of systematic uncertainties can be found in [130], where the strongest limit $\xi^{(3)} < 47$ is found by looking at a very strong flare in the TeV band of the AGN Markarian 501.

One way to alleviate systematic uncertainties—available only in the context of birefringent theories, such as the one with $n = 3$ in EFT—would be to measure the velocity difference between the two polarization states at a single energy, corresponding to

$$\Delta t = 2|\xi^{(3)}|kd/M_{\text{Pl}}. \qquad (13.67)$$

This bound would require that both polarizations be observed and that no spurious helicity-dependent mechanism (such as, for example, propagation through a birefringent medium) affects the relative propagation of the two polarization states.

Let us stress that Eq. (13.66) is no longer valid in birefringent theories. In fact, photon beams generally are not circularly polarized; thus, they are a superposition of fast and slow modes. Therefore, the net effect of this superposition may partially or completely erase the time-delay effect. To compute this effect on a generic photon beam in a birefringent theory, let us describe a beam of light by means of the associated electric field, and let us assume that this beam has been generated with a Gaussian width

$$\mathbf{E} = A\left(e^{i(\Omega_0 t - k^+(\Omega_0)z)} e^{-(z-v_g^+ t)^2 \delta\Omega_0^2} \hat{e}_+ + e^{i(\Omega_0 t - k^-(\Omega_0)z)} e^{-(z-v_g^- t)^2 \delta\Omega_0^2} \hat{e}_-\right), \qquad (13.68)$$

where Ω_0 is the wave frequency, $\delta\Omega_0$ is the Gaussian width of the wave, $k^\pm(\Omega_0)$ is the "momentum" corresponding to the given frequency according to (13.64) and $\hat{e}_\pm \equiv (\hat{e}_1 \pm i\hat{e}_2)/\sqrt{2}$ are the helicity eigenstates. Note that by complex conjugation

$\hat{e}_+^* = \hat{e}_-$. Also, note that $k^\pm(\omega) = \omega \mp \xi \omega^2/M_{\rm Pl}$. Thus,

$$\mathbf{E} = A e^{i\Omega_0(t-z)} \left(e^{i\xi\Omega_0^2/M_{\rm Pl} z} e^{-(z-v_g^+ t)^2 \delta\Omega_0^2} \hat{e}_+ + e^{-i\xi\Omega_0^2/M_{\rm Pl} z} e^{-(z-v_g^- t)^2 \delta\Omega_0^2} \hat{e}_- \right). \tag{13.69}$$

The intensity of the wave beam can be computed as

$$\mathbf{E} \cdot \mathbf{E}^* = |A|^2 \left(e^{2i\xi\Omega_0^2/M_{\rm Pl} z} + e^{-2i\xi\Omega_0^2/M_{\rm Pl} z} \right) e^{-\delta\Omega_0^2((z-v_g^+ t)^2 + (z-v_g^- t)^2)}$$

$$= 2|A|^2 e^{-2\delta\Omega_0^2(z-t)^2} \cos\left(2\xi \frac{\Omega_0}{M_{\rm Pl}} \Omega_0 z \right) e^{-2\xi^2 \frac{\Omega_0^2}{M^2}(\delta\Omega_0 t)^2}. \tag{13.70}$$

This shows that there is an effect even on a linearly-polarised beam. The effect is a modulation of the wave intensity that depends quadratically on the energy and linearly on the distance of propagation. In addition, for a Gaussian wave packet, there is a shift of the packet centre, that is controlled by the square of $\xi^{(3)}/M_{\rm Pl}$ and hence is strongly suppressed with respect to the cosinusoidal modulation.

13.7.2 Vacuum Birefringence

The fact that electromagnetic waves with opposite "helicities" have slightly different group velocities, in EFT LV with $n = 3$, implies that the polarisation vector of a linearly polarised plane wave with energy k rotates, during the wave propagation over a distance d, through the angle [33][6]

$$\theta(d) = \frac{\omega_+(k) - \omega_-(k)}{2} d \simeq \xi^{(3)} \frac{k^2 d}{2 M_{\rm Pl}}. \tag{13.72}$$

Observations of polarized light from a distant source can then lead to a constraint on $|\xi^{(3)}|$ that, depending on the amount of available information—both on the observational and on the theoretical (i.e. astrophysical source modeling) side—can be cast in two different ways [131]:

1. Because detectors have a finite energy bandwidth, Eq. (13.72) is never probed in real situations. Rather, if some net amount of polarization is measured in the band $k_1 < E < k_2$, an order-of-magnitude constraint arises from the fact that if the angle of polarization rotation (13.72) differed by more than $\pi/2$ over this band, the

---

[6]Note that for an object located at cosmological distance (let z be its redshift), the distance d becomes

$$d(z) = \frac{1}{H_0} \int_0^z \frac{1+z'}{\sqrt{\Omega_\Lambda + \Omega_m(1+z')^3}} dz', \tag{13.71}$$

where $d(z)$ is not exactly the distance of the object as it includes a $(1+z)^2$ factor in the integrand to take into account the redshift acting on the photon energies.

detected polarization would fluctuate sufficiently for the net signal polarization to be suppressed [93, 132]. From (13.72), this constraint is

$$\xi^{(3)} \lesssim \frac{\pi M_{Pl}}{(k_2^2 - k_1^2)d(z)}, \tag{13.73}$$

This constraint requires that any intrinsic polarization (at source) not be completely washed out during signal propagation. It thus relies on the mere detection of a polarized signal; there is no need to consider the observed polarization degree. A more refined limit can be obtained by calculating the maximum observable polarization degree, given the maximum intrinsic value [133]:

$$\Pi(\xi) = \Pi(0)\sqrt{\langle\cos(2\theta)\rangle_{\mathscr{P}}^2 + \langle\sin(2\theta)\rangle_{\mathscr{P}}^2}, \tag{13.74}$$

where $\Pi(0)$ is the maximum intrinsic degree of polarization, θ is defined in Eq. (13.72) and the average is weighted over the source spectrum and instrumental efficiency, represented by the normalized weight function $\mathscr{P}(k)$ [132]. Conservatively, one can set $\Pi(0) = 100\,\%$, but a lower value may be justified on the basis of source modeling. Using (13.74), one can then cast a constraint by requiring $\Pi(\xi)$ to exceed the observed value.

2. Suppose that polarized light measured in a certain energy band has a position angle θ_{obs} with respect to a fixed direction. At fixed energy, the polarization vector rotates by the angle (13.72);[7] if the position angle is measured by averaging over a certain energy range, the final net rotation $\langle\Delta\theta\rangle$ is given by the superposition of the polarization vectors of all the photons in that range:

$$\tan(2\langle\Delta\theta\rangle) = \frac{\langle\sin(2\theta)\rangle_{\mathscr{P}}}{\langle\cos(2\theta)\rangle_{\mathscr{P}}}, \tag{13.75}$$

where θ is given by (13.72). If the position angle at emission θ_i in the same energy band is known from a model of the emitting source, a constraint can be set by imposing

$$\tan(2\langle\Delta\theta\rangle) < \tan(2\theta_{obs} - 2\theta_i). \tag{13.76}$$

Although this limit is tighter than those based on Eqs. (13.73) and (13.74), it clearly hinges on assumptions about the nature of the source, which may introduce significant uncertainties.

In conclusion the fact that polarised photon beams are indeed observed from distant objects imposes strong constraints on LV in the photon sector (i.e. on $\xi^{(3)}$), as we shall see later on.

---

[7] Faraday rotation is negligible at high energies.

Table 13.1 Values of p_{th}, according to Eq. (13.77), for different particles involved in the reaction: neutrinos, electrons and proton. Here we assume $\eta^{(n)} \simeq 1$

	$m_\nu \simeq 0.1$ eV	$m_e \simeq 0.5$ MeV	$m_p \simeq 1$ GeV
$n = 2$	0.1 eV	0.5 MeV	1 GeV
$n = 3$	500 MeV	14 TeV	2 PeV
$n = 4$	33 TeV	74 PeV	3 EeV

13.7.3 Threshold Reactions

An interesting phenomenology of threshold reactions is introduced by LV in EFT; also, threshold theorems can be generalized [32]. Sticking to the present case of rotational invariance and monotonic dispersion relations (see [134] for a generalization to more complex situations), the main conclusions of the investigation into threshold reactions are that [31]

- Threshold configurations still corresponds to head-on incoming particles and parallel outgoing ones
- The threshold energy of existing threshold reactions can shift, and upper thresholds (i.e. maximal incoming momenta at which the reaction can happen in any configuration) can appear
- Pair production can occur with unequal outgoing momenta
- New, normally forbidden reactions can be viable

LV corrections are surprisingly important in threshold reactions because the LV term (which as a first approximation can be considered as an additional mass term) should be compared not to the momentum of the involved particles, but rather to the (invariant) mass of the particles produced in the final state. Thus, an estimate for the threshold energy is

$$p_{\text{th}} \simeq \left(\frac{m^2 M_{\text{Pl}}^{n-2}}{\eta^{(n)}} \right)^{1/n}, \qquad (13.77)$$

where m is the typical mass of particles involved in the reaction. Interesting values for p_{th} are discussed, e.g., in [31] and given in Table 13.1. Reactions involving neutrinos are the best candidate for observation of LV effects, whereas electrons and positrons can provide results for $n = 3$ theories but can hardly be accelerated by astrophysical objects up to the required energy for $n = 4$. In this case reactions of protons can be very effective, because cosmic-rays can have energies well above 3 EeV. Let us now briefly review the main reaction used so far in order to casts constraints.

13.7.3.1 LV-Allowed Threshold Reactions: γ-Decay

The decay of a photon into an electron/positron pair is made possible by LV because energy-momentum conservation may now allow reactions described by the basic

QED vertex. This process has a threshold that, if $\xi \simeq 0$ and $n = 3$, is set by the condition [33]

$$k_{th} = \left(6\sqrt{3} m_e^2 M / |\eta_\pm^{(3)}|\right)^{1/3}. \qquad (13.78)$$

Noticeably, as already mentioned above, the electron-positron pair can now be created with slightly different outgoing momenta (asymmetric pair production). Furthermore, the decay rate is extremely fast above threshold [33] and is of the order of $(10 \text{ ns})^{-1}$ ($n = 3$) or $(10^{-6} \text{ ns})^{-1}$ ($n = 4$).

13.7.3.2 LV-Allowed Threshold Reactions: Vacuum Čerenkov and Helicity Decay

In the presence of LV, the process of Vacuum Čerenkov (VC) radiation $e^\pm \to e^\pm \gamma$ can occur. If we set $\xi \simeq 0$ and $n = 3$, the threshold energy is given by

$$p_{VC} = \left(m_e^2 M / 2\eta^{(3)}\right)^{1/3} \simeq 11 \text{ TeV } \eta^{-1/3}. \qquad (13.79)$$

Just above threshold this process is extremely efficient, with a time scale of order $\tau_{VC} \sim 10^{-9}$ s [33].

A slightly different version of this process is the Helicity Decay (HD, $e_\mp \to e_\pm \gamma$). If $\eta_+ \neq \eta_-$, an electron/positron can flip its helicity by emitting a suitably polarized photon. This reaction does not have a real threshold, but rather an effective one [33]—$p_{HD} = (m_e^2 M / \Delta \eta)^{1/3}$, where $\Delta \eta = |\eta_+^{(3)} - \eta_-^{(3)}|$—at which the decay lifetime τ_{HD} is minimized. For $\Delta \eta \approx O(1)$ this effective threshold is around 10 TeV. Note that below threshold $\tau_{HD} > \Delta \eta^{-3} (p/10 \text{ TeV})^{-8} 10^{-9}$ s, while above threshold τ_{HD} becomes independent of $\Delta \eta$ [33].

Apart from the above mentioned examples of reactions normally forbidden and now allowed by LV dispersion relations, one can also look for modifications of normally allowed threshold reactions especially relevant in high energy astrophysics.

13.7.3.3 LV-Allowed Threshold Reactions: Photon Splitting and Lepton Pair Production

It is rather obvious that once photon decay and vacuum Čerenkov are allowed also the related relations in which respectively the our going lepton pair is replaced by two or more photons, $\gamma \to 2\gamma$ and $\gamma \to 3\gamma$, etc., or the outgoing photons is replaced by an electron-positron pair, $e^- \to e^- e^- e^+$, are also allowed.

Photon Splitting This is forbidden for $\xi^{(n)} < 0$ while it is always allowed if $\xi^{(n)} > 0$ [31]. When allowed, the relevance of this process is simply related to its rate. The most relevant cases are $\gamma \to \gamma\gamma$ and $\gamma \to 3\gamma$, because processes with more photons in the final state are suppressed by more powers of the fine structure constant.

The $\gamma \to \gamma\gamma$ process is forbidden in QED because of kinematics and C-parity conservation. In LV EFT neither condition holds. However, we can argue that this process is suppressed by an additional power of the Planck mass, with respect to $\gamma \to 3\gamma$. In fact, in LI QED the matrix element is zero due to the exact cancellation of fermionic and anti-fermionic loops. In LV EFT this cancellation is not exact and the matrix element is expected to be proportional to at least $(\xi E/M_{\text{Pl}})^p$, $p > 0$, as it is induced by LV and must vanish in the limit $M_{\text{Pl}} \to \infty$.

Therefore we have to deal only with $\gamma \to 3\gamma$. This process has been studied in [31, 135]. In particular, in [135] it was found that, if the "effective photon mass" $m_\gamma^2 \equiv \xi E_\gamma^n / M_{\text{Pl}}^{n-2} \ll m_e^2$, then the splitting lifetime of a photon is approximately $\tau^{n=3} \simeq 0.025 \xi^{-5} f^{-1} (50\,\text{TeV}/E_\gamma)^{14}$ s, where f is a phase space factor of order 1. This rate was rather higher than the one obtained via dimensional analysis in [31] because, due to integration of loop factors, additional dimensionless contributions proportional to m_e^8 enhance the splitting rate at low energy.

This analysis, however, does not apply for the most interesting case of ultra high energy photons around 10^{19} eV (see below Sect. 13.8) given that at these energies $m_\gamma^2 \gg m_e^2$ if $\xi^{(3)} > 10^{-17}$ and $\xi^{(4)} > 10^{-8}$. Hence the above mentioned loop contributions are at most logarithmic, as the momentum circulating in the fermionic loop is much larger than m_e. Moreover, in this regime the splitting rate depends only on m_γ, the only energy scale present in the problem. One then expects the analysis proposed in [31] to be correct and the splitting time scale to be negligible at $E_\gamma \simeq 10^{19}$ eV.

Lepton Pair Production The process $e^- \to e^-e^-e^+$ is similar to vacuum Čerenkov radiation or helicity decay, with the final photon replaced by an electron-positron pair. Various combinations of helicities for the different fermions can be considered individually. If we choose the particularly simple case (and the only one we shall consider here) where all electrons have the same helicity and the positron has the opposite helicity, then the threshold energy will depend on only one LV parameter. In [31] was derived the threshold for this reaction, finding that it is a factor ~ 2.5 times higher than that for soft vacuum Čerenkov radiation. The rate for the reaction is high as well, hence constraints may be imposed using just the value of the threshold.

13.7.3.4 LV-Modified Threshold Reactions: Photon Pair-Creation

A process related to photon decay is photon absorption, $\gamma\gamma \to e^+e^-$. Unlike photon decay, this is allowed in Lorentz invariant QED and it plays a crucial role in making our universe opaque to gamma rays above tents of TeVs.

If one of the photons has energy ω_0, the threshold for the reaction occurs in a head-on collision with the second photon having the momentum (equivalently energy) $k_{\text{LI}} = m^2/\omega_0$. For example, if $k_{\text{LI}} = 10$ TeV (the typical energy of inverse Compton generated photons in some active galactic nuclei) the soft photon threshold ω_0 is approximately 25 meV, corresponding to a wavelength of 50 microns.

In the presence of Lorentz violating dispersion relations the threshold for this process is in general altered, and the process can even be forbidden. Moreover, as firstly noticed by Kluźniak [136] and mentioned before, in some cases there is an upper threshold beyond which the process does not occur. Physically, this means that at sufficiently high momentum the photon does not carry enough energy to create a pair and simultaneously conserve energy and momentum. Note also, that an upper threshold can only be found in regions of the parameter space in which the γ-decay is forbidden, because if a single photon is able to create a pair, then *a fortiori* two interacting photons will do [31].

Let us exploit the above mentioned relation $\eta^{e^-}_\pm = (-)^n \eta^{e^+}_\mp$ between the electron-positron coefficients, and assume that on average the initial state is unpolarized. In this case, using the energy-momentum conservation, the kinematics equation governing pair production is the following [33]

$$\frac{m^2}{k^n y(1-y)} = \frac{4\omega_b}{k^{n-1}} + \tilde{\xi} - \tilde{\eta}\left(y^{n-1} + (-)^n(1-y)^{n-1}\right) \tag{13.80}$$

where $\tilde{\xi} \equiv \xi^{(n)}/M^{n-2}$ and $\tilde{\eta} \equiv \eta^{(n)}/M^{n-2}$ are respectively the photon's and electron's LV coefficients divided by powers of M, $0 < y < 1$ is the fraction of momentum carried by either the electron or the positron with respect to the momentum k of the incoming high-energy photon and ω_b is the energy of the target photon. In general the analysis is rather complicated. In particular it is necessary to sort out whether the thresholds are lower or upper ones, and whether they occur with the same or different pair momenta

13.7.4 Synchrotron Radiation

Synchrotron emission is strongly affected by LV, however for Planck scale LV and observed energies, it is a relevant "window" only for dimension four or five LV QED. We shall work out here the details of dimension five QED ($n = 3$) for illustrative reasons (see e.g. [137] for the mSME case).

In both LI and LV cases [33], most of the radiation from an electron of energy E is emitted at a critical frequency

$$\omega_c = \frac{3}{2} eB \frac{\gamma^3(E)}{E} \tag{13.81}$$

where $\gamma(E) = (1 - v^2(E))^{-1/2}$, and $v(E)$ is the electron group velocity.

However, in the LV case, and assuming specifically $n = 3$, the electron group velocity is given by

$$v(E) = \frac{\partial E}{\partial p} = \left(1 - \frac{m_e^2}{2p^2} + \eta^{(3)} \frac{p}{M}\right). \tag{13.82}$$

Therefore, $v(E)$ can exceed 1 if $\eta > 0$ or it can be strictly less than 1 if $\eta < 0$. This introduces a fundamental difference between particles with positive or negative LV coefficient η.

If η is negative the group velocity of the electrons is strictly less than the (low energy) speed of light. This implies that, at sufficiently high energy, $\gamma(E)_- < E/m_e$, for all E. As a consequence, the critical frequency $\omega_c^-(\gamma, E)$ is always less than a maximal frequency $\omega_c^{\max}$ [33]. Then, if synchrotron emission up to some frequency ω_{obs} is observed, one can deduce that the LV coefficient for the corresponding leptons cannot be more negative than the value for which $\omega_c^{\max} = \omega_{\text{obs}}$. Then, if synchrotron emission up to some maximal frequency ω_{obs} is observed, one can deduce that the LV coefficient for the corresponding leptons cannot be more negative than the value for which $\omega_c^{\max} = \omega_{\text{obs}}$, leading to the bound [33]

$$\eta^{(3)} > -\frac{M}{m_e}\left(\frac{0.34 eB}{m_e \omega_{\text{obs}}}\right)^{3/2}. \tag{13.83}$$

If η is instead positive the leptons can be superluminal. One can show that at energies $E_c \gtrsim 8\text{ TeV}/\eta^{1/3}$, $\gamma(E)$ begins to increase faster than E/m_e and reaches infinity at a finite energy, which corresponds to the threshold for soft VC emission. The critical frequency is thus larger than the LI one and the spectrum shows a characteristic bump due to the enhanced ω_c.

13.8 Current Constraints on the QED Sector

Let us now come to a brief review of the present constraints on LV QED and in other sectors of the standard model. We shall not spell out the technical details here. These can be found in dedicated, recent, reviews such as [138].

13.8.1 mSME Constraints

It would be cumbersome to summarize here the constraints on the minimal Standard Model extension (dimension there and four operators) as many parameters characterize the full model. A summary can be found in [128]. One can of course restrict the mSME to the rotational invariant subset. In this case the model basically coincides with the Coleman-Glashow one [28]. In this case the constraints are quite strong, for example on the QED sector one can easily see that the absence of gamma decay up to 50 TeV provides a constraint of order 10^{-16} on the difference between the limit speed of photons and electrons [94]. Constraint up to $O(10^{-22})$ can be achieved on other mSME parameters for dimension four LV terms via precision experiments like Penning traps.

13.8.2 Constraints on QED with $O(E/M)$ LV

It is quite remarkable that a single object can nowadays provide the most stringent constraints for LV QED with $O(E/M)$ modified dispersion relations, this object is the Crab Nebula (CN). The CN is a source of diffuse radio, optical and X-ray radiation associated with a Supernova explosion observed in 1054 A.D. Its distance from Earth is approximately 1.9 kpc. A pulsar, presumably a remnant of the explosion, is located at the centre of the Nebula. The Nebula emits an extremely broad-band spectrum (21 decades in frequency, see [139] for a comprehensive list of relevant observations) that is produced by two major radiation mechanisms. The emission from radio to low energy γ-rays ($E < 1$ GeV) is thought to be synchrotron radiation from relativistic electrons, whereas inverse Compton (IC) scattering by these electrons is the favored explanation for the higher energy γ-rays. From a theoretical point of view, the current understanding of the whole environment is based on the model presented in [140], which accounts for the general features observed in the CN spectrum.

Recently, a claim of $|\xi^{(3)}| \lesssim 2 \times 10^{-7}$ was made using UV/optical polarisation measures from GRBs [141]. However, the strongest constraint to date comes from a local object. In [131] the constraint $|\xi^{(3)}| \lesssim 6 \times 10^{-10}$ at 95 % Confidence Level (CL) was obtained by considering the observed polarization of hard-X rays from the CN [142] (see also [143]).

13.8.2.1 Synchrotron Constraint

How the synchrotron emission processes at work in the CN would appear in a "LV world" has been studied in [139, 144]. There the role of LV in modifying the characteristics of the Fermi mechanism (which is thought to be responsible for the formation of the spectrum of energetic electrons in the CN [145]) and the contributions of vacuum Čerenkov and helicity decay were investigated for $n = 3$ LV. This procedure requires fixing most of the model parameters using radio to soft X-rays observations, which are basically unaffected by LV.

Given the dispersion relations (13.48) and (13.49), clearly only two configurations in the LV parameter space are truly different: $\eta_+ \cdot \eta_- > 0$ and $\eta_+ \cdot \eta_- < 0$, where η_+ is assumed to be positive for definiteness. The configuration wherein both $\eta_\pm$ are negative is the same as the $(\eta_+ \cdot \eta_- > 0, \eta_+ > 0)$ case, whereas that whose signs are scrambled is equivalent to the case $(\eta_+ \cdot \eta_- < 0, \eta_+ > 0)$. This is because positron coefficients are related to electron coefficients through $\eta_\pm^{af} = -\eta_\mp^{f}$ [33]. Examples of spectra obtained for the two different cases are shown in Fig. 13.1.

A χ^2 analysis has been performed to quantify the agreement between models and data [139]. From this analysis, one can conclude that the LV parameters for the leptons are both constrained, at 95 % CL, to be $|\eta_\pm| < 10^{-5}$, as shown by the red vertical lines in Fig. 13.3. Although the best fit model is not the LI one, a careful statistical analysis (performed with present-day data) shows that it is statistically indistinguishable from the LI model at 95 % CL [139].

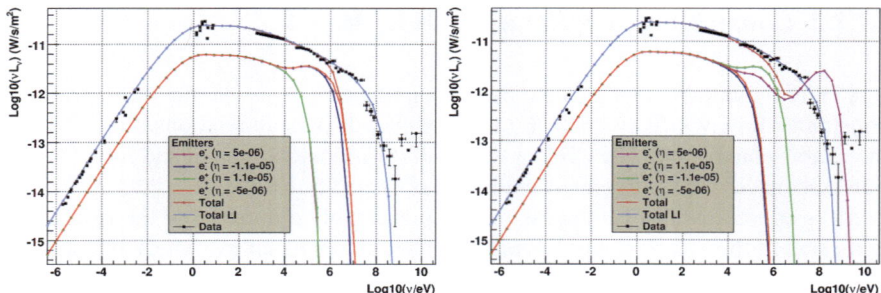

Fig. 13.1 Comparison between observational data, the LI model and a LV one with $\eta_+ \cdot \eta_- < 0$ (*left*) and $\eta_+ \cdot \eta_- > 0$ (*right*). The values of the LV coefficients, reported in the insets, show the salient features of the LV modified spectra. The leptons are injected according to the best fit values $p = 2.4$, $E_c = 2.5$ PeV. The individual contribution of each lepton population is shown

13.8.2.2 Birefringence Constraint

In the case of the CN a (46 ± 10) % degree of linear polarization in the 100 keV–1 MeV band has recently been measured by the INTEGRAL mission [142, 146]. This measurement uses all photons within the SPI instrument energy band. However the convolution of the instrumental sensitivity to polarization with the detected number counts as a function of energy, $\mathcal{P}(k)$, is maximized and approximately constant within a narrower energy band (150 to 300 keV) and falls steeply outside this range [147]. For this reason we shall, conservatively, assume that most polarized photons are concentrated in this band. Given $d_{\text{Crab}} = 1.9$ kpc, $k_2 = 300$ keV and $k_1 = 150$ keV, Eq. (13.73) leads to the order-of-magnitude estimate $|\xi| \lesssim 2 \times 10^{-9}$. A more accurate limit follows from (13.74). In the case of the CN there is a robust understanding that photons in the range of interest are produced via the synchrotron process, for which the maximum degree of intrinsic linear polarization is about 70 % (see e.g. [148]). Figure 13.2 illustrates the dependence of Π on ξ (see Eq. (13.74)) for the distance of the CN and for $\Pi(0) = 70$ %. The requirement $\Pi(\xi) > 16$ % (taking account of a 3σ offset from the best fit value 46 %) leads to the constraint (at 99 % CL)

$$|\xi| \lesssim 6 \times 10^{-9}. \tag{13.84}$$

It is interesting to notice that X-ray polarization measurements of the CN already available in 1978 [149], set a constraint $|\xi| \lesssim 5.4 \times 10^{-6}$, only one order of magnitude less stringent than that reported in [141].

Constraint (13.84) can be tightened by exploiting the current astrophysical understanding of the source. The CN is a cloud of relativistic particles and fields powered by a rapidly rotating, strongly magnetized neutron star. Both the *Hubble Space Telescope* and the *Chandra* X-ray satellite have imaged the system, revealing a jet and torus that clearly identify the neutron star rotation axis [150]. The projection of this axis on the sky lies at a position angle of 124.0° ± 0.1° (measured from North

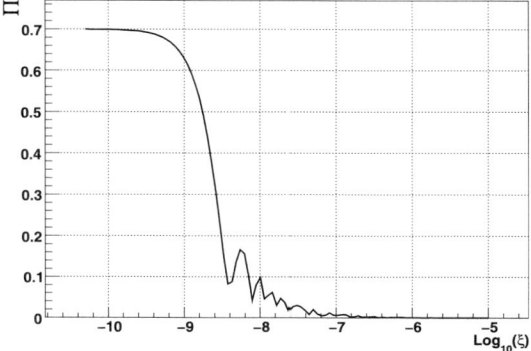

Fig. 13.2 Constraint for the polarization degree. Dependence of Π on ξ for the distance of the CN and photons in the 150–300 keV range, for a constant instrumental sensitivity $\mathscr{P}(k)$

in anti-clockwise). The neutron star itself emits pulsed radiation at its rotation frequency of 30 Hz. In the optical band these pulses are superimposed on a fainter steady component with a linear polarization degree of 30 % and direction precisely aligned with that of the rotation axis [151]. The direction of polarization measured by INTEGRAL-SPI in the γ-rays is $\theta_{\text{obs}} = 123° \pm 11°$ (1σ error) from the North, thus also closely aligned with the jet direction and remarkably consistent with the optical observations.

This compelling (theoretical and observational) evidence allows us to use Eq. (13.76). Conservatively assuming $\theta_i - \theta_{\text{obs}} = 33°$ (i.e. 3σ from θ_i, 99 % CL), this translates into the limit

$$\left|\xi^{(3)}\right| \lesssim 9 \times 10^{-10}, \qquad (13.85)$$

and $|\xi^{(3)}| \lesssim 6 \times 10^{-10}$ for a 2σ deviation (95 % CL).

Polarized light from GRBs has also been detected and given their cosmological distribution they could be ideal sources for improving the above mentioned constraints from birefringence. Attempts in this sense were done in the past [93, 152] (but later on the relevant observation [153] appeared controversial) but so far we do not have sources for which the polarization is detected and the spectral redshift is precisely determined. In [154] this problem was circumvented by using indirect methods (the same used to use GRBs as standard candles) for the estimate of the redshift. This leads to a possibly less robust but striking constraints $|\xi^{(3)}| \lesssim 2.4 \times 10^{-14}$.

Remarkably this constraint was recently further improved by using the INTEGRAL/IBIS observation of the GRB 041219A, for which a luminosity distance of 85 Mpc ($z \approx 0.02$) was derived thanks to the determination of the GRB's host galaxy. In this case a constraint $|\xi^{(3)}| \lesssim 1.1 \times 10^{-14}$ was derived [155].[8]

---

[8]The same paper claims also a strong constraint on the parameter $\xi^{(4)}$. Unfortunately, such a claim is based on the erroneous assumption that the EFT order six operators responsible for this term imply opposite signs for opposite helicities of the photon. We have instead seen that the CPT evenness of the relevant dimension six operators imply a helicity independent dispersion relation for the photon (see Eq. (13.54)).

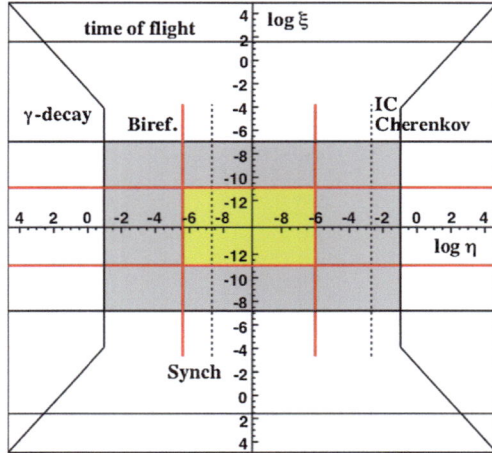

Fig. 13.3 Summary of the constraints on LV QED at order $O(E/M)$. The red lines are related to the constraints derived from the detection of polarized synchrotron radiation from the CN as discussed in the text. For further reference are also shown the constraints that can be derived from the detection of 80 TeV photons from the CN: the *solid black lines* symmetric w.r.t. the ξ axis are derived from the absence of gamma decay, the dashed vertical line cutting the η axis at about 10^{-3} refers to the limit on the vacuum Čerenkov effect coming from the inferred 80 TeV inverse Compton electrons. The *dashed vertical line* on the negative side of the η axis is showing the first synchrotron based constraint derived in [144]

13.8.2.3 Summary

Constraints on LV QED $O(E/M)$ are summarized in Fig. 13.3 where also the constraints—coming from the observations of up to 80 TeV gamma rays from the CN [156] (which imply no gamma decay for these photons neither vacuum Cherenkov at least up to 80 TeV for the electrons producing them via inverse Compton scattering)—are plotted for completeness.

13.8.3 Constraints on QED with $O(E/M)^2$ LV

Looking back at Table 13.1 it is easy to realize that casting constraints on dimension six LV operators in QED requires accessing energies beyond 10^{16} eV. Due to the typical radiative processes characterizing electrons and photons it is extremely hard to directly access these kind of energies. However, the cosmic rays spectrum does extend in this ultra high energy region and it is therefore the main (so far the only) channel for probing these kind of extreme UV LV.

One of the most interesting features related to the physics of Ultra-High-Energy Cosmic Rays (UHECRs) is the Greisen-Zatsepin-Kuzmin (GZK) cut off [41, 42], a suppression of the high-energy tail of the UHECR spectrum arising from interactions with CMB photons, according to $p\gamma \rightarrow \Delta^+ \rightarrow p\pi^0(n\pi^+)$. This process has a (LI) threshold energy $E_{\text{th}} \simeq 5 \times 10^{19} \, (\omega_b/1.3 \text{ meV})^{-1}$ eV (ω_b is the target photon energy). Experimentally, the presence of a suppression of the UHECR flux was claimed only recently [20, 21]. Although the cut off could be also due to the finite acceleration power of the UHECR sources, the fact that it occurs at the expected energy favors the GZK explanation. The results presented in [157] seemed to further strengthen this hypothesis (but see further discussion below).

Rather surprisingly, significant limits on ξ and η can be derived by considering UHE photons generated as secondary products of the GZK reaction [158, 159]. This can be used to further improve the constraints on dimension 5 LV operators and provide a first robust constraint of QED with dimension 6 CPT even LV operators.

These UHE photons originate because the GZK process leads to the production of neutral pions that subsequently decay into photon pairs. These photons are mainly absorbed by pair production onto the CMB and radio background. Thus, the fraction of UHE photons in UHECRs is theoretically predicted to be less than 1 % at 10^{19} eV [160]. Several experiments imposed limits on the presence of photons in the UHECR spectrum. In particular, the photon fraction is less than 2.0 %, 5.1 %, 31 % and 36 % (95 % CL) at $E = 10, 20, 40, 100$ EeV respectively [161, 162].

The point is that pair production is strongly affected by LV. In particular, the (lower) threshold energy can be slightly shifted and in general an upper threshold can be introduced [31]. If the upper threshold energy is lower than 10^{20} eV, then UHE photons are no longer attenuated by the CMB and can reach the Earth, constituting a significant fraction of the total UHECR flux and thereby violating experimental limits [158, 159, 163].

Moreover, it has been shown [159] that the γ-decay process can also imply a significant constraint. Indeed, if some UHE photon ($E_\gamma \simeq 10^{19}$ eV) is detected by experiments (and the Pierre Auger Observatory, PAO, will be able to do so in few years [161]), then γ-decay must be forbidden above 10^{19} eV.

In conclusion we show in Fig. 13.4 the overall picture of the constraints of QED dimension 6 LV operators, where the green dotted lines do not correspond to real constraints, but to the ones that will be achieved when AUGER will observe, as expected, some UHE photon.

Let us add that the same reasoning can be used to further strength the available constraints in dimension 5 LV QED. In this case the absence of relevant UHE photon flux strengthen by at most two order of magnitude the constraint on the photon coefficient while the eventual detection of the expected flux of UHE photons would constrain the electron positron coefficients down to $|\eta^{(3)}| \lesssim 10^{-16}$ (see [138, 159] for further details) by limiting the gamma decay process (note however, that in this case one cannot exclude that only one photon helicity survives and hence a detailed flux reconstruction would be needed).

13.9 Other SM Sectors Constraints

While QED constraints are up to date the more straightforward from a theoretical as well observational point of view, it is possible to cast constraints also on other

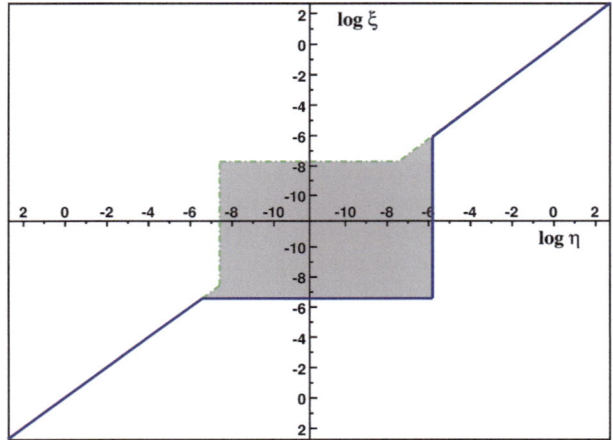

Fig. 13.4 LV induced by dimension 6 operators. The LV parameter space is shown. The allowed regions are *shaded grey*. *Green dotted lines* represent values of (η, ξ) for which the γ-decay threshold $k_{\gamma-dec} \simeq 10^{19}$ eV. *Solid, blue lines* indicate pairs (η, ξ) for which the pair production upper threshold $k_{up} \simeq 10^{20}$ eV

sectors of the SM, most noticeably on the hadronic and neutrino sectors. Let us review them very briefly here.

13.9.1 Constraints on the Hadronic Sector

Being an ultra high energy threshold process, the aforementioned GZK photopion production is strongly affected by LV. Several authors have studied the constraints implied by the detection of this effect [31, 38, 164–167]. However, a detailed LV study of the GZK feature is hard to perform, because of the many astrophysical uncertainties related to the modeling of the propagation and the interactions of UHE-CRs.

As a consequence of LV, the mean free path for the GZK reaction is modified. The propagated UHECR spectrum can therefore display features, like bumps at specific energies, suppression at low energy, recovery at energies above the cutoff, such that the observed spectrum cannot be reproduced. Moreover, the emission of Cherenkov γ-rays and pions in vacuum would lead to sharp suppression of the spectrum above the relevant threshold energy. After a detailed statistical analysis of the agreement between the observed UHECR spectrum and the theoretically predicted one in the presence of LV and assuming pure proton composition, the final constraints implied by UHECR physics are (at 99 % CL) [168]

$$-10^{-3} \lesssim \eta_p^{(4)} \lesssim 10^{-6},$$
$$-10^{-3} \lesssim \eta_\pi^{(4)} \lesssim 10^{-1} \quad (\eta_p^{(4)} > 0) \quad \text{or} \quad \lesssim 10^{-6} \quad (\eta_p^{(4)} < 0). \tag{13.86}$$

Of course for dimension five operators much stronger constraints can be achieved by a similar analysis (order $O(10^{-14})$).

13.9.2 Constraints on the Neutrino Sector

LV can affect the speed of neutrinos with respect to light, influence possible threshold reactions or modified the oscillations between neutrinos flavors. Unfortunately we have a wealth of information only about the latter phenomenon which however constraints only the differences among LV coefficients of different flavors. In this case, the best constraint to date comes from survival of atmospheric muon neutrinos observed by the former IceCube detector AMANDA-II in the energy range 100 GeV to 10 TeV [169], which searched for a generic LV in the neutrino sector [170] and achieved $(\Delta c/c)_{ij} \lesssim 2.8 \times 10^{-27}$ at 90 % confidence level assuming maximal mixing for some of the combinations i, j. Given that IceCube does not distinguish neutrinos from antineutrinos, the same constraint applies to the corresponding antiparticles. The IceCube detector is expected to improve this constraint to $(\Delta c/c)_{ij} \lesssim 9 \times 10^{-28}$ in the next few years [171]. The lack of sidereal variations in the atmospheric neutrino flux also yields comparable constraints on some combinations of SME parameters [172].

For what regards the time of flight constraints we have to date only a single event to rely on, the supernova SN1987a. This was a peculiar event which allowed to detect the almost simultaneous (within a few hours) arrival of electronic antineutrinos and photons. Although only few electronic antineutrinos at MeV energies was detected by the experiments KamiokaII, IMB and Baksan, it was enough to establish a constraint $(\Delta c/c)^{TOF} \lesssim 10^{-8}$ [173] or $(\Delta c/c)^{TOF} \lesssim 2 \times 10^{-9}$ [174] by looking at the difference in arrival time between antineutrinos and optical photons over a baseline distance of 1.5×10^5 ly. Further analyses of the time structure of the neutrino signal strengthened this constraint down to $\sim 10^{-10}$ [175, 176].

The scarcity of the detected neutrino did not allow the reconstruction of the full energy spectrum and of its time evolution in this sense one should probably consider constraints purely based on the difference in the arrival time with respect to photons more conservative and robust. Unfortunately adopting $\Delta c/c \lesssim 10^{-8}$, the SN constraint implies very weak constraints, $\xi_\nu^{(3)} \lesssim 10^{13}$ and $\xi_\nu^{(4)} \lesssim 10^{34}$.

Threshold reactions also can be used to cast constraints on the neutrinos sector. In the literature have been considered several processes most prominently the neutrino Čerekov emission $\nu \to \gamma \nu$, the neutrino splitting $\nu \to \nu \nu \bar{\nu}$ and the neutrino electron/positron pair production $\nu \to \nu e^- e^+$. Let us consider for illustration the latter process. Neglecting possible LV modification in the electron/positron sector (on which we have seen we have already strong constraints) the threshold energy is for arbitrary n

$$E_{th,(n)}^2 = \frac{4m_e^2}{\delta_{(n)}}, \qquad (13.87)$$

with $\delta_{(n)} = \xi_\nu (E_{th}/M)^{n-2}$.

Table 13.2 Summary of typical strengths of the available constrains on the SME at different orders

Order	Photon	e^-/e^+	Hadrons	Neutrinos[a]
$n=2$	N.A.	$O(10^{-16})$	$O(10^{-27})$	$O(10^{-8})$
$n=3$	$O(10^{-14})$ (GRB)	$O(10^{-16})$ (CR)	$O(10^{-14})$ (CR)	$O(30)$
$n=4$	$O(10^{-8})$ (CR)	$O(10^{-8})$ (CR)	$O(10^{-6})$ (CR)	$O(10^{-4})*$ (CR)

GRB = gamma rays burst, CR = cosmic rays

[a] From neutrino oscillations we have constraints on the difference of LV coefficients of different flavors up to $O(10^{-28})$ on dim 4, $O(10^{-8})$ and expected up to $O(10^{-14})$ on dim 5 (ICE3), expected up to $O(10^{-4})$ on dim 6 op. * Expected constraint from future experiments

The rate of this reaction was firstly computed in [47] for $n = 2$ but can be easily generated to arbitrary n [48] (see also [177]). The generic energy loss time-scale then reads (dropping purely numerical factors)

$$\tau_{\nu-\text{pair}} \simeq \frac{m_Z^4 \cos^4 \theta_w}{g^4 E^5} \left(\frac{M}{E}\right)^{3(n-2)}, \qquad (13.88)$$

where g is the weak coupling and θ_w is Weinberg's angle.

The observation of upward-going atmospheric neutrinos up to 400 TeV by the experiment IceCube implies that the free path of these particles is at least longer than the Earth radius implies a constraint $\eta_\mu^{(3)} \lesssim 30$. No effective constraint can be optioned for $n = 4$ LV, however in this case neutrino splitting (which has the further advantage to be purely dependent on LV on the neutrinos sector) could be used on the "cosmogenic" neutrino flux. This is supposedly created via the decay of charged pions produced by the aforementioned GZK effect. The neutrino splitting should modify the spectrum of the ultra high energy neutrinos by suppressing the flux at the highest energies and enhancing it at the lowest ones. In [178] it was shown that future experiments like ARIANNA [179] will achieve the required sensitivity to cast a constraint of order $\eta_\nu^{(4)} \lesssim 10^{-4}$. Note however, that the rate for neutrino splitting computed in [178] was recently recognized to be underestimated by a factor $O(E/M)^2$ [180]. Hence the future constraints here mentioned should be recomputed and one should be able to strengthened them by few orders of magnitude.

13.10 Summary and Perspectives

We can summarize the current status of the constraints for the LV SME in Table 13.2.

A special caveat it is due in the case of $n = 4$ constraints. As we have seen, they mostly rely (in the QED and Hadronic sector) on the actual detection of the GZK feature of the UHECR spectrum. More specifically, UHECR constraints have relied so far on the hypothesis, not in contrast with any previous experimental evidence, that protons constituted the majority of UHECRs above 10^{19} eV. Recent PAO [181] and Yakutsk [182] observations, however, showed hints of an increase of the average mass composition with rising energies up to $E \approx 10^{19.6}$ eV, although still with large uncertainties mainly due to the proton-air cross-section at ultra high energies. Hence, experimental data suggests that heavy nuclei can possibly account for a substantial fraction of UHECR arriving on Earth.

Furthermore the evidence for correlations between UEHCR events and their potential extragalactic sources [157]—such as active galactic nuclei (mainly Blasars)—has not improved with increasing statistics. This might be interpreted as a further hint that a relevant part of the flux at very high enrages should be accounted for by heavy ions (mainly iron) which are much more deviated by the extra and inter galactic magnetic fields due to their larger charge with respect to protons (an effect partially compensated by their shorter mean free path at very high energies).

If consequently one conservatively decides to momentarily suspend his/her judgment about the evidence for a GZK feature, then he/she would lose the constraints at $n = 4$ on the QED sector[9] as well as very much weaken the constraints on the hadronic one.

Assuming that current hints for a heavy composition at energies $E \sim 10^{19.6}$ eV [181] may be confirmed in the future, that some UHECR is observed up to $E \sim 10^{20}$ eV [184], and that the energy and momentum of the nucleus are the sum of energies and momenta of its constituents (so that the parameter in the modified dispersion relation of the nuclei is the same of the elementary nucleons, specifically η_p) one could place a first constraint on the absence of spontaneous decay for nuclei which could not spontaneously decay without LV.[10]

It will place a limit on $\eta_p < 0$, because in this case the energy of the emitted nucleon is lowered with respect to the LI case until it "compensates" the binding energy of the nucleons in the initial nucleus in the energy-momentum conservation. An upper limit for $\eta_p > 0$ can instead be obtained from the absence of vacuum Cherenkov emission. If UHECR are mainly iron at the highest energies the constraint is given by $\eta_p \lesssim 2 \times 10^2$ for nuclei observed at $10^{19.6}$ eV (and $\eta_p \lesssim 4$ for 10^{20} eV), while for helium it is $\eta_p \lesssim 4 \times 10^{-3}$ [185].

So, in conclusion, we can see that the while much has been done still plenty is to be explored. In particular, all of our constraints on $O(E/M)^2$ LV EFT (the most interesting order from a theoretical point of view) are based on the GZK effect (more or less directly) whose detection is still uncertain. It would be nice to be able

---

[9] This is a somewhat harsh statement given that it was shown in [183] that a substantial (albeit reduced) high energy gamma ray flux is still expected also in the case of mixed composition, so that in principle the previously discussed line of reasoning based on the absence of upper threshold for UHE gamma rays might still work.

[10] UHE nuclei suffer mainly from photo-disintegration losses as they propagate in the intergalactic medium. Because photo-disintegration is indeed a threshold process, it can be strongly affected by LV. According to [185], and in the same way as for the proton case, the mean free paths of UHE nuclei are modified by LV in such a way that the final UHECR spectra after propagation can show distinctive LV features. However, a quantitative evaluation of the propagated spectra has not been performed yet.

to cast comparable constrains using more reliable observations, but at the moment it is unclear what reaction could play this role. Similarly, new ideas like the one of gravitational confinement [89] presented in Sect. 13.4.1.2, seems to call for much deeper investigation of LV phenomenology in the purely gravitational sector.

We have gone along way into exploring the possible phenomenology of Lorentz breaking physics and pushed well beyond expectations the tests of this fundamental symmetry of Nature, however still much seems to await along the path.

Acknowledgements I wish to that Luca Maccione and David Mattingly for useful insights, discussions and feedback on the manuscript preparation.

References

1. Mavromatos, N.E.: Lect. Notes Phys. **669**, 245 (2005). arXiv:gr-qc/0407005
2. Weinberg, S.: Phys. Rev. D **72**, 043514 (2005). arXiv:hep-th/0506236
3. Damour, T., Polyakov, A.M.: Nucl. Phys. B **423**, 532 (1994). arXiv:hep-th/9401069
4. Barrow, J.D.: arXiv:gr-qc/9711084 (1997)
5. Bleicher, M., Hofmann, S., Hossenfelder, S., Stoecker, H.: Phys. Lett. B **548**, 73 (2002). arXiv:hep-ph/0112186
6. Kostelecky, V.A.: Phys. Rev. D **69**, 105009 (2004). arXiv:hep-th/0312310
7. Mattingly, D.: Living Rev. Relativ. **8**, 5 (2005). arXiv:gr-qc/0502097
8. Dirac, P.A.M.: Nature **168**, 906 (1951)
9. Bjorken, J.D.: Ann. Phys. **24**, 174 (1963)
10. Phillips, P.R.: Phys. Rev. **146**, 966 (1966)
11. Blokhintsev, D.I.: Sov. Phys. Usp. **9**, 405 (1966)
12. Pavlopoulos, T.G.: Phys. Rev. **159**, 1106 (1967)
13. Rédei, L.B.: Phys. Rev. **162**, 1299 (1967)
14. Nielsen, H.B., Ninomiya, M.: Nucl. Phys. B **141**, 153 (1978)
15. Ellis, J., Gaillard, M.K., Nanopoulos, D.V., Rudaz, S.: Nucl. Phys. B **176**, 61 (1980)
16. Zee, A.: Phys. Rev. D **25**, 1864 (1982)
17. Nielsen, H.B., Picek, I.: Phys. Lett. B **114**, 141 (1982)
18. Chadha, S., Nielsen, H.B.: Nucl. Phys. B **217**, 125 (1983)
19. Nielsen, H.B., Picek, I.: Nucl. Phys. B **211**, 269 (1983)
20. Auger, P., Roth, M.: arXiv:0706.2096 [astro-ph] (2007)
21. HiRes, Abbasi, R., et al.: arXiv:astro-ph/0703099 (2007)
22. Kostelecky, V.A., Samuel, S.: Phys. Rev. D **39**, 683 (1989)
23. Colladay, D., Kostelecky, V.: Phys. Rev. D **58**, 116002 (1998). arXiv:hep-ph/9809521
24. Kostelecky, E., Alan, V.: Prepared for 3rd Meeting on CPT and Lorentz Symmetry (CPT 04), Bloomington, Indiana, 8–11 Aug 2007
25. Amelino-Camelia, G., Ellis, J.R., Mavromatos, N.E., Nanopoulos, D.V., Sarkar, S.: Nature **393**, 763 (1998). arXiv:astro-ph/9712103
26. Coleman, S.R., Glashow, S.L.: Phys. Lett. B **405**, 249 (1997). arXiv:hep-ph/9703240
27. Coleman, S.R., Glashow, S.L.: arXiv:hep-ph/9808446 (1998)
28. Coleman, S.R., Glashow, S.L.: Phys. Rev. D **59**, 116008 (1999). arXiv:hep-ph/9812418
29. Gonzalez-Mestres, L.: arXiv:hep-ph/9610474 (1996)
30. Gonzalez-Mestres, L.: physics/9712005 (1997)
31. Jacobson, T., Liberati, S., Mattingly, D.: Phys. Rev. D **67**, 124011 (2003). arXiv:hep-ph/0209264
32. Mattingly, D., Jacobson, T., Liberati, S.: Phys. Rev. D **67**, 124012 (2003). arXiv:hep-ph/0211466

33. Jacobson, T., Liberati, S., Mattingly, D.: Ann. Phys. **321**, 150 (2006). arXiv:astro-ph/0505267
34. Anselmi, D.: J. High Energy Phys. **0802**, 051 (2008). arXiv:0801.1216 [hep-th]
35. Horava, P.: Phys. Rev. D **79**, 084008 (2009). arXiv:0901.3775 [hep-th]
36. Visser, M.: arXiv:0912.4757 [hep-th] (2009)
37. Myers, R.C., Pospelov, M.: Phys. Rev. Lett. **90**, 211601 (2003). arXiv:hep-ph/0301124
38. Mattingly, D.: arXiv:0802.1561 [gr-qc] (2008)
39. Riess, A.G., et al. (Supernova Search Team): Astron. J. **116**, 1009 (1998). arXiv:astro-ph/9805201
40. Perlmutter, S., et al. (Supernova Cosmology Project): Astrophys. J. **517**, 565 (1999). arXiv:astro-ph/9812133
41. Greisen, K.: Phys. Rev. Lett. **16**, 748 (1966)
42. Zatsepin, G.T., Kuz'min, V.A.: On the interaction of cosmic rays with photons. In: Cosmic rays, Moscow, vol. 11, pp. 45–47 (1969)
43. Takeda, M., et al.: Phys. Rev. Lett. **81**, 1163 (1998). arXiv:astro-ph/9807193
44. Protheroe, R., Meyer, H.: Phys. Lett. B **493**, 1 (2000). arXiv:astro-ph/0005349
45. Adam, T., et al. (OPERA Collaboration): arXiv:1109.4897 (2011). http://press.web.cern.ch/press/pressreleases/Releases2011/PR19.11E.html
46. Amelino-Camelia, G., et al.: Int. J. Mod. Phys. D **20**, 2623 (2011). arXiv:1109.5172 [hep-ph]
47. Cohen, A.G., Glashow, S.L.: Phys. Rev. Lett. **107**, 181803 (2011). arXiv:1109.6562 [hep-ph]
48. Maccione, L., Liberati, S., Mattingly, D.M.: arXiv:1110.0783 [hep-ph] (2011)
49. Carmona, J., Cortes, J.: arXiv:1110.0430 [hep-ph] (2011)
50. Antonello, M., et al. (ICARUS Collaboration): arXiv:1203.3433 [hep-ex] (2012)
51. Amelino-Camelia, G., Ellis, J.R., Mavromatos, N.E., Nanopoulos D.V.: Int. J. Mod. Phys. A **12**, 607 (1997). arXiv:hep-th/9605211
52. Ellis, J.R., Mavromatos, N.E., Nanopoulos, D.V.: Phys. Rev. D **61**, 027503 (2000). arXiv:gr-qc/9906029
53. Ellis, J.R., Mavromatos, N.E., Nanopoulos, D.V.: Phys. Rev. D **62**, 084019 (2000). arXiv:gr-qc/0006004
54. Ellis, J.R., Mavromatos, N.E., Sakharov, A.S.: Astropart. Phys. **20**, 669 (2004). arXiv:astro-ph/0308403
55. Gambini, R., Pullin, J.: Phys. Rev. D **59**, 124021 (1999). arXiv:gr-qc/9809038
56. Carroll, S.M., Harvey, J.A., Kostelecky, V.A., Lane, C.D., Okamoto, T.: Phys. Rev. Lett. **87**, 141601 (2001). arXiv:hep-th/0105082
57. Lukierski, J., Ruegg, H., Zakrzewski, W.J.: Ann. Phys. **243**, 90 (1995). arXiv:hep-th/9312153
58. Amelino-Camelia, G., Majid, S.: Int. J. Mod. Phys. A **15**, 4301 (2000). arXiv:hep-th/9907110
59. Burgess, C.P., Cline, J., Filotas, E., Matias, J., Moore, G.D.: J. High Energy Phys. **03**, 043 (2002). arXiv:hep-ph/0201082
60. Gasperini, M.: Phys. Lett. B **163**, 84 (1985)
61. Gasperini, M.: Phys. Lett. B **180**, 221 (1986)
62. Gasperini, M.: Class. Quantum Gravity **4**, 485 (1987)
63. Gasperini, M., D'Azeglio, C.M.: Lorentz noninvariance and the universality of free fall in quasi-riemannian gravity. In: Gravitational Measurements, Fundamental Metrology and Constants, vol. 1, pp. 181–190. Springer, Berlin (1988)
64. Gasperini, M.: Gen. Relativ. Gravit. **30**, 1703 (1998). arXiv:gr-qc/9805060
65. Mattingly, D., Jacobson, T.: arXiv:gr-qc/0112012 (2001)
66. Eling, C., Jacobson, T., Mattingly, D.: arXiv:gr-qc/0410001 (2004)
67. Jacobson, T.: arXiv:0801.1547 [gr-qc] (2008)
68. Sotiriou, T.P., Visser, M., Weinfurtner, S.: J. High Energy Phys. **0910**, 033 (2009). arXiv:0905.2798 [hep-th]
69. Blas, D., Pujolas, O., Sibiryakov, S.: Phys. Rev. Lett. **104**, 181302 (2010). arXiv:0909.3525 [hep-th]

70. Jacobson, T.: Phys. Rev. D **81**, 101502 (2010). arXiv:1001.4823 [hep-th]
71. Barcelo, C., Liberati, S., Visser, M.: Living Rev. Relativ. **8**, 12 (2005). arXiv:gr-qc/0505065
72. Unruh, W.: Phys. Rev. Lett. **46**, 1351 (1981)
73. Visser, M.: arXiv:gr-qc/9311028 (1993)
74. Garay, L., Anglin, J., Cirac, J., Zoller, P.: Phys. Rev. Lett. **85**, 4643 (2000). arXiv:gr-qc/0002015
75. Barcelo, C., Liberati, S., Visser, M.: Class. Quantum Gravity **18**, 1137 (2001). arXiv:gr-qc/0011026
76. Balbinot, R., Fagnocchi, S., Fabbri, A., Procopio, G.P.: Phys. Rev. Lett. **94**, 161302 (2005). arXiv:gr-qc/0405096
77. Barcelo, C., Liberati, S., Sonego, S., Visser, M.: Phys. Rev. D **77**, 044032 (2008). arXiv:0712.1130
78. Barcelo, C., Liberati, S., Visser, M.: Phys. Rev. A **68**, 053613 (2003). arXiv:cond-mat/0307491
79. Weinfurtner, S., Visser, M., Jain, P., Gardiner, C.: PoS **QG-PH**, 044 (2007). arXiv:0804.1346 [gr-qc]
80. Weinfurtner, S., Jain, P., Visser, M., Gardiner, C.: Class. Quantum Gravity **26**, 065012 (2009). arXiv:0801.2673 [gr-qc]
81. Jain, P., Weinfurtner, S., Visser, M., Gardiner, C.: arXiv:0705.2077 [cond-mat.other] (2007)
82. Parentani, R.: PoS **QG-PH**, 031 (2007). arXiv:0709.3943 [hep-th]
83. Collins, J., Perez, A., Sudarsky, D., Urrutia, L., Vucetich, H.: Phys. Rev. Lett. **93**, 191301 (2004). arXiv:gr-qc/0403053
84. Groot Nibbelink, S., Pospelov, M.: Phys. Rev. Lett. **94**, 081601 (2005). arXiv:hep-ph/0404271
85. Bolokhov, P.A., Nibbelink, S.G., Pospelov, M.: Phys. Rev. D **72**, 015013 (2005). arXiv:hep-ph/0505029
86. ATLAS-Collaboration, https://twiki.cern.ch/twiki/bin/view/AtlasPublic/CombinedSummaryPlots
87. Liberati, S., Visser, M., Weinfurtner, S.: Phys. Rev. Lett. **96**, 151301 (2006). arXiv:gr-qc/0512139
88. Liberati, S., Visser, M., Weinfurtner, S.: Class. Quantum Gravity **23**, 3129 (2006). arXiv:gr-qc/0510125
89. Pospelov, M., Shang, Y.: Phys. Rev. D **85**, 105001 (2012). arXiv:1010.5249 [hep-th]
90. Kostelecky, V.A., Mewes, M.: Phys. Rev. D **66**, 056005 (2002). arXiv:hep-ph/0205211
91. Yao, W.M., et al. (Particle Data Group): J. Phys. G **33**, 1 (2006)
92. Bolokhov, P.A., Pospelov, M.: Phys. Rev. D **77**, 025022 (2008). arXiv:hep-ph/0703291
93. Jacobson, T.A., Liberati, S., Mattingly, D., Stecker, F.W.: Phys. Rev. Lett. **93**, 021101 (2004). arXiv:astro-ph/0309681
94. Stecker, F., Glashow, S.L.: Astropart. Phys. **16**, 97 (2001). arXiv:astro-ph/0102226
95. Ellis, J.R., Mavromatos, N.E., Nanopoulos, D.V.: Phys. Lett. B **293**, 37 (1992). arXiv:hep-th/9207103
96. Ellis, J.R., Mavromatos, N.E., Nanopoulos, D.V., Sakharov, A.S.: Int. J. Mod. Phys. A **19**, 4413 (2004). arXiv:gr-qc/0312044
97. Polchinski, J.: arXiv:hep-th/9611050 (1996)
98. von Ignatowsky, W.: Verh. Dtsch. Phys. Ges. **12**, 788 (1910)
99. von Ignatowsky, W.: Phys. Z. **11**, 972 (1910)
100. von Ignatowsky, W.: Arch. Math. Phys. **3**(17), 1 (1911)
101. von Ignatowsky, W.: Arch. Math. Phys. **3**(18), 17 (1911)
102. von Ignatowsky, W.: Phys. Z. **12**, 779 (1911)
103. Liberati, S., Sonego, S., Visser, M.: Ann. Phys. **298**, 167 (2002). arXiv:gr-qc/0107091
104. Sonego, S., Pin, M.: J. Math. Phys. **50**, 042902 (2009). arXiv:0812.1294 [gr-qc]
105. Baccetti, V., Tate, K., Visser, M.: J. High Energy Phys. **1205**, 119 (2012). arXiv:1112.1466 [gr-qc]
106. Cohen, A.G., Glashow, S.L.: Phys. Rev. Lett. **97**, 021601 (2006). arXiv:hep-ph/0601236

107. Bogoslovsky, G.Y.: arXiv:math-ph/0511077 (2005)
108. Bogoslovsky, G.Y.: Phys. Lett. A **350**, 5 (2006). arXiv:hep-th/0511151
109. Gibbons, G.W., Gomis, J., Pope, C.N.: Phys. Rev. D, Part. Fields **76**, 081701 (2007)
110. Amelino-Camelia, G.: Int. J. Mod. Phys. D **11**, 35 (2002). arXiv:gr-qc/0012051
111. Magueijo, J., Smolin, L.: Phys. Rev. Lett. **88**, 190403 (2002). arXiv:hep-th/0112090
112. Magueijo, J., Smolin, L.: Phys. Rev. D **67**, 044017 (2003). arXiv:gr-qc/0207085
113. Amelino-Camelia, G.: Int. J. Mod. Phys. D **12**, 1211 (2003). arXiv:astro-ph/0209232
114. Lukierski, J.: arXiv:hep-th/0402117 (2004)
115. Amelino-Camelia, G.: arXiv:0806.0339 [gr-qc] (2008)
116. Carmona, J., Cortes, J., Indurain, J., Mazon, D.: Phys. Rev. D **80**, 105014 (2009). arXiv:0905.1901 [hep-th]
117. Judes, S., Visser, M.: Phys. Rev. D **68**, 045001 (2003). arXiv:gr-qc/0205067
118. Ahluwalia, D.V.: arXiv:gr-qc/0212128 (2002)
119. Rembielinski, J., Smolinski, K.A.: Bull. Soc. Sci. Lett. Lodz **53**, 57 (2003). arXiv:hep-th/0207031
120. Schutzhold, R., Unruh, W.G.: JETP Lett. **78**, 431 (2003). arXiv:gr-qc/0308049
121. Hossenfelder, S.: Phys. Rev. Lett. **104**, 140402 (2010). arXiv:1004.0418 [hep-ph]
122. Amelino-Camelia, G., et al.: Phys. Rev. D **78**, 025005 (2008). arXiv:0709.4600 [hep-th]
123. Amelino-Camelia, G., Freidel, L., Kowalski-Glikman, J., Smolin, L.: Phys. Rev. D **84**, 084010 (2011). arXiv:1101.0931 [hep-th]
124. Smolin, L.: arXiv:0808.3765 [hep-th] (2008)
125. Rovelli, C.: arXiv:0808.3505 [gr-qc] (2008)
126. Amelino-Camelia, G., Freidel, L., Kowalski-Glikman, J., Smolin, L.: Phys. Rev. D **84**, 087702 (2011). arXiv:1104.2019 [hep-th]
127. Hossenfelder, S.: arXiv:1202.4066 [hep-th] (2012)
128. Kostelecky, V.A., Russell, N.: arXiv:0801.0287 (2008)
129. Ellis, J.R., Mavromatos, N.E., Nanopoulos, D.V., Sakharov, A.S., Sarkisyan, E.K.G.: Astropart. Phys. **25**, 402 (2006). arXiv:astro-ph/0510172
130. Magic, J.A., et al.: arXiv:0708.2889 [astro-ph] (2007)
131. Maccione, L., Liberati, S., Celotti, A., Kirk, J.G., Ubertini, P.: arXiv:0809.0220 [astro-ph] (2008)
132. Gleiser, R.J., Kozameh, C.N.: Phys. Rev. D **64**, 083007 (2001). arXiv:gr-qc/0102093
133. McMaster, W.H.: Rev. Mod. Phys. **33**, 8 (1961)
134. Baccetti, V., Tate, K., Visser, M.: J. High Energy Phys. **1203**, 087 (2012). arXiv:1111.6340 [hep-ph]
135. Gelmini, G., Nussinov, S., Yaguna, C.E.: J. Cosmol. Astropart. Phys. **0506**, 012 (2005). arXiv:hep-ph/0503130
136. Kluzniak, W.: Astropart. Phys. **11**, 117 (1999)
137. Altschul, B.: Phys. Rev. D **74**, 083003 (2006). arXiv:hep-ph/0608332
138. Liberati, S., Maccione, L.: Annu. Rev. Nucl. Part. Sci. **59**, 245 (2009). arXiv:0906.0681
139. Maccione, L., Liberati, S., Celotti, A., Kirk, J.: J. Cosmol. Astropart. Phys. **2007**, 013 (2007). arXiv:0707.2673 [astro-ph]
140. Kennel, C.F., Coroniti, F.V.: Astrophys. J. **283**, 694 (1984)
141. Fan, Y.-Z., Wei, D.-M., Xu, D.: Mon. Not. R. Astron. Soc. **376**, 1857 (2006). arXiv:astro-ph/0702006
142. Dean, A.J., et al.: Science **321**, 1183 (2008)
143. Forot, M., Laurent, P., Grenier, I.A., Gouiffes, C., Lebrun, F.: arXiv:0809.1292 [astro-ph] (2008)
144. Jacobson, T., Liberati, S., Mattingly, D.: Nature **424**, 1019 (2003). arXiv:astro-ph/0212190
145. Kirk, J.G., Lyubarsky, Y., Petri, J.: arXiv:astro-ph/0703116 (2007)
146. Parmar, A.N., et al.: INTEGRAL mission. In: Truemper, J.E., Tananbaum, H.D. (eds.) X-Ray and Gamma-Ray Telescopes and Instruments for Astronomy. Proceedings of the SPIE, vol. 4851, pp. 1104–1112 (2003)
147. McGlynn, S., et al.: arXiv:astro-ph/0702738 (2007)

148. Petri, J., Kirk, J.G.: Astrophys. J. **627**, L37 (2005). arXiv:astro-ph/0505427
149. Weisskopf, M.C., Silver, E.H., Kestenbaum, H.L., Long, K.S., Novick, R.: Astrophys. J. Lett. **220**, L117 (1978)
150. Ng, C.Y., Romani, R.W.: Astrophys. J. **601**, 479 (2004). arXiv:astro-ph/0310155
151. Kanbach, G., Slowikowska, A., Kellner, S., Steinle, H.: AIP Conf. Proc. **801**, 306 (2005). arXiv:astro-ph/0511636
152. Mitrofanov, I.G.: Nature **426**, 139 (2003)
153. Coburn, W., Boggs, S.E.: Nature **423**, 415 (2003). arXiv:astro-ph/0305377
154. Stecker, F.W.: Astropart. Phys. **35**, 95 (2011). arXiv:1102.2784 [astro-ph]
155. Laurent, P., Gotz, D., Binetruy, P., Covino, S., Fernandez-Soto, A.: arXiv:1106.1068 [astro-ph.HE] (2011)
156. Aharonian, F., et al. (The HEGRA): Astrophys. J. **614**, 897 (2004). arXiv:astro-ph/0407118
157. Auger, P., Abraham, J., et al.: Science **318**, 938 (2007). arXiv:0711.2256 [astro-ph]
158. Galaverni, M., Sigl, G.: Phys. Rev. Lett. **100**, 021102 (2008). arXiv:0708.1737 [astro-ph]
159. Maccione, L., Liberati, S.: J. Cosmol. Astropart. Phys. **0808**, 027 (2008). arXiv:0805.2548 [astro-ph]
160. Gelmini, G., Kalashev, O., Semikoz, D.V.: arXiv:astro-ph/0506128 (2005)
161. Auger, P., Abraham, J., et al.: Astropart. Phys. **29**, 243 (2008). arXiv:0712.1147 [astro-ph]
162. Rubtsov, G.I., et al.: Phys. Rev. D **73**, 063009 (2006). arXiv:astro-ph/0601449
163. Galaverni, M., Sigl, G.: Phys. Rev. D **78**, 063003 (2008). arXiv:0807.1210 [astro-ph]
164. Aloisio, R., Blasi, P., Ghia, P.L., Grillo, A.F.: Phys. Rev. D **62**, 053010 (2000). arXiv:astro-ph/0001258
165. Alfaro, J., Palma, G.: Phys. Rev. D **67**, 083003 (2003). arXiv:hep-th/0208193
166. Scully, S., Stecker, F.: Astropart. Phys. **31**, 220 (2009). arXiv:0811.2230 [astro-ph]
167. Stecker, F.W., Scully, S.T.: New J. Phys. **11**, 085003 (2009). arXiv:0906.1735 [astro-ph.HE]
168. Maccione, L., Taylor, A.M., Mattingly, D.M., Liberati, S.: J. Cosmol. Astropart. Phys. **0904**, 022 (2009). arXiv:0902.1756 [astro-ph.HE]
169. Kelley, J. (IceCube Collaboration): Nucl. Phys. A **827**, 507C (2009)
170. Gonzalez-Garcia, M., Maltoni, M.: Phys. Rev. D **70**, 033010 (2004). arXiv:hep-ph/0404085
171. Huelsnitz, W., Kelley, J. (IceCube Collaboration): Search for quantum gravity with IceCube and high energy atmospheric neutrinos. In: International Cosmic Ray Conference (ICRC 2009) (2009)
172. Abbasi, R., et al. (IceCube Collaboration): Phys. Rev. D **82**, 112003 (2010). arXiv:1010.4096 [astro-ph.HE]
173. Stodolsky, L.: Phys. Lett. B **201**, 353 (1988)
174. Longo, M.J.: Phys. Rev. Lett. **60**, 173 (1988)
175. Ellis, J.R., Harries, N., Meregaglia, A., Rubbia, A., Sakharov, A.: Phys. Rev. D **78**, 033013 (2008). arXiv:0805.0253 [hep-ph]
176. Sakharov, A., Ellis, J., Harries, N., Meregaglia, A., Rubbia, A.: J. Phys. Conf. Ser. **171**, 012039 (2009). arXiv:0903.5048 [hep-ph]
177. Carmona, J., Cortes, J., Mazon, D.: Phys. Rev. D **85**, 113001 (2012). arXiv:1203.2585
178. Mattingly, D.M., Maccione, L., Galaverni, M., Liberati, S., Sigl, G.: J. Cosmol. Astropart. Phys. **1002**, 007 (2010). arXiv:0911.0521 [hep-ph]
179. Barwick, S.W.: J. Phys. Conf. Ser. **60**, 276 (2007). arXiv:astro-ph/0610631
180. Ward, B.: Phys. Rev. D **85**, 073007 (2012). arXiv:1201.1322 [hep-ph]
181. Abraham, J., et al. (Pierre Auger Observatory Collaboration): Phys. Rev. Lett. **104**, 091101 (2010). arXiv:1002.0699 [astro-ph.HE]
182. Glushkov, A., et al.: JETP Lett. **87**, 190 (2008). arXiv:0710.5508 [astro-ph]
183. Hooper, D., Taylor, A.M., Sarkar, S.: Astropart. Phys. **34**, 340 (2011). arXiv:1007.1306 [astro-ph.HE]
184. Abraham, J., et al. (The Pierre Auger Collaboration): Phys. Lett. B **685**, 239 (2010). arXiv:1002.1975 [astro-ph.HE]
185. Saveliev, A., Maccione, L., Sigl, G.: J. Cosmol. Astropart. Phys. **1103**, 046 (2011). arXiv:1101.2903 [astro-ph.HE]

Chapter 14
The Topology of the Quantum Vacuum

Grigorii E. Volovik

Abstract Topology in momentum space is the main characteristic of the ground state of a system at zero temperature, the quantum vacuum. The gaplessness of fermions in bulk, on the surface or inside the vortex core is protected by topology, and is not sensitive to the details of the microscopic physics (atomic or trans-Planckian). Irrespective of the deformation of the parameters of the microscopic theory, the energy spectrum of these fermions remains strictly gapless. This solves the main hierarchy problem in particle physics: for fermionic vacua with Fermi points the masses of elementary particles are naturally small. The quantum vacuum of the Standard Model is one of the representatives of topological matter alongside with topological superfluids and superconductors, topological insulators and semimetals, etc. There is a number of topological invariants in momentum space of different dimensions. They determine the universality classes of the topological matter and the type of the effective theory which emerges at low energy. In many cases they also give rise to emergent symmetries, including the effective Lorentz invariance, and emergent phenomena such as effective gauge and gravitational fields. The topological invariants in extended momentum and coordinate space determine the bulk-surface and bulk-vortex correspondence. They connect the momentum space topology in bulk with the real space. These invariants determine the gapless fermions living on the surface of a system or in the core of topological defects (vortices, strings, domain walls, solitons, monopoles, etc.). The momentum space topology gives some lessons for quantum gravity. In effective gravity emerging at low energy, the collective variables are the tetrad field and spin connections, while the metric is the composite object of tetrad field. This suggests that the Einstein-Cartan-Sciama-Kibble theory with torsion field is more relevant. There are also several scenarios of Lorentz invariance violation governed by topology, including splitting of Fermi point and development of the Dirac points with quadratic and cubic spectrum. The latter leads to the natural emergence of the Hořava-Lifshitz gravity.

G.E. Volovik (✉)
Low Temperature Laboratory, Aalto University, P.O. Box 15100, 00076 Aalto, Finland
e-mail: volovik@boojum.hut.fi

G.E. Volovik
L.D. Landau Institute for Theoretical Physics, Kosygina 2, 119334 Moscow, Russia

14.1 Introduction

14.1.1 Symmetry vs Topology

There is a fundamental interplay of symmetry and topology in physics, both in condensed matter and relativistic quantum fields. Traditionally the main role was played by symmetry: gauge symmetry of Standard Model and GUT; symmetry classification of condensed matter systems such as solid and liquid crystals, magnets, superconductors and superfluids; universality classes of spontaneously broken symmetry phase transitions; etc. The last decades demonstrated the opposite tendency in which topology is becoming primary being the main characteristics of quantum vacua—ground states of the system at $T = 0$, see reviews [1–4] and earlier papers [5–19].

Topology describes the properties of a system, which are insensitive to the details of the microscopic physics. It determines universality classes of the topological matter and the type of the effective theory which emerges at low energy and low temperature, and gives rise to emergent symmetry. Examples are provided by the point nodes in the energy spectrum, which are protected by topology. Close to the nodes the effective Lorentz invariance emerges: the fermionic spectrum forms a relativistic Dirac cone and the fermions behave as Weyl, Dirac or Majorana particles. The bosonic collective modes give rise to effective gauge field and effective metric. All this is the consequence of the topological theorem—the Atiyah-Bott-Shapiro construction [19].

Among the existing and potential representatives of topological materials one can find those which have gap in their fermionic spectrum. These are 3D topological band insulators [3]; fully gapped superfluid ^{3}He-B [15, 20–22]; and 2D materials exhibiting intrinsic (i.e. without external magnetic field) quantum Hall and spin-Hall effects, such as gapped graphene [16]; thin film of superfluid ^{3}He-A and quasi 2D planar phase of triplet superfluid [17, 18, 23]; and chiral superconductor Sr_2RuO_4 [24]. These materials have the topological properties similar to that of the quantum vacuum of Standard Model in its massive "insulating" phase [25].

The gapless topological media are represented by superfluid ^{3}He-A [1]; topological semi-metals [26–30]; gapless graphene [2, 31–34]; nodal cuprate [31] and non-centrosymmetric [35, 36] superconductors. These materials are similar to the quantum vacuum of the Standard Model in its massless "semi-metal" phase [1, 25].

14.1.2 Green's Function vs Order Parameter

Topology operates in particular with integer numbers—topological charges—which do not change under small deformations of the system. The conservation of these topological charges protects the Fermi surface and another object in momentum space—the Fermi point—from destruction. They survive when the interaction between the fermions is introduced and modified. When the momentum of a particle approaches the Fermi surface or the Fermi point its energy necessarily vanishes.

Thus the topology is the main reason why there are gapless quasi-particles in condensed matter and (nearly) massless elementary particles in our Universe.

The momentum-space topological invariants are in many respects similar to the real-space invariants, which describe topological defects in condensed matter systems, and such topological objects as cosmic strings, magnetic monopoles and solitons in particle physics (see Fig. 14.1). While the real-space invariants describes the topologically nontrivial configurations of the order parameter fields in spacetime, the momentum-space invariants describe the nontrivial momentum-space configuration of the Green's function $G(\mathbf{p}, \omega)$ or other response function, which characterizes the ground state of a system (the vacuum state) [1, 19, 37, 38]. In particular, the Fermi surface in metals is topologically stable, because it is analogous to the vortex loop in superfluids or superconductors. In the same way, the Fermi point (the Weyl point) corresponds to the real-space point defects, such as a hedgehog in ferromagnets or a magnetic monopole in particle physics. The fully gapped topological matter, such as topological insulators and fully gapped topological superfluids represent skyrmions in momentum space: they have no nodes in their spectrum or any other singularities, and they correspond to non-singular objects in real space—textures or skyrmions (Fig. 14.1 *top right*).

The topology of Green's function $G(\omega, \mathbf{p}; t, \mathbf{r})$ in the phase-space allows us to consider topologically protected spectrum of fermions living on topological objects such as domain walls, strings and monopoles [1, 15, 37, 41–43].

14.1.3 The Fermi Surface as a Topological Object

Let us start with gapless vacua. For the topological classification of the gapless vacua, the Green's function is considered on the imaginary frequency axis $p_0 = i\omega$. This allows us to consider only the relevant singularities in the Green's function and to avoid the singularities on the mass shell, which exist in any vacuum, gapless or fully gapped. The Green's function is generally a matrix with spin indices. In addition, it may have the band indices (in the case of electrons in the periodic potential of crystals).

We start with zeroes of co-dimension 1. By co-dimension we denote the dimension of $\mathbf{p}$-space minus dimension of the nodes. That is why co-dimension 1 refers to two-dimensional Fermi surface in three-dimensional metal, 1D Fermi line in 2D systems and Fermi point in 1D systems. The general analysis [19] demonstrates that topologically stable nodes of co-dimension 1 are described by the group Z of integers. The corresponding winding number N is expressed analytically in terms of the Green's function [1]:

$$N = \mathbf{tr} \oint_C \frac{dl}{2\pi i} G(\omega, \mathbf{p}) \partial_l G^{-1}(\omega, \mathbf{p}). \tag{14.1}$$

Here the integral is taken over an arbitrary contour C around the Green's function singularity in the $D+1$ momentum-frequency space. See Fig. 14.2 for $D=2$. Example of the Green's function in any dimension D is scalar function $G^{-1}(\omega, \mathbf{p}) =$

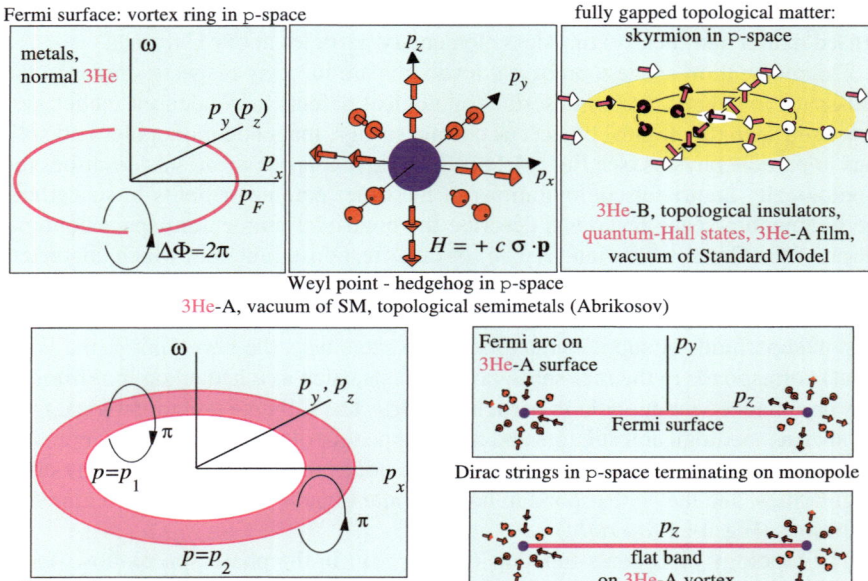

Fig. 14.1 Topological matter, represented in terms of topological objects in momentum space. (*top left*): Fermi surface is the momentum-space analogue of the vortex line: the phase of the Green's function changes by 2π around the element of the line in $(\omega, \mathbf{p})$-space. (*top middle*): Fermi point (Weyl point) is the counterpart of a hedgehog and a magnetic monopole. The hedgehog in this figure has integer topological charge $N = +1$, and close to this Fermi point the fermionic quasiparticles behave as Weyl fermions. Nontrivial topological charges in terms of Green's functions support the stability of the Fermi surfaces and Weyl points with respect to perturbations including interactions [1, 37]. In terms of the Berry phase [39] the Fermi point represents the $\mathbf{p}$-space counterpart of Dirac magnetic monopole with unobservable Dirac string (see Ref. [40] and Fig. 11.4 in [1]). (*top right*): Topological insulators and fully gapped topological superfluids/superconductors are textures in momentum space: they have no singularities in the Green's function and thus no nodes in the energy spectrum in the bulk. This figure shows a skyrmion in the two-dimensional momentum space, which characterizes two-dimensional topological insulators exhibiting intrinsic quantum Hall or spin-Hall effect. (*bottom left*): Flat band emerging in strongly interacting systems [44]. This dispersionless Fermi band is analogous to a soliton terminated by half-quantum vortices: the phase of the Green's function changes by π around the edge of the flat band [45]. (*bottom right*): Fermi arc on the surface of ^{3}He-A [46] and of topological semi-metals with Weyl points [28–30] and flat band inside the vortex core of ^{3}He-A [47] serve as the momentum-space analogue of a Dirac string terminating on a monopole. The Fermi surface formed by the surface bound states terminates on the points where the spectrum of zero energy states merge with the continuous spectrum in the bulk, i.e. with the Weyl points

$i\omega - v_F(|\mathbf{p}| - p_F)$. For $D = 2$, the singularity with winding number $N = 1$ is on the line $\omega = 0$, $p_x^2 + p_y^2 = p_F^2$, which represents the one-dimensional Fermi surface.

Due to the nontrivial topological invariant, the Fermi surface survives the perturbative interaction and exists even in marginal and Luttinger liquids without poles in the Green's function, where quasiparticles are not defined.

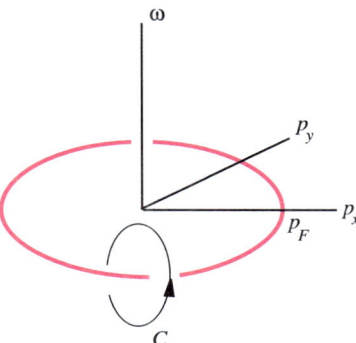

Fig. 14.2 Nodes of co-dimension 1 in 2 + 1 systems. Green's function has singularities on line $\omega = 0$, $p_x^2 + p_y^2 = p_F^2$ in the three-dimensional space (ω, p_x, p_y). The stability of the Fermi surface is protected by the invariant (14.1) which is represented by an integral over an arbitrary contour C around the Green's function singularity. This is applicable to nodes of co-dimension 1 in any $D + 1$ dimension. For $D = 3$ the nodes form conventional Fermi surface in metals and in normal ^{3}He

14.2 Vacuum in a Semi-metal State

If the quantum vacuum of Standard Model obeys the Lorentz invariance, then the relevant object in momentum space is either the Fermi point of chiral type, in which fermionic excitations behave as left-handed or right-handed Weyl fermions [1, 48], or the class of vacua with the nodal point obeying Z_2 topology, where fermionic excitations behave as massless Majorana neutrinos [19].

14.2.1 Fermi Points in 3 + 1 Vacua

The Fermi point is the Green's function singularity described by the following topological invariant expressed via integer valued integral over the surface σ around the singular point in the 4-momentum space $p_\mu = (\omega, \mathbf{p})$ [1]:

$$N = \frac{e_{\alpha\beta\mu\nu}}{24\pi^2} \operatorname{tr} \int_\sigma dS^\alpha\, G\partial_{p_\beta} G^{-1} G\partial_{p_\mu} G^{-1} G\partial_{p_\nu} G^{-1}. \tag{14.2}$$

If the invariant is nonzero, the Green's function has point singularity inside the surface σ—the Fermi point. If the topological charge is $N = +1$ or $N = -1$, the Fermi point represents the so-called conical Dirac point, but actually describes the chiral Weyl fermions. This is the consequence of the so-called Atiyah-Bott-Shapiro construction [19], which leads to the following general form of expansion of the inverse fermionic propagator near the Fermi point with $N = +1$ or $N = -1$:

$$G^{-1}(p_\mu) = e_\alpha^\beta \Gamma^\alpha \left(p_\beta - p_\beta^{(0)}\right) + \cdots. \tag{14.3}$$

Here $\Gamma^\mu = (1, \sigma_x, \sigma_y, \sigma_z)$ are Pauli matrices (or Dirac matrices in the more general case); the expansion parameters are the vector $p_\beta^{(0)}$ indicating the position of the Fermi point in momentum space where the Green's function has a singularity, and the matrix e_α^β; ellipsis denote higher order terms in expansion.

The Fermi or Weyl point represents the exceptional (conical) point of level crossing analyzed by von Neumann and Wigner [49], which takes place in momentum space [9, 40]. Topological invariants for points at which the branches of spectrum merge were introduced by Novikov [50].

14.2.2 Emergent Relativistic Fermionic Matter

Equation (14.3) can be continuously deformed to the simple one, which describes the relativistic Weyl fermions

$$G^{-1}(p_\mu) = i\omega + N\boldsymbol{\sigma} \cdot \mathbf{p} + \cdots, \quad N = \pm 1, \qquad (14.4)$$

where the position of the Fermi point is shifted to $p_\beta^{(0)} = 0$ and ellipsis denote higher order terms in ω and $\mathbf{p}$; the matrix e_α^β is deformed to unit matrix. This means that close to the Fermi point with $N = +1$, the low energy fermions behave as right handed relativistic particles, while the Fermi point with $N = -1$ gives rise to the left handed particles.

Equation (14.4) suggests the effective Weyl Hamiltonian

$$H_{\text{eff}} = N\boldsymbol{\sigma} \cdot \mathbf{p}, \quad N = \pm 1. \qquad (14.5)$$

However, the infrared divergences may violate the simple pole structure of the propagator in Eq. (14.4). In this case in the vicinity of Fermi point one has

$$G(p_\mu) \propto \frac{-i\omega + N\boldsymbol{\sigma} \cdot \mathbf{p}}{(p^2 + \omega^2)^\gamma}, \quad N = \pm 1, \qquad (14.6)$$

with $\gamma \neq 1$. This modification does not change the topology of the propagator: the topological charge of singularity is N for arbitrary parameter γ [2]. For fermionic unparticles one has $\gamma = 5/2 - d_U$, where d_U is the scale dimension of the quantum field [51, 52].

The main property of the vacua with Dirac points is that according to (14.4), close to the Fermi points the massless relativistic fermions emerge. This is consistent with the fermionic content of our Universe, where all the elementary particles—left-handed and right-handed quarks and leptons—are Weyl fermions. Such a coincidence demonstrates that the vacuum of Standard Model is the topological medium of the Fermi point universality class. This solves the hierarchy problem, since the value of the masses of elementary particles in the vacua of this universality class is zero.

Let us suppose for a moment, that there is no topological invariant which protects massless fermions. Then the Universe is fully gapped and the natural masses of fermions must be on the order of Planck energy scale: $M \sim E_P \sim 10^{19}$ GeV. In such a natural Universe, where all masses are of order E_P, all fermionic degrees of freedom are completely frozen out because of the Bolzmann factor $e^{-M/T}$, which is about $e^{-10^{16}}$ at the temperature corresponding to the highest energy reached in accelerators. There is no fermionic matter in such a Universe at low energy. That we survive in our Universe is not the result of the anthropic principle (the latter chooses the Universes which are fine-tuned for life but have an extremely low probability). Our Universe is also natural and its vacuum is generic, but it belongs to a different universality class of vacua—the vacua with Fermi points. In such vacua, the masslessness of fermions is protected by topology (combined with symmetry, see below).

14.2.3 Emergent Gauge Fields

The vacua with Fermi-point suggest a particular mechanism for emergent symmetry. The Lorentz symmetry is simply the result of the linear expansion: this symmetry becomes better and better when the Fermi point is approached and the non-relativistic higher order terms in Eq. (14.4) may be neglected. This expansion demonstrates the emergence of the relativistic spin, which is described by the Pauli matrices. It also demonstrates how gauge fields and gravity emerge together with chiral fermions. The expansion parameters $p_\beta^{(0)}$ and e_α^β may depend on the space and time coordinates and they actually represent collective dynamic bosonic fields in the vacuum with Fermi point. The vector field $p_\beta^{(0)}$ in the expansion plays the role of the effective $U(1)$ gauge field A_β acting on fermions.

For the more complicated Fermi points with $|N| > 1$ the shift $p_\beta^{(0)}$ becomes the matrix field; it gives rise to effective non-Abelian (Yang-Mills) $SU(N)$ gauge fields emerging in the vicinity of Fermi point, i.e. at low energy [1]. For example, the Fermi point with $N = 2$ may give rise to the effective $SU(2)$ gauge field in addition to the effective $U(1)$ gauge field

$$G^{-1}(p_\mu) = e_\alpha^\beta \Gamma^\alpha (p_\beta - g_1 A_\beta - g_2 \mathbf{A}_\beta \cdot \boldsymbol{\tau}) + \text{higher order terms}, \quad (14.7)$$

where $\boldsymbol{\tau}$ are Pauli matrices corresponding to the emergent isotopic spin.

14.2.4 Emergent Gravity

The matrix field e_α^β in (14.7) acts on the (quasi)particles as the field of vierbein, and thus describes the emergent dynamical gravity field. As a result, close to the Fermi point, matter fields (all ingredients of Standard Model: chiral fermions and Abelian and non-Abelian gauge fields) emerge together with geometry, relativistic spin, Dirac matrices, and physical laws: Lorentz and gauge invariance, equivalence principle, etc. In such vacua, gravity emerges together with matter. If this Fermi point mechanism of emergence of physical laws works for our Universe, then the so-called "quantum gravity" does not exist. The gravitational degrees of freedom can be separated from all other degrees of freedom of quantum vacuum only at low energy.

In this scenario, classical gravity is a natural macroscopic phenomenon emerging in the low-energy corner of the microscopic quantum vacuum, i.e. it is a typical and actually inevitable consequence of the coarse graining procedure. It is possible to quantize gravitational waves to obtain their quanta—gravitons, since in the low energy corner the results of microscopic and effective theories coincide. It is also possible to obtain some (but not all) quantum corrections to Einstein equation and to extend classical gravity to the semiclassical level. But one cannot obtain "quantum gravity" by quantization of Einstein equations, since all other degrees of freedom of quantum vacuum will be missed in this procedure.

14.2.5 Topological Invariant Protected by Symmetry in the Standard Model

We assume that the Standard Model contains an equal number of right and left Weyl fermions, $n_R = n_L = 8n_g$, where n_g is the number of generations (we do not consider Standard Model with Majorana fermions, and assume that in the insulating state of Standard Model neutrinos are Dirac fermions). For such a Standard Model the topological charge in (14.2) vanishes, $N = 8n_g - 8n_g = 0$. Thus the masslessness of the Weyl fermions is not protected by the invariant (14.2), and an arbitrary weak interaction may result in massive particles.

However, there is another topological invariant, which takes into account the symmetry of the vacuum. The gapless state of the vacuum with $N = 0$ can be protected by the following integral [1]:

$$N_K = \frac{e_{\alpha\beta\mu\nu}}{24\pi^2} \text{tr}\left[K \int_\sigma dS^\alpha G \partial_{p_\beta} G^{-1} G \partial_{p_\mu} G^{-1} G \partial_{p_\nu} G^{-1} \right], \quad (14.8)$$

where K_{ij} is the matrix of some symmetry transformation, which either commutes or anticommutes with the Green's function matrix. In the Standard Model there are two relevant symmetries, both are the Z_2 groups, $K^2 = 1$. One of them is the center subgroup of $SU(2)_L$ gauge group of weak rotations of left fermions, where the element K is the gauge rotation by angle 2π, $K = e^{i\pi \check{\tau}_{3L}}$. The other one is the group of the hypercharge rotation be angle 6π, $K = e^{i6\pi Y}$. In the $G(224)$ Pati-Salam extension of the $G(213)$ group of Standard Model, this symmetry comes as combination of the Z_2 center group of the $SU(2)_R$ gauge group for right fermions, $e^{i\pi \check{\tau}_{3R}}$, and the element $e^{3\pi i(B-L)}$ of the Z_4 center group of the $SU(4)$ color group—the P_M parity (on the importance of the discrete groups in particle physics see [53, 54] and references therein). Each of these two Z_2 symmetry operations changes sign of left spinor, but does not influence the right particles. Thus these matrices are diagonal,

$K_{ij} = \mathrm{diag}(1, 1, \ldots, -1, -1, \ldots)$, with eigenvalues 1 for right fermions and -1 for left fermions.

In the symmetric phase of Standard Model, both matrices commute with the Green's function matrix G_{ij}, as a result N_K in (14.8) is topological invariant: it is robust to deformations of Green's function which preserve the symmetry K. The value of this invariant $N_K = 16n_g$, which means that all $16n_g$ fermions are massless.

14.2.6 Higgs Mechanism vs Splitting of Fermi Points

The gapless vacuum of the Standard Model is supported by a combined action of topology and symmetry K, and also by the Lorentz invariance which keeps all the Fermi points at $\mathbf{p} = 0$.

Explicit violation or spontaneous breaking of one of these symmetries transforms the vacuum of the Standard Model into one of the two possible vacua. If, for example, the K symmetry is broken, the invariant (14.8) supported by this symmetry ceases to exist, and the Fermi point disappears. All $16n_g$ fermions become massive (Fig. 14.3 *bottom left*). This is assumed to happen below the symmetry breaking electroweak transition caused by Higgs mechanism where quarks and charged leptons acquire the Dirac masses.

If, on the other hand, the Lorentz symmetry is violated, the marginal Fermi point splits into topologically stable Fermi points with non-zero invariant N, which protects massless chiral fermions (Fig. 14.3 *bottom right*). Since the invariant N does not depend on symmetry, the further symmetry breaking cannot destroy the nodes. One can speculate that in the Standard Model the latter may happen with the electrically neutral leptons, the neutrinos [55]. Most interestingly, Fermi-point splitting of neutrinos may provide a new source of T and CP violation in the leptonic sector, which may be relevant for the creation of the observed cosmic matter-antimatter asymmetry [57].

14.2.7 Splitting of Fermi Points and Problem of Generations

An example of the multiple splitting is provided by the model Hamiltonian for fermions in superconductors/superfluids in the state which belongs to $O(D_2)$ symmetry class [58]:

$$H = \frac{1}{\sqrt{2}}(p^2 - p_F^2)\tau_3 + \frac{1}{2}(2p_x^2 - p_y^2 - p_z^2)\tau_1 + \frac{\sqrt{3}}{2}(p_y^2 - p_z^2)\tau_2. \qquad (14.9)$$

At $p_F^2 < 0$ the energy spectrum is fully gapped, for $p_F^2 = 0$ the node in the spectrum appears at $\mathbf{p} = 0$ which at $p_F^2 > 0$ splits into 8 Fermi points at the vertices of cube in momentum space (see Fig. 14.4):

$$\mathbf{p}^{(n)} = \frac{p_F}{\sqrt{3}}(\pm\hat{\mathbf{x}} \pm \hat{\mathbf{y}} \pm \hat{\mathbf{z}}), \quad n = 1, \ldots, 8. \qquad (14.10)$$

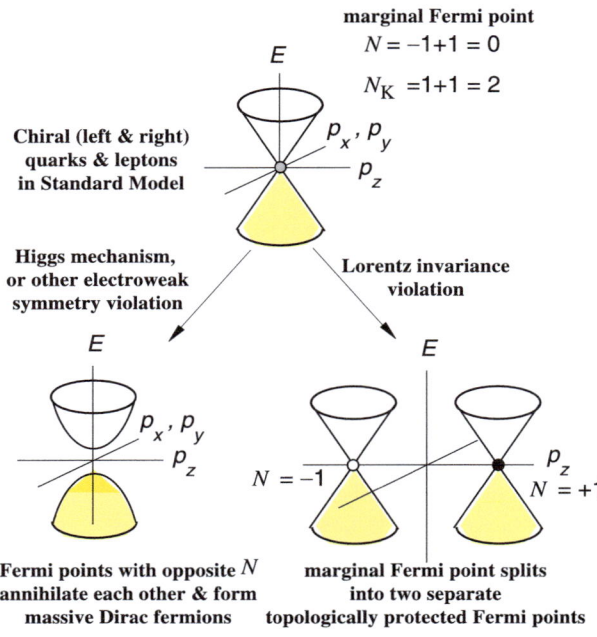

Fig. 14.3 (*top*): In Standard Model the Fermi points with positive $N = +1$ and negative $N = -1$ topological charges are at the same point $\mathbf{p} = 0$, forming the marginal Fermi point with $N = 0$. Symmetry K between the Fermi points prevents their mutual annihilation giving rise to the topological invariant (14.8) with $N_K = 2$. (*bottom left*): If symmetry K is violated or spontaneously broken, Fermi points annihilate each other and Dirac mass is formed. (*bottom right*): If Lorentz invariance is violated or spontaneously broken, the marginal Fermi point splits [55]. The topological quantum phase transition between the state with Dirac mass and the state with splitted Dirac points have been observed in cold Fermi gas [56]

These nodes have topological charges $N = \pm 1$ in Eq. (14.2), and as a result, close to each of 8 nodes the Hamiltonian is reduced to the Hamiltonian describing Weyl fermions:

$$H^{(n)} = \mathbf{e}_1^{(n)} \cdot \left(\mathbf{p} - \mathbf{p}^{(n)}\right)\tau_1 + \mathbf{e}_2^{(n)} \cdot \left(\mathbf{p} - \mathbf{p}^{(n)}\right)\tau_2 + \mathbf{e}_3^{(n)} \cdot \left(\mathbf{p} - \mathbf{p}^{(n)}\right)\tau_3. \quad (14.11)$$

Each Weyl fermion has its own triad (dreibein). Choosing for simplicity $p_F = \sqrt{3}$ one has

$$\mathbf{e}_3^{(n)} = \sqrt{2}(\pm\hat{\mathbf{x}} \pm \hat{\mathbf{y}} \pm \hat{\mathbf{z}}), \quad (14.12)$$

$$\mathbf{e}_1^{(n)} = \pm 2\hat{\mathbf{x}} \mp \hat{\mathbf{y}} \mp \hat{\mathbf{z}}, \quad (14.13)$$

$$\mathbf{e}_2^{(n)} = \sqrt{3}(\pm\hat{\mathbf{y}} \mp \hat{\mathbf{z}}). \quad (14.14)$$

All triads can be transformed to each other by rotations and/or reflection. So in this model one obtains four identical copies of right and left relativistic Weyl fermions. They may be considered as analogues of the generation of Standard Model fermions,

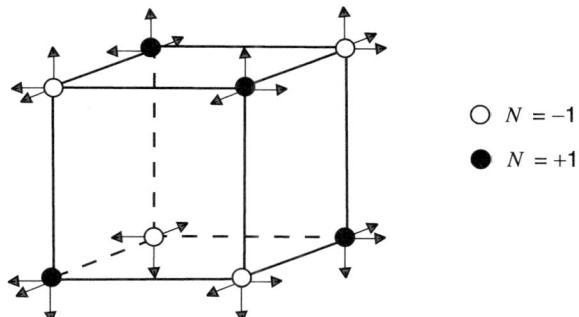

Fig. 14.4 Sketch of topologically protected point nodes in momentum space with topological charge $N = \pm 1$ in Eq. (14.2) in superconductors/superfluids of $O(D_2)$ symmetry class [58]. Chiral fermions emerge in the vicinity of each Fermi points. They have identical emergent Minkowski metric, but different orientations of dreibein. The simplest realization of dreibein for each of eight chiral fermions is shown by arrows. The vierbein orientations in $O(D_2)$ symmetry class superconductors are more complicated than in figure: one of the vectors in each vierbein is along the cube main diagonal, see Eq. (14.12). All four pairs of left and right Weyl fermions have the same quantum numbers, while their triads can be transformed to each other by rotations and reflection. They are analogous to the generation of the Standard Model fermions

but with $n_g = 4$. A different, but related mechanism for the origin of generations is suggested in [59].

14.3 Exotic Fermions

In many systems (including condensed matter and relativistic quantum vacua), the Fermi points with elementary charges $N = \pm 1$ may merge together forming either the neutral point with $N = 0$ or point with multiple N (i.e. $|N| > 1$ [60]). In this case topology and symmetry become equally important, because it is the symmetry which may stabilize the degenerate node. An example is provided by the Standard Model of particle physics, where 16 fermions of one generation have degenerate Dirac point at $\mathbf{p} = 0$ with the trivial total topological charge $N = 8 - 8 = 0$. In the symmetric phase of Standard Model the nodes in the spectrum survive due to a discrete symmetry between the fermions and they disappear in the non-symmetric phase forming the fully gapped vacuum [1]. In the case of degenerate Fermi point with $|N| > 1$, situation is more diverse. Depending on symmetry, the interaction between fermionic flavors may lead to a splitting of the multiple Fermi point to elementary Dirac points [55]; or gives rise to the essentially non-relativistic energy spectrum $E_\pm(p \to 0) \to \pm p^N$, which corresponds to different scaling for space and time in the infrared: $\mathbf{r} \to b\mathbf{r}$, $t \to b^N t$. The particular case of anisotropic scaling with $N = 3$ was suggested by Hořava for quantum gravity at short distances, the so-called Hořava-Lifshitz gravity [61–63], while the anisotropic scaling in the infrared was suggested in Ref. [64]. The topology of the multiple Dirac point provides another possible realization of anisotropic gravity, which is different from the scenario based on Lifshitz point in the theory of phase transitions [65, 66].

14.3.1 Dirac Fermions with Quadratic Spectrum

The nonlinear spectrum arising near the Fermi point with $N=2$ has been discussed for different systems including graphene, double cuprate layer in high-T_c superconductors, surface states of topological insulators and neutrino physics [1, 2, 32, 67–74]. The spectrum of (quasi)particles in the vicinity of the doubly degenerate node depends on symmetry. Let us consider the node with topological charge $N=+2$ in $2+1$ system. Such Fermi point of co-dimension 2 takes place in bilayered graphene. According to general classification [19], the topology alone cannot protect the gapless fermions in 2D: the topological invariant takes place only in the presence of a symmetry. In particular, if we restrict consideration only to real (Majorana) fermions, the nodes obey Z_2 topology with summation law $1+1=0$ [19]. To make the multiple Fermi point possible we need an additional symmetry K which extends the group Z_2 to the full group of integers Z. The relevant symmetry protected topological invariant is [2, 32, 75, 76]:

$$N = \frac{1}{4\pi i}\,\text{tr}\oint_C dl K G(\omega=0,\mathbf{p})\partial_l G^{-1}(\omega=0,\mathbf{p}) = \text{tr}\oint_C dl K \mathscr{H}^{-1}(\mathbf{p})\partial_l \mathscr{H}(\mathbf{p}), \quad (14.15)$$

where C is contour around the Dirac point in 2D momentum space (p_x, p_y); K is the relevant symmetry operator; G is the Green's function matrix at zero frequency, which can be used as the effective Hamiltonian, $\mathscr{H}(\mathbf{p}) = G^{-1}(\omega=0,\mathbf{p})$; the operator K commutes or anticommutes with the effective Hamiltonian.

Provided the symmetry K is preserved and thus the summation law for N takes place, one finds several scenarios of the behavior of the system with the total topological charge $N=+2$.

(i) One may have two fermions with the linear Dirac spectrum, with the nodes being at the same point of momentum space. This occurs if there is some special symmetry, such as the fundamental Lorentz invariance.

(ii) Exotic massless fermions emerge. In the 2D systems, these are fermions with parabolic energy spectrum, which emerge at low energy:

$$E_\pm(\mathbf{p}) \approx \pm p^2. \quad (14.16)$$

They are described by the following effective Hamiltonian

$$\mathscr{H}(p_x, p_y) = \begin{pmatrix} 0 & (p_x+ip_y)^2 \\ (p_x-ip_y)^2 & 0 \end{pmatrix} = (p_x^2 - p_y^2)\sigma_1 - 2p_x p_y \sigma_2, \quad (14.17)$$

where σ_1 and σ_2 are Pauli matrices. The topological charge N of the node at the point $p_x = p_y = 0$ is given by Eq. (14.15), where the symmetry operator

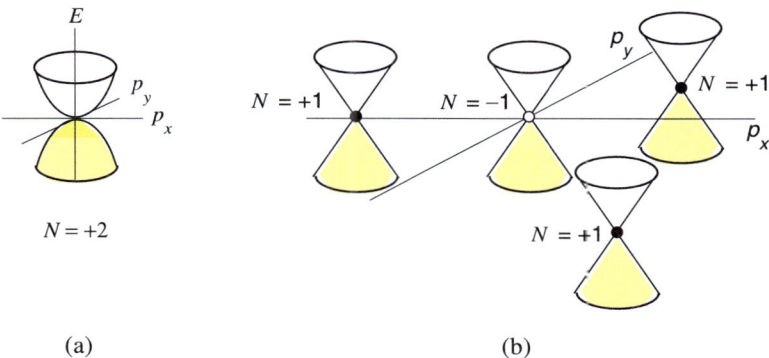

Fig. 14.5 Illustration of summation rule for momentum-space topological invariant. Splitting of $N = 2$ point with quadratic dispersion (**a**) into four Dirac points: $N = 2 = 1 + 1 + 1 - 1$ (**b**)

K is represented by the Pauli matrix σ_3. With effective Hamiltonian (14.17) one obtains that the node with the quadratic spectrum has the charge $N = 2$. In 3D systems, the corresponding fermions with the topological charge $N = 2$ in Eq. (14.2) are the semi-Dirac fermions, with linear dispersion in one direction and quadratic dispersion in the other [67]:

$$E_\pm(\mathbf{p}) \approx \pm\sqrt{c^2 p_z^2 + p_\perp^4}. \quad (14.18)$$

(iii) The Weyl point with $N = +2$ may split either into two Weyl points each with $N = +1$ (see [55] for the relativistic $3 + 1$ system) or into four Weyl points (three with $N = +1$ and one with $N = -1$, see Fig. 14.5). The effective Hamiltonian for the latter case is [77, 78]:

$$\mathcal{H}(p_x, p_y) = \begin{pmatrix} 0 & (p_x + ip_y)^2 + s(p_x - ip_y) \\ (p_x - ip_y)^2 + s(p_x + ip_y) & 0 \end{pmatrix}. \quad (14.19)$$

The energy spectrum of this Hamiltonian has four Dirac points: the node at $\mathbf{p} = 0$ has the topological charge $N = -1$, while three nodes at $p_x + ip_y = -se^{2\pi ki/3}$ with integer k have charges $N = +1$ each, so that the summation rule $N = 1 + 1 + 1 - 1 = 2$ does hold.

Note that in options (ii) and (iii) the (effective) Lorentz invariance of the Dirac point is violated. This suggests that the topological mechanism of splitting of the Dirac point [55] or of the formation of the nonlinear dispersion [67] may lead to the spontaneous breaking of Lorentz invariance in the relativistic quantum vacuum, which in principle may occur in the neutrino sector of the quantum vacuum [79–81].

Let us consider the Fermi point with higher degeneracy, described by the symmetry protected topological invariant $N > 2$.

14.3.2 Dirac Fermions with Cubic and Quadratic Spectrum

Let us consider first the case with $N = 3$ [82]. Examples are three families of right-handed Weyl 2-component fermions in particle physics; three cuprate layers in high-T_c superconductors; three graphene layers, etc. If the Fermi point is topologically protected, i.e. there is a conserved topological invariant N, the node in the spectrum cannot disappear even in the presence of interaction, but it can split into N nodes with elementary charge $N = 1$. The splitting can be prevented if there is a symmetry in play, such as rotational symmetry. Here we provide an example of such a symmetry.

We consider 3 species (families or flavors) of fermions, each of them being described by the invariant $N = +1$ in Eq. (14.15) and an effective relativistic Hamiltonian emerging in the vicinity of the Fermi point

$$\mathcal{H}_0(\mathbf{p}) = \boldsymbol{\sigma} \cdot \mathbf{p} = \sigma_x p_x + \sigma_y p_y. \tag{14.20}$$

The matrix $K = \sigma_z$ anticommutes with the Hamiltonian. This supports the topologically protected node in spectrum, which is robust to interactions. The position of the node here is chosen at $\mathbf{p} = 0$:

$$E^2 = p^2. \tag{14.21}$$

The total topological charge of three nodes at $\mathbf{p} = 0$ of three fermionic species is $N = +3$. Let us now introduce matrix elements which mix the fermions. If these elements violate symmetry K, the topological invariant cannot be constructed and point node will be destroyed, so let the matrix elements obey the symmetry K. For the general case of the matrix elements, but still obeying the symmetry K, the multiple node will split into 3 or more elementary nodes, obeying the summation rule: $3 = 1 + 1 + 1 = 1 + 1 + 1 + 1 - 1 = \cdots$ (see Fig. 14.6). However, in the presence of some extra symmetry, which prevents splitting, the branch with the cubic spectrum emerges.

An example is provided by the following Hamiltonian [82]

$$\mathcal{H}(\mathbf{p}) = \begin{pmatrix} \boldsymbol{\sigma} \cdot \mathbf{p} & g_{12}\sigma^+ & g_{13}\sigma^+ \\ g_{21}\sigma^- & \boldsymbol{\sigma} \cdot \mathbf{p} & g_{23}\sigma^+ \\ g_{31}\sigma^- & g_{32}\sigma^- & \boldsymbol{\sigma} \cdot \mathbf{p} \end{pmatrix}, \tag{14.22}$$

where $\sigma^\pm = \frac{1}{2}(\sigma_x \pm i\sigma_y)$ are ladder operators. This Hamiltonian anti-commutes with $K = \sigma_z$ and thus mixing preserves the topological charge N in (14.15). At $p_x = p_y = 0$ it is independent of the spin rotations up to a global phase of the coupling constants. Under spin rotation by angle θ all elements in the upper triangular matrix are multiplied by $e^{i\theta}$, while all elements in the lower triangular matrix are multiplied by $e^{-i\theta}$. This symmetry of triangular matrices generates does not allow the multiple Fermi point to split at $p = 0$, as a result the gapless branch of spectrum in Fig. 14.7 has the cubic form at low energy, $E \to 0$, which corresponds to the topological charge $N = +3$:

$$E^2 \approx \gamma_3^2 p^6, \quad \gamma_3 = \frac{1}{|g_{12}||g_{23}|}. \tag{14.23}$$

Fig. 14.6 Splitting of Fermi point with $N = 3$ into three Fermi points with $N = 1$ in the model discussed in [82]

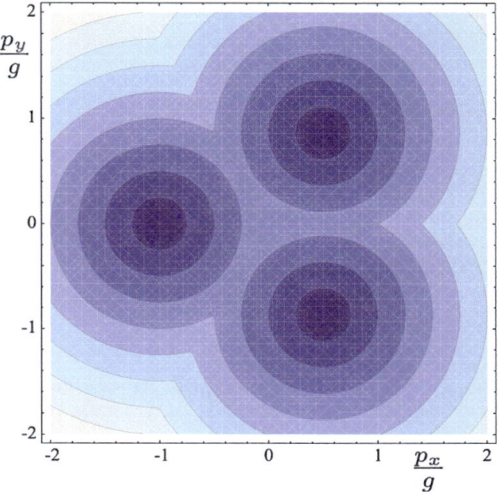

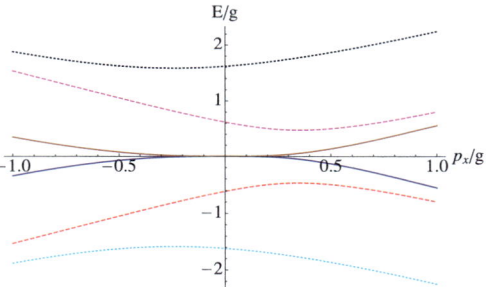

Fig. 14.7 Spectrum of the Hamiltonian (14.22) showing cubic dispersion for the lowest two eigenvalues around the point $\mathbf{p} = 0$. The spectrum has been calculated with equal coupling strengths $g_{12} = g_{13} = g_{23} = g$, but cubic spectrum characterized by topological charge $N = 3$ preserves for any nonzero values of coupling strengths. Spectrum is shown as function of p_x at $p_y = 0$

In the low-energy limit the spectrum in the vicinity of the multiple Fermi point (14.23) is symmetric under rotations. But in general the spectrum is not symmetric as demonstrated in Fig. 14.7. There is only the symmetry with respect to reflection, $(p_x, p_y) \to (p_x, -p_y)$. The rotational symmetry of spectrum (14.23) is an emergent phenomenon, which takes place only in the limit $p \to 0$. For trilayer graphene this spectrum has been discussed in Ref. [83].

In case of four fermionic species, the mixing which does not produce splitting of the Fermi point is obtained by the same principle as in (14.22): all matrix elements above the main diagonal contain only σ^+ (or σ^-):

Fig. 14.8 Spectrum of the Hamiltonian (14.24) showing quartic dispersion for the lowest two eigenvalues around the point $\mathbf{p} = 0$. Different colors correspond to different eigenvalues, spectrum is shown as function of p_x at $p_y = 0$. The spectra have been calculated with equal coupling strengths $g_{12} = g_{13} = g_{23} = g_{14} = g_{24} = g_{34} = g$

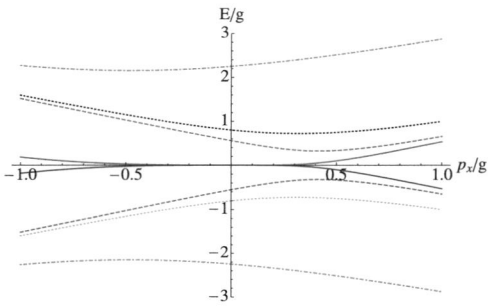

$$\mathcal{H}(\mathbf{p}) = \begin{pmatrix} \boldsymbol{\sigma} \cdot \mathbf{p} & g_{12}\sigma^+ & g_{13}\sigma^+ & g_{14}\sigma^+ \\ g_{21}\sigma^- & \boldsymbol{\sigma} \cdot \mathbf{p} & g_{23}\sigma^+ & g_{24}\sigma^+ \\ g_{31}\sigma^- & g_{32}\sigma^- & \boldsymbol{\sigma} \cdot \mathbf{p} & g_{34}\sigma^+ \\ g_{41}\sigma^- & g_{42}\sigma^- & g_{42}\sigma^- & \boldsymbol{\sigma} \cdot \mathbf{p} \end{pmatrix}. \tag{14.24}$$

Then again under spin rotation by angle θ all elements in the upper triangular matrix are multiplied by $e^{i\theta}$, while all elements in the lower triangular matrix are multiplied by $e^{-i\theta}$. As a result the multiple Fermi point with $N = 4$ is preserved giving rise to the quartic spectrum in vicinity of the Fermi point (see Fig. 14.8):

$$E^2 = \gamma_4^2 p^8, \quad \gamma_4 = \frac{1}{|g_{12}||g_{23}||g_{34}|}. \tag{14.25}$$

For the tetralayer graphene this spectrum was suggested in Ref. [84]. Again the rotational symmetry emerges new the Fermi point, but spectrum is not symmetric under permutations. In fact, there is no such 4×4 matrix that would consist of couplings described by ladder operators and which would be symmetric under permutations.

In general the Fermi point with arbitrary N may give rise to the spectrum

$$E^2 = \gamma_N^2 p^{2N}. \tag{14.26}$$

Such spectrum emerges in multilayered graphene [32, 85]. The discussed symmetry of matrix elements g_{mn} extended to $2N \times 2N$ matrix gives (14.26) with the prefactor

$$\gamma_N = \frac{1}{|g_{12}||g_{23}|\ldots|g_{N-1,N}|}. \tag{14.27}$$

Violation of this symmetry may lead to splitting of the multiple Fermi point into N elementary Fermi points—Dirac points with $N = 1$ and 'relativistic' spectrum $E^2 \propto p^2$.

The effective Hamiltonian describing fermions in the vicinity of multiple Fermi point is $H = \sigma^- p_+^N + \sigma^+ p_-^N$, see [32, 34]. An example of the effective Hamiltonian describing the multiple Fermi point with topological charge N in $3+1$ systems is [60]

$$H = \sigma_z p_z + \sigma^- p_+^N + \sigma^+ p_-^N. \tag{14.28}$$

This Hamiltonian has the spectrum $E^2 = p_z^2 + p_\perp^{2N}$, which has linear dispersion in one direction and non-linear dispersion in the others.

This is also applicable to the vacuum of particle physics. The Lorentz symmetry prohibits both the splitting of the Dirac points and the non-linear non-relativistic spectrum. The situation changes if the Lorentz symmetry is viewed as an emergent phenomenon, which arises near the Dirac point (Fermi points with $N = \pm 1$). In this case both splitting of Dirac points and formation of nonlinear non-relativistic spectrum in the vicinity of the multiple Fermi point are possible, an the choice depends on symmetry. In both cases the mixing of fermions violates the effective Lorentz symmetry in the low-energy corner. This phenomenon, called the reentrant violation of special relativity [67], has been discussed for $N_F = 3$ fermion families in relation to neutrino oscillations [57]. The influence of possible discrete flavor symmetries on neutrino mixing has been reviewed in Ref. [87].

14.4 Flat Bands

Let us turn to the limit case $N \to \infty$ [86]. If the layers are equivalent and interact only via nearest neighbor couplings, i.e. the non-zero matrix elements are

$$g_{12} = g_{23} = \cdots = g_{N-1,N} \equiv t, \tag{14.29}$$

the low-energy spectrum becomes

$$E = t\left(\frac{p}{t}\right)^N. \tag{14.30}$$

In the $N \to \infty$ limit one obtains that in the lowest energy branch all the fermions within circumference $|\mathbf{p}| = t$ have zero energy.

$$E(N \to \infty, |\mathbf{p}| < t) = 0. \tag{14.31}$$

14.4.1 Topological Origin of Surface Flat Band

To understand the topological origin of this branch and its structure let us consider the spectrum in the continuous limit. The effective Hamiltonian in the 3-dimensional bulk system which emerges in the limit of infinite number of layers is the following 2×2 matrix

$$H = \begin{pmatrix} 0 & f \\ f^* & 0 \end{pmatrix}, \quad f = p_x - ip_y - te^{-iap_z}. \tag{14.32}$$

Here t is the magnitude of the hopping matrix element between the layers in Eq. (14.29). The energy spectrum of the bulk system

$$E^2 = \left[p_x - t\cos(ap_z)\right]^2 + \left[p_y + t\sin(ap_z)\right]^2, \tag{14.33}$$

has zeroes on line (see Fig. 14.9):

$$p_x = t\cos(ap_z), \quad p_y = -t\sin(ap_z), \tag{14.34}$$

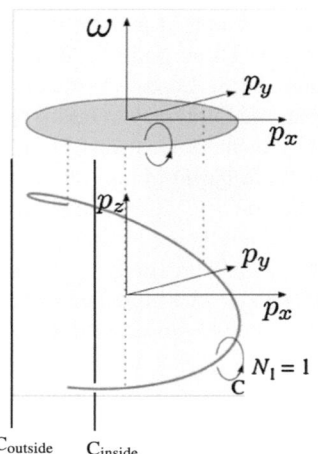

Fig. 14.9 Nodal spiral generates topologically protected flat band on the surface. Projection of spiral on the surface determines boundary of flat band. At each (p_x, p_y) except the boundary of circle $p_x^2 + p_y^2 = t^2$ the system represents the 1D gapped state (insulator). At each (p_x, p_y) inside the circle, the insulator is topological being described by non-zero topological invariant (14.36) and thus one has a gapless edge state. The manifold of these zero-energy edge state inside the circle forms the flat band found in Eq. (14.31)

which forms a spiral, the projection of this spiral on the plane $p_z = $ const being the circle $\mathbf{p}_\perp^2 \equiv p_x^2 + p_y^2 = t^2$. This nodal line is topologically protected by the same topological invariant as in Eq. (14.15)

$$N = \frac{1}{4\pi i} \operatorname{tr} \oint_C dl\, \sigma_z H^{-1} \nabla_l H, \tag{14.35}$$

where the integral is now along the loop C around the nodal line in momentum space, see Fig. 14.9. The winding number around the element of the nodal line is $N = 1$.

Let us consider now the momentum $\mathbf{p}_\perp$ as a parameter of the 1D system, then for $|\mathbf{p}_\perp| \neq t$ the system represents the fully gapped system—1D insulators. This insulator can be described by the same invariant as in Eq. (14.35) with the contour of integration chosen parallel to p_z, i.e. along the 1D Brillouin zone at fixed $\mathbf{p}_\perp$ (due to periodic boundary conditions, the points $p_z = \pm\pi/a$ are equivalent and the contour of integrations forms the closed loop):

$$N(\mathbf{p}_\perp) = \frac{1}{4\pi i} \operatorname{tr} \int_{-\pi/a}^{+\pi/a} dp_z \sigma_z H^{-1} \nabla_{p_z} H. \tag{14.36}$$

For $|\mathbf{p}_\perp| < t$ the 1D insulator is topological, since $N(|\mathbf{p}_\perp| < t) = 1$, while for $|\mathbf{p}_\perp| > t$ one has $N_1(\mathbf{p}_\perp) = 0$ and the 1D insulator is the trivial band insulator. The line $|\mathbf{p}_\perp| = t$ thus marks the topological quantum phase transition between the topological and non-topological 1D insulators.

Topological invariant $N(\mathbf{p}_\perp)$ in (14.36) determines the property of the surface bound states of the 1D system at each $\mathbf{p}_\perp$. Due to the bulk-edge correspondence, the topological 1D insulator must have the surface state with exactly zero energy. Since such states exist for any parameter within the circle $|\mathbf{p}_\perp| = t$, one obtains the flat band of surface states with exactly zero energy, $E(|\mathbf{p}_\perp| < t) = 0$, which is protected by topology. This is the origin of the unusual branch of spectrum in Eq. (14.31): it represents the band of topologically protected surface states with exactly zero energy. Such states do not exist for parameters $|\mathbf{p}_\perp| > t$, for which the 1D insulator is non-topological.

The zero energy bound states on the surface of the system can be obtained directly from the Hamiltonian:

$$\hat{H} = \sigma_x\big(p_x - t\cos(a\hat{p}_z)\big) + \sigma_y\big(p_y + t\sin(a\hat{p}_z)\big), \quad \hat{p}_z = -i\partial_z, \ z < 0. \quad (14.37)$$

We assumed that the system occupies the half-space $z < 0$ with the boundary at $z = 0$. This Hamiltonian has the bound state with exactly zero energy, $E(\mathbf{p}_\perp) = 0$, for any $|\mathbf{p}_\perp| < t$, with the eigenfunction concentrated near the surface:

$$\Psi \propto \begin{pmatrix} 0 \\ 1 \end{pmatrix} (p_x - ip_y) \exp\frac{z\ln(t/(p_x + ip_y))}{a}, \quad |\mathbf{p}_\perp| < t. \quad (14.38)$$

The normalizable wave functions with zero energy exist only for $\mathbf{p}_\perp$ within the circle $|\mathbf{p}_\perp| \leq t$, i.e. the surface flat band is bounded by the projection of the nodal spiral onto the surface. Such correspondence between the flat band on the surface and lines of zeroes in the bulk has been also found in Ref. [35] for superconductors without inversion symmetry.

14.4.2 Dimensional Crossover in Topological Matter: Formation of the Flat Band in Multi-layered Systems

The discrete model with a finite number N of layers has been considered in Ref. [86]. It is described by the $2N \times 2N$ Hamiltonian with the nearest neighbor interaction between the layers in the form:

$$H_{ij}(\mathbf{p}_\perp) = \boldsymbol{\sigma} \cdot \mathbf{p}_\perp \delta_{ij} - t\sigma^+ \delta_{i,j+1} - t\sigma^- \delta_{i,j-1}, \quad 1 \leq i \leq N, \ \mathbf{p}_\perp = (p_x, p_y). \quad (14.39)$$

In the continuous limit of infinite number of layers (14.39) transforms to (14.32) with the nodal line in the spectrum, while for finite N the spectrum contains the Dirac point with multiple topological charge equal to N. Figures 14.10, 14.11, 14.12 and 14.13 demonstrate how this crossover from 2D to 3D occurs. When the number N of layers increases, the dispersionless surface band evolves from the gapless branch of the spectrum $E = \pm|\mathbf{p}_\perp|^N$. Simultaneously, the gapped branches of the spectrum of the finite-N system give rise to the nodal line in bulk. This scenario of formation of the surface flat band takes place if the symmetry does not allow the splitting of the multiple Dirac point, and it continuously evolves to the dispersionless spectrum.

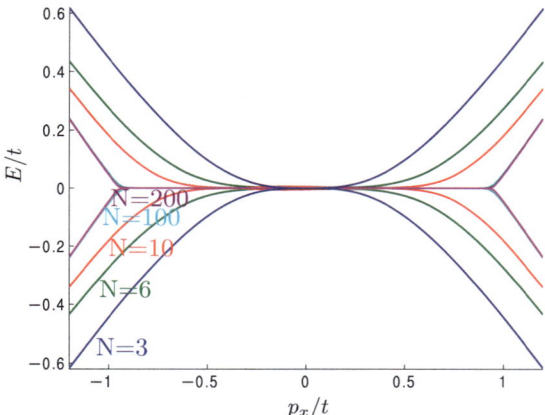

Fig. 14.10 Formation of the surface flat band. When the number N of layers increases, the dispersionless band evolves from the gapless branch of the spectrum, which has the form $E = \pm|\mathbf{p}_\perp|^N$ in the vicinity of multiple Dirac point. The spectrum is shown as a function of p_x for $p_y = 0$. The curves for $N = 100$ and $N = 200$ are almost on top of each other. Asymptotically the spectrum $E = \pm|\mathbf{p}_\perp|^N$ transforms to the dispersionless band within the projection of the nodal line to the surface

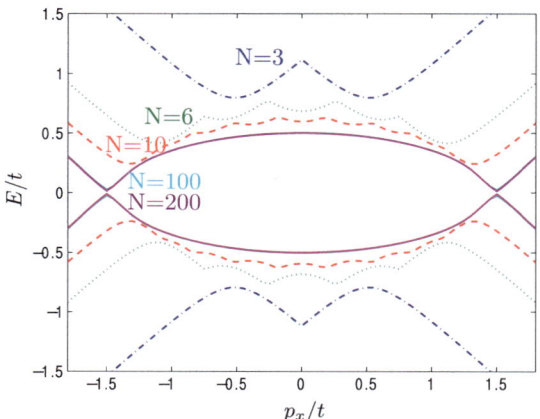

Fig. 14.11 Formation of the nodal line from the evolution of the gapped branch of the spectrum of the multilayered system, when the number N of layers increases. The spectrum is shown as function of p_x for $p_y = 0$. The curves for $N = 100$ and $N = 200$ lie almost on top of each other, indicating the bulk limit. Asymptotically the nodal line $p_x = t\cos(ap_z)$, $p_y = -t\sin(ap_z)$ is formed (two points on this line are shown, which correspond to $p_y = 0$)

If this symmetry is absent, but the symmetry supporting the topological charge persists, the scenario of the flat band formation is different but still is governed by topology. Figure 14.14 demonstrates the formation of the flat band in this situation.

Fig. 14.12 The lowest energy states for different p_x and p_y and arbitrary p_z in the bulk limit. The flat band of surface states is formed in the region $p_x^2 + p_y^2 < t^2$. Outside this region only the bulk states exist. In a simple model considered here flat band comes from the degenerate Dirac point with nonlinear dispersion. However, this is not necessary condition: the flat band emerges whenever the nodal line appears in the bulk

Fig. 14.13 Result of the transformation of the gapped states in the process of dimensional crossover. They form the nodal line in bulk $p_x = t\cos(ap_z)$, $p_y = -t\sin(ap_z)$ whose projection to the (p_x, p_y)-plane is shown

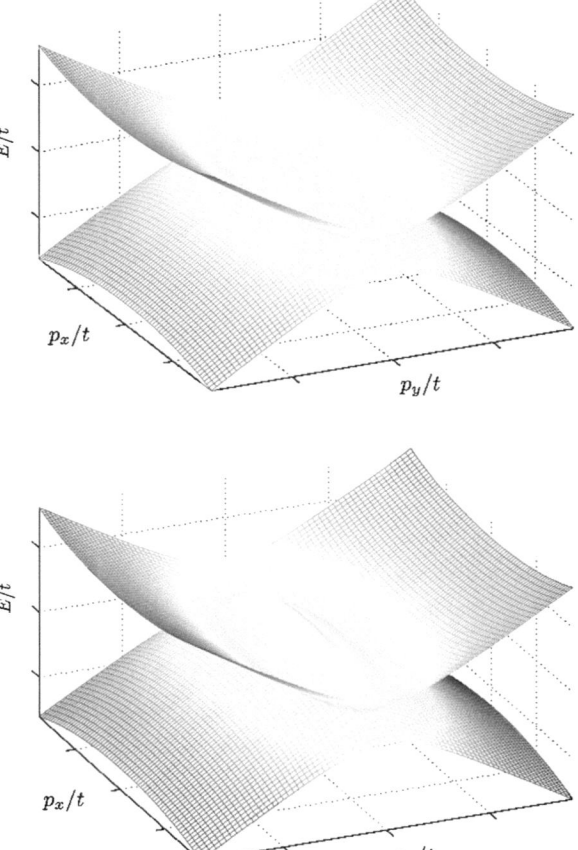

The reason, why the flat band emerges is the formation of the topologically protected nodal line in bulk. The projection of the nodal line on a surface gives the boundary of the flat band emerging on this surface. This is the realization of the bulk-surface correspondence in systems with the nodal lines in bulk. The other examples can be found in Ref. [88].

14.5 Anisotropic Scaling and Hořava Gravity

14.5.1 Effective Theory Near the Degenerate Dirac Point

We know that in the vicinity of the Weyl point with elementary topological charge $N = +1$ or $N = -1$ the quantum electrodynamics and gravity emerge as effective

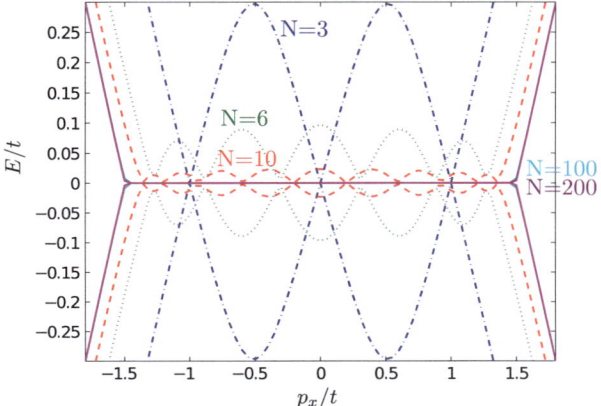

Fig. 14.14 Evolution of the spectrum in a different model, in which the multiple Dirac point is absent, but the surface flat band is formed together with the formation of the nodal line in bulk

fields. What happens near node with higher N. Let us consider again the $2+1$ system with the multiple Dirac point and write the following effective Hamiltonian:

$$\mathcal{H}_N = \frac{\sigma^x + i\sigma^y}{2}\left((\mathbf{e}_1 + i\mathbf{e}_2)\cdot(\mathbf{p} - e\mathbf{A})\right)^{|N|} + \frac{\sigma^x - i\sigma^y}{2}\left((\mathbf{e}_1 - i\mathbf{e}_2)\cdot(\mathbf{p} - e\mathbf{A})\right)^{|N|}. \tag{14.40}$$

For a single layer ($N=1$) the Hamiltonian (14.40) is reduced to the conventional Weyl-Dirac Hamiltonian for massless particles in $2+1$ dimension:

$$\mathcal{H}_{N=1} = \sigma^x \mathbf{e}_1 \cdot (\mathbf{p} - e\mathbf{A}) + \sigma^y \mathbf{e}_2 \cdot (\mathbf{p} - e\mathbf{A}) = e_a^i \sigma^a (p_i - eA_i), \quad a = (1,2). \tag{14.41}$$

Here the vectors $\mathbf{e}_1$ and $\mathbf{e}_2$ play the role of zweibein (in the ground state they are mutually orthogonal). The vector $\mathbf{A}$ is either the vector potential of the effective electromagnetic field which comes from the shifts of the node, or the real electromagnetic field, as it takes place for electrons in graphene. The whole dreibein e_a^μ with $a=(1,2,3)$ and $\mu=(0,1,2)$ emerges for the Green's function and gives the effective metric $2+1$ metric as a secondary object:

$$g^{\mu\nu} = \eta^{ab} e_a^\mu e_b^\nu. \tag{14.42}$$

For a general $|N|$ the situation is somewhat different. While the action for the relativistic fermions is invariant under rescaling $\mathbf{r} = b\mathbf{r}'$, $t = bt'$, the action for the fermions living in the vicinity of the multiple Dirac point is invariant under anisotropic rescaling $\mathbf{r} = b\mathbf{r}'$, $t = b^{|N|}t'$. The anisotropic scaling is in the basis of the Hořava gravity, which is described by the space components of metric are separated from the time component and have different scaling laws [61–64]. The square of Hamiltonian (14.40) gives the space metric in terms of zweibein:

$$\mathcal{H}_N^2 = E_N^2 = \left(g^{ij}(p_i - eA_i)(p_j - eA_j)\right)^{|N|}, \quad g^{ij} = e_1^i e_1^j + e_2^i e_2^j. \tag{14.43}$$

14.5.2 Effective Electromagnetic Action

If the effective action for fields e_a^μ and A_μ is obtained by integration over fermions in the vicinity of the multiple Dirac point, this bosonic action (actually the terms in action which mostly come from these fermions) inherits the corresponding conformal symmetry of the massless fermions. For anisotropic scaling the conformal invariance means invariance under $g^{ik} \to b^2 g^{ik}$ and $g^{00} \to b^{2|N|} g^{00}$; $\sqrt{-g} \to b^{-(|N|+D)}$ (where D is space dimension); while g^{0i} is not considered.

14.5.2.1 Single Layered Graphene and Relativistic Fields

For $D \neq 2$ the spectrum of multiple Fermi point becomes more complicated, and in general is not isotropic, see Eq. (14.28). For general D the spectrum is isotropic only for $|N| = 1$, where one obtains effective relativistic massless $D+1$ quantum electrodynamics, which is Lorentz invariant. This implies the following nonlinear action

$$S_{em}(|N|=1, D) = \int d^D x dt \left[B^2 - E^2\right]^{\frac{D+1}{4}}. \tag{14.44}$$

For $D = 3$ the action is proportional to $(B^2 - E^2)\ln(B^2 - E^2)$, and is imaginary at $B^2 < E^2$ giving rise to Schwinger pair production in massless quantum electrodynamics. The similar imaginary action takes place for $B^2 < E^2$ for $D \neq 3$. For example, a single layer graphene ($D = 2$, $|N| = 1$) reproduces the relativistic $2+1$ QED which gives rise to Lagrangian $(B^2 - E^2)^{3/4}$ [89] with the running coupling constant $1/\alpha = \sqrt{2}\zeta(3/2)/8\pi^2$. The action is imaginary at $B^2 < E^2$ which corresponds to Schwinger pair production with the rate $E^{3/2}$ at $B = 0$.

14.5.2.2 Bilayer Graphene

For bilayered graphene, assuming the quadratic dispersion $|N| = D = 2$, the expected conformal invariant Heisenberg-Euler action for the constant in space and time electromagnetic field, which is obtained by the integration over the $2+1$ fermions with quadratic dispersion, is the function of the scale invariant combination μ [90]:

$$S \sim \int d^2 x dt\, B^2 g(\mu), \quad \mu = \frac{E^2}{B^3}. \tag{14.45}$$

The asymptotical behavior in two limit cases, $g(\mu \to 0) \sim const$ and $g(\mu \to \infty) \sim \mu^{2/3}$, gives the effective actions for the constant in space and time magnetic and electric fields:

$$S_B = a \int d^2 x dt\, B^2, \quad S_E = (b + ic) \int d^2 x dt\, E^{4/3}. \tag{14.46}$$

The parameter a is the logarithmic coupling constant; the parameter b describes the vacuum electric polarization; and the parameter c describes the instability of the vacuum with respect to the Schwinger pair production in the electric field, which leads to the imaginary part of the action. The action also contains the linear non-local term

$$S_{non\text{-}local}(|N| = D = 2) = \int d^2x dt \sqrt{-g} g^{00} g^{kn} F_{0k} \frac{1}{g^{ip} \nabla_i \nabla_p} F_{0n}, \quad (14.47)$$

which corresponds to the polarization operator

$$\Pi_{00} \propto \frac{k^2}{\sqrt{k^4 - \omega^2}}, \quad |N| = D = 2. \quad (14.48)$$

14.5.2.3 $D = 2$ Systems with Nodes with Topological Charge N

In general case of a 2D system with N-th order touching point in spectrum (a kind of multilayered graphene) the Heisenberg-Euler action contains among the other terms the following nonlinear terms in the actions for magnetic and electric fields [90]:

$$S_B(N, D = 2) \sim \int d^2x dt\, B^{\frac{2+|N|}{2}}, \quad S_E(N, D = 2) \sim \int d^2x dt (-E^2)^{\frac{2+|N|}{2(1+|N|)}}, \quad (14.49)$$

where imaginary part of the action is responsible for the pair production in electric field [90, 91], while the linear action of the type (14.47) corresponds to the polarization operator

$$\Pi_{00} \propto \frac{k^2}{\sqrt{k^{2N} - \omega^2}}, \quad D = 2. \quad (14.50)$$

14.5.2.4 Effective Action for Gravity

The expected action for gravitational field is

$$S_{grav} = \int d^2x dt \sqrt{-g} \left[K_1 R^2 + K_2 g^{ik} g^{mn} g^{00} \partial_t g_{im} \partial_t g_{kn} + \cdots \right], \quad (14.51)$$

where $K_1, K_2, \ldots$ are dimensionless quantities, and the other terms of that type are implied.

14.6 Fully Gapped Topological Media

Examples of the fully gapped topological media are the so-called topological band insulators in crystals [3]. Examples are Bi_2Se_3, Bi_2Te_3 and Sb_2Te_3 compounds which are predicted to be $3 + 1$ topological insulators [92]. But the first discussion of the $3 + 1$ topological insulators can be found in Refs. [93, 94]. The main feature of such materials is that they are insulators in bulk, where electron spectrum has a gap, but there are $2+1$ gapless edge states of electrons on the surface or at the interface between topologically different bulk states as discussed in Ref. [94]. The similar properties are shared by the fully gapped 3D topological superfluids and superconductors. The spin triplet p-wave superfluid ^{3}He-B represents the fully gapped superfluid with nontrivial topology. It has $2+1$ gapless quasiparticles living at interfaces between vacua with different values of the topological invariant describing the bulk states of ^{3}He-B [15, 95]. The quantum vacuum of Standard Model below the electroweak transition, i.e. in its massive phase, also shares the properties of the topological insulators and gapped topological superfluids and is actually the relativistic counterpart of ^{3}He-B [25].

Examples of the $2+1$ topological fully gapped systems are provided by the films of superfluid ^{3}He-A with broken time reversal symmetry [23, 96] and by the planar phase which is time reversal invariant [23, 96]. The topological invariants for $2+1$ vacua give rise to quantization of the Hall and spin-Hall conductivity in these films in the absence of external magnetic field (the so-called intrinsic quantum and spin-quantum Hall effects) [23, 97].

14.6.1 $2+1$ Fully Gapped Vacua

14.6.1.1 ^{3}He-A Film: $2+1$ Chiral Superfluid

The gapped ground states (vacua) in $2+1$ or quasi $2+1$ thin films of ^{3}He-A are characterized by the invariant obtained by dimensional reduction from the topological invariant describing the nodes of co-dimension 3. This is the invariant N for the Fermi point in (14.2), which is now over the $(2+1)$-dimensional momentum-frequency space (p_x, p_y, ω):

$$N = \frac{e_{ijk}}{24\pi^2} \mathbf{tr}\left[\int d^2p \, d\omega \, G\partial_{p_i}G^{-1}G\partial_{p_j}G^{-1}G\partial_{p_k}G^{-1}\right]. \quad (14.52)$$

This Eq. (14.52) was introduced in relativistic $2+1$ theories [12–14] and for the film of ^{3}He-A in condensed matter [17, 23], where it was inspired by the dimensional reduction from the Fermi point (see [96] and also Fig. 14.16). In simple case of the 2×2 matrix, the Green's function can be expressed in terms of the three-dimensional vector $\mathbf{d}(p_x, p_y)$,

$$G^{-1}(\omega, p_x, p_y) = i\omega + \boldsymbol{\tau} \cdot \mathbf{d}(p_x, p_y), \quad (14.53)$$

where $\boldsymbol{\tau}$ are the Pauli matrices. Example of the $\mathbf{d}$-vector configuration, which corresponds to the topologically nontrivial vacuum is presented in Fig. 14.15. This is the momentum-space analogue of the topological object in real space—skyrmion. In real space, skyrmions are described by the relative homotopy groups [98]; they have been investigated in detail both theoretically and experimentally in the A phase of ^{3}He, see Sect. 16.2 in [1] and the review paper [99].

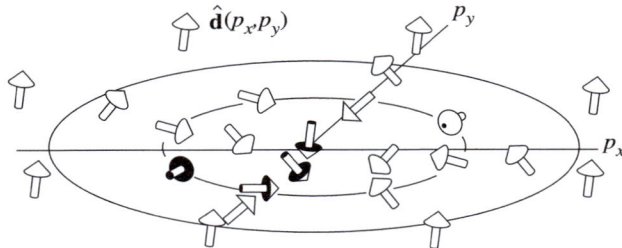

Fig. 14.15 Skyrmion in **p**-space with momentum space topological charge $N = -1$ in (14.54). It describes topologically non-trivial vacua in $2+1$ systems with a fully gapped non-singular Green's function. Vacua with nonzero N have topologically protected gapless edge states. The non-zero topological charge leads also to quantization of Hall and spin Hall conductance

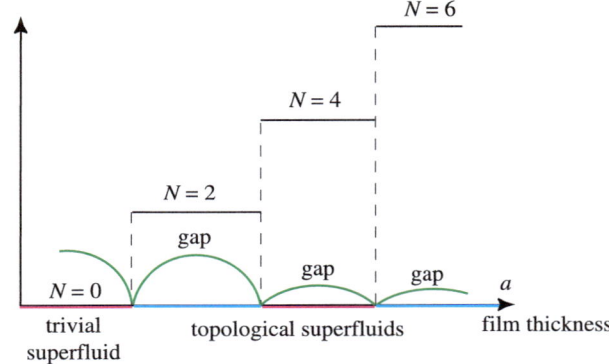

Fig. 14.16 Dependence of the topological invariant (14.52) on the thickness of ^{3}He-A film. The even values of N result from the spin degeneracy. At the topological phase transitions between the states with different N the gap in the spectrum of fermions is nullified

For the Green's function in (14.53) the winding number of the momentum-space skyrmion in Eq. (14.52) is reduced to [17]

$$N = \frac{1}{4\pi} \int d^2 p \, \hat{\mathbf{d}} \cdot \left(\frac{\partial \hat{\mathbf{d}}}{\partial p_x} \times \frac{\partial \hat{\mathbf{d}}}{\partial p_y} \right), \quad (14.54)$$

where $\hat{\mathbf{d}} = \mathbf{d}/|\mathbf{d}|$ is unit vector. For a single layer of the ^{3}He-A film and for one spin projection, the simplified Green's function has the form:

$$G^{-1}(\omega, \mathbf{p}) = i\omega + \boldsymbol{\tau} \cdot \mathbf{d}(\mathbf{p}) = i\omega + \tau_3 \left(\frac{p_x^2 + p_y^2}{2m} - \mu \right) + \tau_1 p_x + \tau_2 p_y. \quad (14.55)$$

For $\mu > 0$ the topological charge $N = 1$ and for $\mu < 0$ the topological charge is $N = 0$. That is why at $\mu = 0$ there is a topological quantum phase transition between the topological superfluid at $\mu > 0$ and non-topological superfluid at $\mu < 0$ [96].

In general case of multilayered ^{3}He-A, topological charge N may take any integer value of group Z. This charge determines quantization of Hall and spin-Hall conductance and the quantum statistics of the topological objects—real-space skyrmions [17, 23, 96, 97]. For $N = 4k + 1$ and $N = 4k + 3$, skyrmion is anyon; for $N = 4k + 2$ it is fermion; and for $N = 4k$ it is boson [96]. This demonstrates the importance of the Z_2 and Z_4 subgroups of the group Z in classification of topological matter; and also provides an example of the interplay of momentum-space and real-space topologies.

14.6.1.2 Planar Phase: Time Reversal Invariant Gapped Vacuum

In case when some symmetry is present, additional invariants appear, which correspond to dimensional reduction of invariant N_K in (14.8):

$$N_K = \frac{e_{ijk}}{24\pi^2} \mathbf{tr}\left[\int d^2 p\, d\omega\, K G \partial_{p_i} G^{-1} G \partial_{p_j} G^{-1} G \partial_{p_k} G^{-1}\right], \quad (14.56)$$

where as before, the matrix K commutes or anticommutes with the Green's function matrix. Example of the symmetric $2 + 1$ gapped state with N_K is the film of the planar phase of superfluid ^{3}He [23, 96]. In the single layer case, the simplest expression for the Green's function is

$$G^{-1}(\omega, p_x, p_y) = i\omega + \tau_3\left(\frac{p_x^2 + p_y^2}{2m} - \mu\right) + \tau_1(\sigma_x p_x + \sigma_y p_y), \quad (14.57)$$

with $K = \tau_3 \sigma_z$ commuting with the Green's function. This state is time reversal invariant. It has $N = 0$ and $N_K = 2$. For the general case of the quasi 2D film with multiple layers of the planar phase, the invariant N_K belongs to the group Z. The magnetic solid state analogue of the planar phase is the 2D time reversal invariant topological insulator, which experiences the quantum spin Hall effect without external magnetic field [3].

14.7 Relativistic Quantum Vacuum and Superfluid ^{3}He-B

Let us now turn to the class of $3 + 1$ fully gapped systems, which is represented by Standard Model in its massive phase and superfluid ^{3}He-B.

In the asymmetric phase of the Standard Model, there is no mass protection by topology and all the fermions become massive, i.e. the Standard Model vacuum becomes the fully gapped insulator. In quantum liquids, the fully gapped three-dimensional system with time reversal symmetry and nontrivial topology is represented by another phase of superfluid ^{3}He—the ^{3}He-B. Its topology is also supported by symmetry and gives rise to the 2D gapless quasiparticles living at interfaces between vacua with different values of the topological invariant or on the surface of ^{3}He-B [15, 95, 100, 101].

14.7.1 Superfluid ^{3}He-B

^{3}He-B belongs to the same topological class as the vacuum of Standard Model in its present insulating phase [25]. The topological classes of the ^{3}He-B states can be represented by the following simplified Green's function:

$$G^{-1}(\omega, \mathbf{p}) = i\omega + \tau_3\left(\frac{p^2}{2m} - \mu\right) + \tau_1 c^B \boldsymbol{\sigma} \cdot \mathbf{p}. \tag{14.58}$$

In the limit $1/m = 0$ this model ^{3}He-B transforms to the vacuum of massive Dirac particles with speed of light $c = c^B$.

In the fully gapped systems, the Green's function has no singularities in the whole 4-dimensional space $(\omega, \mathbf{p})$. That is why we are able to use the Green's function at $\omega = 0$. The topological invariant relevant for ^{3}He-B and for quantum vacuum with massive Dirac fermions is:

$$N_K = \frac{e_{ijk}}{24\pi^2} \operatorname{tr}\left[\int_{\omega=0} d^3 p K G \partial_{p_i} G^{-1} G \partial_{p_j} G^{-1} G \partial_{p_k} G^{-1}\right], \tag{14.59}$$

with matrix $K = \tau_2$ which anti-commutes with the Green's function at $\omega = 0$. In ^{3}He-B, the τ_2 symmetry is combination of time reversal and particle-hole symmetries; for Standard Model the matrix $\tau_2 = \gamma_5\gamma^0$. Note that at $\omega = 0$ the symmetry of the Green's function is enhanced, and thus there are more matrices K, which commute or anti-commute with the Green's function, than at $\omega \neq 0$.

Figure 14.17 shows the phase diagram of topological states of ^{3}He-B in the plane $(\mu, 1/m)$. The line $1/m = 0$ corresponds to the vacuum of Dirac fermions with the mass parameter $M = -\mu$, its topological charge

$$N_K = \operatorname{sign}(M). \tag{14.60}$$

The real superfluid ^{3}He-B lives in the weak-coupling corner of the phase diagram: $\mu > 0$, $m > 0$, $\mu \gg mc_B^2$. However, in the ultracold Fermi gases with triplet pairing the strong coupling limit is possible near the Feshbach resonance [102]. When μ crosses zero the topological quantum phase transition occurs, at which the topological charge N_K changes from $N_K = 2$ to $N_K = 0$.

There is an important difference between ^{3}He-B and Dirac vacuum. The space of the Green's function of free Dirac fermions is non-compact: G has different asymptotes at $|\mathbf{p}| \to \infty$ for different directions of momentum $\mathbf{p}$. As a result, the topological charge of the interacting Dirac fermions depends on the regularization at large momentum. ^{3}He-B can serve as regularization of the Dirac vacuum, which can be made in the Lorentz invariant way [25]. One can see from Fig. 14.17, that the topological charge of free Dirac vacuum has intermediate value between the charges of the ^{3}He-B vacua with compact Green's function. On the marginal behaviour of free Dirac fermions see Refs. [1, 16, 20, 21, 100].

The vertical axis separates the states with the same asymptote of the Green's function at infinity. The abrupt change of the topological charge across the line, $\Delta N_K = 2$, with fixed asymptote shows that one cannot cross the transition line adiabatically. This means that all the intermediate states on the line of this QPT

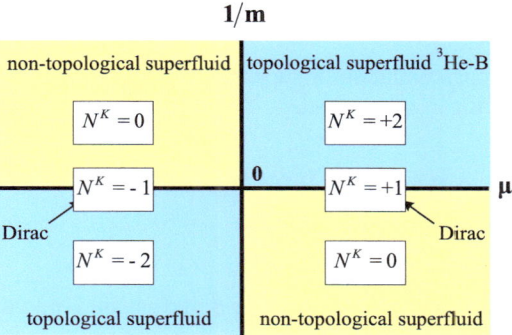

Fig. 14.17 Phase diagram of topological states of ^{3}He-B in Eq. (14.66) in the plane $(\mu, 1/m)$. States on the line $1/m = 0$ correspond to the Dirac vacua, which Hamiltonian is non-compact. Topological charge of the Dirac fermions is intermediate between charges of compact ^{3}He-B states. The line $1/m = 0$ separates the states with different asymptotic behavior of the Green's function at infinity: $G^{-1}(\omega = 0, \mathbf{p}) \to \pm \tau_3 p^2/2m$. The line $\mu = 0$ marks topological quantum phase transition, which occurs between the weak coupling ^{3}He-B (with $\mu > 0$, $m > 0$ and topological charge $N_K = 2$) and the strong coupling ^{3}He-B (with $\mu < 0$, $m > 0$ and $N_K = 0$). This transition is topologically equivalent to quantum phase transition between Dirac vacua with opposite mass parameter $M = \pm|\mu|$, which occurs when μ crosses zero along the line $1/m = 0$. The interface which separates two states contains single Majorana fermion in case of ^{3}He-B, and single chiral fermion in case of relativistic quantum fields. Difference in the nature of the fermions is that in Fermi superfluids and in superconductors the components of the Bogoliubov-Nambu spinor are related by complex conjugation. This reduces the number of degrees of freedom compared to Dirac case

are necessarily gapless. For the intermediate state between the free Dirac vacua with opposite mass parameter M this is well known. But this is applicable to the general case with or without relativistic invariance: the gaplessness is protected by the difference of topological invariants on two sides of transition.

14.7.2 From Superfluid Relativistic Medium to ^{3}He-B

Fully gapped 3-dimensional fermionic systems may arise also in relativistic quantum field theories. In particular, the Dirac vacuum of massive Standard Model particles has also the nontrivial topology, and the domain wall separating vacua with opposite signs of the mass parameter M contains fermion zero modes [103]. Topologically nontrivial states may arise in dense quark matter, where chiral and color superconductivity is possible. The topological properties of such fermionic systems have been recently discussed in Ref. [104]. In particular, in some range of parameters the isotropic triplet relativistic superconductor is topological and has the fermion zero modes both at the boundary and in the vortex core. On the other hand, there is a range of parameters, where this triplet superconductor is reduced to the non-relativistic superfluid ^{3}He-B [105]. That is why the analysis in Ref. [104] is applicable to ^{3}He-B and becomes particularly useful when the fermions living in the vortex core are discussed.

In a relativistic superconductor or superfluid with the isotropic pairing—such as a color superconductor in quark matter—the fermionic spectrum is determined by the Hamiltonian

$$H = \tau_3(c\boldsymbol{\alpha} \cdot \mathbf{p} + \beta M - \mu_R) + \tau_1 \Delta, \tag{14.61}$$

for spin singlet pairing, and by Hamiltonian

$$H = \tau_3(c\boldsymbol{\alpha} \cdot \mathbf{p} + \beta M - \mu_R) + \gamma_5 \tau_1 \Delta, \tag{14.62}$$

for spin triplet pairing [104, 105]. Here α^i, β and γ_5 are Dirac matrices, which in standard representation are

$$\boldsymbol{\alpha} = \begin{pmatrix} 0 & \boldsymbol{\sigma} \\ \boldsymbol{\sigma} & 0 \end{pmatrix}, \quad \beta = \begin{pmatrix} 1 & 0 \\ 0 & -1 \end{pmatrix}, \quad \gamma_5 = \begin{pmatrix} 0 & 1 \\ 1 & 0 \end{pmatrix}; \tag{14.63}$$

M is the rest energy of fermions; μ_R is their relativistic chemical potential as distinct from the non-relativistic chemical potential μ; τ_a are matrices in Bogoliubov-Nambu space; and Δ is the gap parameter.

In the non-relativistic limit the low-energy Hamiltonian is obtained by standard procedure, see e.g. [106]. The non-relativistic limit is determined by the conditions

$$cp \ll M \tag{14.64}$$

and

$$\left| M - \sqrt{\mu_R^2 + \Delta^2} \right| \ll M. \tag{14.65}$$

Under these conditions the Hamiltonian (14.61) reduces to the Bogoliubov–de Gennes (BdG) Hamiltonian for fermions in spin-singlet s-wave superconductors, while (14.62) transforms to the BdG Hamiltonian relevant for fermions in isotropic spin-triplet p-wave superfluid ^{3}He-B:

$$H = \tau_3 \left(\frac{p^2}{2m} - \mu \right) + c^B \tau_1 \boldsymbol{\sigma} \cdot \mathbf{p}, \quad m = \frac{M}{c^2}, \quad c^B = c\frac{\Delta}{M}, \tag{14.66}$$

where the nonrelativistic chemical potential $\mu = \sqrt{\mu_R^2 + \Delta^2} - M$.

The Dirac-BdG system in Eq. (14.62) has the following spectrum

$$\varepsilon = \pm \sqrt{M^2 + c^2 p^2 + \Delta^2 + \mu_R^2 \pm 2\sqrt{M^2(\mu_R^2 + \Delta^2) + \mu_R^2 c^2 p^2}}. \tag{14.67}$$

This spectrum is plotted in Fig. 14.18. Depending on the value of the parameters μ_R, Δ, M the spectral branches have different configurations.

There is a soft quantum phase transition, at which the position of the minimum of energy $E(p)$ shifts from the origin $\mathbf{p} = 0$, and the energy profile forms the Mexican hat in momentum space. This momentum-space analogue of the Higgs transition [2] occurs when the relativistic chemical potential μ_R exceeds the critical value

$$\mu_R^* = \left(\frac{M^2}{2} + \sqrt{\frac{M^4}{4} + M^2 \Delta^2} \right)^{1/2}. \tag{14.68}$$

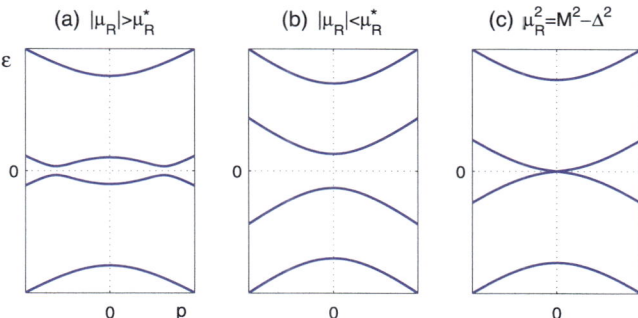

Fig. 14.18 Plot of the spectrum of relativistic Hamiltonian (14.62) for two generic cases: (a) $|\mu_R| > \mu_R^*$ when the minimum in the energy spectrum is away from the origin and (b) $|\mu_R| < \mu_R^*$ when the minimum in the energy spectrum is at $\mathbf{p} = 0$. At $|\mu_R| > \mu_R^*$ there is a soft quantum phase transition between these two vacua. This transition is not topological, and thus the gap in the energy spectrum does not close at the transition. The gap closes at the topological transition occurring at $\mu_R^2 = M^2 - \Delta^2$ as shown in plot (c)

Figures 14.18(a) and (b) demonstrate two generic cases: $|\mu_R| > \mu_R^*$ when there are extremums of function $\varepsilon(p)$ at $p \neq 0$ and $|\mu_R| < \mu_R^*$ when all extremums are at the point $p = 0$. The formation of the Mexican hat at $|\mu_R| = \mu_R^*$ is an example of non-topological quantum phase transition. Let us turn to the topological quantum phase transitions, at which the topological invariant changes and the gap closes as is shown in Fig. 14.18(c).

14.7.3 Topology of Relativistic Medium and ^{3}He-B

Figure 14.19 shows the phase diagram of the vacuum states of relativistic triplet superconductors. Different vacuum states are characterized by different values of the topological invariant N_K in Eq. (14.59), where the Green's function matrix at zero frequency $G^{-1}(\omega = 0, \mathbf{p})$ is equivalent to effective Hamiltonian. For ^{3}He-B in Eq. (14.66) and for triplet relativistic superconductor in Eq. (14.62) the relevant matrix $K = \tau_2$, which anti-commutes with the Hamiltonian. The vacuum states with different N_K cannot be adiabatically connected and thus at the phase transition lines the states are gapless. The circle $\mu_R^2 + \Delta^2 = M^2$ is an example of the line of topological quantum phase transition. In non-relativistic limit this corresponds to the line $\mu = 0$ in Fig. 14.17. The states inside the circle $\mu_R^2 + \Delta^2 = M^2$ are topologically trivial, while the states outside this circle represent topological superconductivity [104]. The vacuum states with $\mu_R^2 + \Delta^2 > M^2$ and $\mu_R^2 + \Delta^2 < M^2$ cannot be adiabatically connected which leads to the gap closing in Fig. 14.18(c). Discontinuity in the topological charge across the transition induces discontinuity in energy across the transition. For example, for the $2 + 1$ $p_x + ip_y$ superfluid/superconductor the phase transition is of third order, meaning that the third-order derivative of the ground state energy is discontinuous [107].

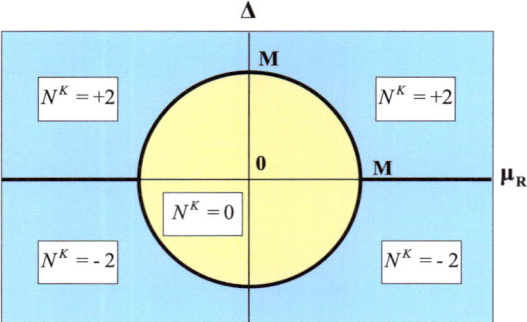

Fig. 14.19 Phase diagram of ground states of relativistic triplet superfluid in Eq. (14.62) in the plane (μ_R, Δ). Topological quantum phase transitions are marked by *thick lines*. The states inside the circle $\mu_R^2 + \Delta^2 = M^2$ are topologically trivial. The states outside this *circle* represent topological superconductors. The states on the *lines* of topological quantum phase transition are gapless

14.8 Fermions in the Core of Strings in Topological Materials

14.8.1 Vortices in ^{3}He-B and Relativistic Strings

In relativistic theories there is an index theorem which relates the number of fermion zero modes localized on a vortex with the vortex winding number [108]. We know that the Dirac vacuum considered in Ref. [108] has non-zero topological charge. This suggests that the existence of zero energy states in the core is sensitive not only to the real-space topological charge of a vortex, but also to the momentum-space topological charge of the quantum vacuum in which the vortex exists, and if so the index theorem can be extended to vortices in any fully gapped systems, including the non-relativistic superfluid ^{3}He-B. Since both the momentum-space topology of bulk state and the real-space topology of the vortex or other topological defects are involved, the combined topology of the Green's function in the coordinate-momentum space $(\omega, \mathbf{p}, \mathbf{r})$ [1, 41, 109–111] seems to be relevant. However, though the bulk-vortex correspondence does evidently exist, the explicit index theorem which relates the existence of the fermion zero modes to the topological charge of the bulk state and the vortex winding number is still missing. The existing index theorems are applicable only to particular cases, see e.g. [104, 109, 111, 112]. There is also a special index theorem for superconductors/superfluids with a small gap $\Delta \ll \mu$. Spectrum of fermions in these superconductors has branches which cross zero energy as a function discrete quantum number—angular momentum L [113]. The index theorem relates the number of such branches with the vortex winding number [114]. Here we are interested in the true fermion zero modes—the branches of spectrum $E(p_z)$, which cross zero as function of p_z.

An example, which demonstrates that the connection between the topological charge N_K and the existence of Majorana fermions—fermion zero modes on

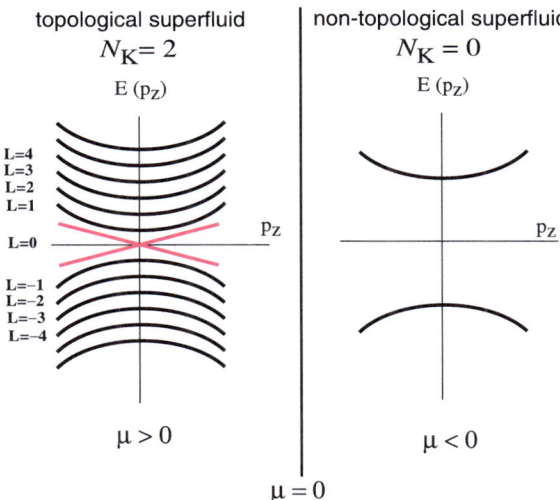

Fig. 14.20 Schematic illustration of spectrum of the fermionic bound states in the core of the most symmetric vortex with $n = 1$, the so-called o-vortex [99], in fully gapped spin triplet superfluid/superconductor of ^{3}He-B type. (*left*): Spectrum of bound state in the ^{3}He-B, vortex which corresponds to the weak coupling limit with non-zero topological charge $N_K = 2$ [115]. L is the azimuthal quantum number of fermions in the vortex core. There are two fermion zero modes, which cross zero energy in the opposite directions. (*right*): The same vortex but in the topologically trivial state of the liquid, $N_K = 0$, does not have fermion zero modes. The spectrum of bound states is fully gapped. Fermion zero modes disappear at the topological quantum phase transition, which occurs in bulk liquid at $\mu = 0$. A similar situation may take place for strings in color superconductors in quark matter [104]

vortices—is in Fig. 14.20. For ^{3}He-B, which lives in the range of parameters where $N_K \neq 0$, the gapless fermions in the core have been found in Ref. [115]. On the other hand, in the BEC limit, when μ is negative and the Bose condensation of molecules takes place, there are no gapless fermions. Thus in the BCS-BEC crossover region the spectrum of fermions localized on vortices must be reconstructed. The topological reconstruction of the fermionic spectrum in the vortex core cannot occur adiabatically. It should occur only during the topological quantum phase transition in bulk, when the bulk gapless state is crossed. Such topological transition occurs at $\mu = 0$, see Fig. 14.17. At $\mu < 0$ the topological charge N_K nullifies and simultaneously the gap in the spectrum of core fermions arises, see Fig. 14.20. This is similar to the situation discussed in Ref. [116] for the other type of p-wave vortices, and in Refs. for Majorana fermions in semiconductor quantum wires [117, 118]. A review on Majorana fermions in superconductors is in Ref. [119].

Another example is provided by the fermions on relativistic vortices in Dirac vacuum discussed in Ref. [108]. The Dirac vacuum has the nonzero topological invariant, $N_K = \pm 1$, see Fig. 14.17. This is consistent with the existence of the fermion zero modes on vortices, found in Ref. [108]. The index theorem for fermion zero modes on these vortices can be derived using the topology in combined coordinate and momentum space. The number of fermion zero modes on the vortex can be expressed via the 5-form topological invariant in terms of Green's function [43, 120]

$$N = \frac{1}{4\pi^3 i} \text{tr} \left[\int d^3 p \, d\omega \oint_C dl \, G \partial_{p_x} G^{-1} G \partial_{p_y} G^{-1} G \partial_{p_z} G^{-1} G \partial_\omega G^{-1} G \partial_l G^{-1} \right]. \tag{14.69}$$

The space integral is along the closed contour C around the vortex line. For the vortex in Dirac vacuum, Eq. (14.69) gives $N = n$, where n is the winding number which reproduces the index theorem discussed in Ref. [108]: the algebraic number of fermion zero modes N_{zm} equals the vortex winding number n.

For a vortex in ^{3}He-B one obtains $N = 0$, which is consistent with Fig. 14.20: two branches of zero modes have opposite signs of velocity v_z and thus produce zero value for the algebraic sum of zero modes, $N_{zm} = 0$. To resolve the fermion zero modes in the systems where the branches cancel each other due to symmetry, the index theorem for the zero modes must be complemented by symmetry consideration.

The 5-form topological invariant similar to Eq. (14.69) has been discussed also in [1, 121]. In particular, it is responsible for the topological stability of the $3 + 1$ chiral fermions emerging in the core of the domain wall separating topologically different vacua in $4 + 1$ systems (see Sect. 22.2.4 in [1]). The topological invariant for the general $2n + 1$ insulating relativistic vacua and the bound chiral fermion zero modes emerging there have been considered in [59, 122, 123]. Application of the 5-form topological invariant to the states in lattice chromodynamics can be found in [38, 124].

14.8.2 Flat Band in a Vortex Core: Analogue of Dirac String Terminating on Monopole

The topological bulk-vortex correspondence exists also for vortices in gapless vacua. The topological protection of fermion zero modes is provided by the nontrivial topology of three-dimensional Weyl points in the bulk. This bulk-vortex correspondence [101] is illustrated in Fig. 14.21. In bulk there is a pair of Weyl points with opposite topological charges $N = \pm 1$ in Eq. (14.2). The projections of these Weyl points on the direction of the vortex line determine the boundaries of the region where the spectrum of fermions bound to the vortex core is exactly zero, $E(p_z) = 0$. Such flat band first obtained in Ref. [125] for the noninteracting model is not destroyed by interactions. The spectrum of bound states in a singly quantized vortex in ^{3}He-A is shown in Fig. 14.21. The 1D flat band terminates at points where the spectrum of bound state merges with zeroes in the bulk, i.e. with Weyl points.

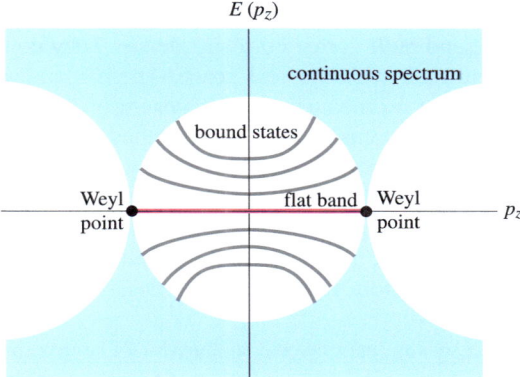

Fig. 14.21 Schematic illustration of the spectrum of bound states $E(p_z)$ in the vortex core. The branches of bound states terminate at points where their spectrum merges with the continuous spectrum in the bulk. The flat band terminates at points where the spectrum has zeroes in the bulk, i.e. when it merges with Weyl points. It is the **p**-space analogue of a Dirac string terminating on a monopole, another analogue is given by the Fermi arc in Fig. 14.1 *bottom right*

14.9 Discussion

The last decades demonstrated that topology is a very important tool in physics. Topology in momentum space is the main characteristic of the ground states of a system at zero temperature ($T = 0$), in other words it is the characteristic of the quantum vacuum. The gaplessness of fermions in bulk on the surface or inside the vortex core is protected by topology, and thus is not sensitive to the details of the microscopic physics (atomic or trans-Planckian). Irrespective of the deformation of the parameters of the microscopic theory, the value of the gap (mass) in the energy spectrum of these fermions remains strictly zero. This solves the main hierarchy problem in particle physics: for fermionic vacua with Fermi points the masses of elementary particles are naturally small.

The vacua, which have nontrivial topology in momentum space, are called the topological matter, and the quantum vacuum of Standard Model is the representative of the topological matter alongside with topological superfluids and superconductors, topological insulators and semi-metals, etc. There is a number of topological invariants in momentum space of different dimensions. They determine universality classes of the topological matter and the type of the effective theory which emerges at low energy and low temperature. In many cases they also give rise to emergent symmetries, including the effective Lorentz invariance and probably all the symmetries of Standard Model, and emergent phenomena such as gauge and gravitational fields. The symmetry appears to be the secondary factor, which emerges in the low-energy corner due to topology, and it is possible that it is the topology of the quantum vacuum, which is responsible for the properties of the fermionic matter in the present low-energy Universe.

The topological invariants in extended momentum and coordinate space determine the bulk-surface and bulk-vortex correspondence. They connect the momentum space topology in bulk with the real space. These invariants determine the fermion zero modes living on the surface of a system or in the core of topological defects (vortices, strings, domain walls, solitons, hedgehogs, etc.).

In respect to gravity, the momentum space topology delivers some lessons. First, in the effective gravity emerging at low energy, the collective variables represent the tetrad field and spin connections. In this approach the metric field emerges as the composite object of a tetrad field, and thus the Einstein-Cartan-Sciama-Kibble theory with torsion field is more relevant for the description of gravity (see also [126–129]).

Secondly, the topology suggests several scenarios of Lorentz invariance violation governed by topology. Among them the splitting of Fermi point and development of the Dirac points with quadratic and cubic spectrum. The latter leads to the natural emergence of the Hořava-Lifshitz gravity.

Acknowledgements This work is supported in part by the Academy of Finland and its COE program 2006–2011.

References

1. Volovik, G.E.: The Universe in a Helium Droplet. Clarendon, Oxford (2003)
2. Volovik, G.E.: Quantum phase transitions from topology in momentum space. In: Unruh, W.G., Schützhold, R. (eds.) Quantum Analogues: From Phase Transitions to Black Holes and Cosmology. Springer Lecture Notes in Physics, vol. 718, pp. 31–73 (2007). arXiv: cond-mat/0601372
3. Hasan, M.Z., Kane, C.L.: Topological insulators. Rev. Mod. Phys. **82**, 3045–3067 (2010)
4. Qi, X.-L., Zhang, S.-C.: Topological insulators and superconductors. Rev. Mod. Phys. **83**, 1057–1110 (2011)
5. Nielsen, H.B., Ninomiya, M.: Absence of neutrinos on a lattice, I: proof by homotopy theory. Nucl. Phys. B **185**, 20 (1981)
6. Nielsen, H.B., Ninomiya, M.: Absence of neutrinos on a lattice, II: intuitive homotopy proof. Nucl. Phys. B **193**, 173 (1981)
7. Volovik, G.E., Mineev, V.P.: Current in superfluid Fermi liquids and the vortex core structure. J. Exp. Theor. Phys. **56**, 579–586 (1982)
8. Nielsen, H.B., Ninomiya, M.: The Adler-Bell-Jackiw anomaly and Weyl fermions in a crystal. Phys. Lett. B **130**, 389–396 (1983)
9. Avron, J.E., Seiler, R., Simon, B.: Homotopy and quantization in condensed matter physics. Phys. Rev. Lett. **51**, 51–53 (1983)
10. Semenoff, G.W.: Condensed-matter simulation of a three-dimensional anomaly. Phys. Rev. Lett. **53**, 2449–2452 (1984)
11. Niu, Q., Thouless, D.J., Wu, Y.-S.: Quantized Hall conductance as a topological invariant. Phys. Rev. B **31**, 3372–3377 (1985)
12. So, H.: Induced topological invariants by lattice fermions in odd dimensions. Prog. Theor. Phys. **74**, 585–593 (1985)
13. Ishikawa, K., Matsuyama, T.: Magnetic field induced multi component QED in three-dimensions and quantum Hall effect. Z. Phys. C **33**, 41–45 (1986)
14. Ishikawa, K., Matsuyama, T.: A microscopic theory of the quantum Hall effect. Nucl. Phys. B **280**, 523–548 (1987)

15. Salomaa, M.M., Volovik, G.E.: Cosmiclike domain walls in superfluid ^{3}He-B: instantons and diabolical points in (**k**, **r**) space. Phys. Rev. B **37**, 9298–9311 (1938)
16. Haldane, F.D.M.: Model for a quantum Hall effect without landau levels: condensed-matter realization of the "parity anomaly". Phys. Rev. Lett. **61**, 2015–2018 (1988)
17. Volovik, G.E.: Analogue of quantum Hall effect in superfluid ^{3}He film. J. Exp. Theor. Phys. **67**, 1804–1811 (1988)
18. Yakovenko, V.M.: Spin, statistics and charge of solitons in $(2 + 1)$-dimensional theories. Fizika (Zagreb) **21**(suppl. 3), 231 (1989). arXiv:cond-mat/9703195
19. Hořava, P.: Stability of Fermi surfaces and K-theory. Phys. Rev. Lett. **95**, 016405 (2005)
20. Schnyder, A.P., Ryu, S., Furusaki, A., Ludwig, A.W.W.: Classification of topological insulators and superconductors in three spatial dimensions. Phys. Rev. B **78**, 195125 (2008)
21. Schnyder, A.P., Ryu, S., Furusaki, A., Ludwig, A.W.W.: Classification of topological insulators and superconductors. AIP Conf. Proc. **1134**, 10 (2009). arXiv:0905.2029
22. Kitaev, A.: Periodic table for topological insulators and superconductors. AIP Conf. Proc. **1134**, 22–30 (2009). arXiv:0901.2686
23. Volovik, G.E., Yakovenko, V.M.: Fractional charge, spin and statistics of solitons in superfluid ^{3}He film. J. Phys. Condens. Matter **1**, 5263–5274 (1989)
24. Mackenzie, A.P., Maeno, Y.: The superconductivity of Sr_2RuO_4 and the physics of spin-triplet pairing. Rev. Mod. Phys. **75**, 657–712 (2003)
25. Volovik, G.E.: Topological invariants for Standard Model: from semi-metal to topological insulator. JETP Lett. **91**, 55–61 (2010). arXiv:0912.0502
26. Abrikosov, A.A., Beneslavskii, S.D.: Possible existence of substances intermediate between metals and dielectrics. Sov. Phys. JETP **32**, 699 (1971)
27. Abrikosov, A.A.: Quantum magnetoresistance. Phys. Rev. B **58**, 2788 (1998)
28. Burkov, A.A., Balents, L.: Weyl semimetal in a topological insulator multilayer. Phys. Rev. Lett. **107**, 127205 (2011)
29. Burkov, A.A., Hook, M.D., Balents, L.: Topological nodal semimetals. Phys. Rev. B **84**, 235126 (2011)
30. Wan, X., Turner, A.M., Vishwanath, A., Savrasov, S.Y.: Topological semimetal and Fermi-arc surface states in the electronic structure of pyrochlore iridates. Phys. Rev. B **83**, 205101 (2011)
31. Ryu, S., Hatsugai, Y.: Topological origin of zero-energy edge states in particle-hole symmetric systems. Phys. Rev. Lett. **89**, 077002 (2002)
32. Manes, J.L., Guinea, F., Vozmediano, M.A.H.: Existence and topological stability of Fermi points in multilayered graphene. Phys. Rev. B **75**, 155424 (2007)
33. Vozmediano, M.A.H., Katsnelson, M.I., Guinea, F.: Gauge fields in graphene. Phys. Rep. **496**, 109 (2010)
34. Cortijo, A., Guinea, F., Vozmediano, M.A.H.: Geometrical and topological aspects of graphene and related materials. arXiv:1112.2054
35. Schnyder, A.P., Ryu, S.: Topological phases and flat surface bands in superconductors without inversion symmetry. Phys. Rev. B **84**, 060504(R) (2011). arXiv:1011.1438
36. Schnyder, A.P., Brydon, P.M.R., Timm, C.: Types of topological surface states in nodal noncentrosymmetric superconductors. arXiv:1111.1207
37. Essin, A.M., Gurarie, V.: Bulk-boundary correspondence of topological insulators from their Green's functions. Phys. Rev. B **84**, 125132 (2011)
38. Zubkov, M.A.: Generalized unparticles, zeros of the Green function, and momentum space topology of the lattice model with overlap fermions. arXiv:1202.2524
39. Berry, M.V.: Quantal phase factors accompanying adiabatic changes. Proc. R. Soc. Lond. A **392**, 45–57 (1984)
40. Volovik, G.E.: Zeros in the fermion spectrum in superfluid systems as diabolical points. Pis'ma Zh. Eksp. Teor. Fiz. **46**, 81–84 (1987). JETP Lett. **46**, 98–102 (1987)
41. Grinevich, P.G., Volovik, G.E.: Topology of gap nodes in superfluid ^{3}He: π_4 homotopy group for ^{3}He-B disclination. J. Low Temp. Phys. **72**, 371–380 (1988)

42. Volovik, G.E.: Gapless fermionic excitations on the quantized vortices in superfluids and superconductors. JETP Lett. **49**, 391–395 (1989)
43. Silaev, M.A., Volovik, G.E.: Topological superfluid ^{3}He-B: fermion zero modes on interfaces and in the vortex core. J. Low Temp. Phys. **161**, 460–473 (2010). arXiv:1005.4672
44. Khodel, V.A., Shaginyan, V.R.: Superfluidity in system with fermion condensate. JETP Lett. **51**, 553 (1990)
45. Volovik, G.E.: A new class of normal Fermi liquids. JETP Lett. **53**, 222 (1991)
46. Tsutsumi, Y., Ichioka, M., Machida, K.: Majorana surface states of superfluid ^{3}He A and B phases in a slab. Phys. Rev. B **83**, 094510 (2011)
47. Volovik, G.E.: Flat band in the core of topological defects: bulk-vortex correspondence in topological superfluids with Fermi points. JETP Lett. **93**, 66 (2011)
48. Froggatt, C.D., Nielsen, H.B.: Origin of Symmetry. World Scientific, Singapore (1991)
49. von Neumann, J., Wigner, E.P.: Über das Verhalten von Eigenwerten bei adiabatischen Prozessen. Phys. Z. **30**, 467–470 (1929)
50. Novikov, S.P.: Magnetic Bloch functions and vector bundles. Typical dispersion laws and their quantum numbers. Sov. Math. Dokl. **23**, 298–303 (1981)
51. Georgi, H.: Another odd thing about unparticle physics. Phys. Lett. B **650**, 275–278 (2007). arXiv:0704.2457
52. Luo, M., Zhu, G.: Some phenomenologies of unparticle physics. Phys. Lett. B **659**, 341 (2008)
53. Pepea, M., Wieseb, U.J.: Exceptional deconfinement in $G(2)$ gauge theory. Nucl. Phys. B **768**, 21–37 (2007)
54. Kadastik, M., Kannike, K., Raidal, M.: Dark matter as the signal of grand unification. Phys. Rev. D **80**, 085020 (2009)
55. Klinkhamer, F.R., Volovik, G.E.: Emergent CPT violation from the splitting of Fermi points. Int. J. Mod. Phys. A **20**, 2795–2812 (2005). arXiv:hep-th/0403037
56. Tarruell, L., Greif, D., Uehlinger, T., Jotzu, G., Esslinger, T.: Creating, moving and merging Dirac points with a Fermi gas in a tunable honeycomb lattice. arXiv:1111.5020
57. Klinkhamer, F.R.: Possible new source of T and CP violation in neutrino oscillations. Phys. Rev. D **73**, 057301 (2006)
58. Volovik, G.E., Gorkov, L.P.: Superconductivity classes in the heavy fermion systems. J. Exp. Theor. Phys. **61**, 843–854 (1985)
59. Kaplan, D.B., Sun, S.: Spacetime as a topological insulator: mechanism for the origin of the fermion generations. Phys. Rev. Lett. **108**, 181807 (2012)
60. Volovik, G.E., Konyshev, V.A.: Properties of the superfluid systems with multiple zeros in fermion spectrum. JETP Lett. **47**, 250–254 (1988)
61. Hořava, P.: Spectral dimension of the Universe in quantum gravity at a Lifshitz point. Phys. Rev. Lett. **102**, 161301 (2009)
62. Hořava: Quantum gravity at a Lifshitz point. Phys. Rev. D **79**, 084008 (2009)
63. Hořava, P.: Membranes at quantum criticality. J. High Energy Phys. **0903**, 020 (2009). arXiv:0812.4287
64. Xu, C., Hořava, P.: Emergent gravity at a Lifshitz point from a Bose liquid on the lattice. Phys. Rev. D **81**, 104033 (2010)
65. Lifshitz, E.M.: On the theory of second-order phase transitions I. Zh. Eksp. Teor. Fiz. **11**, 255 (1941)
66. Lifshitz, E.M.: On the theory of second-order phase transitions II. Zh. Eksp. Teor. Fiz. **11**, 269 (1941)
67. Volovik, G.E.: Reentrant violation of special relativity in the low-energy corner. JETP Lett. **73**, 162–165 (2001). arXiv:hep-ph/0101286
68. Dietl, P., Piechon, F., Montambaux, G.: New magnetic field dependence of Landau levels in a graphenelike structure. Phys. Rev. Lett. **100**, 236405 (2008)
69. Montambaux, G., Piechon, F., Fuchs, J.-N., Goerbig, M.O.: A universal Hamiltonian for motion and merging of Dirac points in a two-dimensional crystal. Eur. Phys. J. B **72**, 509–520 (2009). arXiv:0907.0500

70. deGail, R., Fuchs, J.-N., Goerbig, M.O., Piechon, F., Montambaux, G.: Manipulation of Dirac points in graphene-like crystals. Physica B **407**, 1948–1952 (2012)
71. Chong, Y.D., Wen, X.G., Soljacic, M.: Effective theory of quadratic degeneracies. Phys. Rev. B **77**, 235125 (2008)
72. Banerjee, S., Singh, R.R., Pardo, V., Pickett, W.E.: Tight-binding modeling and low-energy behavior of the semi-Dirac point. Phys. Rev. Lett. **103**, 016402 (2009)
73. Sun, K., Yao, H., Fradkin, E., Kivelson, S.A.: Topological insulators and nematic phases from spontaneous symmetry breaking in 2D Fermi systems with a quadratic band crossing. Phys. Rev. Lett. **103**, 046811 (2009)
74. Fu, L.: Topological crystalline insulators. Phys. Rev. Lett. **106**, 106802 (2011)
75. Wen, X.G., Zee, A.: Gapless fermions and quantum order. Phys. Rev. B **66**, 235110 (2002)
76. Beri, B.: Topologically stable gapless phases of time-reversal invariant superconductors. Phys. Rev. Lett. **103**, 016402 (2009)
77. McCann, E., Fal'ko, V.I.: Landau-level degeneracy and quantum Hall effect in a graphite bilayer. Phys. Rev. Lett. **96**, 086805 (2006)
78. Koshino, M., Ando, T.: Transport in bilayer graphene: calculations within a self-consistent Born approximation. Phys. Rev. B **73**, 245403 (2006)
79. Klinkhamer, F.R., Volovik, G.E.: Superluminal neutrino and spontaneous breaking of Lorentz invariance. Pis'ma Zh. Eksp. Teor. Fiz. **94**, 731–733 (2011). arXiv:1109.6624
80. Klinkhamer, F.R.: OPREA's superluminal muon-neutrino velocity and a Fermi-point-splitting model of Lorentz violation. arXiv:1109.5671
81. Klinkhamer, F.R.: Superluminal neutrino, flavor, and relativity. arXiv:1110.2146
82. Heikkilä, T.T., Volovik, G.E.: Fermions with cubic and quartic spectrum. Pis'ma Zh. Eksp. Teor. Fiz. **92**, 751–756 (2010). JETP Lett. **92**, 681–686 (2010). arXiv:1010.0393
83. Guinea, F., Castro Neto, A.H., Peres, N.M.R.: Electronic states and Landau levels in graphene stacks. Phys. Rev. B **73**, 245426 (2006)
84. Mak, K.F., Shan, J., Heinz, T.F.: Electronic structure of few-layer graphene: experimental demonstration of strong dependence on stacking sequence. Phys. Rev. Lett. **104**, 176404 (2010)
85. Castro Neto, A.H., Guinea, F., Peres, N.M.R., Novoselov, K.S., Geim, A.K.: The electronic properties of graphene. Rev. Mod. Phys. **81**, 109–162 (2009)
86. Heikkilä, T.T., Volovik, G.E.: Dimensional crossover in topological matter: evolution of the multiple Dirac point in the layered system to the flat band on the surface. Pis'ma Zh. Eksp. Teor. Fiz. **93**, 63–68 (2011). JETP Lett. **93**, 59–65 (2011). arXiv:1011.4185
87. Altarelli, G., Feruglio, F.: Discrete flavor symmetries and models of neutrino mixing. Rev. Mod. Phys. **82**, 2701–2729 (2010)
88. Heikkilä, T.T., Kopnin, N.B., Volovik, G.E.: Flat bands in topological media. Pis'ma Zh. Eksp. Teor. Fiz. **94**, 252–258 (2011). JETP Lett. **94**, 233–239 (2011). arXiv:1012.0905
89. Andersen, J.O., Haugset, T.: Magnetization in $(2+1)$-dimensional QED at finite temperature and density. Phys. Rev. D **51**, 3073–3080 (1995)
90. Katsnelson, M.I., Volovik, G.E.: Quantum electrodynamics with anisotropic scaling: Heisenberg-Euler action and Schwinger pair production in the bilayer graphene. JETP Lett. **95**, 411–415 (2012). arXiv:1203.1578
91. Zubkov, M.A.: Schwinger pair creation in multilayer graphene. Pis'ma Zh. Eksp. Teor. Fiz. **95**, 540–543 (2012). arXiv:1204.0138
92. Kane, C.L., Mele, E.: Z_2 topological order and the quantum spin Hall effect. Phys. Rev. Lett. **95**, 146802 (2005)
93. Volkov, B.A., Gorbatsevich, A.A., Kopaev, Yu.V., Tugushev, V.V.: Macroscopic current states in crystals. J. Exp. Theor. Phys. **54**, 391–397 (1981)
94. Volkov, B.A., Pankratov, O.A.: Two-dimensional massless electrons in an inverted contact. JETP Lett. **42**, 178–181 (1985)
95. Volovik, G.E.: Fermion zero modes at the boundary of superfluid ^{3}He-B. Pis'ma Zh. Eksp. Teor. Fiz. **90**, 440–442 (2009). JETP Lett. **90**, 398–401 (2009). arXiv:0907.5389
96. Volovik, G.E.: Exotic Properties of Superfluid ^{3}He. World Scientific, Singapore (1992)

97. Volovik, G.E.: Fractional statistics and analogs of quantum Hall effect in superfluid ^{3}He films. In: Ihas, G.G., Takano, Y. (eds.) Quantum Fluids and Solids. AIP Conference Proceedings, vol. 194, pp. 136–146 (1989)
98. Mineev, V.P., Volovik, G.E.: Planar and linear solitons in superfluid ^{3}He. Phys. Rev. B **18**, 3197–3203 (1978)
99. Salomaa, M.M., Volovik, G.E.: Quantized vortices in superfluid ^{3}He. Rev. Mod. Phys. **59**, 533–613 (1987)
100. Volovik, G.E.: Topological invariant for superfluid ^{3}He-B and quantum phase transitions. Pis'ma Zh. Eksp. Teor. Fiz. **90**, 639–643 (2009). JETP Lett. **90**, 587–591 (2009). arXiv:0909.3084
101. Volovik, G.E.: Topological superfluid ^{3}He-B in magnetic field and Ising variable. JETP Lett. **91**, 201–205 (2010). arXiv:1001.1514
102. Gurarie, V., Radzihovsky, L.: Resonantly-paired fermionic superfluids. Ann. Phys. **322**, 2–119 (2007)
103. Jackiw, R., Rebbi, C.: Solitons with fermion number 1/2. Phys. Rev. D **13**, 3398–3409 (1976)
104. Nishida, Y.: Is a color superconductor topological? Phys. Rev. D **81**, 074004 (2010)
105. Ohsaku, T.: BCS and generalized BCS superconductivity in relativistic quantum field theory: formulation. Phys. Rev. B **65**, 024512 (2001)
106. Nishida, Y., Santos, L., Chamon, C.: Topological superconductors as nonrelativistic limits of Jackiw-Rossi and Jackiw-Rebbi models. arXiv:1007.2201
107. Rombouts, S.M.A., Dukelsky, J., Ortiz, G.: Quantum phase diagram of the integrable $p_x + ip_y$ fermionic superfluid. arXiv:1008.3406
108. Jackiw, R., Rossi, P.: Zero modes of the vortex-fermion system. Nucl. Phys. B **190**, 681–691 (1981)
109. Volovik, G.E.: Localized fermions on quantized vortices in superfluid ^{3}He-B. J. Phys. Condens. Matter **3**, 357–368 (1991)
110. Teo, J.C.Y., Kane, C.L.: Majorana fermions and non-Abelian statistics in three dimensions. Phys. Rev. Lett. **104**, 046401 (2010)
111. Teo, J.C.Y., Kane, C.L.: Topological defects and gapless modes in insulators and superconductors. Phys. Rev. B **82**, 115120 (2010)
112. Lu, C.-K., Herbut, I.F.: Pairing symmetry and vortex zero-mode for superconducting Dirac fermions. Phys. Rev. B **82**, 144505 (2010)
113. Caroli, C., de Gennes, P.G., Matricon, J.: Phys. Lett. **9**, 307 (1964)
114. Volovik, G.E.: Vortex motion in Fermi superfluids and Callan-Harvey effect. JETP Lett. **57**, 244–248 (1993)
115. Misirpashaev, T.Sh., Volovik, G.E.: Fermion zero modes in symmetric vortices in superfluid ^{3}He. Physica B **210**, 338–346 (1995)
116. Mizushima, T., Machida, K.: Vortex structures and zero-energy states in the BCS-to-BEC evolution of p-wave resonant Fermi gases. Phys. Rev. A **81**, 053605 (2010)
117. Lutchyn, R.M., Sau, J.D., Das Sarma, S.: Majorana fermions and a topological phase transition in semiconductor-superconductor heterostructures. Phys. Rev. Lett. **105**, 077001 (2010)
118. Oreg, Y., Refael, G., von Oppen, F.: Helical liquids and Majorana bound states in quantum wires. Phys. Rev. Lett. **105**, 177002 (2010)
119. Beenakker, C.W.J.: Search for Majorana fermions in superconductors. arXiv:1112.1950
120. Shiozaki, K., Fujimoto, S.: Green's function method for line defects and gapless modes in topological insulators: beyond semiclassical approach. arXiv:1111.1685
121. Wang, Z., Qi, X.-L., Zhang, S.-C.: General theory of interacting topological insulators. arXiv:1004.4229
122. Kaplan, D.B.: Method for simulating chiral fermions on the lattice. Phys. Lett. B **288**, 342–347 (1992). arXiv:hep-lat/9206013
123. Golterman, M.F.L., Jansen, K., Kaplan, D.B.: Chern-Simons currents and chiral fermions on the lattice. Phys. Lett. B **301**, 219–223 (1993). arXiv:hep-lat/9209003
124. Zubkov, M.A., Volovik, G.E.: Topological invariants for the $4D$ systems with mass gap. Nucl. Phys. B **860**(2), 295–309 (2012). arXiv:1201.4185

125. Kopnin, N.B., Salomaa, M.M.: Mutual friction in superfluid ^{3}He: effects of bound states in the vortex core. Phys. Rev. B **44**, 9667–9677 (1991)
126. Akama, K.: An attempt at pregeometry—gravity with composite metric. Prog. Theor. Phys. **60**, 1900 (1978)
127. Volovik, G.E.: Superfluid ^{3}He-B and gravity. Physica B **162**, 222 (1990)
128. Wetterich, C.: Gravity from spinors. Phys. Rev. D **70**, 105004 (2004)
129. Diakonov, D.: Towards lattice-regularized quantum gravity. arXiv:1109.0091

Chapter 15
Einstein^2: Brownian Motion Meets General Relativity

Matteo Smerlak

Abstract Blending general relativity with Brownian motion theory leads to an interesting prediction: the classical root-of-time behavior of the RMS displacement of a Brownian particle is affected by gravity. These corrections to the diffusion law are not just a theoretical curiosity: they provide an opportunity for a new kind of "metamaterials", where diffusive transport—as opposed to ray propagation, as in transformation optics—can be tailored with suitably designed effective metrics. What is more, this effect force us to reconsider the formulation of the second law of (non-equilibrium) thermodynamics in gravitational analogues, as tracers diffusing in curved effective metrics can sometimes accumulate (instead of spreading), thereby decreasing their Gibbs entropy, without any force being applied to them.

15.1 Introduction

It is hard to tell which of Einstein's insights has been the most fecund of them all. His reappraisal of the concepts of space and time has given birth to astrophysics and cosmology as we know them today. His understanding of Brownian motion has demonstrated the reality of atoms and opened the path to the theory of equilibrium and non-equilibrium fluctuations. His explanation of the photoelectric effect has prompted the rise of quantum mechanics. And so forth—the whole twentieth century of physics, in one way or another, is his.

Yet, many have stressed that Einstein's revolution is "unfinished": as of today, we still have little clue how to merge general relativity with quantum mechanics. Attempts in this direction have proved fruitful in unexpected ways, resulting for instance in an entirely new branch of mathematics known as "topological quantum field theory" [1, 23] or "quantum topology" [7]. This notwithstanding, we must confess that, in our quest to a theory of quantum gravity, we remain—and perhaps for a long time to come—"half-way through the woods" [13].

M. Smerlak (✉)
Max-Planck-Institut für Gravitationsphysik (Albert-Einstein-Institut), Am Mühlenberg 1,
14476 Golm, Germany
e-mail: smerlak@aei.mpg.de

The purpose of this chapter is to discuss *another* way in which Einstein's revolution is unfinished: at the interface between general relativity and Brownian motion theory. The former teaches that spacetime is curved; the latter that the RMS displacement of a Brownian particle grows with the square root of time. Are these fundamental laws consistent with each other?

They are not. Indeed, the diffusion square-root law follows from the *diffusion equation* for the probability density p of Brownian motion[1]

$$\frac{\partial p}{\partial t} = \Delta p, \qquad (15.1)$$

which conflicts with general relativity (GR) in two ways.

- It is parabolic and hence propagates signals faster than light, while GR demands that no physical process can be used to signal faster than light.
- It relies on the assumption that space is flat, while GR states that, where there is matter, there is spacetime curvature.

The first point is definitely not a blocker: Brownian motion is a coarse-grained approximation to the microscopic dynamics of Brownian particles, which holds at local equilibrium, in a regime where the gradients $\nabla_a p$ are weak. To improve this approximation and account for the short-time dynamics of the particle in a causal way, we can replace the diffusion equation by a kinetic (Boltzmann, Klein-Kramers or else) equation in *phase space* [5].

What about the second point? How can the effect of gravity and spacetime curvature be included in the diffusion equation? Answering this question is interesting from a theoretical perspective, as dissipative processes are seldom studied within the framework of general relativity and could reveal interesting aspects thereof. But, more importantly, it is also of practical significance: as emphasized in the "analogue gravity" paradigm, curved spacetimes are good models of many inhomogeneous condensed-matter systems, ranging from fluid flows to Bose-Einstein condensates and gradient-index dielectrics. Finding a generalization of diffusion theory to arbitrary spacetimes could thus have very concrete applications. It is the purpose of this chapter to discuss such a generalization.

Our plan is as follows. After some technical preliminaries in Sect. 15.2, we describe in Sect. 15.3 the modifications to the classical theory of stochastic processes required by the presence of a non-trivial gravitational field. In Sect. 15.4, we consider in more detail the special case of Brownian motion, and discuss possible applications in Sect. 15.5. We close our conclusions in Sect. 15.6.

15.2 Preliminaries

Throughout this paper, we consider a $(D + 1)$-dimensional spacetime M with signature $(-++\cdots)$. (We keep D unspecified to include lower dimensional analogue

---

[1] We choose units where the diffusion coefficient is 1.

spacetimes in the discussion.) We denote ∇ the spacetime Levi-Civita connection, and $a, b, c, \ldots, i, j, \ldots$ are abstract indices. (The standard references for general relativity and the $D+1$ formalism are [10, 22]; stochastic processes and Fokker-Planck equations are exposed in [12, 20].)

15.2.1 The $D+1$ Formalism

Consider a relativistic fluid with velocity u^a. Assume its flow is *irrotational*, viz.

$$u_{[a}\nabla_b u_{c]} = 0. \tag{15.2}$$

Then (according to the Frobenius theorem) there is a foliation of spacetime by hypersurfaces Σ_t orthogonal to u^a. Furthermore, the slices Σ_t are the level sets of a time functions $t : M \to \mathbb{R}$ such that

$$u_a = -N\nabla_a t \tag{15.3}$$

for some non-negative function N. The function N is called the *lapse function*, and the slices Σ_t have the interpretation of "instantaneous space" relative to observers comoving with the fluid. In the following, we will denote σ a flow line of u^a (a "spatial point"), and σ_t its intersection with Σ_t.

The intrinsic geometry of the spatial hypersurfaces Σ_t is coded by the induced metric

$$h_{ij} = g_{ij} + u_i u_j, \tag{15.4}$$

and its associated covariant derivative[2] D_a and Laplace-Beltrami operator Δ, while their embedding in spacetime is measured by the (symmetric) extrinsic curvature tensor

$$K_{ij} = D_i u_j. \tag{15.5}$$

Its trace $\theta = D_a u^a = \nabla_a u^a$ is the *expansion scalar*, and measures the fractional rate of change of an infinitesimal volume δV about a spatial point along the flow, viz.

$$\theta = u^a \nabla_a \ln \delta V = \frac{1}{N} \frac{1}{\delta V} \frac{d(\delta V)}{dt}. \tag{15.6}$$

The factor $1/N$ above converts the proper time along the flow to the global time coordinate t.

A situation of particular interest is the *hydrostatic equilibrium*: the vector $\xi^a = \nabla^a t = -u^a/N$ is then *Killing*, i.e. generates timelike isometries. In this context,

---

[2]The covariant derivative D_a associated to h_{ab} acts on a tensor field $T^{a_1 \cdots a_n}{}_{b_1 \cdots b_m}$ according to

$$D_c T^{a_1 \cdots a_n}{}_{b_1 \cdots b_m} = h^{a_1}_{e_1} \cdots h^{d_m}_{b_m} h^{f}_{c} \nabla_f T^{d_1 \cdots d_n}{}_{e_1 \cdots e_m}.$$

the lapse function N is usually denoted χ, and called the *redshift factor*. It satisfies $u^a \nabla_a \chi = 0$, and gives the acceleration $a^i = u^c \nabla_c u^i$ of the flow by

$$a^i = \nabla^i \ln \chi. \tag{15.7}$$

Moreover, the time-time component of the Ricci tensor $E = R_{ab} u^a u^b$ (sometimes called the Raychaudhuri scalar), is given in this case by

$$E = D_i a^i + a_i a^i. \tag{15.8}$$

In general relativity, this scalar is closely related to the local mass-energy density, by virtue of the Einstein equation. We will see that E also plays an interesting role in diffusion phenomena.

15.2.2 Markov Processes

Let Σ be a Riemannian manifold with metric h_{ij} and covariant derivative D_i, representing a curved *space*, and denote $\sigma_t \in \Sigma$ the instantaneous position of a random walker at time t. In the Markovian setup, we assume that σ_t completely determines its later positions $\sigma_{t'}$ ($t' > t$), according to *transition rates* $\Gamma(\sigma \to \sigma')$. By definition, these are such that the elementary probability for the walker to jump from a volume $dV(\sigma)$ about $\sigma \in \Sigma$ to a volume $dV(\sigma')$ about $\sigma' \in \Sigma$ in time dt is given by

$$\Gamma(\sigma \to \sigma') dV(\sigma) dV(\sigma') dt. \tag{15.9}$$

These transition rates are implicit functions of the metric h_{ij}.

Let $p(\sigma, t)$ denote the probability density that the walker is in neighborhood of σ at time t, i.e. $\sigma_t = \sigma$, and

$$j(\sigma \to \sigma', t) = p(\sigma, t) \Gamma(\sigma \to \sigma'). \tag{15.10}$$

the corresponding *probability fluxes*. Balancing the incoming and outgoing fluxes at σ, we can write the evolution equation for p as

$$\partial_t p_t(\sigma) = \int_\Sigma dV(\sigma') \big(j(\sigma' \to \sigma, t) - j(\sigma \to \sigma', t)\big), \tag{15.11}$$

i.e.

$$\partial_t p(\sigma, t) = \int_\Sigma dV(\sigma') \big(p(\sigma', t) \Gamma(\sigma' \to \sigma) - p(\sigma, t) \Gamma(\sigma \to \sigma')\big), \tag{15.12}$$

where $dV(\sigma_t)$ is the Riemannian volume element on Σ_t. This integro-differential equation is known as the *master equation*, and the operator $\mathcal{M}$ such that $\partial_t p = \mathcal{M} p$ as the *master operator*.

In this stochastic framework, *equilibrium states* are defined as follows. A steady-state solution p^* is an *equilibrium distribution* if the corresponding probability fluxes cancel pairwise, i.e.

$$p^*(\sigma)\Gamma(\sigma \to \sigma') = p^*(\sigma')\Gamma(\sigma' \to \sigma). \tag{15.13}$$

This condition is known as the *detailed balance condition*.

Under certain regularity conditions for the rates Γ, one can show that the paths (σ_t) are discontinuous: for this reason one often speaks of *jump processes* in this case. The situation changes in the limit where the jumps become infinitely frequent and short-ranged (with respect to some relevant coarse-graining scale). Then Γ becomes *distributional*, and the master operator $\mathcal{M}$ reduces to its second-order truncation $\mathcal{L}$ in a moment expansion,

$$\mathcal{L}p = -D_i\left(w_1^i p\right) + \mathbf{f}_{12} D_i D_j\left(w_2^{ij} p\right). \tag{15.14}$$

Here w_1^i is a vector field on Σ, the *drift vector*, and w_2^{ij} a symmetric and positive-definite rank-2 tensor field, the *diffusion tensor*. Note that the transition rates Γ are related to $\mathcal{L}$ according to

$$\Gamma(\sigma' \to \sigma) = \mathcal{L}\delta(\sigma',\sigma), \tag{15.15}$$

where δ is the Dirac distribution on Σ and $\mathcal{L}$ acts on the σ' variable. Stochastic processes described by such Fokker-Planck equations are called *diffusion processes*.

The simplest example of such a diffusion process is *Brownian motion*, for which (by definition) $w_1^j = 0$ and $w_2^{ij} = 2\kappa h^{ij}$ for some positive constant κ. The corresponding Fokker-Planck equation $\partial_t p = \mathcal{L}p$ is the well-known *diffusion equation*

$$\partial_t p = \kappa \Delta p. \tag{15.16}$$

15.3 Master and Fokker-Planck Equations in Curved Spacetimes

In this section we describe the curved spacetime generalization of the master and Fokker-Planck equations for Markov processes.

15.3.1 Markovian Setup

Consider a Markov process defined by stationary transition rates $\Gamma(\sigma' \to \sigma)$, depending parametrically on a Riemannian metric h_{ij}. In the case of Brownian motion, for instance, $\Gamma(\sigma' \to \sigma) = \kappa \Delta \delta(\sigma,\sigma')$, with Δ the Laplace-Beltrami operator associated to h_{ij}.

Assume that this process defines the *instantaneous* dynamics of a random walker in spacetime, in *proper time*. In other words, given an irrotational flow u^a, consider the associated orthogonal foliation (Σ_t), evaluate Γ on the induced metric h_{ab},[3] and assume that the probability that a random walker carried by the flow u^a will jump from the position σ_t to the position σ'_t in proper time $ds(\sigma_t)$ is given that

$$\Gamma(\sigma_t \to \sigma'_t)dV(\sigma_t)dV(\sigma'_t)ds(\sigma_t), \tag{15.17}$$

where $ds(\sigma_t)$ is the proper time along σ.

15.3.2 Master Equation

Now, to write the corresponding probability equation, we must convert the proper time $s(\sigma_t)$ in (15.17) into the time coordinate t. This is achieved by means of the lapse function N, as

$$ds(\sigma_t) = N(\sigma_t)dt. \tag{15.18}$$

Hence, we can rewrite (15.17) as

$$\Gamma(\sigma_t \to \sigma'_t)dV(\sigma_t)dV(\sigma'_t)N(\sigma_t)dt. \tag{15.19}$$

Denoting $p(\sigma_t)$ the probability density of the stochastic process, the probability flux is therefore

$$j(\sigma_t \to \sigma'_t) = N(\sigma_t)p(\sigma_t)\Gamma(\sigma_t \to \sigma'_t). \tag{15.20}$$

This expression is physically intuitive: *where proper time runs faster (high N), the walker jumps more frequently (high j)*.

From this simple argument, we see that, if $\mathcal{M}$ is the master operator associated to the rates Γ, the right-hand side of the curved-spacetime master equation should be $\mathcal{M}(Np)$, i.e.

$$\int_{\Sigma_t} dV(\sigma'_t)(N(\sigma'_t)p_t(\sigma'_t)\Gamma(\sigma'_t \to \sigma_t) - N(\sigma_t)p_t(\sigma)\Gamma(\sigma_t \to \sigma'_t)). \tag{15.21}$$

A moment of reflection shows that the left-hand side of the master equation should also be modified in the presence of gravity. Indeed, recall that in a curved spacetime, the time-variation of an integrated density does not coincide with the integral of the time-derivative of the density: it V_t is a region in Σ_t, then

$$\mathbf{f}\frac{d}{dt}\int_{V_t} dV(\sigma_t)p_t(\sigma_t) \neq \int_{V_t} dV(\sigma_t)\partial_t p_t(\sigma_t). \tag{15.22}$$

---

[3] If spacetime is not static, this makes the transition rates implicit functions of time.

This is due to the fact that the volume element $dV(\sigma_t)$ itself depends on time. The correct formula follows from the relationship (15.6) defining the expansion scalar, and reads

$$\frac{d}{dt}\int_{V_t} dV(\sigma_t)p_t(\sigma_t) = \int_{V_t} dV(\sigma_t)\big(\partial_t p_t(\sigma_t) + N\theta p_t\big). \tag{15.23}$$

Shrinking the volume V_t down to zero, we thus find that the left-hand side of the master equation should be $\partial_t p + N\theta p$ instead of $\partial_t p_t$.

Combining both insights, we find that the master equation in a curved spacetime with lapse N and expansion θ is

$$\partial_t p + N\theta p = \mathscr{M}(Np). \tag{15.24}$$

It is easy to check that this equation conserves the total probability $\int_{\Sigma_t} dV(\sigma_t)p_t(\sigma_t)$, as it should.

Note that, in the case of static spacetimes ($\theta = 0$ and $N = \chi$ is the redshift factor), we can read off from (15.20) the generalized *detailed balance condition*: for an equilibrium distribution p^*, the probability fluxes cancel pairwise if

$$\Gamma(\sigma' \to \sigma)\chi(\sigma')p^*(\sigma') = \Gamma(\sigma \to \sigma')\chi(\sigma)p^*(\sigma). \tag{15.25}$$

Hence, the *product* χp^* must satisfy the usual detailed balance condition defined by the rates $\Gamma(\sigma \to \sigma')$, instead of p^* itself, as in the non-relativistic case.

15.3.3 Diffusive Limit: The Generalized Fokker-Planck Equation

Assume from now on that the stochastic process is of diffusive type (or can be approximated by one[4]) and denote $\mathscr{L}$ the Fokker-Planck operator defined by the rates Γ, as in (15.14). Then from (15.24) it follows immediately that the Fokker-Planck equation reads

$$\partial_t p + N\theta p = \mathscr{L}(Np), \tag{15.26}$$

i.e.

$$\partial_t p + N\theta p = -D_i\big(w_1^i Np\big) + \frac{1}{2}D_i D_j\big(w_2^{ij} Np\big) \tag{15.27}$$

where w_1^a and w_2^{ab} are the drift vector and diffusion tensor associated to the rates Γ, as in Sect. 15.2.2. This is the curved-spacetime generalization of the usual Fokker-Planck equation.

---

[4] I recommend van Kampen's note [18] for a discussion of the applicability of this approximation.

Note that (15.27) can be given a more hydrodynamical flavor, by replacing the unphysical derivative ∂_t by the convective derivative $u^a \nabla_a$, which evolves the probability distribution in proper time rather than in coordinate time; it then becomes

$$u^a \nabla_a p + \theta p = -\frac{D_i(w_1^i Np)}{N} + \frac{1}{2}\frac{D_i D_j(w_2^{ij} Np)}{N}. \quad (15.28)$$

15.4 The Case of Brownian Motion

In this section we focus on the properties of *Brownian motion* in curved spacetimes, computing in particular the gravitational corrections to the classical diffusion square-root law.

15.4.1 The Generalized Diffusion Equation

We saw in Sect. 15.2 that Brownian motion is characterized among diffusion processes by the vanishing of the drift vector, $w_1^a = 0$, and by $w_2^{ab} = 2\kappa h^{ab}$, with κ the diffusivity. The corresponding Fokker-Planck equation is therefore

$$\partial_t p + N\theta p = \kappa \Delta(Np) \quad (15.29)$$

or equivalently

$$u^a \nabla_a p + \theta p = \kappa \frac{\Delta(Np)}{N}. \quad (15.30)$$

In the hydrostatic limit, this equation reduces to[5]

$$u^a \nabla_a p = \kappa \frac{\Delta(\chi p)}{\chi}. \quad (15.31)$$

Using the relation $a_b = D_b \log \chi$ between the acceleration of the congruence a_b and the spatial gradient of the redshift factor, we can rewrite (15.31) as

$$\left(u^b - 2\kappa a^b\right)\nabla_b p = \kappa \Delta p + \kappa E p, \quad (15.32)$$

where u^b is the hydrostatic velocity (Killing vector divided by its norm) and E is the Raychaudhuri scalar. In addition to the usual diffusion term Δp, we see that the generalized diffusion equation contains two new terms.

---

[5]It is interesting to note that this equation is the same as the one postulated by Eckart [6] for thermal diffusion in his attempt to formulate a general-relativistic theory of dissipative hydrodynamics.

- *Drift.* The term $2\kappa a^b \nabla_b p$ is a *drift term*. Unlike the drift term in the classical Fokker-Planck equation (15.14), it vanishes in the limit $\kappa \to 0$, and is therefore a genuine stochastic effect.
- *Source.* The term $\kappa E p$, where $E = D_b a^b + a_b a^b$, is a *source term*. It implies that the probability density appears to comoving observers as sourced by (κ times) the Raychaudhuri scalar E.[6]

Both terms, which result from the non-homogeneity of χ in space, can be interpreted as stochastic *gravitational redshift* effects.

15.4.2 Gravitational Corrections to the Mean Square Displacement

The most significant observable of Brownian motion is the *mean square displacement* (MSD). Consider for simplicity a static spacetime, which is furthermore radially symmetric about a distinguished spatial point o. If $\rho(\sigma)$ denotes the Riemannian distance between σ and o, the MSD of a Brownian particle starting at o at time $t = 0$ is defined by

$$\langle \rho^2 \rangle_t = \int_\Sigma dV(\sigma) K_t(\sigma) \rho^2(\sigma). \tag{15.33}$$

Here K_t is the Green function (or heat kernel) of the generalized diffusion equation

$$\frac{\partial p}{\partial t} = \kappa \Delta(\chi p), \tag{15.34}$$

namely the solution with initial condition

$$\lim_{t \to 0} K_t(\sigma) = \delta(\sigma, o), \tag{15.35}$$

where $\delta(\sigma, o)$ is the Dirac distribution on Σ with support at o. (Note that, with the definition (15.33), the MSD is measured as a function of the t coordinate, and not proper time: unlike the non-relativistic situation, there is no global physical time parameter in a curved spacetime.)

Let us denote $\mathscr{D}$ the differential operator $\kappa \Delta_q(\chi \cdot)$. Then Eqs. (15.34)–(15.35) can be solved formally as

$$K_t(\sigma) = e^{t\mathscr{D}} \delta(\sigma, o) = \sum_{n=0}^{\infty} \frac{t^n}{n!} \mathscr{D}^n \delta(\sigma, o). \tag{15.36}$$

---

[6]That is *not* to say that the total probability is not conserved; we saw that it is.

The MSD, in turn, can be computed by evaluating this distribution the squared distance function ρ^2. This gives

$$\langle \rho^2 \rangle_t = \sum_{n=0}^{\infty} \frac{t^n}{n!} (\mathscr{D}^\dagger)^n \rho^2(o), \tag{15.37}$$

where $\mathscr{D}^\dagger = \kappa \chi \Delta_q$ is the formal adjoint of $\mathscr{D}$, i.e.

$$\langle \rho^2 \rangle_t = \sum_{n=0}^{\infty} \frac{(\kappa t)^n}{n!} (\chi \Delta)^n \rho^2(o). \tag{15.38}$$

This formula provides the asymptotic expansion of the MSD in the small time limit $t \to 0$. To second order in t, we get [14][7]

$$\langle \rho^2 \rangle_t = 2\kappa D \chi(o) t \left\{ 1 + \left(\frac{\Delta \chi(o)}{2\chi(o)} - \frac{R^{(D)}(o)}{3D} \right) \kappa t + \mathcal{O}(t^2) \right\}. \tag{15.39}$$

Here $R^{(D)}(o)$ is the Ricci scalar curvature of Σ at o. (To arrive at this expression we used the geometric identities $\Delta \rho^2(o) = 2D$ and $\Delta^2 \rho^2(o) = -4R^{(D)}(o)/3$.[8]) At this order, we thus find that diffusion is enhanced by a convex lapse profile about o and by negative spatial curvature. For instance, in the case of a constant-density Schwarzschild star with mass M and radius R, one finds that the curvature correction term to the MSD is positive and grows with $2GM/R$, showing that diffusion is actually *enhanced* by gravity, at least at short times. This is a somewhat counterintuitive result, given that gravity is supposed to be "attractive"; the point is that a Brownian particle is not free-falling, but collides constantly with the (constant-density) stellar fluid.

15.5 Application: Tailored Diffusion in Gravitational Analogues

In this section we discuss possible applications of the theory of Brownian motion in curved spacetimes to condensed-matter physics, via the "analogue gravity" paradigm.

15.5.1 Dissipation in Gravitational Analogues

It is now well known that certain condensed-matter systems are best thought of as "analogue spacetimes", meaning that their inhomogeneous structure can be sub-

---

[7] The original article [14] contains an error in this formula, pointed out by James Bonifacio at the University of Canterbury (New Zealand). I thank him for that.

[8] The higher order terms involve higher derivatives of the squared distance function, which can also be expressed in terms of local curvature invariants [4, 11].

sumed under an effective, non-Minkowskian metric—this book provides many examples. The theory of Brownian motion in curved spacetimes sketched above allows us to consider the issue of *dissipation* in these analogue spacetimes.

Consider the diffusion of tracers in a material medium with space-dependent diffusivity $D(\sigma)$, such that the density of tracers p satisfies

$$\frac{\partial p}{\partial t} = \Delta(Dp). \tag{15.40}$$

Not all inhomogeneous media lead to this diffusion equation [19], but some do, one simple example being fluids with space-varying viscosity [2]. Comparing (15.40) with the diffusion equation (15.34) for static spacetimes, we see that the diffusion coefficient $D(\sigma)$ can be interpreted as (a constant times) a *redshift factor*. This is not surprising, since—heuristically—a smaller diffusion coefficient means slower diffusion, hence, in general-relativistic terms, smaller redshift factor. Hence, just like certain inhomogeneous dielectrics behave as curved spacetimes with respect to the propagation of light waves [9], certain inhomogeneous media also behave as curved spacetimes with respect to diffusive transport. This observation suggest that the concept of "metamaterials", materials with tuned properties according to desired applications, can in fact be extended from the realm of conservative physics (where it has been exclusively applied so far) to the realm of dissipative physics (where it has not) [16].

15.5.2 From Diffusion to Antidiffusion

Here is a simple example of the way diffusion can be tailored using the effective redshift factor i.e. diffusion coefficient $D(\sigma)$. Consider a medium with a trap-like redshift profile, with $D(\sigma)$ monotonously increasing about a global minimum. (This profile of course recalls the redshift profile of a star, where χ is an increasing function of the distance to the star: Newton's law of gravitation.) Suppose the tracers are initially diluted within the medium, viz. the initial distribution $p_0(\sigma)$ has a large variance. What evolution does the diffusion equation (15.40) dictate?

On the one hand, we may expect that the effective gravitational field will tend to attract the tracers towards the center of the trap; on the other hand, the effect of diffusion is always to smooth out density gradients. The answer is the first one: it turns out that, in this particular case, Brownian motion will *concentrate* the tracers at the bottom of the trap, see Fig. 15.1. This effect shows that effective gravitational fields can be used to control diffusion to the point of reversing its effect, from the spontaneous dilution to the spontaneous concentration of tracers.

15.5.3 On the Second Law of Thermodynamics

This "antidiffusion" effect raises a puzzle concerning the second law of thermodynamics. In its standard (non-equilibrium) formulation, the latter states that "the

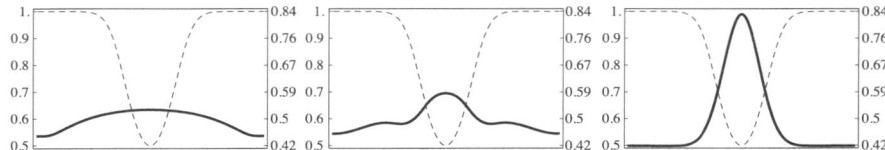

Fig. 15.1 Numerical solution of the generalized diffusion equation (15.40) in one space dimension (*horizontal axis*) with Neumann boundary conditions, at three times $t_0 < t_1 < t_2$ (left to right). The dashed curve (*left axis*) is the local diffusivity $D(x)$ normalized to its maximal value; the thick curve (*right axis*) are the probability densities $p(t_i, x)$, with x in arbitrary units. The amplitude of variation of $D(x)$ is consistent with the fluid experiment of Ref. [21]

entropy of an isolated system at local equilibrium can never decrease". Is this principle respected by Eq. (15.40)?

To answer this question, one should first identify the relevant thermodynamic entropy S. So long as we are dealing with non-interacting tracers diffusing within a thermal bath, this entropy can be decomposed into the *internal entropy* S_{int}, accounting for the uncertainty on the microscopic velocity of each tracer, and the *positional entropy* S_{pos}, measuring the uncertainty on the position of each tracer. In the context of Brownian motion, the assumption of local thermal equilibrium (Maxwellian velocity distribution at each point) implies that the former depends only on the local temperature $T(\sigma)$; as for the latter, it is by definition a functional of the probability density $p(\sigma, t)$, usually taken to be

$$S_{\text{pos}}[p(\sigma, t)] = -\int_\Sigma dV(\sigma) p(\sigma, t) \ln p(\sigma, t). \tag{15.41}$$

In a case where the temperature of bath is *constant* but $\chi(\sigma)$ is not, one has therefore

$$\frac{dS}{dt} = -\frac{d}{dt} \int_\Sigma dV(\sigma) p(\sigma, t) \ln p(\sigma, t). \tag{15.42}$$

Using the diffusion equation (15.40), one readily computes [17]

$$\frac{dS}{dt} = \kappa \int_\Sigma dV(\sigma) \chi(\sigma) \left(\frac{\nabla p(\sigma)}{p(\sigma)} + \frac{\nabla \chi(\sigma)}{\chi(\sigma)} \right). \tag{15.43}$$

This quantity is clearly *not* always non-negative: it suffices that ∇p and $\nabla \chi$ have opposite directions, and $\nabla \chi$ is sufficiently large, for this quantity to be negative and thus for S to decrease. This is the consequence of the fact, already discussed in the previous section, that certain χ profiles can force tracers to accumulate in a given region of space, even when they were initially perfectly diluted. Does this mean that *gravitational analogues can violate the second law of thermodynamics*?

No, but this result does force us to reconsider the notion of "positional entropy" in an inhomogeneous context, where $\chi(\sigma)$ actually depends on space. In this case, the above argument shows that the "Gibbs" expression (15.41) is not a suitable definition of positional entropy. Consider instead the *relative entropy* associated to

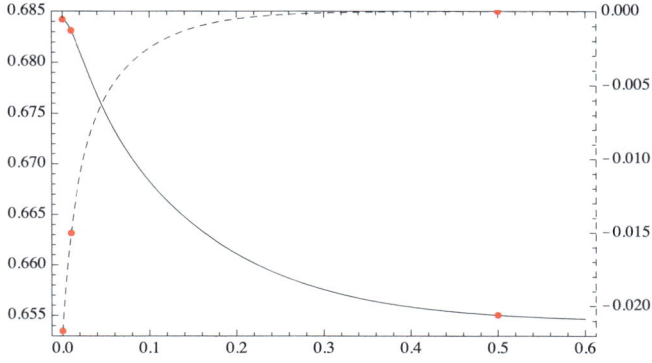

Fig. 15.2 The positional entropy (*continuous curve, left axis*) and relative positional entropy (*dashed curve, right axis*) in the "trap" of Fig. 15.1 as functions of time (*horizontal axis*, arbitrary units). The *red dots* indicate the times t_0, t_1, t_2 plotted in Fig. 15.1

a medium with equilibrium probability density $p^* \propto \chi^{-1}$, defined by

$$S_{\rm rel} = -\int_\Sigma dV(\sigma) p(\sigma,t) \ln \frac{p(\sigma,t)}{p^*(\sigma)}. \tag{15.44}$$

This quantity, also known as the Kullback-Leibler divergence [8], measures the "distance" between the instantaneous probability density $p(\sigma,t)$ and the equilibrium $p^*(\sigma)$. (The "Gibbs" entropy (15.41) is a special case of this, with $p^* = $ constant.) Now, due to the dissipative nature of diffusion, this "distance" can only increase in time: from (15.40) we find [17]

$$\frac{dS_{\rm rel}}{dt} = \kappa \int_\Sigma dV(\sigma)\chi(\sigma) \frac{(\nabla(\chi p)(\sigma))^2}{p(\sigma)} \geq 0. \tag{15.45}$$

This shows that a proper formulation of the second law of thermodynamics, applicable to inhomogeneous media as well as to homogeneous media, should rely on relative entropy—and not "Gibbs" entropy. (This statement can be sharpened within the framework of "stochastic thermodynamics" in the form of a "fluctuation theorem", see [15].)

15.6 Conclusion

The arguments developed in this chapter boil down to a very simple observation: if time runs at different rates in different places (due to an actual or effective gravitational field), then a Brownian particle will be accelerated correspondingly, and this affects its diffusive motion—Einstein 1905 amended by Einstein 1912, really! Yet, I have argued that this observation could have interesting consequences for condensed-matter physics, in effect creating an opportunity for a new class of metamaterials where diffusion could be tailored. No doubt some of the harvests of Einstein's revolutionary ideas are still to be gathered!

Acknowledgements The title of this chapter is inspired by the article [3], in which P. Castro-Villarreal computes the corrections of the RMS displacement of a Brownian particle in a curved *space* (as opposed to a curved *spacetime*, as in this chapter).

References

1. Atiyah, M.: Topological quantum field theories. Inst. Hautes Etudes Sci. Publ. Math. **68**, 175–186 (1989)
2. Bringuier, E.: Particle diffusion in an inhomogeneous medium. Eur. J. Phys. **32**(4), 975–992 (2011)
3. Castro-Villarreal, P.: Brownian motion meets Riemann curvature. J. Stat. Mech. **2010**(08), P08006 (2010)
4. DeWitt, B.S.: The Global Approach to Quantum Field Theory. Oxford University Press, Oxford (2003)
5. Dunkel, J., Hänggi, P.: Relativistic Brownian motion. Phys. Rep. **471**(1), 1–73 (2009)
6. Eckart, C.: The thermodynamics of irreversible processes, III: relativistic theory of the simple fluid. Phys. Rev. **58**(10), 919–924 (1940)
7. Kauffman, L.H., Baadhio, R.A. (eds.): Quantum Topology. World Scientific, Singapore (1993)
8. Kullback, S., Leibler, R.A.: On information and sufficiency. Ann. Math. Stat. **22**(1), 79–86 (1951)
9. Leonhardt, U., Philbin, T.: Geometry and Light: The Science of Invisibility. Dover, New York (2010)
10. Misner, C.W., Thorne, K.S., Wheeler, J.A.: Gravitation. Freeman, San Francisco (1973)
11. Ottewill, A.C., Wardell, B.: A transport equation approach to calculations of Green functions and HaMiDeW coefficients
12. Risken, H.: The Fokker-Planck Equation: Methods of Solution and Applications. Springer, Berlin (1989)
13. Rovelli, C.: Halfway through the woods: contemporary research on space and time. In: Earman, J., Norton, J. (eds.) The Cosmos of Science. University of Pittsburgh Press, Pittsburgh (1997)
14. Smerlak, M.: Diffusion in curved spacetimes. New J. Phys. **14**, 023019 (2012)
15. Smerlak, M.: Stochastic thermodynamics for inhomogeneous media. arXiv:1207.1026v4 (2012)
16. Smerlak, M.: Tailoring diffusion in analog spacetimes. Phys. Rev. E **85**(4), 041134 (2012)
17. Smerlak, M., Youssef, A.: Self-organization without heat: the geometric ratchet effect. arXiv:1206.3441 (2012)
18. Van Kampen, N.G.: The diffusion approximation for Markov processes. In: Lamprecht, I., Zotin, A.I. (eds.) Thermodynamics & Kinetics of Biological Processes, pp. 181–195. de Gruyter, Berlin (1982)
19. Van Kampen, N.G.: Diffusion in inhomogeneous media. J. Phys. Chem. Solids **49**, 673–677 (1988)
20. Van Kampen, N.G.: Stochastic Processes in Physics and Chemistry. North Holland, Amsterdam (1992)
21. van Milligen, B.P., Bons, P.D., Carreras, B.A., Sanchez, R.: On the applicability of Fick's law to diffusion in inhomogeneous systems. Eur. J. Phys. **26**, 913–925 (2005)
22. Wald, R.M.: General Relativity. Lecture Notes in Physics, vol. 721. University of Chicago Press, Chicago (1984)
23. Witten, E.: Topological quantum field theory. Commun. Math. Phys. **117**(3), 353–386 (1988)

Chapter 16
Astrophysical Black Holes: Evidence of a Horizon?

Monica Colpi

Abstract In this Lecture Note we first follow a short account of the history of the black hole hypothesis. We then review on the current status of the search for *astrophysical black holes* with particular attention to the black holes of stellar origin. Later, we highlight a series of observations that reveal the albeit indirect presence of supermassive black holes in galactic nuclei, with mention to forthcoming experiments aimed at testing directly the black hole hypothesis. We further focus on evidences of a black hole event horizon in cosmic sources.

16.1 The Black Hole Hypothesis

In General Relativity (GR), an *event horizon* is a boundary in spacetime, defined with respect to the external universe, inside of which events cannot affect any external observer. Anything that passes through the event horizon from the observer's side (regardless it is matter or light) can not return to the outside world.

The most common case of an event horizon is that surrounding a *black hole*. Within the event horizon of a black hole all paths that light could take are warped so as to fall farther into the black hole.

The first black hole solution of Einstein's field equations was discovered by Karl Schwarzschild [1].[1] It represents the exact solution for the metric tensor of a *point mass* M, in otherwise empty space. This solution describes the static, isotropic gravitational field generated by an uncharged point mass M. In the weak field limit, the solution recovers the static limit of a Newtonian point mass. The Schwarzschild solution also describes the gravitational field generated by any spherically symmetric star of mass M and surface radius R in vacuum, i.e. exterior to the star itself, at radii $r > R$.

---

[1] The references have been limited to a minimum and are focussed on mentioning main key papers supplemented by specific reviews and some recent publications, which do include more extensive references.

M. Colpi (✉)
Dipartimento di Fisica G. Occhialini, Università degli Studi di Milano Bicocca, Piazza della Scienza 3, 20126 Milano, Italy
e-mail: monica.colpi@unimib.it

In Schwarzschild coordinates, the surface of no return, i.e. the horizon of the Schwarzschild solution appears as a critical surface where spacetime is singular: more precisely where the radial component g_{rr} of the metric tensor diverges to infinity. In the usually adopted Schwarzschild coordinates, this occurs at $r = 2GM/c^2$, the Schwarzschild's radius R_S of the point mass M. For the Sun the Schwarzschild radius is 2.95 km, deep in the solar interior where Einstein's equations in matter space exhibits no Schwarzschild singularity, nor any other singularity. The singularity clearly emerges only for a hypothetical massive object (the black hole) so small that the radius R_S lies outside its hypothetical surface, in empty space. In these exotic circumstances, the Schwarzschild solution holds down to R_S, the radius that Finkelstein named event horizon.

Is this singularity real? It is now regonized that the Schwarzschild singularity is an *apparent* singularity, i.e. an artifact of the coordinate system used. The *curvature scalar* is perfectly well behaved and regular at the Schwarzschild radius. If any one of the curvature invariants had been singular at R_S, then this singularity would of course have been present in all coordinate systems. The non singular nature of the Schwarzschild singularity is evident in a set of coordinates, the Kruskal–Szekeres coordinates [2], that cover the entire spacetime manifold of the maximally extended Schwarzschild solution and that are well-behaved everywhere outside the physical singularity, present at the origin. While in Schwarzschild coordinates an infalling test particle is seen to take an infinite coordinate time to reach the horizon, a free-falling particle will only take a finite proper time (time as measured by its own clock) to cross the event horizon, and if the particle's world line is drawn in the Kruskal–Szekeres diagram this will also only take a finite coordinate time in Kruskal–Szekeres coordinates. Furthermore, and again in Kruskal–Szekeres coordinates, any event inside the black hole interior region (defined by the horizon) will have a future light cone that remains in this region such that any world line within the event's future light cone will eventually hit the black hole central, physical singularity where spacetime curvature is truly infinite.

The Schwarzschild solution is a limiting case of a more general solution of the Einstein's field equations found by Roy Kerr [3], describing the spacetime metric of an *axially-symmetric* point mass M, in empty space. This solution is endowed by an event horizon and describes an uncharged, *rotating black hole*. The event horizon of a Kerr black hole, which is *topologically a sphere*, is where g_{rr} diverges to infinity, in Boyer Lindquist coordinates. This inner spherical surface is now determined by the mass M and angular momentum $\mathbf{J}$ of the black hole. The Schwarzschild black hole is simply a *non* rotating black hole, i.e. black hole with $\mathbf{J} = 0$, and the Kerr metric approaches the Schwarzschild metric as $\mathbf{J} \to 0$.

Kerr black holes posses a further critical surface where $g_{tt} = 0$, defining the boundary where g_{tt} changes sign from negative (at larger radii) to positive (at smaller radii), and that is exterior to the event horizon. This outer surface can be visualized as an oblate spheroid. This surface together with the event horizon encloses a region called *ergosphere*: within this region, the purely temporal component g_{tt} is positive, i.e. acts like a purely spatial metric component. Consequently, particles within this ergosphere must co-rotate with the inner mass, if they are to retain their time-like character. Penrose demonstrated that within that region energy can be extracted at the expenses of the rotational energy and this may have observable consequences [4].[2]

Kerr black holes, and thus Schwarzschild black holes that are a limiting case, hide at their center a singularity that can not be eliminated by any change of coordinates. There, curvature scalars that are invariant under coordinate transformations diverge to infinity. This divergence is generally a sign for a missing piece in the theory, a failure of GR without quantum mechanics which forbids wavelike particles to inhabit a space smaller than their wavelength.

The *cosmic censorship conjecture* is the hypothesis that no *naked* singularities form in Nature. It asserts that singularities (should they be present) need to be hidden from an observer at infinity by the event horizon of a black hole. This conjecture is fundamental when studying black hole equilibrium states. Theorems of Israel, Carter, Hawking and Robinson [5–8], gave proof of the remarkable result that Kerr back holes are the *only* possible stationary vacuum black holes, paving the way to a further conjecture, known as *no-hair theorem*. The no-hair theorem postulates that all black hole solutions of the Einstein-Maxwell equations of gravitation and electromagnetism in GR can be completely characterized by only three externally observable parameters: the mass M, angular momentum $\mathbf{J}$, and electric charge. All other information (for which "hair" is a metaphor) about the matter which formed a black hole or falling into it, disappears behind the black hole event horizon and is therefore permanently not accessible to external observers. A corollary of the no-hair conjecture asserts that the only deformations that black holes admit are those obtained by a change of mass, angular momentum, and charge.

The black hole solution to the Einstein's field equation defines a *classical* black hole, i.e. a black hole for which *vacuum quantum fluctuation* near the event horizon are not accounted for. Mass and angular momentum $\mathbf{J}$ (or spin hereon) are conserved quantities, for isolated *classical* black holes. When *Hawking radiation* is included (as it should) isolated black holes evaporate [9]. The radiation is as if it were emitted by a black body with a temperature that is inversely proportional to the black hole's mass. Black holes of one solar mass have a temperature of only one hundred nano-Kelvin, and evaporate on a timescale of 10^{63} years, much longer than the age of the universe. Furthermore, these black holes would absorb far more radiation than they emit in their interaction with the cosmic microwave background radiation. Only when the black hole mass is much smaller than 10^{15} g is the evaporation timescale shorter than the Hubble time. Such *mini-black holes* formed as a result of fluctuations or phase transitions in the early universe would long since have evaporated. Those weighing $\sim 10^{15}$ g would just now be exploding. Since we have observational evidence of black holes heavier than a few solar masses in the universe, evaporation is of no concern in this Lecture Note.

---

[2]Black holes can also carry a finite electric charge: they are described by the known Kerr-Newman metric which is the solution of the Einstein-Maxwell field equation in GR for a rotating, charged point mass. As any electric charge excess is shorted by opposite charges in the cosmic environment, charged black holes have no astrophysical relevance, in this Lecture Note.

16.2 Gravitational Collapse

A black hole forms when a star or gas cloud collapses under its own self-gravity, having lost its dynamical stability. Ideally, the simplest case is the *spherical collapse* of a *dust, pressure-free cloud* of mass M and radius a_o. Oppenheimer and Snyder [10] demonstrated that a cloud of initial uniform density ρ_o and zero pressure collapses from rest to a state of *infinite proper energy density* in a finite time $T_{\rm col} = (3\pi/32G\rho_o)^{1/2}$. The collapsing cloud reduces its radius a with time until it develops a trapped surface, i.e. an event horizon at a radius equal to the Schwarzschild radius of the mass M. When the radius of the sphere a approaches R_S, light signals emitted from the surface take an infinite time to reach a distant observer and are infinitely redshifted. Since the gravitational redshift increases exponentially as $a \to R_S$, with an e-folding time $4GM/c^3$ (corresponding to the light travel time across the horizon), the cloud dims and fades out of sight. In practice, after an early and rapid contraction of the cloud on the dynamical time $\sim (a_o^3/GM)^{1/2}$, where the gravitational redshift $z_G = (1 - a/R_S)^{-1/2} - 1 \ll 1$, collapse appears to relent as redshift z_G increases to infinity. As the Schwarzschild metric is the only metric outside a spherically symmetric body, the only external parameter accessible for a measurement is M the mass of the star or cloud.

Similarly, an axially symmetric collapsing dust cloud settles down to a stationary state of infinite proper energy density, and as the only metric is the Kerr metric, according to theorems of uniqueness, a Kerr black hole forms as endpoint of axisymmetric gravitational collapse. As a Kerr black hole has only two measurable parameters, i.e. the mass M and the angular momentum $\mathbf{J}$, any information about the matter distribution in the collapsing body is lost [11].

Black holes have no-hair, according to the no-hair conjecture, thus even collapse under *generic* initial conditions inevitably form a Kerr black hole. Gravitational waves during collapse carry away asymmetries seeded in the collapsing star so that a Kerr black hole eventually forms. Thus, if gravitational collapse is the inevitable fate of massive bodies, black holes, i.e. hidden singularities, should be ubiquitous, in the universe. But how inevitable is gravitational collapse to a black hole in stars, star's clusters and even in larger scale structures such as galaxies? In order to learn about instabilities toward collapse we need to learn about gravitational equilibria and their stability.

16.3 Gravitational Equilibria and Stability

The idea that black holes do form in nature developed as soon as it was recognized that *stars* can *not* remain in *stable equilibrium* when the pressure support against gravity, determined by the *microphysical* properties of matter, drops to the point that the total energy E of the star (composed of N baryons) is no longer a *minimum*. The minimum is computed here with respect to all variations in the density profile $\rho(r)$ that leave the number of particles N unchanged, and unchanged and uniform the entropy per nucleon and the chemical composition.

16 Astrophysical Black Holes: Evidence of a Horizon?

Loss of dynamical stability occurs under a variety of conditions: when the star is either supported by the pressure of *degenerate relativistic electrons* or *neutrons* in cold, dense matter, or by *radiation* pressure in a hot, tenuous medium. In addition, the instability is seeded in any star and regardless the equation of state of matter (under currently known conditions) when the *non-linear nature* of the gravitational interaction becomes important, in the strong field limit.

A star in Newtonian equilibrium has a gravitational energy Ω in full balance with $-3 \int P dV$, i.e. three times the integral of the pressure P over the star's volume V (with minus sign). This is known as *virial relation*. For non-relativistic (relativistic) particles, P is equal to [2/3] ([1/3]) times the internal (nuclear) energy u. If E is the total energy of the star $E = \Omega + U$, the virial relation gives $E = -U = (1/2)\Omega$ for non-relativistic particles ($E = 0$ for relativistic particles), where U is the internal energy (i.e., $U = \int u dV$ over the star's volume). E is negative relative to a state of matter at rest at infinity, and for a star dynamically stable E can not be lowered by any other way (i.e. E is a minimum for constant entropy perturbations). By contrast, equilibria with $E \sim 0$ correspond to states of marginal stability.

If the fluid making the star is a classical perfect fluid, the virial relation yields $U \approx N k_B \langle T \rangle \approx G M^2 / R$, where $\langle T \rangle$ is the mean temperature inside the star of mass M and radius R: $\langle T \rangle \propto M/R$. Stars radiate away their energy at a rate L, and thus contract their radius if there is no internal energy source. As energy is lost in the form of radiation, the star (evolving along equilibrium sequences) gets smaller, denser, and hotter according to the virial relation, because gravity has *negative* specific heat.

Stars tend to develop degeneracy as they evolve away from the main sequence, since they contract faster than they can heat-up. Electrons are the first to become degenerate being the lightest particles in ordinary matter. When quantum effects in the fluid particles become important, the star internal structure gets simpler as temperature does not play any role. Electrons carry a large conductivity so that the star's interior is isothermal and radiative transfer is confined just in the tiny atmosphere.

Since for a degenerate non-relativistic gas of particles $P = (2/3)u$, Newtonian equilibria of degenerate non-relativistic matter have $E < 0$, and are thus stable. In addition they follow a mass-radius relation, $M R^3 = $ constant, since $P \propto \rho^{5/3}$. The relation is simple as there is only a single parameter, the pressure at the center of the star that determines uniquely its equilibrium property.[3] The heavier the degenerate star, the denser and smaller the star is. If we let M to increase to larger and larger values, there will be a critical mass above which the degenerate particles inside the star enter the relativistic dominated regime (when the Fermi energy exceeds the rest mass energy of the particles). When this occurs, $P = (1/3)u$ and $E \sim 0$ is no longer a minimum. This signals a turning point for stability on an equilibrium sequence.

---

[3] We recall that degenerate particles of mass m_e (i.e. electrons for white dwarfs) at a given density n have a distribution in the momentum space p which is flat up to the maximum, known as Fermi momentum, $p_F = (3h^3 n/8\pi)^{1/3}$, where h the Planck constant. Pressure in this fluid scales as $P \propto n(p/m_e)p \propto n p_F^2 \propto n^{5/3}$ when the particles are non-relativistic; for ultra-relativistic electrons, $P \propto n c p_F \propto n^{4/3}$.

White dwarfs and neutron stars do exist in nature, the first supported against gravity by the pressure of degeneracy of the electrons, at densities of millions $g\,cm^{-3}$, the second by the degeneracy pressure of neutrons, at nuclear densities of 10^{14} $g\,cm^{-3}$. However these stars exist over a limited range of masses.

As soon as electrons become relativistic, white dwarfs loose their stability and start collapsing under their own gravitational pull. This occurs when the white dwarf mass exceeds the *Chandrasekhar mass limit* of $M_{CH} \approx (\hbar c)^{3/2}/m^2 G^{3/2}$ equal to $1.46 M_\odot$, for a composition of heavy ions (of atomic weight A, charge Z and mass Am) and electrons (with mean electron molecular weight $\mu_e \sim A/Z \sim 2$). M_{CH} is the *maximum mass* for a white dwarf to remain in stable equilibrium [12]. At M_{CH} a white dwarf is on the verge of collapsing. Collapse can further progress or can be halted and reversed depending on the white dwarf chemical composition. If composed of carbon and oxygen nuclei the star ultimately explodes: energy deposition by runaway nuclear reactions drives a detonation wave that unbinds the entire star, leaving no remnant.[4] By contrast, the collapse continues if the white dwarf (the core of a very massive star) is composed of iron nuclei, the most bound nuclei in nature. Via endothermic dissociation of iron nuclei into α-particles and free neutrons, and deleptonization of matter via inverse β-decay ($e^- + p \to n + \nu$), collapse continues unhalted until the star's core reaches nuclear densities. At this point, a neutron star forms composed mostly of free degenerate neutrons. The mass of the neutron star is close to the Chandrasekhar mass limit as it results from the collapse of a stellar core at M_{CH}. A fundamental question then arises. Can neutron stars carry a mass arbitrarily large so that very massive stars end their life as neutron stars, always?

Neutron stars are (crudely) like white dwarfs, except that they consist of clod degenerate neutrons and are more compact having a radius m_e/m_n times smaller, i.e. $R_{NS} \approx 10$ km. Since the electron mass m_e does not enter the expression of M_{CH} (only the baryon mass m is involved there, and the neutron mass m_n is close to the proton mass m), we would expect that degenerate neutrons will become relativistic (and so the star unstable) at just such mass. A critical maximum mass exists also for neutron stars but with some slight complication as neutron stars are *relativistic stars*.

In GR, the density ρ of a star in hydrostatic equilibrium is the mass-energy density and as energy is equivalent to mass, any form of energy counterbalancing gravity ultimately becomes a source of gravity. The key finding by Oppenheimer and Volkoff was that neutron stars composed of an ideal-degenerate gas of (non-relativistic) neutrons can not remain in stable equilibrium when their mass exceeds a maximum mass at $0.7 M_\odot$ [13]. This occurs well before relativistic degeneracy for the neutrons is encountered inside the star. This remarkable finding is due to GR corrections of non-linear gravity inside the star. No stars with a mass higher than $0.7 M_\odot$ can exist in nature that are supported by neutron degeneracy. This maximum mass is known as *Oppenheimer-Volkoff limit*.

But how much should we trust the upper mass limit of Oppenheimer and Volkoff? Above nuclear matter densities, i.e. above a few 10^{14} $g\,cm^{-3}$, the physics is not very

---

[4]These explosions are identified as Type Ia supernovae.

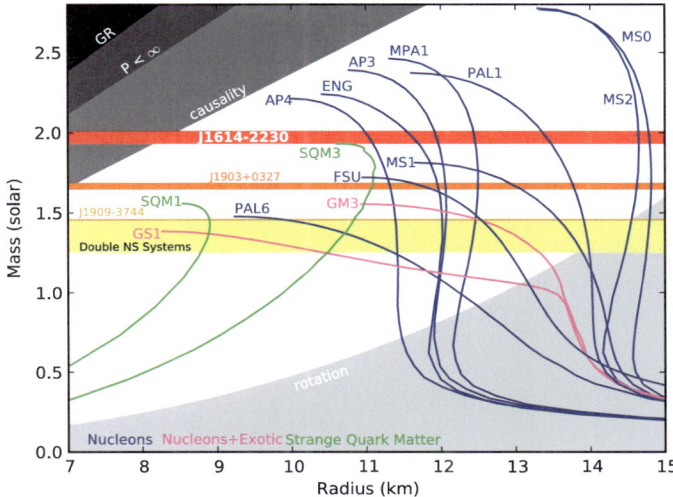

Fig. 16.1 Equilibrium sequences of neutron stars for selected equations of state [14]. The figure shows the non-rotating mass versus physical radius. The horizontal bands show the observational constraints from the mass of J1614-2230 of $1.97 \pm 0.04 M_\odot$ (*red*, [15]), of J1903+0327 (*orange*, [16]) and of the double neutron star binaries [14]. Any equation of state line that does not intersect the J1614-2230 band is ruled out by the measurement by Demorest et al. [15]. In particular, most equations of state involving exotic matter, such as kaon condensates or hyperons, tend to predict maximum neutron star masses well below $2.0 M_\odot$, and are therefore ruled out. *Green lines* refer to strange star models. The *upper left gray areas* of different intensity refer to regions excluded by GR and causality. The figure is from [15]

well understood and is poorly constrained by experimental data. No unique model exists for nuclear forces, understood as a residual coupling of the more fundamental force among quarks. This causes huge uncertainties on the correct equation of state and thus on the true value of the maximum neutron star's mass M_{max}. Nonetheless, the existence of a maximum mass is inevitable. This was considered to be an alarming result as the concept of a maximum mass for a neutron star is conducive to the concept that black holes unavoidably form during core collapse of very massive stars when their core's mass exceeds M_{max}.

The Oppenheimer-Volkoff limit was discovered much earlier than the discovery of the Kerr solution of a rotating black hole, and for long times it was believed that stars, in their latest stages of evolution, manage to lose significant amounts of mass in order to "avoid" hitting the Oppenheimer-Volkoff limit, and that rotation could be a cause of such mass loss.

There is indeed a considerable range of values for the neutron star maximum mass M_{max} today, in relation to the equation of state adopted [14]. It can vary from $1.6 M_\odot$ to $2.5 M_\odot$ as illustrated Fig. 16.1 where the mass versus radius relation is plotted for a set of equations of state. Soft equations of state (those with a lower pressure for a given mass-energy density) predict lower-mass maximum masses, and are those where hyperons, mesons, and/or pion condensates and/or are present. A phase transition to quark matter in the core is also considered to be a possibility.

As any star is endowed of rotation, then centrifugal support against gravity would enhance the maximum mass, but studies by [17, 18] indicate a mass increase of only 20 %. Since rotation is limited by the condition of break up and mass shedding at the equator, the most rapidly spinning neutron stars can be used to infer the correct equation of state for nuclear matter if their mass is known, and thus to infer the true maximum mass of a neutron star. Soft equations of state allow for the lowest spin periods ever possible (0.4 milliseconds for a neutron star of $1.4 M_\odot$, as shown in the works by Cook et al. [17, 18]). An inequality can be derived for the spin period

$$P_{\rm rot} > P_{\rm rot, min} \simeq 0.96 \pm 0.03 \left(\frac{M}{M_\odot}\right)^{-1/2} \left(\frac{R}{10\ {\rm km}}\right)^{3/2}\ {\rm ms}, \quad (16.1)$$

where M and R are the mass and radius of the spherical static neutron star solution [14]. The apparent lack of neutron stars spinning with periods below a millisecond is a hint that at nuclear densities matter does not follow a soft equation of state.

To circumvent uncertainties in the equation of state of cold matter at ultra-high densities Rhoades and Ruffini in 1974 presented an argument to derive a *firm upper mass limit* [19]. If (i) gravity is described by GR; (ii) the equation of state is well known below a threshold density ρ_o; (iii) matter is microscopically stable (i.e., $dP/d\rho \geq 0$); and (iv) the equation of state satisfies the causality condition, that the sound speed is less than the speed of light (i.e., $c_{\rm sound}^2 = dP/d\rho < c^2$): then

$$M_{\rm max} \simeq 3.2 \left(\frac{\rho_o}{4.6 \times 10^{14}\ {\rm g\,cm^{-3}}}\right)^{-1/2}\ M_\odot. \quad (16.2)$$

Thus, if there is a maximum mass for a (rotating) neutron star, any dense star heavier than $M_{\rm max}$ *must be/become* a Kerr back hole.[5]

A key question rises: Do laws of nature manage to protect any star from collapsing to a black hole by some mean? To answer that question we need to consider how stars form and evolve. This will be explored in Sect. 16.5. Meantime, observations are providing compelling evidence that white dwarfs and neutron stars do form in nature as well as *stellar-mass black holes*, and this topic is shortly reviewed in the incoming section.

---

[5] A radically different viewpoint was presented by Witten with the introduction of the strange star model. The idea of the strange star rests on the hypothesis that strange quark matter, composed by equal number of up, down and strange quarks, could be the absolute ground state of matter [20]. The simplest model of self-bound strange quark matter is the MIT bag model for which $P = (\rho c^2 - 4B)/3$ where B is the Bag constant. Interestingly, strange stars exist only below a maximum mass of about $2.033 M_\odot$ for $B = 56\ {\rm MeV\,fm^{-3}}$.

16.4 Neutron Stars and Stellar-Mass Black Holes in the Realm of Observations

While Sirius B, the white dwarf companion to Sirius A was discovered in 1844, the discovery of neutron stars and stellar-mass black holes had to wait until the mid 1960s. Neutron stars were discovered as *Pulsars*, i.e., as *rotation-powered* highly magnetized neutron stars in 1967 [21], and as *accretion-powered X-ray sources* in binary systems a year later, in 1968 [22]. Similarly, stellar-mass black hole candidates were discovered as *accretion-powered X-ray sources*. The chief argument used in discriminating between a stellar-mass black hole or a neutron star was and still is the *mass*. Remarkably, not only we do observe these compact stars as cosmic sources but we witness the moment of their formation. *Core-collapse supernovae* and *γ ray burst* (those called long GRBs) that are among the most powerful sources in the universe are the events that accompany their birth.

Neutron stars, besides their strong gravitational field (nearly as strong as that of a black hole having a radius $R_{NS} \sim 3R_S$), are endowed by rapid rotation, and often by intense magnetic fields. They further have a *surface* and a *crust* that can be a site of nuclear explosions. By contrast black holes are endowed by a very strong gravitational field, and by rotation only. As a consequence neutron stars display a much richer phenomenology than black holes.

As sources of extreme gravity, both neutron stars and black holes can be observed in the universe, as *accreting sources* [23–27]. According to the *accretion paradigm*, energy can be extracted in the form of radiation with high efficiency just outside the event horizon of the black hole and outside the surface of the neutron star. A test particle of mass m_p in free fall, hitting the surface of a neutron star releases a kinetic energy in form of radiation as large as $(GM/R_{NS}c^2)m_pc^2 \sim 0.1m_pc^2$ given the high surface gravity of the star. Similarly a test particle of mass m_p releases a comparable radiation energy $(0.1m_pc^2)$ in order to rich the event horizon of a black hole after spiralling inward along a sequence of nearly circular orbits. (Note that free-fall onto a black hole does not require any release of radiation since black holes do not hold a surface.)

Customarily, accretion occurs via a *geometrically thin, optically thick accretion disc* as matter in the vicinity of the black hole likely carries a non-vanishing angular momentum [30]. Gas moves on nearly Keplerian orbits around the compact object forming a disc. As cooling is rapid, the gas settles in a geometrically thin disc and flows inward under the action of viscous stress. During the slow drift, the gas heats up reaching typical temperatures of several million degrees (a few keV), each annulus emitting as a black body at an effective temperature that increases closer to the black hole or neutron star. Under these conditions more than 10 % of the rest mass energy is released at the expense of the gravitational energy by matter prior crossing the event horizon or prior touching the surface of the neutron star. This high efficiency ($\varepsilon \sim 10$ %), much higher than nuclear reactions, makes black holes and neutron stars sites of large energy production. Having a horizon, i.e. not a surface, accretion is the main and only known vehicle for highlighting the presence of a black hole in the universe. A disc forms around a neutron star, similarly to the case of a black hole: the disc however interacts with the star's magnetosphere and later with the surface producing additional dissipation.

Besides accretion, the intense magnetic fields that neutron stars can carry give origin to a phenomenology that is unique of neutron stars and that categorize them as *rotation powered pulsars*. The emission, distributed over a wide range of spectral energies (often from radio to γ-rays) results from a complex and degrading cascade of high energy photons and electron-positron pairs interacting with the magnetic field that drain energy from the rotation of the neutron star [28, 29].

After nearly 50 years of continuous discoveries and continuous monitoring of the sources, the picture that emerges is as follows:

(i) Neutron stars are ubiquitous in the Milky Way (more than a billion), and so in any other galaxy. Similarly, black holes of stellar origin are ubiquitous in the galaxies (more than a few millions) as in our own Galaxy.

(ii) More than ~ 1500 rotation-powered pulsars has been discovered in our own Galaxy, from radio surveys, and now in increasing number from surveys at MeV energies.

(iii) Rotation-powered pulsars come into two favours: (1) The *young highly magnetized* (with magnetic fields $B \gtrsim 10^{12}$ G) isolated pulsars. Their lifetime is about 10^7 yr as they fade away when crossing the death line [29]. They map the young stellar population in the spiral arms of the Milky Way. (2) The *old, weakly magnetized* ($B \lesssim 10^9$ G) *millisecond* pulsars, spinning with periods below 10 milliseconds, which (mostly) live in binaries with a white dwarf as a companion. In the Galaxy they are found in the bulge, in the galactic disc, and in large numbers in globular clusters. They trace the old stellar population. In binaries, pulsar masses can be measured with high accuracy, and so they offer the possibility of studying the mass distribution of neutron stars (those formed in binary systems). Intermediate to the two classes are the rotation-powered pulsars that have as companion a second neutron star, referred to as *double neutron star binaries*. These in binaries are the most exquisite laboratories for testing GR.

(iv) Accretion-powered X-ray sources housing either a neutron star or a stellar-mass black hole are ubiquitous in the Milky Way and are currently observed in large numbers also in nearby galaxies [37]. They are broadly divided into two classes: (1) The *High Mass X-ray Binaries* where a *massive star* (more massive than $\sim 10 M_\odot$) feeds the compact star through an accretion disc, via its powerful wind. Dissipation in the viscous disc heats the fluid on its way to the compact object, and a spectrum at hard X-ray energies is emitted by the in-flowing plasma onto the large magnetosphere. Given the short lifetime of the massive star, these sources trace the *young* stellar populations, and are present in large numbers in star-forming galaxies. (2) The *Low Mass X-ray Binaries* where a *low-mass star* (less massive than the sun) feeds the compact companion star through an extended accretion disc, via Roche lobe overflow when the star exists the main sequence and ascends the red-giant branch. Dissipation in the viscous disc heats the fluid on its way to the compact object, and a multi-black body spectrum is emitted at soft X-ray energies. These sources trace the *old* stellar populations, and are present in large numbers in bright, massive elliptical galaxies. In both classes, X-ray emission can be either persistent or transient (i.e. varying on timescale of weeks, months to years).

(v) In both classes of accreting X-ray sources, a *lower limit* to the mass of the compact star can be inferred.

(vi) The compact object in accreting X-ray binaries can be identified as a neutron star, if the source is *pulsating* (i.e., if there is a periodicity in the X-ray light curve), or in the absence of any pulsation, if a runaway thermonuclear flash is observed resulting in the so called Type I X-ray burst [31]. Both features are signature of the presence in the source of a *hard surface*; in other words the *absence of an event horizon*. Sources that lack of these two features may host a stellar-mass black hole. In this case, this is not a sufficient condition to identify a black hole.

(vii) Low Mass X-ray Binaries which host a neutron star (instead of a black hole) are the progenitors of the rotation-powered Millisecond Pulsars in binaries as in these low mass systems the neutron star is weakly magnetized. The B-field has likely decayed from high values ($\sim 10^{12}$ G) due to long-lived episodes of accretion down to $\sim 10^8$ G. During these episodes the mass of the accreting object can increase sizably, as well as its angular momentum in its interaction with the accretion disc. This process is termed *recycling*. As soon as accretion ceases the now rapidly rotating neutron star can turn on as rotation powered pulsar [26]. Similarly, High Mass X-ray Binaries housing a neutron star are the progenitor of double neutron star binaries [26]. The second neutron star forms when the high mass star evolves away from the main sequence. The evolutionary link among these classes is firmly confirmed by the observations, and the remarkable finding is that many of these binary systems housing the two compact stars are tight enough to emit gravitational waves. The Hulse-Taylor binary pulsar PSR 1913+16 , and the double-pulsar PSR J0737-3039 (where the two neutron stars are both active as rotation-powered radio-pulsars) are the show-case binaries where there is compelling, albeit indirect, evidence that gravitational waves are emitted from the stars in the binary, due to the time varying quadrupolar mass distribution [32–35].

(viii) There exists a class of accreting X-ray sources, named *Ultra Luminous X-ray Sources*, that have luminosities $L_X \sim 10^{39}$–10^{41} erg s^{-1} is excess of the maximum luminosity for steady accretion onto a 10M$_\odot$ stellar-mass black hole. These are sources found close to young, forming star's clusters, and for many of these sources there is evidence of a companion star. Ultra Luminous X-ray Sources may not be a homogeneous sample but those which show a binary nature likely host the *heaviest* black holes of stellar origin [36].

(ix) No radio pulsar has been observed yet to orbit around a stellar-mass black hole, in a binary. This represents one of the biggest challenges of incoming radio observatories like SKA [38]. The pulsar in such systems could be used as a powerful probe of the spacetime around the black hole. This occurs by tracing the "in situ" propagation of the radio signal when climbing up the potential well of the invisible black hole.

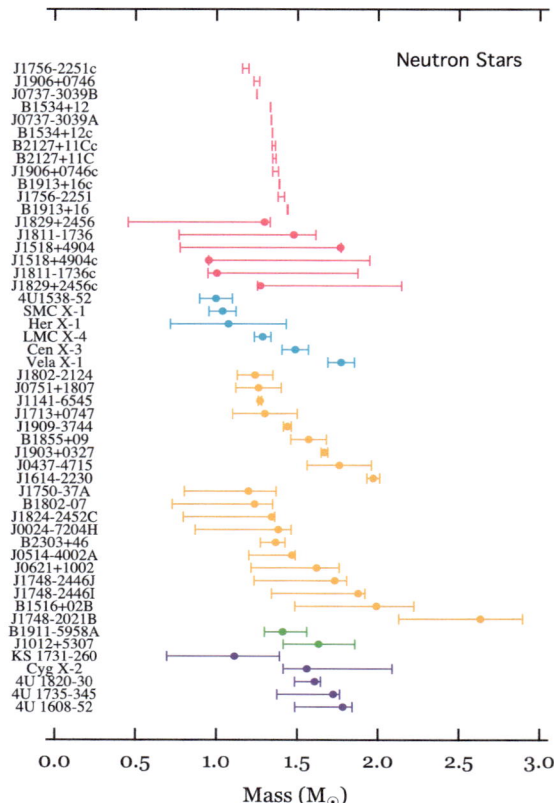

Fig. 16.2 The masses of neutron star from a variety of sources: from rotation powered pulsars to accretion powered neutron stars. In particular, mass measurements refer to neutron stars in double neutron star binary systems (*magenta*); in High Mass eclipsing X-ray Binaries (*cyan*); in Low Mass X-ray Binaries, with white dwarf companions (*gold*), with optical observations of the white dwarf companions (*green*), and in accreting bursters (*purple*). The figure is from [39]

In this well explored and studied field of galactic binaries there is now compelling evidence for the existence of stellar-mass black holes, and the guiding argument rests in the measure of the mass of the compact object. Figure 16.2 shows data grouping measured neutron star masses [39], and Fig. 16.3 data for selected stellar-mass black hole candidates [40].

The masses of compact objects are measured in binaries by tracing the systematic Doppler shifts observed in the spectrum of the companion ordinary star, consistent with it moving on a binary orbit under the influence of the unseen companion. From the Doppler shift data, a radial velocity curve can be constructed, giving the variation with time of the component of the star's velocity along the line of sight. Using Kepler's laws of orbital motion one can obtain a quantity called *mass function*: $f(M) = (M \sin i)^3/(M + m_*)^2$, where M is the mass of the compact object and m_* that of the visible, ordinary star. If $m_* \ll M$ (as in Low Mass X-ray Binaries) a firm lower mass limit can be inferred, even if the inclination i of the binary is unknown. More accurate mass measurements can be inferred in binary systems housing a pulsating neutron star, through accurate timing of the pulsar signals (as shown in Fig. 16.2, magenta data points).

Fig. 16.3 The mass of galactic black hole candidates in Low Mass X-ray Binaries, as presented and discussed in [40]. The two colors refer to two different categories based on the amount of information available on the mass ratios and inclinations of binaries. Arrows indicate the measured mass function

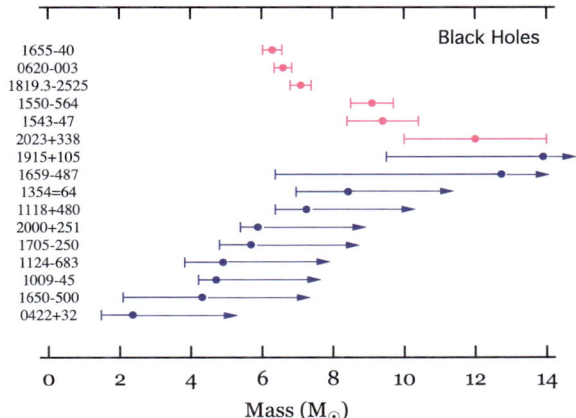

Figure 16.2 collects an inventory of measured neutron star masses, from a variety of different sources: from the young rotation-powered neutron stars in non-recycled binary systems to the old ones recycled in low mass binaries. As illustrated in the figure, neutron stars inhabiting young non-recycled high mass binary systems, all believed near their birth masses, have a mean mass of $1.28 M_\odot$ and a dispersion of $0.24 M_\odot$, in agreements with expectations that neutron stars from in core collapse supernovae. The mass of the neutron star can rise over M_{CH} if *fall-back* of material occurs during the supernova explosion. This depends sensitively on the energetic of the explosion, and the initial mass of the progenitor star. Double neutron star binaries may have evolved in systems where fall-back was important and for this reason they carry a slightly higher mass of $1.33 M_\odot$. The neutron star mass can further increase during *accretion episodes*. In Low Mass X-ray Binaries in particular, accretion can be long lived. Again Fig. 16.2 indicates that the mass distribution of neutron stars that have been recycled in Low Mass X-ray Binaries has a mean of $1.48 M_\odot$ and a dispersion of $0.2 M_\odot$, consistent with the expectation that they have experienced extended mass accretion episodes.

Recently, radio timing observations of the binary millisecond pulsar PSR J1614-2230 (also reported in Figs. 16.1 and 16.2) have led to the discovery of the heaviest neutron star known of $1.97 \pm 0.04 M_\odot$ measured with such certainty [15]. This finding effectively rules out the presence of hyperons, bosons, or free quarks at the center of neutron stars, and thus of equations of state with a high degree of softness [41]. Though the sample of Fig. 16.2 is likely biased toward target neutron stars that formed and evolved in binary systems, i.e. in peculiar galactic environments, $1.97 \pm 0.04 M_\odot$ can be taken as a *firm lower limit to the maximum mass of a neutron star*. Little is known about the mass of isolated neutron stars, i.e. born from progenitor stars evolved in isolation, but in the future there will be the prospect of measuring their masses and this will be discussed in Sect. 16.9.

Figure 16.3 shows a collection of data from galactic Low Mass X-ray transient sources housing *stellar-mass black hole candidates*. Following [40], observations are best described with a narrow mass distribution peaked around $7.8 \pm 1.2 M_\odot$. The

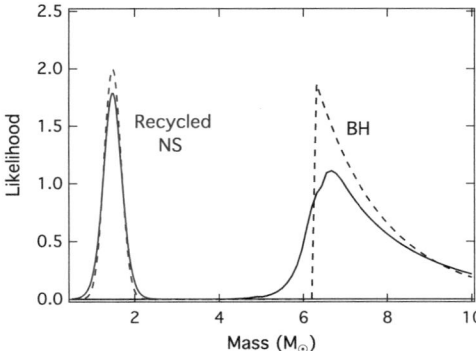

Fig. 16.4 The inferred mass distributions for the class of recycled neutron stars (the heaviest of the population of neutron stars as a whole in the Galaxy) and for black hole candidates in Low Mass X-ray Binaries (for clarity the black hole distribution has been scaled up by a factor of three). The dashed lines correspond to the most likely values of the parameters while the solid lines represent the weighted mass distribution. The figure shows the natural *divide* existing between neutron stars and stellar-mass black holes. The figure is from [39]

heaviest black hole has been observed in a distant galaxy: identified as M33 X-7 it has a mass of $15.65 \pm 1.45 M_\odot$.

According to our theoretical expectation, *the maximum mass of a neutron star establishes a divide between neutron stars and black holes of stellar origin*. In principle a black hole can carry any mass, thus even a mass as small as M_{CH} or even smaller. However, as we are guided by our knowledge on stellar evolution, any collapsing iron core in a massive star will evolve into a neutron star and not a black hole as long as its mass is below M_{max}. With this strong argument (or prejudice) the dividing line between neutron stars and black holes of stellar origin is represented by M_{max}. Thus, M_{max} could be viewed as the *minimum mass for a stellar-mass black hole*. With this premise, it is quite remarkable that the mass distributions of neutron stars and of black hole candidates shows this divide, as clearly illustrated in Fig. 16.4.

In the future, constraining both the low-mass and high-mass ends of the black hole mass distribution will bring clues on the way black holes form, in relation to the explosion energy in core collapse supernovae, to the degree of rotation of the collapsing core, and to the metallicity. An example of such analysis is presented in the incoming section.

16.5 Black Holes of Stellar Origin: A Maximum Mass?

One of the key, yet unanswered, question is: *what is the maximum mass of a black hole of stellar origin?* Pilot studies by Heger et al. [42] highlighted the key role played by metallicity in affecting the final fate of massive stars (with mass on the main sequence in excess of 10M$_\odot$) and in determining the black hole mass at the end of star's core collapse [42].

Stars form following the fragmentation of cold (10–100 K) molecular gaseous cores of initial density $\rho \sim 10^{-24}$ g cm^{-3}, corresponding to the densest phase of the interstellar medium. Stellar embryons generally evolve to increasing central densities and temperatures as they grow by accretion of surrounding gas, and contraction and growth halt when burning of hydrogen into helium starts in their cores. Any star has contracted up to a mean density $\rho \sim 1$ g cm^{-3}, when entering the main sequence phase of hydrogen ignition. In the case of massive stars, burning of helium into carbon and oxygen, then carbon, neon, oxygen into silicon continues until finally iron is produced in the core. The dense growing core is progressively dominated by electron degeneracy, at temperature of $\sim 10^9$ K, and when the star has built up a large enough iron mass exceeding its Chandrasekhar mass limit M_{CH}, collapse continues unhalted and either a neutron star (of mean density $\gtrsim 10^{14}$ g cm^{-3}) or a black hole forms. The relic core weight no less than $\approx M_{CH}$ or slightly more, depending on the dynamics of the collapse, i.e. whether material has fallen back and a luminous or dim supernova explosion or no-explosion has occurred.

Figure 16.5 from Heger et al. (2003) shows, in a rough scheme, the fate of single stars as a function of the initial stellar mass and metallicity Z (defined as the logarithm in power of ten of the iron to hydrogen abundance ratio and often expressed in units of the solar metallicity $Z_\odot$). Primordial (or Pop III) stars (at the bottom of the diagram) are stars resulting from the collapse and fragmentation of pristine gas formed from the composition as made in the Big Bang, and thus lacking of any metal ($Z = 0$). Modern stars (at the top of the diagram) are stars formed out of gas clouds with solar metallicity or even larger, so that on the y-axis the span of metallicity covers more than seven orders of magnitude (near to pristine with $Z = 10^{-6} Z_\odot$ up to $10Z_\odot$ for the very metal rich stars).

Figure 16.5 captures the basic findings on the end-states of stellar evolution, showing that the fate of stars is in close relation to their metallicity at birth, beside their mass. Stars evolve always in neutron stars if their initial mass is below 25M$_\odot$, regardless the value of Z. The fate is a neutron star, regardless the initial star's mass (i.e. even when the initial mass exceeds 25M$_\odot$) if they are very metal rich: this is due to the dramatic mass loss by metal-induced winds that these metal rich stars experience even during the main sequence phase. Mass loss cause these stars evolving *as if* they were initially much lighter. Above 25M$_\odot$, stellar black holes form over a wide range of metallicities provided the values of Z are not very large. In this regime, we can distinguish black holes resulting (i) from *fall-back*: initially a neutron star forms in the collapsing star launching a shock wave that drives matter out (with supernova display). Later, fall-back of outflowing gas, bound to the forming neutron star core, drives the central core above M_{max} resulting in the formation of a black hole; and (ii) from *direct collapse* if the collapse of the stellar core is un-halted (with no or dim supernova display). Black hole formed along the first channel weight less than those formed via direct collapse, and their mass is a fraction of the mass at birth. As noted by Heger et al. and illustrated in Fig. 16.5 the relative importance of the fall-back and direct channels depends on the metallicity of the progenitor star.

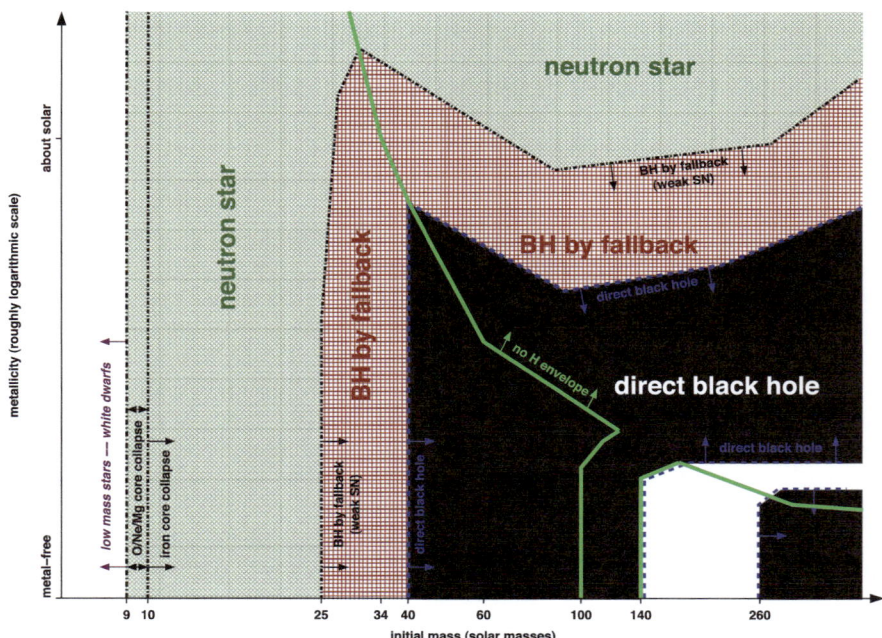

Fig. 16.5 Remnants of massive single stars as a function of initial metallicity (y-axis; qualitatively) and initial mass (x-axis; solar masses). The *thick green line* separates the regimes where the stars keep their hydrogen envelope (left and lower right) from those where the hydrogen envelope is lost (upper right and small strip at the bottom between $100 M_\odot$ and $140 M_\odot$). The *dashed blue line* indicates the border of the regime of direct black hole formation (*black*). This domain is interrupted by a strip of pair-instability supernovae that leave no remnant (*white*). Outside the direct black hole regime, at lower mass and higher metallicity, follows the regime of black hole formation by fall-back (*red cross hatching* and bordered by a *black dash-dotted line*). Outside of this, *green cross hatching* indicates the formation of neutron stars. At even lower mass, the cores do not collapse and only white dwarfs are made (white strip at the very left). The figure is from [42]

A further finding by Heger et al. [42] is that black hole formation in metal free and/or metal poor stars is confined to the (albeit wide) intervals of initial masses in between $40 M_\odot$–$140 M_\odot$ and above $260 M_\odot$. By contrast, stars with masses in between $140 M_\odot$ and $260 M_\odot$ have a different fate: *they explode leaving no black hole remnant*. Gamma-rays produced in their cores become so energetic that they annihilate into particle and anti-particle electron pairs, after central carbon burning. The resulting drop in radiation pressure causes the stellar core to collapse under its own huge gravity and to heat up (according to the virial theorem) to produce further gamma-ray photons via nuclear reaction that cascade back into electron-positron pairs. This triggers runaway burning of oxygen and silicon so that rapid energy deposition blows the star completely apart *leaving no-remnant*. At higher metallicities, stars loose their mass without ever encountering the electron-positron pair instability, and this explains why it is suppressed in stars with increasing Z as shown in Fig. 16.5.

Only above 260M$_\odot$, direct collapse to a black hole of comparable mass is "restored" as photo-disintegration of alpha particles (which themselves are already the result of photo-disintegration of iron group elements which were made in silicon burning) reduces the pressure enough that the collapse of the star, deprived by carbon/helium nuclei ready to burn, is not turned around but directly continues into a black hole.

The diagram in Fig. 16.5 thus shows that nature manages to avoid forming black holes during the collapse of very massive stars, in a certain range of masses and metallicities, illustrating how delicate is the balance between gravity and microphysics when coupled to radiative transport and nuclear energy production.

In summary, the maximum mass of a black hole of stellar origin is undetermined: primordial/metal-poor stars likely end their life as *massive* stellar-mass black holes weighing more than 260M$_\odot$. No black holes of this mass have been ever observed. Indeed, little is known at observational level not only on the true evolution of very massive stars, but also on their statistical inference. The initial mass function of stars in relation to the metallicity of their cosmic environment in not very well constrained, particularly at low metallicities. As a rule o thumb, a canonical Salpeter initial mass function produces one stellar black hole every 10^{3-4} ordinary stars so that the search and identification of black holes of stellar origin is always problematic and challenging. It has been speculated that the initial mass function of Pop III and of metal poor stars is *top-heavy*, i.e. there is a predominance of very massive stars over stars as ordinary as the sun, but theoretical uncertainties and lack of observations make this scenario highly debated.

Recently, it has been suggested that tracing the population of Ultra Luminous X-ray Sources may be of central importance in order to address the issue on the maximum mass of black holes of stellar origin, as these sources (or some of them) may host the heaviest black holes resulting from stellar core collapse [43].

16.6 Black Holes: The Other Flavor

So far, our focus was on stellar-mass black holes of $\gtrsim$3M$_\odot$ up to $\sim$260M$_\odot$ (or more, e.g. $\lesssim$1000M$_\odot$), for which we outlined a clear formation path. Black holes appear however to come in nature in another flavor: they are the *supermassive black holes*, weighing between $\sim 10^6$M$_\odot$ up to $\sim 10^9$M$_\odot$ [44, 45].

The suggestion of the existence of supermassive black holes originated in the early sixties following the discovery of Quasi Stellar Objects (QSOs; or quasars). QSOs are *active nuclei* that are so luminous (with bolometric luminosities in the range of $\gtrsim 10^{44-47}$ erg s^{-1} $\sim 2 \times 10^{10-13}$L$_\odot$) that often outshine the galaxy they inhabit. Their radiation is emitted across a spectrum, almost equally, from X-ray to the far-infrared, and in a fraction of cases from γ-rays to radio waves. Their variability on short timescales reveals that the emitting region is only a few light years across. Correlated variability over broad energy intervals indicates that this mechanism can not be ascribed to any stellar process. Efficiencies of mass-to-light conversion of the order of 10 % are required in order to fulfill all observed requirements and this is suggestive that QSOs and the more broadly called AGN (acronym of Active Galactic Nuclei) are associated with the relativistic potential of a supermassive black hole. A fraction (∼10 %) of the AGN, from radio to γ-rays, shows the presence of collimated *jets* transporting relativistic particles (with bulk Lorentz factors around ten) extending rather coherently for up to millions of light years. Again, this requires the presence of a relativistic potential well and of a stable preferred axis over timescales of hundred million years, properties that hint for the presence of a stable source, i.e. supermassive black hole. The jet power might even be extracted directly from the rotational spin energy of the black hole itself by means of the Blandford-Znajek mechanism [46], or from the accretion disc [47] who's axis of orientation fixes a preferred direction in space. A further fact supporting the black hole conjecture is that the basic properties of AGN and QSOs seem to scale self-similarly over a luminosity range of more than six orders of magnitude indicating that a universal engine is present for a certain time in most if not all galaxies.

16.6.1 Weighing Active Supermassive Black Holes

For long time the chief argument for weighing a black hole in an AGN or QSO has been the *Eddington limit* on the luminosity, corresponding to when the radiation pressure force acting on accreting electrons equals the force of gravity (upon protons). Above this limit accreting gas that would be responsible of the emission can not fall onto the black hole as it is blown away. This occurs when the accretion luminosity exceeds the Eddington luminosity $L_{\rm Edd} = 4\pi G M_{\bullet} m_{\rm p} c/\sigma_{\rm T} \sim 10^{46}(M_{\bullet}/10^8 M_{\odot})$ erg s^{-1} (where $\sigma_{\rm T}$ is the Thomson cross section, $m_{\rm p}$ the proton mass, and $M_{\bullet}$ the black hole mass). The linear correspondence between the Eddington luminosity and $M_{\bullet}$ clearly fixes a lower limit on the black hole mass of an AGN of given luminosity $L_{\rm AGN}$, and the interval of AGN luminosities unambiguously indicate that supermassive black holes of $10^6 M_{\odot}$–$10^9 M_{\odot}$ are required to power their loud emission.

Alternatively, if $\varepsilon \delta M_{\rm acc} c^2$ is the fraction of radiated away energy by an accreting black hole, $(1 - \varepsilon) \delta M_{\rm acc}$ is the mass accreted by the black hole. As $\varepsilon \delta M_{\rm acc} c^2 = L_{\rm AGN} \delta t_{\rm AGN}$ is the radiated away energy for an AGN emitting a luminosity $L_{\rm AGN}$ over a timescale $\delta t_{\rm AGN}$, a mass as large as $6 \times 10^7 M_{\odot}$ is acquired by the active black hole over a time of $\sim 10^8$ yr, if $L_{\rm AGN} = 10^{12} L_{\odot}$, and $\varepsilon \sim 10$ %. This unavoidably lead to the accretion-induced growth of a supermassive black hole.

The mass $M_{\bullet}$ of an active black hole can also be determined by studying the kinematics of gas orbiting around. Warm gas is revealed through the presence of broad emission lines that are observed in the optical spectra of QSOs [44]. Mass estimates require the independent measure of the velocity $V_{\rm gas}$ and distance $R_{\rm gas}$ of the orbiting gas, assumed to move on Keplerian circular orbits ($V_{\rm gas}^2 = G M_{\bullet}/R$). At present, the measurements of the velocity and velocity-width of broad emission lines (Hβ, Mg II and C IV, in particular) originating from gas moving around the supermassive black hole on parsec and even sub-parsec scales provide measurements of their

16 Astrophysical Black Holes: Evidence of a Horizon?

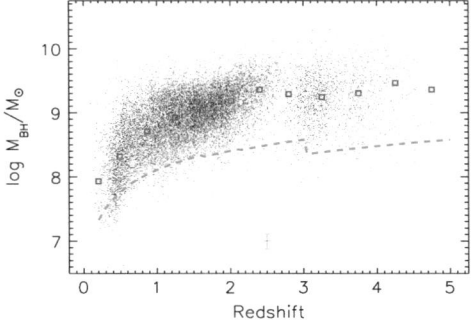

Fig. 16.6 Redshift distribution of the black hole masses of 14,434 QSOs selected from DSDD DR3, by Vestergaard et al. [49]. The median mass in each selected redshift bin is marked with a *square*. The median propagated black hole mass measurement error is shown in the *midst of the lower part of the diagram*. The *dashed line* refers to the faint SDSS flux limit so that at any given redshift observation can not pierce deep into the faint, lower mass end of the black hole mass distribution

mass. The technique uses (besides V_{gas}) the monochromatic continuum luminosity to estimate the position of the emitting lines R_{gas} via an empirical relation that has been inferred for a number of selected sources using the so called reverberation mapping technique [48].[6]

Figure 16.6 from [49] shows the mass distribution of the QSOs selected from the Sloan Digital Sky Survey (SDSS-DR3) over a sample of 15180 QSOs in the redshift range $0.3 < z \lesssim 5$. The median mass in redshift bins varies from a few times $10^7 M_\odot$ (at lower z) up to $10^9 M_\odot$ (at higher z). QSOs (o more generally AGN) are a population of cosmic sources distributed over an ample interval of cosmological redshifts [50]. Observations indicate that AGN are a population that underwent severe cosmic evolution and that had their peak of activity around redshift $2 \lesssim z \lesssim 3$ where also the overall star formation rate had its peak, in the universe [50]. Evidence is accumulating that *the complex, yet understood mechanism of black hole fueling goes hand in hand with the formation and assembly of the baryonic component of galaxies* over the entire cosmic epoch. Gas needs to be accumulated in the nuclear regions of a forming galaxy, and dynamical instabilities on large scales need to reverberate down to the smallest scales to feed the central black hole. This may occur when structures are on their way of forming [51, 52]. Thus, a clear prediction is that as soon as the accretion activity has superseded owing to the lack of fuel, a

---

[6]This technique reminiscent of echo mapping makes use of the intrinsic variability of the continuum source in active galactic nuclei to map out the distribution and kinematics of line-emitting gas from its light travel time-delayed response to continuum changes. These echo mapping experiments yield sizes for the broad line-emitting region that have been studied in about three dozen AGNs. The dynamics of the line-emitting gas appears to be dominated by the gravity of the central black hole, enabling measurement of the black-hole masses in AGN.

dormant supermassive black hole should reside in galaxies that were able to grow a supermassive black hole [53].

16.6.2 Dormant Black Holes in the Local Universe and Their Demography

In recent years, studies of the dynamics of stars and/or gas at the centers of nearby galaxies have revealed the presence of *massive dark objects*, i.e. of a *non-luminous mass* in *excess* of the mass (in stars) resulting from the underlying galactic potential [54]. This is inferred through observations of the galaxy's surface brightness profile at the smallest resolvable spatial scales, able to pierce down to a volume of size comparable to the gravitational sphere of influence of a hypothetical supermassive black hole nested at the center of a typical stellar bulge. The detection of a dark mass comes from the signature of a Keplerian rise in the velocity field of a spatially resolved ensemble of stars (or gas). This implies that the dark massive object is *point-like* on the smallest resolvable scales though not necessarily point-like on the true scale. If the star's velocity dispersion in the bulge is σ_* (typically of 100 km s^{-1} up to 300 km s^{-1}), the influence radius of a black hole of mass $M_\bullet$ is $R_{\rm inf} \sim GM_\bullet/\sigma_*^2 \sim 10(M_\bullet/10^8 M_\odot)(200 \text{ km s}^{-1}/\sigma_*)^2$ pc, corresponding to 0.11 arcseconds of angular size for a galaxy at a distance of 20 Mpc.

Over the last ten years, thanks to the unprecedentedly high angular resolution and sensitivity of the *Hubble Space Telescope* (*HST*), it has been possible to measure the spatial distribution and spectroscopic velocities of stars in the nuclei of several nearby galaxies. At present, there are 17 robust mass determinations from stellar dynamics with *HST*, and among these is the case of the nearby Andromeda galaxy (M31). All observations invariably point towards the presence of a massive dark object with inferred masses in the interval between $10^7 M_\odot$ and $3 \times 10^9 M_\odot$ [54]. Similarly, gas dynamical studies with *HST* have led to the discovery of 11 galaxies (among which M87 in the Virgo cluster) housing dark matter objects of similar mass. By exploiting the dynamics of stars and gas, *HST* is probing regions of size $\sim R_{\rm inf}$ that are roughly a million times bigger than the Schwarzschild radius of the supermassive black hole, should it be present there.

While the motion of stars is directly and almost solely affected by the gravitational potential of the galaxy and of the dark massive object, gas can be influenced by forces other than gravity. However, since internal energy can easily be dissipated whereas angular momentum cannot, the gas at the center of a galaxy plausibly settles into a relatively *cold rotating, massive nuclear disc*. In this context, an independent observational technique, which has provided one of the strongest cases for the presence of supermassive black holes, is the measurement of gas dynamics by means of the H_2O megamaser emission line at the wavelength of 1.35 cm. Radio measurements with the VLBA (Very Long Baseline Array) can achieve angular resolutions 100 times smaller than *HST* (less than half of a milliarcsecond). The text-book case refers to NGC 4258 housing a massive dark object of $M_\bullet \sim 3.9 \times 10^7 \pm 0.1 M_\odot$

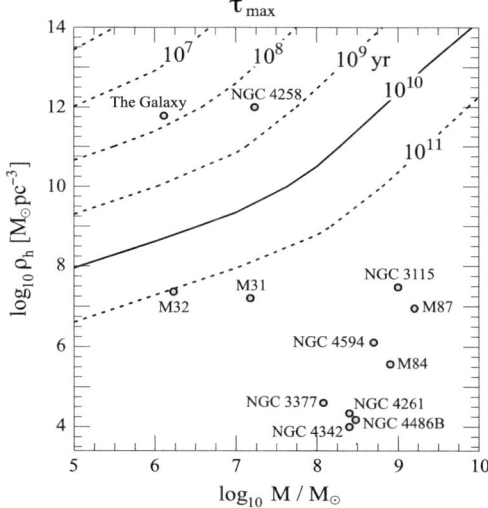

Fig. 16.7 Mass versus density diagram re-adapted from [71]: *Filled squares* represent current data for most of the observed dark massive objects: their half-mass M_h is the measured mass $M_\bullet/2$ and the half-mass density is computed as $\rho_h = 3M_h/4\pi R_h^3$ accounting for the spatial resolution at the time of the measurement. *Lines* refer to loci of constant maximum possible lifetimes for a dark cluster with half-mass M_h and half-mass density ρ_h, against the processes of evaporation and destruction by physical collisions. The lifetime of such hypothetical clusters is much shorter than 10 Gyr, the age of the universe, only in the cases of NGC 4258 and our Galaxy, thus strongly arguing for a point mass in these nuclei

confined within a region as small as 0.03 parsecs, strongly pointing towards the presence of a black hole [69, 70]. Figure 16.7 collects (on the x-axis) the masses $M_\bullet$ of the dormant black holes in terms of $M_h = M_\bullet/2$ (the half-mass of an hypothetical cluster of dark star) that have been collected by Maoz [71].

The phenomenological evidence discussed so far of supermassive black holes in the centres of nearby galaxies has also important astrophysical and cosmological implications (that go beyond the scope of this review but that are worth mentioning). For a long time the AGN phenomenon was depicted as caused by a process exclusively confined to the nuclear region of galaxies, with no relation with the host. This picture of disjoint black hole and galaxy evolution changed soon after data on black hole masses were collected and confronted with properties of the underlying host galaxies, thanks to *HST* that allowed a full kinematical and morphological characterization of the hosts. If was then found [56, 57] that galaxy spheroids with higher stellar velocity dispersions, i.e. with deeper gravitational potential wells and accordingly higher stellar masses and luminosities, host heavier central black holes with little dispersion in the correlation. More massive galaxies thus grow more massive black holes: the black hole "sees" the galaxy that it inhabits, and the galaxy "sees" the black hole at its centre despite its small influence radius [58]. The tightest correlation found is between $M_\bullet$ and the star's velocity dispersion σ_* in the bulge:

$\log(M_{\bullet}/M_{\odot}) = (8.12 \pm 0.08) + (4.24 \pm 0.41)\log(\sigma_*/200 \text{ km s}^{-1})$, for all galaxies and with a scatter of 0.44 ± 0.06 [55]. Consensus is rising that the $M_{\bullet} - \sigma_*$ is fossil evidence of a *co-evolution of black holes and galaxies*. The relation may have been rising along the course of galactic mergers and in episodes of self-regulated accretion during the formation of the early structures, but its true origin and evolution at look-back times is still unclear [59–61].

16.6.3 Supermassive Black Holes: How Do They Form?

There exist a number of key and independent observations that hint for a major growth of supermassive black holes via accretion during their luminous phase as QSOs AGN [62]. What is then the mass of the supermassive black holes at *birth*? This key question has at present no answer being unconstrained by observations. Are they growing from *black hole seeds*? When and where do these seeds form and which is their characteristic mass?

Middleweight black holes of $10^5 M_{\odot}$ (much lighter than the black holes in the AGN, as shown in Fig. 16.6 are now increasingly found in low mass spirals and dwarf galaxies with and without a bulge, and evidence is rising that these lighter black holes *co-habit* dense nuclear star clusters at the center of galaxies. According to the current paradigm of galaxy formation, dwarf galaxies in the galactic field are believed to suffer a quieter evolution than their much brighter analogues, the massive elliptical often found in galaxy's clusters. Thus, dwarf galaxies are the preferred sites for the search of these pristine black holes. However our knowledge is rather incomplete.

Models of hierarchical structure formation predict that galaxy-sized dark matter halos start to become common at redshift $z \sim 10$–20. This is the beginning of the nonlinear phase of density fluctuations in the universe, and hence also the epoch of baryonic collapse leading to star and galaxy formation. This is also believed to be the time of formation of the black hole seeds [63]. In this context, different populations of *seed black holes* have been proposed in the range of $100 M_{\odot}$ up to $10^6 M_{\odot}$.

Small mass seeds of $100 M_{\odot}$ or $1000 M_{\odot}$ may result from the core collapse of the first generation of massive stars (Pop III) that form inside unstable *metal-free* gas clouds, at $z \sim 20$ and in halos of $10^6 M_{\odot}$ [64–66]. Alternatively, *large* mass seeds may form later and in heavier halos of $10^8 M_{\odot}$ from the collapse of unstable gaseous discs [67]. This route leads to the formation of a massive quasi-star able to grow in its core a black hole that keeps on accreting and growing from the dense envelope. A third possibility, among others, is that star collisions in ultra dense nuclear star clusters lead to the formation of a supra-massive star that ultimately collapses into a black hole [68]. It is clear that these pathways are all reminiscent of the way stellar-mass black hole form. Quite intriguingly, no current observations give us any clue on how black hole formed and grew to the giant size characteristic of the supermassive black holes. While it is currently accepted that black hole of stellar origin form in the aftermath of core collapse (dim) supernovae, witnessing the formation of a seed black hole at look back times is still unaccessible to observation, and has no unique signature.

16.6.4 The Supermassive Black Hole at the Galactic Center

The determination of the mass of the dark massive objects at the center of galaxies can give a hint on the mass of the dormant supermassive black holes residing there, but it does not prove that the object is a Kerr black hole endowed by an event horizon. Even in the text-book case of NGC 4258 water maser lines are detected out to distances $R_{obs} \sim 10^4 R_S$, much larger than the size of the event horizon R_S. Combining the mass of NGC 4258 with the resolution scale R_{obs} of the observation one can infer a lower limit to the mass density associated to the dark object: $\rho_{dark-obj} \sim 4M_\bullet/R_{obs}^3$ which for NGC 4258 is $10^{12} M_\odot$ pc^{-3}, for $R_{obs} = 0.03$ pc. This density is much higher than the stellar mass density of a globular cluster, but still far from the *mean black hole mass density* $\rho_{horizon} \sim 4M_\bullet/R_S^3 \sim 2c^6/(G^3 M_\bullet^2)$, equal to $\rho_{horizon} \sim 10^{21} M_\odot$ pc^{-3} for a supermassive black hole of $10^8 M_\odot$. Is there a case where the presence of a supermassive black hole is more compelling?

The Galactic Center offers the closest view to a hypothetical supermassive black hole [72, 73]. The nucleus of the Milky Way is one hundred times closer to Earth than the nearest large external galaxy Andromeda, and one hundred thousand times closer than the nearest AGN. Due to its proximity, it is the *only nucleus in the universe* that can be studied and imaged in great detail. The central few parsecs of the Milky Way house gas cloud complexes in both neutral and hot phases, a dense luminous *nuclear star cluster*, and a faint radio source SgrA* of extreme compactness (3 to 10 light minutes across).

Observations, using diffraction-limited imaging and spectroscopy in the near-infrared, have been able to probe the densest region of the star cluster and measure the stellar dynamics of more than two hundred stars a few light days far from the dynamical center. The latter is coincident, to within 0.1 arcseconds, with the compact radio source SgrA*. The stellar velocities increase toward SgrA* with a Kepler law, implying the presence of a $4 \pm 0.06 \pm 0.35 \times 10^6 M_\odot$ central *point-like* dark mass (the largest uncertainty coming from uncertainties on the distance of the Earth relative to the Galactic Center).

This technique has also led to the discovery of nearly thirty young stars that orbit the innermost region: the so called S0 (or S stars). These young stars are seen to move on Keplerian orbits, all sharing the same focus. Any spherically symmetric extended configuration of dark stars distributed over the S0 star's complex would give rise to planar *precessing* orbits that would deviate from being Keplerian. This implies that the dark object at the Galactic Center has a size that can not exceed the smallest periapsis of the collection of the S0 stars. For this reason, particular attention has been given to S02 (or S2) the *showcase star* orbiting the putative black hole on the closest, highly eccentric (0.88) orbit with a period of 15.9 years. S02 is skimming the hypothetical horizon of the supermassive black hole at a distance which is about 1000 R_S. The periapsis of this orbit imposes a *lower limit* on the density of the dark mass concentration of $\sim 10^{13} M_\odot$ pc^{-3}, larger than that inferred for NGC 4258. Additionally, an even lower "lower-limit" of $10^{18} M_\odot$ pc^{-3} can be inferred from the compactness of the radio source SgrA*.

Are these limits providing compelling evidence that the dark point-mass at SgrA* is a supermassive black hole?

16.6.5 Testing the Black Hole Hypothesis?

To argue convincingly that the massive dark objects discovered at the center of nearby galaxies and in our own Galaxy are dormant black holes, one must rule out alternatives to the black hole hypothesis. The simplest is a *cluster of non luminous objects*, such as stellar remnants [71]. It is clear that SgrA* poses the most stringent limit on the properties of such non luminous clusters. If SgrA* is not a "special" object, the limit inferred from SgrA* should indicate that all other dark objects would satisfy the same limit, if explored with the same angular resolution.

Can dark clusters of relic stars or of failed stars such as planets and/or brown dwarfs of such compactness exist in nature? In other words can clusters with densities much in excess of $10^9 M_\odot$ pc^{-3} and up to $10^{18} M_\odot$ pc^{-3}, remain in dynamical equilibrium for a timescale comparable to the Hubble time?

Black holes in nearby galaxies are likely to exist since the early epoch of galaxy formation when they were outshining as QSOs, and the widespread stability of the AGN phenomenon and of the radio jets in particular argues in favor of a long-lived structure. Thus, given the observed masses $M_\bullet$ and the lower limits of the densities of all the massive dark objects $\rho_{\text{dark-obj}}$, is the characteristic lifetime of a hypothetical cluster longer than 13 Gyrs? If its lifetime is longer than the Hubble time, then hyper-dense star clusters are an alternative to the supermassive black hole hypothesis.

An upper limit to the lifetime of any bound stellar system (such as a star's cluster) is given by its evaporation time. Evaporation is the inevitable outcome of cluster evolution as stars escape from any dynamical system due to weak gravitational encounters that lead the system to reach equipartition of the star's kinetic energy. The evaporation timescale of a cluster consisting of equal-mass (m_*) objects is $t_{\text{evap}} \sim 300 t_{\text{relax}}$ where the relaxation time $t_{\text{relax}} = (0.14 N / \ln N)(R_h^3 / G M_h)^{1/2}$ is given as a function of the number of stars $N = M/m_*$, the half-mass radius R_h and the total mass M of the cluster. Denoting with M_h and ρ_h the cluster's half-mass half-mass-density, one has

$$t_{\text{evap}} \sim \frac{4 \times 10^4 (M_h/m_*)}{\ln[0.8(M_h/m_*)]} \left(\frac{\rho_h}{10^8 M_\odot \text{ pc}^{-3}}\right)^{-1/2} \text{ yr.} \quad (16.3)$$

The other limit on a cluster lifetime comes from the disruption of stars by physical collisions (occurring at extreme star's densities). After a time t_{coll}, every star experiences a collision. Runaway collisions among ordinary stars would lead to the formation of a large star and to severe mass loss via stellar disruption. The large star may then collapse into a black hole. Maoz estimates a collision time of

$$t_{\text{coll}} \sim \left[16\pi^{1/2} n_* \sigma_* r_*^2 \left(1 + \frac{Gm_*}{2\sigma_*^2 r_*} \right) \right]^{-1}, \qquad (16.4)$$

where n_*, r_*, and σ_* is the star's density, radius and velocity dispersion, respectively.

The lifetime $\tau(m_*, r_*)$ of a cluster of stars of given mass and radius is the minimum between t_{evap} and t_{coll}. Accordingly, for every combination of M_h and ρ_h, one can define the maximum cluster lifetime as the one obtained exploring all types of non luminous stars and selecting the longest lived one: $\tau_{\max}(M_h, \rho_h) = \max[\tau(m_*, r_*)]$ over all families of dark stars considered (comprising stellar-mass black holes of $\gtrsim 3 M_\odot$, or neutron stars of $1.4 M_\odot$, or white dwarfs in the mass interval $[0.01 M_\odot, 1.4 M_\odot]$, or very low mass objects of mass $10^{-3} M_\odot$, as planets or mini-black holes).[7]

Following Maoz (1998), Fig. 16.7 shows, in the half-mass and half-mass density plane, the loci of constant cluster maximum lifetimes against the processes of evaporation and destruction due to physical collisions. The squares refer to current data for most of the observed massive dark objects in nearby galaxies. The figure shows that all massive dark objects observed are consistent with being dense star clusters of any plausible form of non-luminous objects *except* for SgrA* at the Galactic Center, and NGC4 258. (Note that a much higher density is currently estimated for the dark object at the Galactic Center from the compactness of the radio source, as inferred from its variability.) Since the dark mass at the Galactic Center is the only mass distribution probed with sufficient space resolution, and since we expect that the dark mass in SgrA* is *not exceptional*, this observation provides strong evidence in support that *all* dark massive objects are point masses, within the limits imposed by SgrA*.

Collision and evaporation arguments can not exclude the presence of dark clusters, if these consist of elementary particles: in the limit of $m_* \to 0$ the collision and evaporation times can be made arbitrarily long [71]. Collisionless dark matter particles is a possibility. However they may not mass segregate to extreme densities ρ_h as they are dissipationless. A viable alternative to a point mass, i.e. to a Kerr black hole, is a *boson star* made of self-gravitating repulsive bosons for which dynamical stability has been proven to hold [74]. Current observations can not exclude this possibility for the dark object at the Galactic Center, and additional and alternative tools are necessary to prove or exclude this hypothesis, tools that will be described in Sect. 16.9.

---

[7] Physical collisions do not affect the lifetime of clusters made of black holes as light as $0.005 M_\odot$, due to their small sizes. For these light mini-black holes the evaporation time can also be made arbitrarily long, owing to the weakness of gravitational encounters, when $m_* \to 0$. These mini-black-holes would however not be of stellar origin, and no current astrophysical scenario predict their existence.

16.7 Black Holes: Are They Spinning?

Astrophysical black holes carry a mass and a spin (which is a vector), and mass and spin determine uniquely the structure of the spacetime. In Boyer Lindquist coordinates, the event horizon of a Kerr black hole is fixed uniquely by the black hole mass M and by the modulus of the spin vector J, customarily expressed in terms of the dimensionless spin parameter a such that: $J = aGM^2/c$ with $a \leq 1$, according to the cosmic censorship conjecture. The event horizon

$$R_{\text{horizon}}(M, a) = \frac{GM}{c^2}\left[1 + \left(1 - a^2\right)^{1/2}\right] \quad (16.5)$$

of a rotating Kerr black hole is decreasing with increasing a, from $2GM/c^2$ for $a = 0$ to GM/c^2 for $a = 1$. Equation (16.5) indicates that a hypothetical measure of the black hole spin can be obtained by pinpointing the position of the event horizon, if the mass is known by some other, independent mean.

Can we measure the spin of a black hole? The spin itself can be viewed as a indirect manifestation of an event horizon, since it is near the horizon that the ergosphere of a Kerr black hole develops, or in other terms since the spacetime is warped in a well defined manner. The spin is also a manifestation of the conservation of angular momentum: cosmic bodies carry often if not always a spin and if they collapse into a black hole, the hole is inevitably rotating and at a much higher frequency owing to the reduction of the moment of inertia of the collapsing star.

The spin of neutron stars can be directly measured from the modulation imprinted in the light curve of either a rotation-powered (isolated) pulsar or of an accreting X-ray pulsar in a binary system, due to the natural misalignment that exists between the spin axis $\hat{\mathbf{l}} \equiv \mathbf{J}/J$ and the axis of the magnetic moment $\mathbf{m}$ of the star (if highly magnetized). For the case of a black hole, the "access" to a spin measurement is much more subtle.

In the simplest, albeit realistic case of an *accreting* black hole, the accretion disc extends from far out distances (corresponding to radii where either the disc self-gravity becomes important [for the case of supermassive black hole], or where the Roche radius is located in a binary [for the case of a stellar-mass black hole]) down to the so called *Innermost Stable Circular Orbit* (ISCO).

In the Kerr metric there exists a critical radius, the radius of the ISCO R_{ISCO}, defined in the equatorial plane of the rotating black hole, below which *no dynamically stable* circular orbit exists for test particles with finite mass. R_{ISCO} is a monotonic decreasing function of a, in the interval $-1 \leq a \leq 1$. For negative values of a, corresponding to counter-rotating discs relative to the hole's spin axis, and in particular for $a = -1$, $R_{\text{ISCO}} = 9GM/c^2$. For $a = 0$ the radius is at $6GM/c^2$, and for $a > 0$, corresponding to co-rotating discs, R_{ISCO} further decreases so that at $a = +1$, $R_{\text{ISCO}} = GM/c^2$ coincides with the horizon.

R_{ISCO} can be viewed as the inner rim of any geometrically thin accretion disc around an active black hole. At R_{ISCO} the gas loses its dynamical stability and falls into the black hole without the intervention of viscous torques. It is before matter reaches R_{ISCO} that there is the highest energy dissipation in the accretion disc, and it is before and around R_{ISCO} that a series of GR effects on light propagation leave the spin signature imprinted in the X-ray spectra of accreting black holes, and in the shape of their emission lines.

The sudden change in the disc structure due to the presence of this dynamical instability in the Kerr spacetime is unique to black holes and gives access to measuring the black hole spin.

Two methods are employed, both of which depend upon identifying the inner radius of the accretion disc as the ISCO whose dimensionless radius $R_{\text{ISCO}}c^2/GM$ is a function of the spin parameter a, only [77]. The first is the *continuum-fitting* method which has so far only been applied to stellar-mass black holes that consists in modeling the thermal X-ray continuum spectrum of the truncated accretion disc. The second is the Fe Kα method, which applies to both classes of black holes (stellar-mass and supermassive), that models the profile of a relativistically broadened fluorescence iron line with special focus on the gravitational *redshifted red wing* of the line. Also in this case the truncation of the disc at R_{ISCO} is a key assumption of the model to infer a.

The GR generalization of the standard accretion disc model [30] by Novikov & Thorne [76] in the late 90s allowed computing the multicolor black-body-spectra from a geometrically thin, optically thick accretion disc. This model provides spectra of the continuum that can be used as templates to fit the data. The guiding line for the continuum-fitting method relies on an analogy: that of measuring the radius of a star R_* from the observed flux F_{obs} and the effective temperature T_{eff} derived from the black body spectrum of the star, i.e. from $L_* = 4\pi D^2 F_{\text{obs}} = 4\pi R_*^2 T_{\text{eff}}^4$, once the distance D to the source is known. From F_{obs} and T_{eff}, one can infer $\pi (R_*/D)^2$, and thus R_* known D.

Similarly, the continuum-fitting model aims at measuring $\pi \cos i\, (R_{\text{ISCO}}/D)^2$ from the observed flux, and T_{eff} near R_{ISCO} by accurate fits of the multicolor blackbody disc spectrum. If the distance D to the source, the black hole mass and the inclination of the disc i to the line of sight are known, then fitting the continuum gives R_{ISCO} in dimenssionless units, and in turn gives a. As T_{eff} scales with the black hole mass as $M^{-1/4}$ in the standard model [30] (i.e. lighter black holes have harder/hotter spectra) knowledge of the black hole mass is necessary to avoid degeneracy in the evaluation of the temperature of the continuum. Since the flux F emitted locally at a distance R by the disc increases with decreasing radius R, the effective radiation temperature T_{eff} varies with R as well. Accordingly, the hottest emitting annuli are those near R_{ISCO}, where T_{eff} is the highest. As spinning black holes (of fixed mass) have smaller $R_{\text{ISCO}}c^2/GM$, they result in a higher effective temperature at the cut-ff radius R_{ISCO}: fastly spinning black holes should have warmer spectra and higher luminosities per unit accreted matter $L/\dot{M}$ (with $\dot{M}$ the mass accretion rate, and L the disc luminosity) than non spinning black holes.

Figure 16.8 shows the logarithmic derivative of $L/\dot{M}$ versus the radius r in units of GM/c^2, for the Novikov & Thorne model (dahsed lines; [75]). The sharp down turn of $d(L/\dot{M})/d\ln r$ in the Novikov & Thorne model mirrors the effect of net truncation of the accretion disc at R_{ISCO}. Arguments have been advanced to suggest that a magnetized accreting gas will indeed have a non-zero shear stress at the

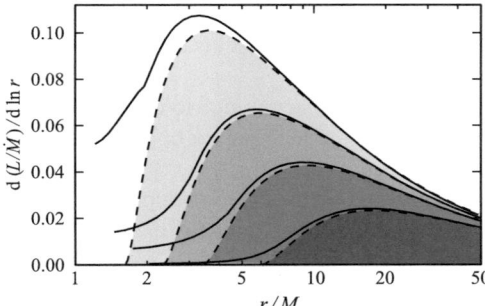

Fig. 16.8 Luminosity profiles (*solid lines*) from GR magnetohydrodynamic simulations by Kulkarni et al. [75] compared with those (*dashed lines*) re-adapted from [76] for a razor thin accretion disc truncated at the innermost stable circular orbit R_{ISCO}. From bottom to top, the spin parameter a increases from 0, 0.7, 0.9, and 0.98

ISCO [44], and that furthermore this stress could be so large that it may completely invalidate the Novikov & Thorne model even in very thin discs. This is clearly an important question that strikes at the heart of the continuum fitting method. A number of recent studies of magnetized discs using three-dimensional general relativistic magnetohydrodynamic (GRMHD) simulations have explored this question [75, 78]. The conclusion of these authors is that the shear stress and the luminosity of the simulated discs do differ from the Novikov & Thorne model, but in a "controlled" way, as illustrated in Fig. 16.8 (solid line). Thus fitting the continuum remains a viable method, providing all these caveats are taken into consideration.

Are the accretion discs in real sources truncated? Observational evidence of truncated discs in a class of highly variable sources, the Soft X-ray Transients, has been inferred via monitoring large changes in the flux of thermal origin. As R_{ISCO} is fixed by the geometry of the underlying Kerr black hole, the disc is expected to extend down to R_{ISCO} regardless the magnitude of the accretion flux and intrinsic luminosity (related to the mass transfer rate). Observations reveal that in these sources the fitted inner disc radius remains stable over changes in the flux by 10–100 factors.[8]

In the Fe Kα method, one determines R_{ISCO} by modeling the profile of the broad and skewed iron line, which is formed in the inner disc by Doppler effects, light bending, relativistic beaming, and gravitational redshift [79–81]. Of central importance is the effect of the gravitational redshift on the red wing of the line. This wing extends to very low energies for a rapidly rotating black hole ($a \lesssim 1$) because in this case gas can continue to orbit down to the event horizon. This is illustrated in Fig. 16.9 where the line profiles predicted in the case of Schwarzschild (red) and maximal Kerr (blue) black holes are shown.

---

[8]Recently, the evidence for a constant inner radius in the thermal state has been presented for a number of sources via plots showing that the bolometric luminosity of the thermal component is approximately proportional to T_{eff}^4. This indicates the stability of the radii of the inner annuli contributing most to the thermal emission.

Fig. 16.9 Line profiles predicted in the case of Schwarzschild (*red*) and maximal Kerr (*blue*) black holes. It is the extent of the red wing and its importance relative to the blue wing that allow black hole spin to be determined with disc lines. Re-adapted from the review by Miller [81]

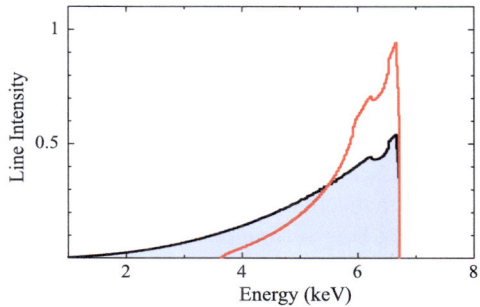

Measuring the extent of the red wing in order to infer the spin a is hindered by the relative faintness of the signal. However, the Fe Kα method has the virtues that it is independent of the mass M of the black hole and distance D to the sources, and that the blue wing of the line permits an estimate of the inclination i. This method, while applicable to both classes of black holes, is presently the only viable approach to measuring the spins of supermassive black holes.

Only for a handful of Galactic stellar-mass black holes and AGN, spins have been measured [77, 81]. The values of the spin are found to cover the whole interval $(0 \lesssim a \lesssim 1)$, with no preferred value for the spin. Nor, measurements can guarantee an accuracy such to yield a measure of the spin of an extreme Kerr black hole. While spins of stellar-mass black holes are genetic since accretion in Galactic binaries has never had time to alter a since birth,[9] spins of supermassive black holes can mirror specific evolutionary paths and accretion histories, given the large amount of gas that active black holes can accrete when shining as QSOs. Thus, information on their spin at birth is erased. No spin statistics is available yet to infer the true spin evolution of supermassive black holes: theoretical models suggest that if accretion occurs in the *coherent mode* in most of the AGN, i.e. it occurs through long-lived episodes of accretion along a preferred plane then black hole should be spinning close to $a \sim 1$. By contrast, if accretion occurs in the *chaotic mode*, i.e. via a sequence of randomly oriented episodes favor very low spins, the black hole spins in AGN should be quite low, owing to the rough balance of prograde and retrograde accretion events. A further key issue is the potential correlation between the black hole spin and the jet power in radio loud AGN. At present this correlation is quite poor, highly debated and extremely controversial [82].

16.8 Black Holes: Evidence of an Event Horizon?

Are there prospects of demonstrating the existence of an event horizon in an alternative way?

---

[9]The spin of a black hole changes sizably only if the black hole accretes a mass of the order of the black hole mass itself.

Consider again the case of an accretion disc around a compact object extending down to the ISCO [83]. Gas, having lost stability, inevitably falls into the event horizon. By contrast, if there is a surface inside ISCO (as for the case of a neutron star whose physical radius is in between R_S and R_{ISCO} for most of the equations of state) then gas that falls in rapidly is shock-heated within a boundary layer on the star's surface. In other terms, under identical accretion conditions a neutron star is brighter than a black hole because it has a surface. If $L_{BH} = L_{disc}$ is the luminosity of a black hole surrounded by an accretion disc, $L_{NS} = L_{disc} + L_{surface} > L_{BH}$ is the luminosity for a neutron star. The key step forward is to find a class of sources housing either a black hole or a neutron star that accrete under the *same* conditions. If so, black holes would be underluminous and this would be a signature of an event horizon. This is the chief argument that was pioneered by Narayan et al. (2008) and his colleagues.

The argument is slightly more subtle: to magnify the effect of a large luminosity mismatch between L_{BH} and L_{NS}, Narayan et al. (2008) searched for sources where the break down of the thin disc approximation occurs. Accretion discs can be considered geometrically thin if gas cools rapidly (on a timescale shorter than the viscous time of the mass inflow). However flows with long cooling times and thus low radiative efficiencies can in principle form under at least two critical circumstances: (i) when the accretion rate is so low that the inflowing gas has low density (and long cooling times), or conversely (ii) when the accretion rate is so large that the gas, optically thick, traps the radiation that is dragged in. In both cases energy is advected inward and these flows are called ADAF, acronym of Advection Dominated Accretion Flows. If the accretor is a black hole the advected energy is captured within the horizon and invariably; if it is a neutron star that energy is ultimately released at the surface.

In the low accretion rate regime, this process has been considered to explain the emission of a class of X-ray binaries (the Soft X-ray Transients) that are transients as they alternate long lived phases of quiescence with short lived phases of intense accretion. During quiescence, the sources are emitting at a very low luminosity compared to the luminosity during the outburst phase. Figure 16.10 shows the quiescent luminosity of a number of Soft X-ray Transients versus the orbital period (expressed in hours). Filled symbols correspond to known black hole candidates and open circles to neutron stars, while shaded bands are to guide the eye. As these binaries house low mass donors of similar mass, binaries with comparable orbital periods likely host similar accretion disc. Thus comparing sources in quiescence having similar orbital periods means selecting sources that accrete under similar conditions.

Figure 16.10 shows that as a group, the neutron stars are a factor a hundred or more brighter than the black hole candidates and this difference can be interpreted as evidence that these sources posses an event horizon, and thus house a black hole. There are caveats, since the interpretation of the quiescent luminosity of neutron stars is not unique. For example, this luminosity can be attributed to heating of the star's crust during outburst followed by cooling in quiescence, and if true the whole picture becomes invalid.

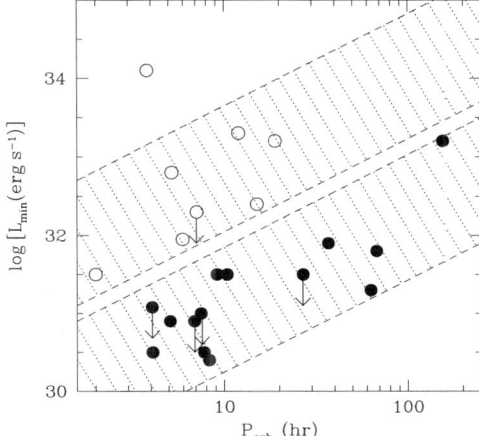

Fig. 16.10 Eddington-scaled luminosities in the energy interval of 0.5–10 keV of black hole transients (*filled circles*) and neutron star transients (*open circles*) versus the orbital period. The plot shows all systems with known orbital periods, which have optical counterparts and good distance estimates. The *diagonal hatched areas* delineate the regions occupied by the two classes of sources and indicate the observed dependence of luminosity on orbital period. Note that the black holes systems are on average nearly 3 orders of magnitude fainter than the neutron star systems with similar orbital periods. The figure is from [83]

16.9 Event Horizons: A New Perspective

Observations indicate that the center-most part of galaxy spheroids and bulges of spirals consists of a cluster of a few 10^7 to a few 10^8 stars and of the dark massive object of mass $M_\bullet$, from now on a supermassive black hole [84]. Star's densities can be in excess of 10^6 pc^{-3}, and up to 10^8 pc^{-3} near the center, suggesting that stars can interact dynamically with the central dark object. Under these exceptional conditions stars can either be *tidally torn apart* and disrupted by the central black hole or, if compact, can get *swallowed*. Thus, stars can probe the innermost regions of the spacetime around a black hole as gas particles do in an accretion disc. If an ordinary star, like the sun, of radius r_* and mass $m_* \sim M_\odot$ happens to pass very close to the massive black hole, i.e. within a distance $r_{\text{tidal}} \sim r_*(M_\bullet/M_\odot)^{1/3}$, some part of it or all of it is teared apart under the action of the tidal field by the black hole, and a fraction of the gas composing the star likely form a ring that is eventually accreted by the black hole. If R_S ($R_{S,*}$) is the Schwartzshild radius of the central black hole (star), the disruption occurs at a distance (in units of the Schwartzshild radius) $r_{\text{tidal}}/R_S \sim (r_*/R_{S,*})(m_*/M_\bullet)^{2/3}$ which is only $\sim 2 \times 10^5 (m_*/M_\bullet)^{2/3} \sim 20$, for a black hole of $10^6 M_\odot$ and $m_* = M_\odot$.

Depending on the distribution and topology of the orbits, stars in galactic nuclei can alternatively be swallowed whole on their first pass by the central supermassive black hole and these are called *direct plunges*. There is however a very interesting possibility (and a range of orbits) corresponding to *gradual inspiral*. If the star is a compact object (a neutron star or a stellar-mass black hole), the star is not tidally disrupted (having r_{tidal} smaller than R_S) and can orbit very close to the hole's event horizon, as if it were a "point" mass (i.e. unaffected by tides). Under these circumstances the compact object orbits deep in the relativistic potential of the supermassive black hole and slowly spirals inward toward the black hole as it emits *gravitational waves*. This process of slow inspiral is referred to as *Extreme Mass Ratio Inspiral* (EMRI), as extreme is the mass ratio between the large, massive black hole $M_\bullet$ and the neutron star or stellar-mass black holes [84]. Owing to the extreme compactness, the star can explore the entire volume of spacetime outside the supermassive black hole before being swallowed. The compact object doomed to death spends many orbits (up to 10^5) around the large black hole and when doing so, it radiates gravitational waves whose waveform is containing detailed information about the spacetime and all the physical parameters which characterize the EMRI.

Gravitational waves are emitted typically at frequencies of $f_{\text{GW}} > 10^{-4}$ Hz, one or two years before the object is swallowed, allowing the detection of EMRIs in galaxies up to cosmological redshift $z \sim 0.7$, corresponding to a volume of several Gpc^3 [85]. If one considers the New Gravitational wave Observatory (NGO, often referred to as extended LISA or eLISA) as reference experiment for the direct detection of low frequency gravitational waves from space [85], a few to a hundred EMRIs could be revealed in two years of operation. Stellar-mass black hole inspirals are expected to dominate the detection rate, both because higher mass means greater gravitational wave amplitude (and hence a larger detection volume) and because dynamical mass segregation in galactic nuclei concentrates the heaviest objects closest to the supermassive black hole. Thus, stellar-mass black holes should be the dominant population near the large black hole.

The highly relativistic orbits of EMRIs, lying within $\sim$5–10 Schwarzschild radii of the supermassive black hole, display extreme versions of both relativistic pericenter precession and Lense-Thirring precession of the orbital plane about the spin axis, as depicted in Fig. 16.11.

The large number of cycles and complexity of the orbits encode wonderfully detailed information concerning the system's physical parameters. The mass of the compact object and of the supermassive black hole, and the eccentricity of the orbit will typically all be determined to a fractional accuracy of 10^{-4} or even less. The analysis of these events will unable us (i) to measure the mass of isolated stellar-mass black holes and neutron stars (a measure not accessible by any other mean), and learn about their maximum mass; (ii) to measure the mass $M_\bullet$ of the supermassive black hole in a mass range $10^4 M_\odot$ and $10^6 M_\odot$ which is complementary to that probed by electromagnetic observations of AGN and nearby dormant black holes; (iii) and to have a clean and *direct* measure of the spin of the supermassive black hole, i.e. a measurement unaffected by systematics introduced by model fitting, and with an absolute error smaller than 10^{-5}. Thus, future experiments for the detection of gravitational waves (as NGO/eLISA) will bring revolutionary information on the nature of the dark massive objects and we further elaborate this topic in the incoming section.

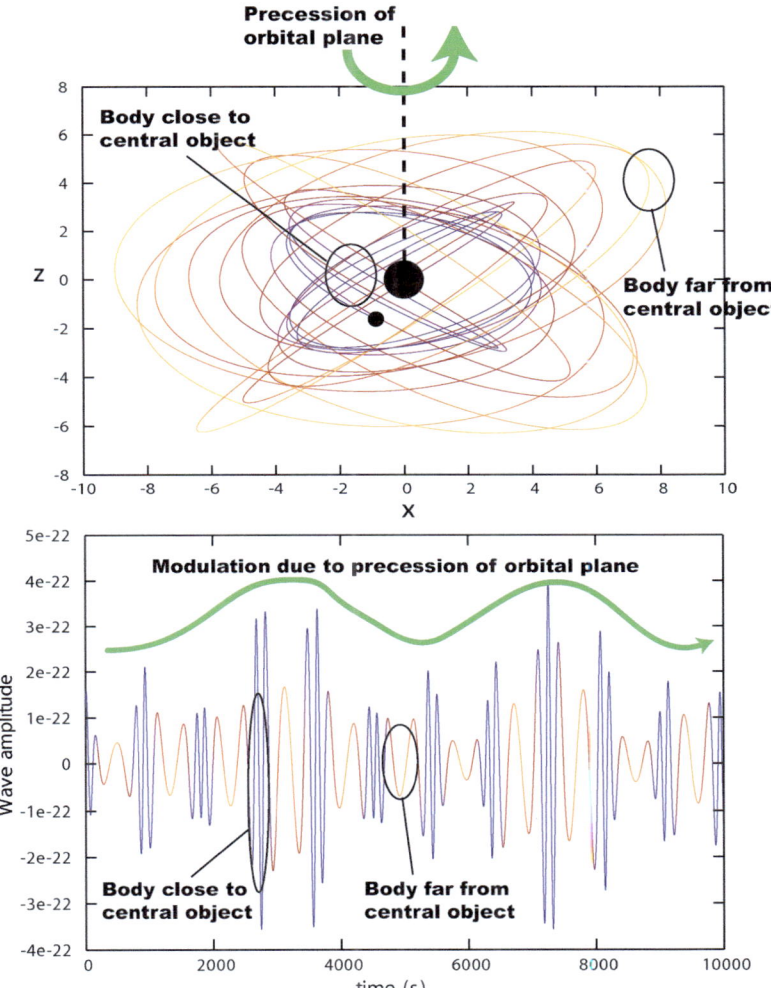

Fig. 16.11 Cartoon of an EMRI orbit, as viewed from the side (*top panel*) and the emitted gravitational wave (*bottom panel*). The gravitational wave is characterized by higher amplitude and frequency radiation associated with extreme pericenter precession when the body is close to the central object, and lower amplitude and frequency radiation when the body is further away. There is an overall modulation due to precession of the orbital plane. The waveform is colored to illustrate this structure. The figure is from [81]

16.9.1 Testing the Kerr-ness of the Spacetime

Are the massive dark objects lurking at the center of galactic nuclei supermassive black holes? No electromagnetic observations can provide a direct test on the *Kerr-*

ness of a massive dark object while gravitational waves do, as EMRIs can be used to probe directly the texture of spacetime in a galactic nucleus.

As already mentioned EMRI's orbits are highly relativistic exhibiting extreme forms of periastron and orbital plane precession, and given the large amount of cycles collected during a typical EMRI observation (1–2 years) a fit of the observed gravitational waveforms to theoretically calculated templates is very sensitive even to small changes in the physical parameters of the system. The level of precision of the mass and spin measurements of the central black hole in an EMRI event can be used as a highly accurate test of the Kerr-ness of the central massive object [85].

The spacetime outside a stationary axisymmetric object is fully determined by its *mass moments* M_l and current multipole moments S_l. These moments fully characterize the spacetime so that the orbits of the smaller (test) object and the gravitational waves it emits are fully determined by the multipolar structure of spacetime. By detecting the gravitational waves with NGO/eLISA one can obtain a map of the spacetime. Extracting the moments from the EMRI signal is analogous to geodesy in which the distribution of the Earth's mass is determined by studying the orbits of satellites.

Black hole geodesy, also known as holiodesy, is very powerful because Kerr black holes have a very special multipolar structure. A Kerr black hole with mass $M_\bullet$ and spin parameter a (in units with $G = c = 1$) has multipole moments given by

$$M_l + i S_l = (ia)^l M_\bullet^{(l+1)}. \tag{16.6}$$

Thus, $M_0 = M_\bullet$, $S_1 = aM_\bullet^2$, and $M_2 = -a^2 M_\bullet^3$. Similarly, all other multipole moments are all completely determined by the first two moments, the black hole mass $M_\bullet$ and spin a. This is nothing more than the black hole *no-hair* theorem: the properties of a black hole are entirely determined by its mass and spin.

For inspiraling trajectories that are slightly eccentric and slightly non-equatorial, in principle all the multipole moments are redundantly encoded in the emitted gravitational waves, through the time-evolution of the three fundamental frequencies of the orbit: the fundamental frequencies associated with the r, θ, and ϕ motions, or, equivalently, the radial frequency and the two precession frequencies.

The mass quadrupole moment M_2 of a Kerr black hole can be measured to within $\Delta M_2 / M_\bullet^3 \sim 10^{-2}$–$10^{-4}$ for EMRI's signals with an SNR of 30, and at the same time $\Delta M_\bullet / M_\bullet$ and $\Delta S_1 / M_\bullet^2$ can be estimated up to an accuracy of 10^{-3}–10^{-4}. Any inconsistency with the Kerr multipole structure could signal a failure of GR, the discovery of a new type of compact object, or a surprisingly strong perturbation from some other material or object [86, 87].

EMRI signals can be used to distinguish definitively between a central massive black hole and a boson star [88]. In the case of a black hole, the gravitational wave signal *shuts off* shortly after the inspiraling compact object reaches the last stable orbit (and then plunges through the event horizon), while for a massive boson star, the signal does not fade, and its frequency derivative changes sign, as the body enters the boson star and spirals toward its centre. Similarly, if the central object's horizon is replaced by some kind of membrane (this is the case for the so-called gravastars) the orbital radiation produced by the orbiting body could resonantly excite the quasi normal modes of the gravastar, with characteristic signatures in the gravitational wave energy spectrum that would be detectable by eLISA [89].

Other studies within GR considered axisymmetric solutions of the Einstein field equations for which the multipole moments can differ from the Kerr metric, such as the Manko-Novikov solution. These studies revealed ergodic orbital motion in some parts of the parameter space [90] as a result of the loss of the third integral of motion. Similar studies suggested that the inspiralling body could experience an extended resonance in the orbital evolution when the ratio of intrinsic frequencies of the system is a rational number. If detected, these features would be a robust signature of a deviation from the Kerr metric.

We remark that, if GR must be modified, the "true" theory of gravity should lead to similar deviations in all observed EMRI. For this reason, statistical studies of many EMRI events to test GR would alleviate possible disturbances that may cause deviations in individual systems, such as interactions with an accretion disc [91, 92] or perturbations due to a second nearby black hole or by a near-by star.

Consensus and appreciation is now rising on experiments such as NGO/eLISA that are designed to bring transformational science results, i.e. new data that could improve significantly our knowledge of black holes as astrophysical sources.

16.10 Conclusions

Astrophysical black holes appear to be ubiquitous in the universe. Owing to their extreme gravity they are hypothesized to power the Galactic X-ray sources and the luminous QSOs. Black holes appear to influence the dynamics of gas and stars in a unique way, giving rise of a variety of astrophysical phenomena that we outlined in this Lecture Note.

There are many unanswered questions that will be addressed in the future, and we here list three, among the compelling: (i) Are the dark massive objects in galactic nuclei supermassive black holes? As highlighted in Sect. 16.9, detection of low frequency gravitational wave from EMRI events in future experiments like NGO/eLISA will provide a direct test of the Kerr-ness of dark massive objects. Incoming electromagnetic observations with elevated spatial and time resolutions will also improve our view on how the gas and stars orbit around the central dark objects in galactic nuclei, providing accurate measurements on their masses and a view of their physical environment. (ii) How, when and where supermassive black holes form in the universe? What is the mass of seed black holes and their physical channel of formation? Observations indicate that black holes evolve in concordance with galaxies. However, little is known on how galaxies assemble: whether via major mergers, or via repeated minor mergers of satellite halos, or from cosmic filaments or from a combination of all these effects. It is likely that supermassive black holes are just a manifestation and a key outcome of this intricate phenomenological process, and major progress is expected in the incoming years from both observations and theory. At present, the mechanism of black hole seed formation is highly debated, and uncertain. (iii) Are powerful jets in radio sources powered by the spin of the black hole? Also this is a question still unattended. Galactic X-ray binaries provide the best laboratory for testing this hypothesis, due to their small size and high variability on scales of milliseconds to years: they could be view as supermassive black holes in miniature. Their continuous monitoring will disclose differences between sources housing a neutron star, i.e. a star with a surface, or a black hole, i.e. a star with an event horizon and an ergosphere. Jets are ubiquitous in cosmic sources, but relativistic jets are somewhat unique as they require the presence of a relativistic potential, but the different role of the disc, the magnetosphere and the spin is difficult to disentangle.

Given all these open issues, black holes will still be offering surprises in the incoming years.

References

1. Schwarzschild, K.: Preuss. Akad. Wiss. Berlin, Sitzber. **189** (1916)
2. Kruskal, M.D.: Phys. Rev. **119**, 1743 (1960)
3. Kerr, R.: Phys. Rev. Lett. **11**, 237 (1963)
4. Penrose, R.: Nuovo Cimento **1**, 252 (1969)
5. Israel, W.: Phys. Rev. **164**, 1776 (1967)
6. Carter, B.: Phys. Rev. **174**, 1559 (1968)
7. Hawking, S.W.: Commun. Math. Phys. **33**, 323 (1972)
8. Robinson, D.C.: Phys. Rev. Lett. **34**, 905 (1975)
9. Hawking, S.W.: Nature **248**, 30 (1974)
10. Oppenheimer, J.R., Snyder, H.: Phys. Rev. **56**, 455 (1939)
11. Harrison, B.K., Thorne, K.S., Wakano, M., Wheeler, J.A.: Gravitation Theory and Gravitational Collapse. University of Chicago Press, Chicago (1965)
12. Chandrasekhar, S.: Astrophys. J. **74**, 81 (1931)
13. Oppenheimer, J.R., Volkoff, G.M.: Phys. Rev. **55**, 734 (1939)
14. Lattimer, J.M., Prakash, M.: Phys. Rep. **442**, 109 (2007)
15. Demorest, P.B., Pennucci, T., Ransom, S.M., Roberts, M.S.E., Hessels, J.W.T.: Nature **467**, 7319 (2010)
16. Jacoby, B.A., Hotan, A., Bailes, M., Ord, S., Kulkarni, S.R.: Astrophys. J. **629**, L113 (2005)
17. Cook, G., Shapiro, S.L., Teukolsky, S.: Astrophys. J. **424**, 823 (1994)
18. Cook, G., Shapiro, S.L., Teukolsky, S.: Astrophys. J. **423**, L117 (1994)
19. Rhoades, C.E., Ruffini, R.: Phys. Rev. Lett. **32**, 324 (1974)
20. Witten, W.: Phys. Rev. D **30**, 272 (1984)
21. Hewish, A., Bell, S.J., Pilkington, J.D., Scott, P.F., Collins, R.A.: Nature **217**(5130), 709 (1968)
22. Giacconi, R., Gursky, H., Paolini, F.R., Rossi, B.B.: Phys. Rev. Lett. **9**, 439 (1962)
23. Salpeter, E.: Astrophys. J. **140**, 796 (1964)
24. Zel'dovich, Ya., Novikov, I.D.: Sov. Phys. J. **9**, 246 (1964)
25. Lynden-Bell, D., Rees, M.J.: Mon. Not. R. Astron. Soc. **152**, 461 (1971)
26. Tauris, T.M., van den Heuvel, E.P.J.: In: Lewin, van der Klis (eds.) Compact Stellar X-Ray Sources. Cambridge Astrophysics Series, vol. 39. Cambridge University Press, Cambridge (2006)
27. King, A.R.: In: Lewin, van der Klis (eds.) Compact Stellar X-Ray Sources. Cambridge Astrophysics Series, vol. 39. Cambridge University Press, Cambridge (2006)

28. Ghosh, P.: Class. Quantum Gravity **25**(5), 059001 (2008)
29. Lorimer, D.R.: Living Rev. Relativ. **8**, 7 (2008)
30. Shakura, N.I., Sunyaev, R.A.: Astron. Astrophys. **24**, 337 (1973)
31. Lewin, W.H.G., van Paradijs, J., Taam, R.E.: Space Sci. Rev. **62**, 223 (1993)
32. Hulse, R.A., Taylor, J.H.: Astrophys. J. **195**, L51 (1975)
33. Burgay, M., et al.: Nature **426**, 6966, 531 (2003)
34. Lyne, A.G., et al.: Science **303**, 5661, 1153 (2004)
35. Kramer, M., Wex, N.: Class. Quantum Gravity **26**(7), 073001 (2009)
36. Mushotzky, A.: Adv. Space Res. **38**(12), 2793 (2006)
37. Fabbiano, G.: Annu. Rev. Astron. Astrophys. **44**(1), 323 (2006)
38. Kramer, M.: New horizons in time-domain astronomy. IAU Symp. **285**, 147 (2012)
39. Ozel, F., Psaltis, D., Narayan, R., Villareal, S.A.: Astrophys. J. **757**, 55 (2012)
40. Ozel, F., Psaltis, D., Narayan, R., McClintock, J.E.: Astrophys. J. **725**, 1918 (2010)
41. Ozel, F., Psaltis Dimitrios, D., Ransom, R., Demorest, P., Alford, M.: Astrophys. J. **724**, L199 (2010)
42. Heger, A., Fryer, C.L., Woosley, S.E., Langer, N., Hartmann, D.H.: Astrophys. J. **591**, 288 (2003)
43. Mapelli, M., Ripamonti, E., Zampieri, L., Colpi, M., Bressan, A.: Mon. Not. R. Astron. Soc. **408**, 234 (2010)
44. Krolik, J.H.: Active Galactic Nuclei: From the Central Black Hole to the Galactic Environment. Princeton University Press, Princeton (1999)
45. Begelman, M.C., Blandford, R.D., Rees, M.J.: Rev. Mod. Phys. **56**, 255 (1984)
46. Blandford, R.D., Znajek, R.L.: Mon. Not. R. Astron. Soc. **179**, 433 (1977)
47. Blandford, R.D., Payne, R.: Mon. Not. R. Astron. Soc. **199**, 883 (1982)
48. Peterson, B.M., Horne, K.: Astron. Nachr. **325**, 248 (2004)
49. Vestergaard, M., Fan, X.X., Tremonti, C.A., Osmer, P., Patrick, S., Richards, G.T.: Astrophys. J. **674**, L1 (2008)
50. Merloni, A., Heinz, S.: In: Keelbook, W. (ed.) Planets, Stars and Stellar Systems, vol. 6 (2012)
51. White, S.D.M., Rees, M.J.: Mon. Not. R. Astron. Soc. **183**, 341 (1978)
52. Mo, H., van den Bosch, F.C., White, S.D.M.: Galaxy Formation and Evolution. Cambridge University Press, Cambridge (2010)
53. Kormendy, J., Richstone, D.O.: Annu. Rev. Astron. Astrophys. **33**, 581 (1995)
54. Ferrarese, L., Ford, H.: Space Sci. Rev. **116**, 523 (2005)
55. Gultekin, K., et al.: Astrophys. J. **698**, 198 (2009)
56. Ferrarese, L., Merritt, D.: Astrophys. J. **539**, 9 (2000)
57. Gebhardt, K., et al.: Astrophys. J. **593**, 13 (2000)
58. Magorrian, J., et al.: Astron. J. **115**, 2285 (1998)
59. Silk, J., Rees, M.J.: Astron. Astrophys. **331**, 1 (1998)
60. King, A.R.: Astrophys. J. **596**, 27 (2003)
61. Di Matteo, T., Springel, V., Hernquist, L.: Nature **433**, 604 (2005)
62. Marconi, A., Risaliti, G., Gilli, R., Hunt, L.K., Maiolino, R., Salvati, M.: Mon. Not. R. Astron. Soc. **351**, 169 (2004)
63. Volonteri, M.: Astron. Astrophys. Rev. **18**, 279 (2010)
64. Tegmark, M., Silk, J., Rees, M.J., Blanchard, A., Abel, T., Palla, F.: Astrophys. J. **474**, 1 (1997)
65. Abel, T., Bryan, G.L., Norman, M.L.: Science **295**, 93 (2002)
66. Omukai, K., Palla, F.: Astrophys. J. **561**, 55 (2001)
67. Begelman, M.C., Volonteri, M., Rees, M.J.: Mon. Not. R. Astron. Soc. **370**, 289 (2006)
68. Devecchi, B., Volonteri, M., Rossi, E., Colpi, M., Portegies Zwart, S.: Mon. Not. R. Astron. Soc. **412**, 1465 (2012)
69. Miyoshi, M., et al.: Nature **373**, 6510, 127 (1995)
70. Siopsis, C., et al.: Astrophys. J. **693**, 946 (2009)
71. Maoz, E.: Astrophys. J. **494**, L181 (1998)
72. Ghez, A.M., et al.: Astrophys. J. **689**, 1044 (2008)

73. Gillessen, S., et al.: Astrophys. J. **692**, 1075 (2009)
74. Colpi, M., Shapiro, S.L., Wasserman, I.: Phys. Rev. Lett. **57**, 2485 (1986)
75. Kulkarni, A.K., et al.: Mon. Not. R. Astron. Soc. **414**, 1183 (2011)
76. Novikov, I.D., Thorne, K.S.: In: Dewiit, C. (ed.) Astrophysics of Black Holes, p. 343 (1973)
77. McClintock, J.E., et al.: Class. Quantum Gravity **28**, 114009 (2011)
78. Penna, R.F., SaDowski, A., McKinney, J.C.: Mon. Not. R. Astron. Soc. **420**, 684 (2011)
79. Fabian, A.C., Iwasawa, K.K., Reynolds, C.S., Young, A.J.: Publ. Astron. Soc. Pac. **112**, 1145 (2000)
80. Reynolds, C.S., Nowak, M.A.: Phys. Rep. **377**, 389 (2003)
81. Miller, J.M.: Annu. Rev. Astron. Astrophys. **45**, 441 (2007)
82. Fender, R.P., Gallo, E., Russell, D.: Mon. Not. R. Astron. Soc. **406**, 1425 (2010)
83. Narayan, R., McClintock, J.E.: New Astron. Rev. **51**, 733 (2008)
84. Amaro Seoane, P., et al.: Living Rev. Relativ. (2012, in press)
85. Amaro Seoane, P., et al.: Class. Quantum Gravity **29**(12), 24016 (2012)
86. Sopuerta, C.F.: Gravit. Wave Not. **4**, 3 (2010)
87. Babak, S., et al.: Class. Quantum Gravity **28**, 114001 (2011)
88. Kesden, M., et al.: Phys. Rev. D **71**, 044015 (2005)
89. Pani, P., Cardoso, V.: Phys. Rev. D **79**, 084031 (2009)
90. Gair, J.R., Li, C., Mandel, I.: Phys. Rev. **77**, 024035 (2008)
91. Barausse, E., Rezzolla, L., Petroff, D., Ansorg, M.: Phys. Rev. **75**, 064026 (2007)
92. Kocsis, B., Yunes, N., Loeb, A.: Phys. Rev. **84**, 024032 (2011)

Index

Symbols
γ-decay, 324

A
Accretion, 407
Acoustic black hole, 183, 201
Acoustic horizon, 183
Adiabatic eigen-states, 56, 57
Adiabatic expansion, 56, 57
Adiabatic limit, 55, 59
Airy interference, 155, 158, 159, 164
Amazing fact, 15
Amplification factor, 75
Analogue models for gravity, 31, 301
Analogue spacetime, 31, 45, 46
Asymptotic expansion, 57

B
Bernoulli equation, 67
Black hole demography, 418
Black hole horizon: observational evidence, 427
Black holes, 169
Black holes and gravitational waves, 429
Black holes of stellar origin, 412
Blocking flow, 75, 77
Blue horizon, 146, 150, 161
Bogoliubov coefficients, 53, 54, 59, 75
Bogoliubov dispersion, 120, 191
Bogoliubov spectrum, 286
Bogoliubov theory, 185
Bogoliubov transformation, 199, 207, 215
Bogoliubov-de Gennes, 186
Boltzmann-distribution, 167, 169, 170
Boost Hamiltonian, 15
Boost Killing vector, 12

Bose-Einstein condensates, 61, 183, 185, 209, 301
Brownian motion, 385, 386, 389, 392

C
Capillary wave, 148
Casimir force, 232, 237
Čerenkov condition, 113
Čerenkov effect, 109
Čerenkov emission, moving charge, 116
Chandrasekhar mass limit, 404
Conformally flat metric, 65
Conserved norm, 74
Correlation functions, 200
Cosmic microwave background, 52, 59, 60
Cosmological particle creation, 51–53, 60
Crab Nebula, 329
Curvature, 394

D
De Sitter metric, 58, 59
Deep fluid, 132
Deep water waves, 74
Density correlations, 184
Density-density correlations, 200, 209
Detailed balance, 389, 391
Diffusion equation, 386, 389, 392, 395
Dispersion, 24
  parabolic, 124
Dispersion relation, 191, 202
Dissipation, 160
Doubly special relativity, 316

E
EFT, 310
Electromagnetic action, 365
Energy conservation, 208

Entanglement, 17
Entanglement entropy, 17
Ergoregion, 9
Ergosurface, 10
Evaporation, 169
Event horizon, 3, 399
Excitation spectrum, 172

F
Fermi surface, 345
Fluctuations, 70, 285, 291, 294
Fokker-Planck equation, 389, 391
Friedmann-Robertson-Walker metric, 52, 58
Froude number, 69, 77

G
Galactic Center, 421
General linearized waves, 75
Generalized Čerenkov effect, 109
Geodesic, 5
Gordon metric, 31–35, 42
Graphene, 365
Gravitational analogue, 386, 395
Gravitational analogy, 184, 187, 188, 191, 211
Gravitational collapse, 402
Gravitational equilibria of stars, 402
Gravitational lensing, 224
Gravity wave, 145, 147, 152
Gross-Pitaevskii equation, 185–187
GZK, 332

H
Hartle-Hawking state, 18
Hawking radiation, 17, 159, 182, 207, 208, 287, 292
Hawking temperature, 18, 288, 294
Healing length, 121
Helicity decay, 325
High energy cosmic rays, 297
Horizon, 146, 147, 151, 152, 159, 161
Horizon entanglement entropy, 17
Horizon generating Killing vector, 11
Hubble parameter, 58
Hybridon, 158
Hydrodynamic approximation, 187
Hyperbolic angle, 12
Hyperbolic rotation, 12

I
Inflation, 52, 57–59
Inflaton field, 57
Inner product, 53, 54
Invisibility cloaks, 227
Irrotational flow, 64

K
Kerr nonlinearity, 276, 280
Killing energy, 10
Killing vector, 8
Killing vector/stationarity, 387, 391, 392
Klein-Gordon equation, 278, 286, 287, 292

L
Laboratory analogues, 52, 60
Landau critical velocity, 119
Laplacian, 65
Laval nozzle, 276, 279, 280
Lorentz boost, 12
Lorentz invariance, 297
Lorentz invariance violation, 378
Lorentz violations, 297
Low-energy effective action, 60
Luminous fluid, 276
Luminous liquid, 282, 285, 287

M
Mach cone, 118
Mach horizon, 275
Madelung transformation, 278
Markov process, 388, 389
Master equation, 388, 391
Maximum neutron star mass, 404
Maxwell equations, 223
Maxwell's fish eye, 235
Mean square displacement, 393, 394
Metamaterials, 395
Mode conversion, 147, 153
Modified dispersion relations, 307
Moving media, 241

N
Naturalness problem, 308
Negative energy states, 211
Negative horizon, 146
Negative refraction, 230
Non-perturbative, 56
Nonlinear Schrödinger equation, 277
Nonlinearity length, 278, 287, 294

O
Oppenheimer-Volkoff limit, 404

P
Painlevé-Gullstrand coordinates, 4
Pair-creation, 326
Parker, 51
Particle creation, 208
Penrose process, 11
Perfect imaging, 230, 236

Phonon partners, 208, 210
Photon splitting, 325
Photon time of flight, 320
Planck length scale, 287

Q
QED, 312
Quantization, 290
Quantum gravity phenomenology, 299
Quantum levitation, 232
Quantum norm, 75
Quantum potential, 278, 279, 286, 289, 294

R
Redshift/lapse function, 387, 388, 390, 392
Relative entropy, 397
Riccati equation, 55
Rindler Hamiltonian, 15
Rindler horizon, 12
Rindler wedge, 12
Rotation powered pulsars, 407

S
Scattering matrix, 192, 215
Schrödinger, 51–53
Schwarzschild coordinates, 3
Schwarzschild spacetime, 2
Second law of thermodynamics, 395
Shallow fluid, 134
Shallow water wave, 73
Ship-wave pattern, 133
Spacetime foam, 315
Spatial geometry, 226
Specific pressure, 64
Spinning black holes, 424
Stationarity, 210
Step configuration, 189
Stimulated emission, 23, 26, 169
Straddled fluctuation, 285, 287, 292, 294
Stream function, 66
Subluminal dispersion, 146, 159, 162
Subsonic flow, 193
Superfluid black/white hole, 137
Superfluids, 119
Superluminal dispersion, 146
Supermassive black holes, 415
Supersonic flow, 201, 203, 212
Surface gravity, 7, 13, 170, 189, 208
Surface waves, 127, 145, 169
    analogues, 139

SUSY, 309
Synchrotron radiation, 327

T
Texture, 8
Threshold reactions, 324
Time dilation, 390, 393
Time slowing, 287
Topology, 344
Trans-Planckian problem, 152, 159, 160, 162, 169, 216
Trans-Planckian question, 23
Transformation media, 228
Transformation optics, 221
Transition rates, 388
Trapped ions, 60
Two dimensional flow, 64
Two-roton decay, 160

U
Ultra-high-energy cosmic rays, 332
Undulation, 27
Unruh effect, 14
Unruh state, 18
Unruh temperature, 15

V
Vacuum
    semi-metal, 347
    Čerenkov, 325
Vacuum birefringence, 322
Vacuum fluctuations, 182
Vacuum state, 207
Vacuum wave functional, 15
Vancouver experiment, 27

W
Wake pattern, 113
White hole, 8, 26, 145, 167
White hole radiation, 26
Windows on quantum gravity, 299
WKB approximation, 158
WKB limit, 55, 59
WKB method, 53, 54, 59
WKB phase, 54
WKB turning points, 55

Z
Zero mode, 159
Zero-point quantum fluctuations, 211

MIX
Papier aus verantwortungsvollen Quellen
Paper from responsible sources
FSC® C105338

If you have any concerns about our products,
you can contact us on
ProductSafety@springernature.com

In case Publisher is established outside the EU,
the EU authorized representative is:
**Springer Nature Customer Service Center GmbH
Europaplatz 3, 69115 Heidelberg, Germany**

Printed by Libri Plureos GmbH
in Hamburg, Germany